Exosomal RNA

Role in Human Diseases and Therapy

EXOSOMAL RNA

ROLE IN HUMAN DISEASES AND THERAPY

Edited by

DEBALINA BHATTACHARYA

Department of Microbiology, Maulana Azad College, Kolkata, West Bengal, India

MAINAK MUKHOPADHYAY

Department of Biosciences, JIS University, Kolkata, West Bengal, India

KURSAD TURKSEN

Ottawa Hospital Research Institute, Ottawa, ON, Canada

ELSEVIER

ACADEMIC PRESS

An imprint of Elsevier

Academic Press is an imprint of Elsevier
125 London Wall, London EC2Y 5AS, United Kingdom
525 B Street, Suite 1650, San Diego, CA 92101, United States
50 Hampshire Street, 5th Floor, Cambridge, MA 02139, United States

ISBN: 978-0-443-14008-2

For information on all Academic Press publications visit our website at https://www.elsevier.com/books-and-journals

Publisher: Stacy Masucci
Acquisitions Editor: Michelle Fisher
Editorial Project Manager: Barbara Makinster
Production Project Manager: Jayadivya Saiprasad
Cover Designer: Matthew Limbert

Typeset by TNQ Technologies

Contents

Contributors

Muhammad Abbas
Department of Neurosurgery, University of Florida, Gainesville, FL, United States

Avik Acharya Chowdhury
Department of Biosciences, JIS University, Kolkata, West Bengal, India

Ravindra V. Badhe
Pharmaceutical Chemistry Department, Marathwada Mitra Mandal's College of Pharmacy, Pune, Maharashtra, India

Kalyanee Bera
Department of Biosciences, JIS University, Kolkata, West Bengal, India

Debalina Bhattacharya
Department of Microbiology, Maulana Azad College, Kolkata, West Bengal, India

Abhishek Bhattacherjee
Jubilant Biosys, Bengaluru, Karnataka, India
Department of Chemistry, Department of AFNS, University of Alberta, Edmonton, AB, Canada

Avirup Chakraborty
Department of Neurosurgery, University of Florida, Gainesville, FL, United States

Shrena Chakraborty
TEAM "DNA Recombination, Replication, and Genome Stability", UMR3348, Institut Curie, Orsay, France

Anusha Chaudhuri
Zentrum für Molekulare Medizin, Institut für kardiovaskuläre Regeneration, Goethe-Universität, Universitätsklinikum, Frankfurt, Germany

Anjali Chauhan
Department of Neurosurgery, University of Florida, Gainesville, FL, United States

Sougata Ghosh Chowdhury
Department of Life Science and Biotechnology, Jadavpur University, Kolkata, West Bengal, India

Juan D. Cortés García
Facultad de Medicina, Universidad Autónoma de San Luis Potosí (UASLP), San Luis Potosí, San Luis Potosí, México

Subhadeep Das
Department of Biochemistry, Purdue University, West Lafayette, IN, United States

Rajarshi Das
Amity Institute of Biotechnology, Amity University, Noida, UP, India

Supriya Das
Department of Biosciences, JIS University, Kolkata, West Bengal, India
Department of Biotechnology, JIS University, Kolkata, West Bengal, India

Yongtao Duan
Henan Provincial Key Laboratory of Pediatric Hematology, Children's Hospital Affiliated to Zhengzhou University, Zhengzhou University, Zhengzhou, China

Nalok Dutta
Biochemical Engineering Department, University College London, London, United Kingdom

Rameesha Fareed
Experimental Therapeutics Group, University of Florida Health Cancer Center, University of Florida, Gainesville, FL, United States

Suparna Ghosh
Department of Biosciences, JIS University, Kolkata, West Bengal, India

Kakali Ghoshal
Department of Medicine, Division of Nephrology and Hypertension, Vanderbilt University School of Medicine, Nashville, TN, United States

Sounak Ghosh Roy
Henry M Jackson Foundation for the Advancement of Military Medicine, Inc., Silver Spring, MD, United States

Ana K. González Palomo
Coordinación para la Innovación y Aplicación de la Ciencia y Tecnología (CIACYT), Universidad Autónoma de San Luis Potosí (UASLP), San Luis Potosí, San Luis Potosí, México

Abhishek Jaiswal
Viral Research and Diagnostic Laboratory (VRDL), Department of Microbiology, Midnapore Medical College and Hospital, Midnapore, West Bengal, India

Jorge A. Jiménez Avalos
Centro de Investigación y Desarrollo Oncológico, S.A. de C.V. (CIDO), San Luis Potosí, San Luis Potosí, México

Parimal Karmakar
Department of Life Science and Biotechnology, Jadavpur University, Kolkata, West Bengal, India

Vinay Kumar
Department of Physiology and Cell Biology, The Ohio State University Wexner Medical Centre, Columbus, OH, United States

Hima Makala
Radiation Oncology Branch, National Cancer Institute, NIH, Bethesda, MD, United States

Moumita Mandal
Department of Microbiology, Maulana Azad College, Kolkata, West Bengal, India

Mainak Mukhopadhyay
Department of Biosciences, JIS University, Kolkata, West Bengal, India
Department of Biotechnology, JIS University, Kolkata, West Bengal, India

Anik Paul
Department of Biochemistry, Purdue University, West Lafayette, IN, United States
Department of Biochemistry and Molecular Biology, University of Dhaka, Dhaka, Bangladesh

Saptarshi Ray
Department of Microbiology, Maulana Azad College, Kolkata, West Bengal, India

Anirban Roychowdhury
Section of Hematology and Oncology, Medicine Service, McGuire Cancer Center, Hunter Holmes McGuire VA Medical Center, Richmond, VA, United States
Division of Hematology, Oncology, and Palliative Care, Department of Internal Medicine, Massey Cancer Center, Virginia Commonwealth University, Richmond, VA, United States

Bidisha Saha
Department of Biosciences, JIS University, Kolkata, West Bengal, India
Department of Biotechnology, JIS University, Kolkata, West Bengal, India

Soumyadipta Saha
Department of Microbiology, Maulana Azad College, Kolkata, West Bengal, India

Soumyadev Sarkar
Center for Fundamental and Applied Microbiomics, Biodesign Institute, Arizona State University, Tempe, AZ, United States

Ananya Sarkar
Department of Microbiology, Maulana Azad College, Kolkata, West Bengal, India

Debasish Sarkar
Fosun Pharma USA Inc., Lexington, MA, United States

Sampriti Sarkar
Department of Biosciences, JIS University, Kolkata, West Bengal, India

Paramjeet Singh
Department of Biosciences, JIS University, Kolkata, West Bengal, India

Eneida Turiján Espinoza
Instituto de Investigación en Biociencias, Centro Universitario de Los Altos, Universidad de Guadalajara, Guadalajara, Jalisco, México

Kursad Turksen
Ottawa Hospital Research Institute, Ottawa, ON, Canada

Preface

Exosomes are extracellular vesicles with a size range of 30–100 nm released from endosomes. Exosomes facilitate communication between cells and tissues by serving as a second messenger. Exosomes contain lipids, proteins, nucleic acids, and metabolites. Exosomal RNA, which includes both coding and noncoding RNAs, is generated depending on the microenvironment of the secreting cells. Exosomal RNA has a crucial role in the development and pathogenesis of human diseases and also has diagnostic and therapeutic potential to inhibit the diseases.

Exosomal RNA: Role in Human Diseases and Therapy presents the most recent findings in scientific research on exosomal RNA and their role in human diseases and therapy. This book shows the latest research on the relationship between different exosomal RNAs and diseases like cancer, type 2 diabetes, cardiovascular diseases, kidney diseases, immune diseases, neurodegenerative diseases, respiratory diseases, and bacterial infections. It also goes into detail about how to diagnose and treat these diseases. This book consists of 14 chapters. Chapter 1 discusses exosomal RNA and its types. Exosomal biogenesis, exosomal composition, and exosomal isolation and characterization techniques are discussed in Chapter 2. Several studies have reported that exosomal RNA has a vital role in the development and pathogenesis of cancer progression. So, in Chapter 3, the role of exosomal RNA in different cancer progressions has been discussed. Exosomal RNA can be easily detected in clinical samples like blood, urine, saliva, or other body fluids. Based on this property, exosomal RNA is used as a biomarker in several cancers. In Chapter 4, exosomal RNA as a biomarker in cancer diagnostics and therapy is discussed. Chapter 5 provides a comprehensive analysis of the molecular and functional interactions between the exosomal RNA and the autophagy pathways. Recent studies indicate that exosome-derived RNAs control insulin resistance, β cell dysfunction, adipose tissue inflammation, endothelial dysfunction, and various other pathophysiological consequences of type 2 diabetes mellitus (T2DM) and related metabolic complications. In Chapter 6, the role of exosome-derived RNAs in the progression and screening of T2DM and related metabolic complications has been discussed, along with their role in novel treatment strategies for T2DM. Recent studies have reported that exosomal RNAs, notably microRNAs and circular RNAs, have been reported to be novel candidates to control immune responses and immunological disorders. Therefore, Chapter 7 emphasizes the role of exosomal RNAs in the immune system and the diseases they regulate. Chapter 8 focuses on the role of exosomal RNA in disease progression, specifically neurodegenerative diseases, brain injury, spinal cord injury, and glioma. Exosomal RNA expression targets the transcription and translation of the molecular pathways essential to human longevity and is linked to ageing mechanisms. Chapter 9 has covered the interactions of exosomal RNA with ageing and age-associated diseases. The knowledge of exosomal RNA-mediated intercellular communication with cardiac injuries, repairs, and regeneration has been discussed in Chapter 10. Moreover, their potential role as predictive and diagnostic biomarkers for cardiovascular illnesses (CVDs) and their possible future use as tools for CVD therapy have also been covered in this chapter. According to recent findings, exosomes may influence the pathogenesis of kidney disorders while also controlling tissue healing. Exosomal RNAs may therefore be a target for pharmaceutical therapy and may be employed therapeutically to treat both acute and long-term kidney injuries. Based on this context, Chapter 11 focuses on the role of exosomal RNA in renal diseases. Chapter 12 has addressed the role of exosomal RNA in host—pathogen communication, especially long noncoding RNAs, in bacterial infection, which is still in its infancy. Chapter 13 focuses on the role of exosomal RNA in tissue repair and wound healing mechanisms. Chapter 14 has covered the role of exosomal RNA in other human diseases such as ocular diseases, chronic obstructive pulmonary diseases (COPD), and other respiratory diseases. This book should be utilized for many applications, such as education, research, and clinical applications. It has been designed to be pedagogical so that even beginners can learn the fundamentals of exosomes and exosomal RNA and their role in human diseases.

CHAPTER ONE

Exosomal RNA—an introduction

Bidisha Saha, Supriya Das, Kalyanee Bera, Mainak Mukhopadhyay
Department of Biosciences, JIS University, Kolkata, West Bengal, India

1. Introduction

Eukaryotic cells as well as prokaryotic cells secrete a variety of membrane-bound structures as a part of their regular physiology known as extracellular vesicles (EVs). Invasion of the plasma membrane causes the formation of multivesicular bodies [1]. The latter can then collaborate with other organelles and vesicles found within cells to produce an extremely broad range of unique shapes [2]. Exosomes and ectosomes are the two primary types of EVs. Ectosomes are a form of EV released by cells via outward budding of the plasma membrane; examples include microvesicles, microparticles, and large vesicles. Exosomes are EVs between 40 and 160 nm in size that are released from endosomes [3]. After being released into the extracellular space, these exosomes can be picked up by other cells [4]. Exosomes facilitate communication between cells and tissues by serving as a second messenger that may travel both short and long distances (Fig. 1.1). Lipids, proteins, nucleic acids, and ribose nucleotides are all components of exosomes [5]. The composition of exosomes is dependent on physiological condition and stimulation of the cells within the tissue where they are produced. During the process of exosome formation, the exosomal RNA, which includes both coding and noncoding RNA, depends on the microenvironment of the secreting cells. The unique heterogenous composition of exosomal RNAs serves as a signature of the originating cell or biological system. The fundamental methods by which RNA is loaded into exosomes remain unspecified, despite the discovery that several RNA-binding proteins may bind and transfer RNA molecules into intracellular vesicles utilizing specific patterns. Exosomal RNAs play a crucial role in intercellular communication and has gained significant attention in recent years. In this chapter, we compile the most up-to-date information on the biological roles of exosomal RNA and their links to key neurological, cardiovascular, metabolic, and immune system problems and diseases. The therapeutic potential of exosomal RNA for diseases associated with chronic inflammation and tissue regeneration is also discussed.

Exosomal RNA
ISBN 978-0-443-14008-2
https://doi.org/10.1016/B978-0-443-14008-2.00014-0

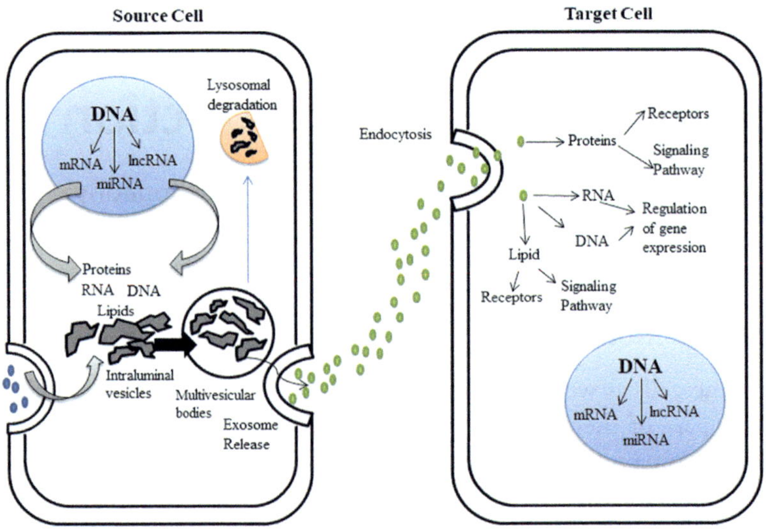

Figure 1.1 The generation of exosomes and their impact on recipient cells, depicted graphically: Cytoplasmic intraluminal vesicles, or exosomes, are created when the endocytic membrane invaginates. When these vesicles reach maturity, they become multivesicular bodies (MVBs) that contain nucleic acids, proteins, and lipids. Exosomes are released into the extracellular environment when MVBs fuse with the plasma membrane. MVBs can also be recycled or sent to lysosomes for destruction. The source cell's exosomal payloads can be transported to recipient cells, where they may alter signaling, gene expression, or immunological response.

2. Characteristics of exosomal RNA

Some distinguishing features of exosomal RNA are as follows:

- *Diversity of RNA Species:* Exosomal RNA encompasses a diverse range of RNA species, including messenger RNA (mRNA), microRNA (miRNA), long non-coding RNA (lncRNA), circular RNA (circRNA), small nucleolar RNA (snoRNA), small nuclear RNA (snRNA), and transfer RNA (tRNA). This variety of RNA molecules reflects the complexity of cellular processes and functions that exosomal RNA can control.
- *Stability:* Exosomal RNA exhibits remarkable stability, even in harsh extracellular conditions. The encapsulation of RNA within the protective lipid bilayer of exosomes helps shield it from degradation by extracellular ribonucleases. This stability enables exosomal RNA to be preserved during transportation between cells and in various biological fluids.
- *Packaging:* Exosomal RNA is selectively packaged into intraluminal vesicles (ILVs) within multivesicular bodies (MVBs) during exosome biogenesis. The packaging process involves specific sorting mechanisms that recognize and select certain RNA species for inclusion in exosomes. The exact mechanisms of RNA sorting into ILVs are still under investigation.

- *Transportability:* Exosomal RNA is transported through body fluids, allowing long-distance communication between cells and tissues. Exosomes, including their RNA cargo, can traverse through various biological barriers and reach recipient cells or target tissues.
- *Cell-specific composition:* The composition of exosomal RNA can vary depending on the cell type, physiological state, and environmental conditions. Different cell types may selectively package specific subsets of RNA molecules into exosomes, resulting in cell-specific RNA profiles.
- *Regulatory functions:* Exosomal RNA, particularly miRNAs and circRNAs, can act as regulatory molecules in recipient cells. They can modulate gene expression by targeting specific messenger RNAs (mRNAs) for degradation or translational repression. This regulatory activity can impact various cellular processes, including proliferation, differentiation, immune response, and metabolism.

The characteristics of exosomal RNA make it a fascinating and dynamic area of research, with significant implications for understanding cell-to-cell communication, disease mechanisms, and the development of diagnostic and therapeutic strategies.

2.1 Exosomal MiRNAs

MicroRNAs (miRNAs) are more abundant in exosomes than in the cells from which they are produced [6]. Because the lipid bilayer shields them from RNase oxidation, exosomal miRNAs are more durable than those in serum. Consequently, ongoing research aims to assess the usefulness of miRNAs carried by exosomes as molecular indicators for diagnosing diseases. The following characteristics of malignant cell exosomal miRNAs make them valuable as potential biological markers for monitoring cancer development and therapy response: (1) cancer-related proteins, RNAs, and DNA fragments can be found in exosomes secreted by cancer cells [7], (2) the structure of exosomal membranes contributes to the stability of miRNAs, which boosts their potential as biomarkers for diseases. (3) miRNAs derived from circulating exosomes have greater resistance to degradation than miRNAs found in cells or plasma [8,9], and (4) the complexity of blood constituents reduces specific cancer-associated proteins, making early detection or low expression difficult. Exosomal miRNAs hold significant promise as molecular markers of cancer, with applications in both diagnosis and prognosis due to the aforementioned characteristics. Expression of exosomal RNAs as biomarkers in several cancers such as respiratory system cancer [10], digestive system cancer [11−14], nervous system cancer [15], and genitourinary system cancers [16,17] has been reported.

2.2 Exosomal circular RNAs

Exosomal circular RNAs (circRNAs) have emerged as potential players in various cellular processes such as cell proliferation, angiogenesis, epithelial−mesenchymal transition, and targeted therapy. These RNAs have been identified in exosomes obtained from serum, urine, and tumors [18–20]. Li was the first to use RNA-sequencing techniques to show the enhancement and durability of numerous circRNAs in exosomes relative to the producer cells [21]. Although interest in exosomal circRNAs is increasing, little is known about their structure, function, or other properties. Exosomes transport molecular information across cells, which are made up of exosomal circRNAs perhaps being influenced by changes in related miRNA concentrations in donor cells. Serum exosomes include circRNAs, which participate in RNA–RNA competitive interactions that contribute to the development of many neoplasias. Zhao demonstrated that exosomal circRNAs may be involved in neuronal growth and repair, transmission of nerve impulses, and modulation of the cGMP-PKG signaling pathway (Fig. 1.2) [22]. Therefore, continued investigation of exosomal RNAs not only has significant implications for our knowledge of exosome functions and properties but also offers a fresh approach to the diagnosis and treatment of numerous diseases.

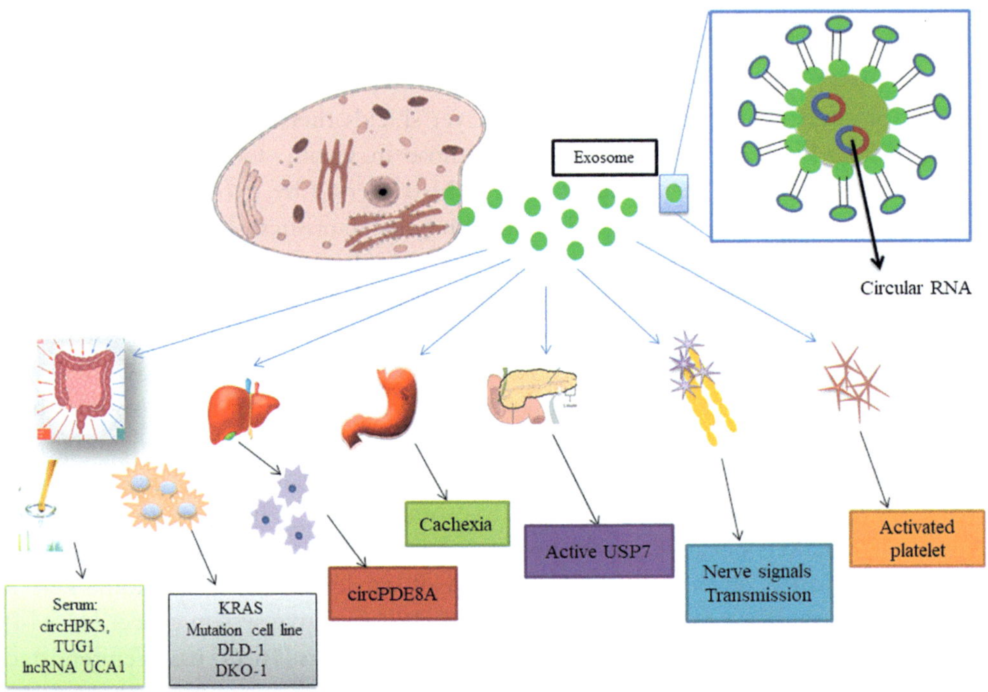

Figure 1.2 The role of exosomal circRNA biogenesis and origin in disorders. Under various physiological and pathological circumstances, most cells produce exosomes. Exosomes allow the transport of circRNAs from donor cells to recipient cells, where they can serve as a messenger in a variety of signaling pathways.

3. Importance of exosomal RNAs

The importance of exosomal RNA stems from its involvement in several biological processes and its potential as a diagnostic and therapeutic tool. Here are some key points highlighting the significance of exosomal RNA [23].

- *Intercellular communication:* The RNA contained in exosomes serves as a form of intercellular communication. These vesicles are capable of ferrying RNA molecules from one cell to another, altering the behavior and function of both cells in the process. Exosomes' ability to transport RNA allows them to affect cellular processes, control gene expression, and aid in tissue homeostasis to regulate gene expression and biological functioning in recipient cells.
- *Disease pathogenesis and progression:* Exosomal RNA has been associated with the development and progression of disease. In conditions such as cancer, exosomal RNA can transfer oncogenic or tumor-suppressive RNA molecules to neighboring or distant cells, influencing tumor growth, and angiogenesis, metastasis, and treatment resistance. Similarly, in neurodegenerative diseases, exosomal RNA can contribute to the spread of pathological proteins and RNA species, impacting neuronal function and survival. Understanding the roles of exosomal RNA in disease processes can provide insights into disease mechanisms and potential therapeutic targets.
- *Disease biomarkers:* Exosomal RNA has emerged as a potential source of biomarkers for various diseases. The RNA content within exosomes reflects the molecular characteristics of the cells from which they are derived, including diseased cells. By analyzing the RNA profiles of exosomes in biofluids such as blood, urine, or cerebrospinal fluid, researchers can identify disease-specific RNA signatures. These signatures can be utilized for early detection, diagnosis, prognosis, and monitoring of diseases, including cancer, neurodegenerative disorders, cardiovascular diseases, and infectious diseases.
- *Liquid biopsy:* The analysis of exosomal RNA in bodily fluids, such as blood or urine, has opened up possibilities for liquid biopsy approaches. Liquid biopsy involves the noninvasive detection and monitoring of diseases by analyzing biomarkers in body fluids. Exosomal RNA analysis offers a potential alternative to traditional tissue biopsies and enables real-time monitoring of disease progression and treatment response.
- *Prognostic indicators and treatment response:* Exosomal RNA profiles can serve as prognostic indicators for disease outcomes and treatment response. By analyzing changes in exosomal RNA content over time, researchers can monitor disease progression, evaluate the effectiveness of therapies, and make informed decisions regarding patient management and personalized treatment approaches.

- *Regenerative medicine:* Exosomal RNA has implications for regenerative medicine and tissue repair. Exosomes derived from stem cells or other regenerative cell types contain specific RNA molecules that can modulate tissue regeneration processes. By delivering exosomal RNA to damaged tissues, researchers aim to promote tissue repair, enhance healing, and potentially develop new approaches for regenerative therapies.
- *Therapeutic applications:* Exosomal RNA holds promise as a therapeutic tool. Exosomes can be harnessed as delivery vehicles for RNA-based therapeutics, including small interfering RNA (siRNA), microRNA (miRNA), and messenger RNA (mRNA). These exosomes protect the RNA cargo from degradation, facilitate targeted delivery, and enhance the efficacy of RNA-based therapies. By engineering exosomes to carry specific RNA molecules, researchers can enhance the stability, target specificity, and therapeutic efficacy of these molecules. Exosomal RNA-based therapies have the potential to treat various diseases, including cardiovascular diseases, genetic disorders, regenerative medicine, and so on.

4. Role of exosomal RNAs in different diseases

4.1 Cancer diagnosis and prognosis

Cancers of the digestive tract, including the esophagus, stomach, liver, pancreas, and colon, have been linked to exosomal RNA. Brain malignancies such as glioblastoma, which develop in the nervous system, have also been linked to exosomal RNA. In hepatocellular carcinoma (HCC), both cell proliferation and cell cycle regulation are influenced by microRNAs (miRNAs) [24]. One such example is exosomal circRNA_100284, which targets EZH2 in various malignancies, including HCC [25,26]. Dai's study revealed that it functions as a sponge for miR-217, regulating the cell cycle and suppressing cell proliferation by causing cell cycle arrest in the G2/M phase. However, the specific effects of miR-217 [27] on malignant transformation and underlying processes are still not well understood. Recent research has shown that exosomes released from arsenite-transformed cells have elevated levels of circRNAs. In normal cells, overexpression of miR-217 reduced the expression of the proliferation biomarker EZH2 [28,29] and the cell cycle regulator cyclin-D1 [30]. This suggests that exosomal RNAs stimulate normal cell proliferation and accelerate the cell cycle. In hepatocellular carcinoma (HCC), exosomal circRNAs released by adipose tissues can regulate deubiquitination through the miR-34a/USP7/CyclinA2 signaling pathway. Overexpressed circ-deubiquitination (circ-DB) significantly downregulates miR-34a, supporting cell expansion and protecting against DNA damage. Another interesting finding is the role of exosomal circRASSF2 in laryngeal squamous cell carcinoma (LSCC) development [31]. CircRASSF2 expression was consistently higher in LSCC tumor tissues compared with controls. Suppression of exosomal circRASSF2 significantly reduced cell proliferation and migration through the miR-302b-3p/IGF-1R axis.

Research has demonstrated that circRNAs can be transported to the exosomes of colorectal cancer cell lines and are markedly downregulated in KRAS mutants [32]. Recently, it was discovered that the miR338/MACC1/MET pathway in pancreatic ductal adenocarcinoma is linked to exosomal circ-PDE8A, which in turn promotes tumor development and lymphatic invasion [33]. Exosomal circRNA has also been shown to increase browning of white adipose tissue by targeting the miR-133/PRDM16 pathway, which offers new insight into the causes of cancer-related cachexia. Endometrial cancer and hepatocellular carcinoma also contribute exosomal RNAs that promote tumor growth by inhibiting deubiquitination of USP7 [34]. Human platelets, unlike other hematopoietic cells, are more closely associated with hemostasis, inflammation, and wound healing, and so are common sources of exosomal RNAs that are not cancer-related [35,36].

4.2 Autophagy-related disorder

Cell autophagy is a process that has been shown to be controlled by exosomal long noncoding RNA. Some miRNAs can be sponged by lncRNA to target autophagy-related proteins [37]. For instance, the autophagy factor ATG7 and microRNA (miR)-188-3p expression can be controlled by the lncRNA APF. TTGFB2-OT1, a long noncoding RNA, can bind to several different types of protein expression regulators, allowing it to control the production of several different autophagy-related proteins. The CERS1, LARP1, and NAT8L are all examples of this. Cancer cells are shielded from cellular stress by autophagy. Autophagy is thought to be a survival mechanism utilized by cancer cells. lncRNAs have been discovered to have a crucial role in regulating autophagy, a process that aids cancer cells in resisting cellular stress and chemotherapy. The lncRNA-CAF is essential for tumor cell maintenance and growth. By inhibiting p62-dependent autophagy, it hinders the breakdown of IL-33, a proinflammatory cytokine. High quantities of lncRNA-CAF, which can stimulate tumor development, are found in normal stromal fibroblasts. Cancer cells release lncRNA-CAF, which is known to promote the development of tumor cells, creating a positive feedback loop in the tumor microenvironment. Increased levels of lncRNA-CAF in OSCC tumor microenvironments predict a worse outcome for OSCC patients. Autophagy is stimulated in endothelial cells by the lncRNA ANRIL, which does so through increasing Beclin1. Angiogenesis and cell proliferation are both aided by this method [38]. Release of the lncRNA ANRIL, which controls autophagy in cardiomyocytes, could be a potential treatment for uremic cardiovascular disease [39]. One study suggests that high glucose levels trigger autophagy by lowering lncRNA H19 levels [40]. A detrimental effect on heart function has been associated with autophagy. Hypoxia-induced damage may be mitigated by H19's ability to activate the PI3K/Akt pathway [41]. The likelihood of heart failure is decreased by overexpressing H19, mitigating this effect. In response to hypoxia, ATG7 and Beclin1 expression can be boosted by downregulating H19. Researchers Wang et al. claim that lncRNA H19 reduces the survival of cultured cells by activating autophagy via the DUSP5-ERK1/2 axis [42]. The results of this investigation lend credence to the hypothesis that H19 is linked to autophagy. An essential transcriptional regulator in the progression of cancer is metastatic lung adenocarcinoma transcript 1 (MALAT1). Pancreatic ductal adenocarcinoma could be diagnosed by testing positive for MALAT1. The expression of MALAT1 had shown a capacity to suppress autophagy and tumor development [43]. Tumor development and metastatic suppression were both correlated with MALAT1 silencing's ability to inhibit autophagy. Reducing MALAT1 protein levels increases p62 and decreases LAMP2 expression, two proteins critical for lysosome-autophagosome fusion, according to a study. Also, it can induce autophagy by preventing the production of the Beclin1 protein. Beclin1 protein expression was shown to be decreased in multiple myeloma tissues after MALAT1 knockdown, according to a recent study. The lncRNA MALAT1 inhibits the ability of cerebral microvascular endothelial cells to reoxygenate. As a result, these cells are less able to recover from the harm induced by ULK2 overexpression [44]. According to research, inhibiting MALAT1 levels reduces miR-30a expression, which in turn reduces neuronal death [45].

4.3 Diabetic and metabolic diseases

It has been discovered that exosomes, which contain noncoding RNA, play an important role in the onset, progression, and complications of both type 1 and type 2 diabetes [46]. Through mechanisms such as the recruitment of epigenetic modifier proteins, the regulation of mRNA degradation and translation, and the DNA sequestration of transcription factors, it has been discovered that these noncoding RNAs can exert their effects on distant organs. Insulin-producing pancreatic cells are targeted and destroyed by the immune system in people with type 1 diabetes (T1D), making it an autoimmune disease. The autoimmune response in T1D has been associated with exosomes. Both rat and human pancreatic islets secrete exosomes that are then taken up by dendritic cells [47]; these include cell autoantigens and proinsulin. T1D and pancreatic cell death originate from the activation of B and T lymphocytes caused by the ingestion of exosomal proteins [46]. On the other hand, exosomes produced by benign pancreatic tumors called insulinomas arouse the innate immune response through proinflammatory signaling pathways. Pancreatic mesenchymal stromal cell exosomes also stimulate T cell activation and inflammation, which play a role in autoimmune reactions that lead to T1D.

High levels of glucose in the blood due to insulin resistance induced by excessive fat accumulation in adipocytes characterize the majority of diagnosed cases of diabetes, known as type 2 diabetes (T2D). Adipose cells, or adipocytes, secrete exosomes that carry substances that promote insulin resistance [48]. T2D exosomes, for instance, include more of the miRNA 20b-5p that inhibits AKT signaling and lead to decreased glycogen storage in skeletal muscle and possibly elevated insulin resistance [49]. Exosomal miRNA-155 from adipose tissue macrophages in diabetic mice has been shown to bind the peroxisome proliferator—activated receptor, a transcription factor that affects lipid metabolism, hence promoting glucose intolerance and insulin resistance. Glucose intolerance can be induced in lean mice using exosomes from obese mice with elevated levels of miRNA-192, miRNA-122, miRNA-27a-3p, and miRNA-27b-3p. These miRNAs show prospects for the diagnosis and treatment of diabetes when they produce glucose intolerance and insulin resistance in lean mice [50]. Noncoding RNA carried by exosomes plays crucial roles in the development of both T1D and T2D. While in T2D they contribute to insulin resistance and glucose intolerance, in T1D they contribute to autoimmune reactions and pancreatic cell death. The development of new diagnostic and perhaps therapeutic approaches to diabetes is made possible by elucidating the role of exosomal noncoding RNAs in this disease.

4.4 Immunological diseases

The advancement of sequencing technology and microarray technology has led to the identification of numerous circular RNAs (circRNAs). These circRNAs have been associated with the pathophysiology and diagnosis of various immunological disorders, including primary immunodeficiency diseases (PIDs), autoimmune diseases (AIDs), and other immune diseases. Additionally, the interplay between circRNAs and miRNAs has been implicated in these immune-related disorders. The discovery and understanding of circRNAs and miRNAs provide valuable insights into the mechanisms underlying immunological diseases and offer potential opportunities for their diagnosis and management. PIDs are a group of disorders that affect the immune system, leading to deficiencies or dysfunctions in immune cells, tissues, and organs. Severe combined immunodeficiency (SCID) and Wiskott—Aldrich syndrome (WAS) [51,52] are examples of PIDs, caused by mutations in specific genes such as adenosine deaminase (ADA) and WAS gene, respectively. Research has shown that the levels of certain circular RNAs (circRNAs) derived from genes such as roundabout guidance receptor 1 (ROBO1), CDC42-binding protein kinase alpha (CDC42BPA), and tumor necrosis factor receptor superfamily member 11a (TNFRSF11A) are significantly increased in patients with SCID and WAS [53] compared with healthy controls. These circRNAs are associated with the pathogenesis of these PIDs. ROBO family members and CDC42BPA are implicated in the development of WAS, while TNFRSF11A is linked to SCID and alterations in various signaling pathways [54,55]. These findings contribute to our understanding of the pathophysiology of PIDs and open avenues for further research and potential treatment approaches. Existing research suggests that noncoding RNAs (ncRNAs) enclosed in exosomes play a significant role in autoimmune diseases, which have complex molecular mechanisms that are currently under extensive investigation.

Rheumatoid arthritis (RA) [56], a chronic autoimmune disease, is characterized by leukocyte infiltration into joints, resulting in the production of inflammatory mediators and subsequent damage to bone and cartilage tissue. Exosomal miR-150 has been found to alleviate RA symptoms by downregulating MMP14 and VEGF, thereby reducing angiogenesis. Additionally, miR-150 is associated with T cell maturation and can modulate angiogenesis in RA. In RA patients, serum levels of C-reactive protein, rheumatoid factor, and erythrocyte sedimentation rate are inversely correlated with miR-548a-3p, which is significantly decreased in serum exosomes of RA patients. The impact of exosomal RNA on RA has been established, with evidence showing that exosomal miR-548a-3p regulates macrophage-mediated inflammation in RA through the TLR4/NF-κB signaling pathway. Moreover, exosomal microRNAs (miRNAs) have been implicated in the pathogenesis of systemic lupus erythematosus (SLE) [57], a multisystem autoimmune disease characterized by persistent inflammation and the production of autoantibodies. These findings highlight the involvement of exosomal ncRNAs, particularly miRNAs, in the development and progression of autoimmune disorders, providing potential avenues for understanding the underlying mechanisms and developing novel therapeutic strategies.

4.5 Neurodegenerative diseases

Exosomal cellular communication plays a crucial role in the development and progression of neurodegenerative disorders such as Alzheimer's disease (AD), Parkinson's disease (PD), and Huntington's disease (HD) [58]. Exosomes are being investigated as potential prognostic indicators for various neurodegenerative conditions, including traumatic brain injury and stroke [59,60]. The protein and RNA contents of exosomes, which can be found in cerebrospinal fluid and peripheral fluids, undergo changes months or even years before the official diagnosis of these disorders, making exosomes a potential diagnostic tool. Importantly, exosomes have the ability to cross the blood—brain barrier and enter the peripheral circulation, allowing for their detection to monitor disease progression or facilitate early detection of neurodegenerative diseases. Ongoing research in this area aims to utilize exosomal biomarkers for improved diagnosis and prognosis of neurodegenerative disorders. Researchers have discovered promising findings regarding exosomal RNA levels in individuals with AD. By comparing the levels of miRNA-384 in exosomes derived from serum, they were able to differentiate between AD dementia, PD dementia, and vascular dementia. Additionally, miRNA-384, miRNA-193b, and miRNA-135a showed potential as early-stage diagnostic markers for AD [61]. This research highlights the potential of exosomal miRNAs as biomarkers for the early detection and diagnosis of AD, offering new insights into the field of AD research. PD is a prevalent neurological disorder, and exosomes have been found to play a role in its pathogenesis and neuroprotection. Exosomes have been shown to transport toxic (misfolded) synuclein, a protein associated with PD, between cells, potentially contributing to the degenerative process [62]. Additionally, exosomes derived from PD patients have been found to induce cellular apoptosis, which can dampen the neural stress response [63]. Moreover, exosomes have demonstrated the ability to remove misfolded proteins [64], suggesting a potential role in neuroprotection. In the context of PD, specific miRNAs targeting genes in the "dopaminergic synapse" and "cholinergic synapse" pathways have been identified in exosomes isolated from cerebrospinal fluid [65–68]. These findings support the use of exosomal RNA molecules as biomarkers to differentiate PD from other neurodegenerative disorders such as AD. Overall, exosomes hold promise for understanding the pathogenesis of PD and exploring potential therapeutic strategies. HD is characterized by the accumulation of mutant Huntingtin protein (mHTT) in the brain, resulting from the abnormal amplification of CAG repeats in the Huntington gene. While exosomes have not been implicated in the transmission of mHTT across the nervous system, other mechanisms such as tunneling nanotubes or vesicle processes have been suggested [69]. However, research has shown that overexpression of an HTT-exon 1 polyQ-GFP construct in cells leads to the secretion of polyQ-GFP protein and amplified RNA in the form of exosomes. Ingestion of these exosomes by striatal mouse neurons results in an upregulation of polyQ-GFP RNA [70]. This demonstrates that exosomes have the ability to clear polyQ proteins from cells, reducing their accumulation in the brain, and can also transmit harmful amplified repetitive RNA. The amplification of RNAs in the bodily fluids of HD patients may serve as a potential biomarker for tracking the progression of the disease. While the role of exosomes in HD transmission is not fully understood, their involvement in the clearance and transmission of mutant proteins and RNA provides insights into the pathogenesis and potential diagnostic strategies for HD.

4.6 Aging-related disorder

The inevitable decline in an organism's biological, physical, and mental capabilities over time is what we mean when we talk about aging. The research encompasses the process of aging and its correlation with age-related degenerative diseases such as osteoporosis (OP), osteoarthritis (OA), and other associated disorders [71]. As people live longer, their risk of developing RA rises. Recent research has shown that ncRNAs can control senescence, suggesting that they may play a role in the control of aging and related disorders. Devastating to the ligaments, cartilage, surrounding bones, and joint lining, osteoarthritis (OA) is a condition of the synovial joints. A combination of factors, including an increase in inflammatory cytokines in aged joints and an imbalance between extracellular matrix (ECM) synthesis and degradation, may contribute to the development of OA. Inflammatory mediators, such as RNA molecules, can be transported via exosomes released by chondrocytes and synovial cells located within the joint. Gene expression in inflammatory pathways can be controlled by exosomal RNAs such as microRNAs (miRNAs). Exosomal miR-146a, for instance, is upregulated in the synovial fluid of OA patients; this miR-146a functions as a negative regulator of the inflammatory nuclear factor kappa B (NF-κB) signaling pathway. Exosomal miRNA dysregulation has been linked to persistent joint inflammation and the development of osteoarthritis. Some of the RNAs that play a role in degrading cartilage and rebuilding the ECM are secreted by chondrocytes in the form of exosomes. Matrix metalloproteinase (MMP) and tissue inhibitor of metalloproteinase (TIMP) genes are among those that can be targeted by exosomal RNAs in the process of preserving cartilage. Disturbance of the equilibrium between matrix synthesis and degradation, which leads to cartilage disintegration in OA, might be caused by an imbalance in the amounts of these RNAs. Age- and OA-associated accumulation of senescent chondrocytes results in exosome release, which can transport miRNAs that regulate cellular senescence and death. Cellular senescence, apoptosis, and compromised cartilage repair are only some of the outcomes of these exosomal miRNAs affecting the behavior of recipient cells. Chronic skeletal sickness characterized by decreased bone strength, poor bone quality, and reduced bone mineral density is called osteoporosis (OP). The aging process has a major role in the pathophysiology of bone deregulation. Bone tissue development is the function of osteoblasts, while bone resorption is the province of osteoclasts. Bone cell exosomes can transport miRNAs (microRNAs) that control osteoblast and osteoclast function and development. For instance, osteoblast-derived exosomal miR-214-3p can inhibit osteoblast activity and degrade bone formation by directly targeting the messenger RNA (mRNA) of runt-related transcription factor 2 (Runx2). Similar to how osteoclast-derived exosomal miR-214-3p can suppress osteoclast development and activity by degrading nuclear factor of activated T cells 1 (NFATc1) mRNA, miR-214-3p is a key transcription factor in the regulation of osteoclastogenesis. The delicate equilibrium between osteoblasts and osteoclasts can be upset by dysregulation of these exosomal miRNAs, which can therefore contribute to the development of osteoporosis. Mesenchymal stem cell—derived exosomes (MSC-derived exosomes) can target the messenger RNA (mRNA) of the receptor activator of nuclear factor kappa-B ligand (RANKL), a critical factor driving osteoclast formation, once delivered to osteoblasts. Bone resorption and osteoclastogenesis are slowed when exosomal miR-23a reduces RANKL expression. The delicate equilibrium between bone formation and resorption might be thrown off by dysregulation of these exosomal miRNAs, leading to osteoporosis. Certain microRNAs (miRNAs) and other RNAs transported by exosomes have been proposed as biomarkers of bone resorption and osteoporosis development. RA is a common and devastating autoimmune disorder that affects the joints. RA is influenced by aging. EVs contribute to the progression of RA because they transport autoantigens, microRNAs, and matrix degradation enzymes. In response to TNF, synovial fibroblasts secrete exosomes, which carry miRNAs that promote and suppress bone growth by primarily targeting inhibitors of bone morphogenetic proteins (BMPs) and Wnt pathways, respectively. In RA, exosomal miRNAs such as miR-155 and miR-146a have been shown to play a role in controlling the expression of inflammatory and immune response genes. Immune cells, including macrophages and T cells, release exosomes into the body that may contain RNA with immunomodulatory properties. Possible contributors to the aberrant immune response seen in RA are exosomal RNAs that govern immune cell activation, differentiation, and polarization. An increase in angiogenesis and synovial hyperplasia, both hallmarks of RA, have been linked to exosomal RNA. Inflamed and hyperplastic synovium can develop in part because exosomes released by synovial fibroblasts and endothelial cells carry proangiogenic proteins and microRNAs that encourage new blood vessel growth. For RA diagnostics, disease activity evaluation, and treatment response monitoring, exosomal RNA has been demonstrated to be a promising biomarker source. Exosomes recovered from synovial fluid or peripheral blood contain various RNA molecules, some of which have been identified as possible biomarkers for disease activity, joint damage, and therapy response. The detection and analysis of exosomal RNAs may aid in understanding the mechanisms behind RA and the creation of targeted therapeutic approaches.

4.7 Cardiovascular diseases

Exosomes, vesicles released by cardiomyocytes (CMs), play a crucial role in maintaining homeostasis in the circulatory system. In a hypoxic environment, such as artery blockage, the release of exosomes from cardiomyocytes increases, leading to a change in their composition [72]. Under stress, cardiomyocytes can produce the proinflammatory cytokine TNF and release it in exosomes, which can trigger apoptosis in nearby healthy cells. This suggests that stressed cells have the ability to adapt to a proinflammatory stimulus by producing exosomes [73]. Exosomes also facilitate communication between different types of cardiac cells. They can transfer DNA and RNA molecules to recipient cells, such as cardiomyocytes and fibroblasts, influencing gene expression in these cells [74]. This intercellular communication mediated by exosomes plays a role in coordinating the function and response of cardiac cells. The understanding of exosomes and their role in cardiac physiology and pathology is still evolving. Further research is needed to uncover the full extent of their functions and potential implications for cardiovascular diseases. Studying exosomal composition and communication in the heart could provide valuable insights into cardiac adaptation, response to stress, and the development of novel therapeutic strategies. There is growing evidence that exosomal microRNAs (miRNAs) can serve as new indicators for cardiovascular disorders [75]. Specific miRNAs, including hsa-miRNA-1 and hsa-miRNA-133a, were shown to be increased in exosomes from individuals with acute coronary syndrome (ACS), suggesting their potential diagnostic usefulness for ACS and acute myocardial infarction (AMI) [76]. Common indicators for AMI, such as troponin and creatine kinase levels, reach their peaks about 12 h after myocardial damage [77]. As their peak levels approach earlier than troponin, exosomal miRNAs show promise as early markers. For instance, miRNA-208a was undetectable in healthy people but present in 10% of AMI patients as early as 4 h after the onset of chest pain. MiRNAs with higher levels detectable within 12 h after AMI have been found as possible biomarkers [78]; they include miRNA-208b, miRNA-1, miRNA-133a, and miRNA-499 [79]. When compared with these miRNAs, troponin is still the gold standard for diagnosis. Both miRNA-208b and miRNA-499-5p were found to have potential as prognostic indicators in heart failure, with greater levels being linked to a higher risk of mortality or heart failure. Further evidence for their usefulness as predictive biomarkers comes from the finding that miRNA-192, miRNA-194, and miRNA-34a were all upregulated in AMI patients with increasing heart failure symptoms within 12 months. Exosomes were extracted from the blood of AMI patients, and these miRNAs were found within them [80]. As a whole, exosomal miRNAs have potential as diagnostic and prognostic indicators in cardiovascular illnesses, with benefits for early identification and risk stratification. To completely establish their therapeutic value and integrate them into ordinary treatment, more research and validation studies are required.

4.8 Renal disease [81]

Exosomal RNA has been found to play a significant role in renal diseases, including acute kidney injury (AKI), chronic kidney disease (CKD), and renal cell carcinoma (RCC). In AKI, exosomes released from injured renal cells contain various types of RNA, including messenger RNA (mRNA), microRNA (miRNA), and long noncoding RNA (lncRNA). These exosomal RNAs can be transferred to neighboring cells, promoting inflammation, apoptosis, and fibrosis, which contribute to the progression of AKI. Studies have shown that exosomal miRNAs, such as miR-146a and miR-155, are involved in regulating the immune response and inflammation during AKI. Additionally, exosomal lncRNAs, such as MALAT1 and NEAT1, have been associated with renal cell apoptosis and fibrosis in AKI. Additionally, exosomal miRNAs have been identified as potential biomarkers for the early detection and prognosis of AKI. In CKD, exosomal RNAs have been implicated in the intercellular communication between renal cells and the development of fibrosis. Exosomes derived from activated fibroblasts or injured tubular epithelial cells carry specific miRNAs and lncRNAs that can regulate the activation of target cells, promote fibroblast-to-myofibroblast transition, and stimulate ECM production. These processes contribute to the progression of renal fibrosis in CKD. Furthermore, exosomal miRNAs have shown promise as diagnostic and prognostic markers for CKD. Exosomes derived from renal cells can transfer profibrotic miRNAs, such as miR-21 and miR-29, to recipient cells, promoting fibrogenesis and contributing to the development of renal fibrosis. Furthermore, exosomal lncRNAs, such as TUG1 and HOTAIR, have been implicated in modulating renal cell proliferation, apoptosis, and fibrosis in CKD. In renal cell carcinoma (RCC), exosomal RNAs have been investigated for their potential role in tumor development, progression, and metastasis.

Exosomes derived from RCC cells carry various types of RNA, including miRNAs, lncRNAs, and mRNA, which can be transferred to recipient cells in the tumor microenvironment. These exosomal RNAs can influence the behavior of recipient cells, such as by promoting angiogenesis, immune evasion, and tumor cell proliferation. Exosomal miRNAs, such as miR-210 and miR-1233, have been identified as promising biomarkers for RCC detection and monitoring disease progression. Additionally, exosomal lncRNAs, such as GAS5 and H19, have shown potential as prognostic markers for RCC patients. Additionally, exosomal miRNAs have been explored as noninvasive biomarkers for RCC diagnosis and monitoring. Several pathogenic processes, such as inflammation, cell death, fibrosis, and tumor growth, are facilitated by exosomal RNA in renal disorders. Insights into disease causes, easier early identification, and new therapy methods could be gained by studying the functions of exosomal RNAs in renal disorders.

4.9 Bacterial infection diseases

Exosomal RNA has been implicated in the modulation of immune responses and intercellular communication during bacterial infections. Here are some important roles for exosomal RNA in bacterial infection:

- Immune response modulation: Exosomes released by immune cells, such as macrophages and dendritic cells, during bacterial infections can contain RNA molecules that modulate the immune response. These exosomal RNAs, including miRNAs and other noncoding RNAs, can be transferred to target cells, such as other immune cells or infected cells, to regulate gene expression and shape the immune response against bacterial pathogens. Exosomal RNA can influence the production of proinflammatory cytokines, antimicrobial peptides, and other immune-related factors, impacting the outcome of the infection.
- Bacterial pathogenesis and virulence: Pathogenic bacteria may enhance their pathogenicity and virulence by using exosomal RNA interactions. Exosomes are one type of bacterial extracellular vesicle found to carry RNA cargo with the ability to regulate host cell responses. These bacterial exosomal RNAs may evade immune responses and promote bacterial survival and dissemination by targeting host immune cells, altering host gene expression, and manipulating immune signaling systems. Exosome-like vesicles carrying bacterial RNA and other components have been demonstrated to be secreted by some bacteria, including *Mycobacterium TB* [82] and *Helicobacter pylori*. Bacterial RNA can be delivered to host cells via these vesicles, where it can influence immune responses, aid in survival, and create long-term infections. By encouraging phagocytosis and boosting the antimicrobial activity of immune cells, exosomal RNA can aid in the elimination of bacterial infections.
- Host—pathogen interactions: Exosomal RNA can mediate bidirectional communication between host cells and bacteria. Exosomes, also known as outer membrane vesicles (OMVs), are RNA-carrying vesicles that can be secreted by bacteria

themselves. These bacterial OMVs are capable of exchanging RNA molecules with other bacteria, which promotes bacterial cooperation and alters the likelihood that a certain strain will become virulent or survive. After interacting with nearby host cells, these exosomes can transport bacterial RNA to trigger an immunological response. Conversely, host cells can influence bacterial growth and pathogenicity by secreting exosomes containing RNA that targets bacterial virulence factors, antimicrobial peptides, or immunological regulators. Gene expression, biofilm development, antibiotic resistance, and other crucial bacterial activities are all influenced by bacterial exosomal RNA. Infected cells can secrete exosomes containing bacterial components such as RNA.

- Horizontal gene transfer: Some bacterial species have been discovered to engage in horizontal gene transfer via exosomal RNA. RNA molecules, including those from transmissible genes such as plasmids and transposons, can be released by bacteria and taken up by recipient bacteria via exosomes. The dissemination of antibiotic-resistant genes and other virulence factors among bacterial populations is facilitated by exosome-mediated gene transfer, which in turn fosters the evolution and adaptation of bacterial pathogens.
- Diagnostic potential: There is hope that exosomal RNA can be used to diagnose bacterial infections. To locate potentially harmful bacteria, it is possible to evaluate the RNA content of exosomes in bodily fluids such as blood or urine. Noninvasive and sensitive detection of bacterial infections using exosomal RNA-based diagnostic techniques could pave the way for early diagnosis and effective treatment.

Overall, exosomal RNA plays a multifaceted role in bacterial infections, including immune response modulation, pathogen clearance, interbacterial communication, and diagnostic and therapeutic potential. Understanding the intricate interactions between exosomal RNA and bacterial pathogens can provide insights into host—pathogen dynamics and open new avenues for the development of diagnostic and therapeutic strategies against bacterial infections. However, more research is needed to unravel the complex mechanisms underlying exosomal RNA-mediated communication during bacterial infections and explore its full potential in clinical applications.

4.10 Wound healing and tissue repairing

Wound healing and tissue repair are significantly facilitated by exosomal RNA [83]. Key functions of exosomal RNA in various cellular activities include the following:

- Cell communication and signaling: Exosomes, including those derived from various cell types involved in wound healing, such as fibroblasts, macrophages, and MSCs, contain diverse RNA molecules, including miRNAs, mRNAs, and other noncoding RNAs. These exosomal RNAs can be transferred to recipient cells at the site of injury, facilitating intercellular communication and signaling. The transferred exosomal RNAs can regulate gene expression and modulate cellular behaviors and influence wound healing processes. Exosomal RNA can regulate gene expression, signaling pathways, and cellular responses, promoting cell migration, proliferation, angiogenesis, and ECM remodeling.

- Angiogenesis and vascularization: Exosomal RNA has been shown to promote angiogenesis, the formation of new blood vessels, which is crucial for delivering oxygen and nutrients to the healing tissues. Exosomes released by endothelial cells, MSCs, and other cell types can carry proangiogenic miRNAs, such as miR-126, miR-130a, and miR-132, which can stimulate endothelial cell proliferation, migration, and tube formation. These exosomal miRNAs can activate angiogenic signaling pathways and facilitate the growth of new blood vessels in the injured area.
- Modulation of inflammation: Exosomal RNA can regulate the inflammatory response during wound healing. Exosomes released by immune cells, such as macrophages, carry miRNAs and other RNA molecules that can target inflammatory mediators, such as cytokines and chemokines, modulating their expression levels and thus influencing the balance between proinflammatory and antiinflammatory signals. Specific miRNAs present in these exosomes can target inflammatory signaling pathways, regulate the recruitment and activation of immune cells, modulate the release of cytokines and growth factors involved in tissue repair, and promote the resolution of inflammation, which is crucial for efficient wound healing.
- ECM remodeling: Wound healing relies on the creation of ECM components such as collagen, fibronectin, elastin, and proteoglycans, and the ability of exosomal RNA molecules to regulate their synthesis and degradation to stimulate ECM production, MMP activity, and tissue remodeling. MMPs and TIMPs control the ECM's remodeling and turnover, and their expression can be influenced by these factors. The equilibrium between ECM deposition and remodeling can be affected by exosomal miRNAs by targeting genes involved in ECM production and degradation. This is necessary for the wound to close and for the newly produced tissue to be properly organized and remodeled.
- Regulation of cell differentiation and stem cell function: Tissue repair stem cells can be influenced in their development and function by exosomal RNA. Cells, such as progenitor cells at the wound site, can take up exosomes generated by stem cells because they contain RNA molecules that regulate their differentiation into specific cell types essential for tissue regeneration. Tissue repair may be assisted by exosomal RNA released by stem cells such as MSCs. Exosomes released by MSCs include a variety of RNAs that can stimulate the community of stem cells or progenitor cells to become activated, migrate, and differentiate. Tissue regeneration is aided by the exosomal RNAs' ability to modulate cellular processes such as cell proliferation, angiogenesis, and ECM development.
- Promotion of cell proliferation and migration: Exosomal RNA can influence cell proliferation and migration, which are essential for tissue regeneration and wound

closure. Certain exosomal miRNAs can enhance cell proliferation by targeting genes involved in cell cycle regulation. Additionally, exosomal RNA can promote cell migration by influencing the expression of genes associated with cell adhesion, cytoskeletal rearrangement, and ECM remodeling.

Exosomal RNA plays a role in the intricate system of intercellular communication and signaling events that control the capacity of the human body to heal wounds and repair damaged tissues. Tissue regeneration and repair is promoted by the exchange of exosomal RNAs between cells, which coordinate processes such as cell proliferation, migration, angiogenesis, immune response modulation, stem cell function, and ECM remodeling. Exosomal RNA is believed to have an aspect in wound healing and tissue repair, although further study is needed to determine its precise function in these processes.

5. Concluding remarks

In summary, further research is needed to explore the therapeutic and biomarker potential of exosomes, which is still in its early stages. Exosomes play a crucial role in various aspects of human biology by facilitating communication processes that are implicated in disorders affecting the central nervous system, metabolism, immune system, cardiovascular system, and more. These vesicles transport a diverse array of information through RNA and proteins. Exosomal RNA contributes to the therapeutic qualities of exosomes formed from stem cells and provides early diagnostic information, although the specific methods by which exosomes affect biological processes still need to be elucidated. Exosomal therapy has a great deal of potential because these vesicles can be designed to carry and transmit genetic material, influencing the progression of diseases and targeting specific organ systems and cells. By rearranging the genetic information carried within exosomes, it becomes possible to modify processes that underlie physiological abnormalities. However, to fully exploit the potential of exosomal therapy, a comprehensive understanding of the intricate cellular interactions mediated by exosomes is necessary. In-depth studies of these interactions are crucial to unraveling the mechanisms of information transmission and utilizing them for diagnostic and therapeutic purposes. Given the importance of exosomes in human biology, it is vital to continue investigating their therapeutic and biomarker potential. Through ongoing research, exosomal therapy has the potential to revolutionize the field of medicine by offering innovative approaches for diagnosis, treatment, and monitoring of various conditions.

References

[1] Doyle LM, Wang MZ. Overview of extracellular vesicles, their origin, composition, purpose, and methods for exosome isolation and analysis. Cells 2019;8(7):727.

[2] Abels ER, Breakefield XO. Introduction to extracellular vesicles: biogenesis, RNA cargo selection, content, release, and uptake. Cellular and Molecular Neurobiology 2016;36(3):301–12.

[3] Raposo G, Stoorvogel W. Extracellular vesicles: exosomes, microvesicles, and friends. The Journal of Cell Biology 2013;200(4):373–83.
[4] Kalluri R, LeBleu VS. The biology, function, and biomedical applications of exosomes. Science 2020;367(6478).
[5] Zhang Y, Liu Y, Liu H, Tang WH. Exosomes: biogenesis, biologic function and clinical potential. Cell and Bioscience 2019;9(19).
[6] Simpson RJ, Kalra H, Mathivanan S. ExoCarta as a resource for exosomal research. Journal of Extracellular Vesicles 2012;1(1):18374.
[7] Kok VC, Yu CC. Cancer-derived exosomes: their role in cancer biology and biomarker development. International Journal of Nanomedicine 2020:8019–36.
[8] Hu C, Meiners S, Lukas C, Stathopoulos GT, Chen J. Role of exosomal microRNAs in lung cancer biology and clinical applications. Cell Proliferation 2020;53(6):e12828.
[9] Manna I, Iaccino E, Dattilo V, Barone S, Vecchio E, Mimmi S, et al. Exosome-associated miRNA profile as a prognostic tool for therapy response monitoring in multiple sclerosis patients. The FASEB Journal 2018;32(8):4241–6.
[10] Xu K, Zhang C, Du T, Gabriel ANA, Wang X, Li X, et al. Progress of exosomes in the diagnosis and treatment of lung cancer. Biomedicine and Pharmacotherapy 2021;134:111111.
[11] Li S, Yao J, Xie M, Liu Y, Zheng M. Exosomal miRNAs in hepatocellular carcinoma development and clinical responses. Journal of Hematology and Oncology 2018;11:1–9.
[12] Machlowska J, Baj J, Sitarz M, Maciejewski R, Sitarz R. Gastric cancer: epidemiology, risk factors, classification, genomic characteristics and treatment strategies. International Journal of Molecular Sciences 2020;21(11):4012.
[13] Santasusagna S, Moreno I, Navarro A, Castellano JJ, Martinez F, Hernández R, et al. Proteomic analysis of liquid biopsy from tumor-draining vein indicates that high expression of exosomal ECM1 is associated with relapse in stage I-III colon cancer. Translational Oncology 2018;11(3):715–21.
[14] Xu YF, Xu X, Bhandari K, Gin A, Rao CV, Morris KT, et al. Isolation of extra-cellular vesicles in the context of pancreatic adenocarcinomas: addition of one stringent filtration step improves recovery of specific microRNAs. PLoS One 2021;16(11):e0259563.
[15] Lugli G, Cohen AM, Bennett DA, Shah RC, Fields CJ, Hernandez AG, et al. Plasma exosomal miRNAs in persons with and without Alzheimer disease: altered expression and prospects for biomarkers. PLoS One 2015;10(10):e0139233.
[16] Bhagirath D, Yang TL, Bucay N, Sekhon K, Majid S, Shahryari V, et al. microRNA-1246 is an exosomal biomarker for aggressive prostate CancerExosomal microRNA in aggressive prostate cancer. Cancer Research 2018;78(7):1833–44.
[17] Bhome R, Del Vecchio F, Lee GH, Bullock MD, Primrose JN, Sayan AE, et al. Exosomal microRNAs (exomiRs): small molecules with a big role in cancer. Cancer Letters 2018;420:228–35.
[18] Zhang L, Zhang S, Yao J, Lowery FJ, Zhang Q, Huang WC, et al. Microenvironment-induced PTEN loss by exosomal microRNA primes brain metastasis outgrowth. Nature 2010;527(7576):100–4.
[19] Whiteside TL. The effect of tumor-derived exosomes on immune regulation and cancer immunotherapy. Future Oncology 2017;13(28):2583–92.
[20] Chen G, Huang AC, Zhang W, Zhang G, Wu M, Xu W, et al. Exosomal PD-L1 contributes to immunosuppression and is associated with anti-PD-1 response. Nature 2018;560(7718):382–6.
[21] Li Y, Zheng Q, Bao C, et al. Circular RNA is enriched and stable in exosomes: a promising biomarker for cancer diagnosis. Cell Research 2015;25:981–4.
[22] Zhao RT, Zhou J, Dong XL, Bi CW, Jiang RC, Dong JF, et al. Circular ribonucleic acid expression alteration in exosomes from the brain extracellular space after traumatic brain injury in mice. Journal of Neurotrauma 2018;35(17):2056–66.
[23] Zhao Z, Zhao G, Yang S, et al. The significance of exosomal RNAs in the development, diagnosis, and treatment of pancreatic cancer. Cancer Cell International 2021;21:364.
[24] Fernández PL, Hernández L, Farré X, Campo E, Cardesa A. Alterations of cell cycle-regulatory genes in prostate cancer. Pathobiology: Journal of Immunopathology, Molecular and Cellular Biology 2002;70(1):1–10.

[25] Xu Y, Xia F, Ma L, Shan J, Shen J, Yang Z, et al. MicroRNA-122 sensitizes HCC cancer cells to adriamycin and vincristine through modulating expression of MDR and inducing cell cycle arrest. Cancer Letters 2011;310(2):160–9.

[26] Fang F, Chang RM, Yu L, Lei X, Xiao S, Yang H, et al. MicroRNA-188-5p suppresses tumor cell proliferation and metastasis by directly targeting FGF5 in hepatocellular carcinoma. Journal of Hepatology 2015;63(4):874–85.

[27] Xue J, Liu Y, Luo F, Lu X, Xu H, Liu X, et al. Circ100284, via miR-217 regulation of EZH2, is involved in the arsenite-accelerated cell cycle of human keratinocytes in carcinogenesis. Biochimica et Biophysica Acta - Molecular Basis of Disease 2017;1863(3):753–63.

[28] Fidler IJ. The pathogenesis of cancer metastasis: the 'seed and soil' hypothesis revisited. Nature Reviews Cancer 2003;3(6):453–8.

[29] Luna J, Boni J, Cuatrecasas M, Bofill-De Ros X, Núñez-Manchón E, Gironella M, et al. DYRK1A modulates c-MET in pancreatic ductal adenocarcinoma to drive tumour growth. Gut 2019;68(8):1465–76.

[30] Zhang X, Wang S, Wang H, et al. Circular RNA circNRIP1 acts as a microRNA-149-5p sponge to promote gastric cancer progression via the AKT1/mTOR pathway. Molecular Cancer 2019;18:20.

[31] Wang J, Zhang Q, Zhou S, Xu H, Wang D, Feng J, et al. Circular RNA expression in exosomes derived from breast cancer cells and patients. Epigenomics 2019;11(4):411–21.

[32] Dou Y, Cha DJ, Franklin JL, Higginbotham JN, Jeppesen DK, Weaver AM, et al. Circular RNAs are down-regulated in KRAS mutant colon cancer cells and can be transferred to exosomes. Scientific Reports 2016;6:37982.

[33] Li Z, Yanfang W, Li J, Jiang P, Peng T, Chen K, et al. Tumor-released exosomal circular RNA PDE8A promotes invasive growth via the miR-338/MACC1/MET pathway in pancreatic cancer. Cancer Letters 2018;432:237–50.

[34] Zhang H, Deng T, Ge S, et al. Exosome circRNA secreted from adipocytes promotes the growth of hepatocellular carcinoma by targeting deubiquitination-related USP7. Oncogene 2019;38:2844–59.

[35] Smyth SS, McEver RP, Weyrich AS, Morrell CN, Hoffman MR, Arepally GM, et al. Platelet colloquium participants (2009). Platelet functions beyond hemostasis. Journal of Thrombosis and Haemostasis 2009;7(11):1759–66.

[36] Preußer C, Hung LH, Schneider T, Schreiner S, Hardt M, Moebus A, et al. Selective release of circRNAs in platelet-derived extracellular vesicles. Journal of Extracellular Vesicles 2018;7(1):1424473.

[37] Xu X, Cui L, Zhong W, Cai Y. Autophagy-associated lncRNAs: promising targets for neurological disease diagnosis and therapy. Neural Plasticity; 2020.

[38] Zeng R, Song XJ, Liu CW, Ye W. LncRNA ANRIL promotes angiogenesis and thrombosis by modulating microRNA- 99a and microRNA-449a in the autophagy pathway. American Journal of Tourism Research 2019;11(12):7441–8.

[39] Xu Y, Chen J, Wang M, Yu R, Zou W, Shen W. Mechanism of lncRNA-ANRIL/miR-181b in autophagy of cardiomyocytes in mice with uremia by targeting ATG5. PLoS One 2021;16(9):e0256734.

[40] Yang L, Wang H, Shen Q, Feng L, Jin H. Long non-coding RNAs involved in autophagy regulation. Cell Death and Disease 2017;8(10):e3073.

[41] Yuan L, Yu L, Zhang J, Zhou Z, Li C, Zhou B, et al. Long noncoding RNA H19 protects H9c2 cells against hypoxia-induced injury by activating the PI3K/AKT and ERK/p38 pathways. Molecular Medicine Reports 2020;21(4):1709–16.

[42] Wang J, Cao B, Han D, Sun M, Feng J. Long non-coding RNA H19 induces cerebral ischemia-reperfusion injury via activation of autophagy. Aging and Disease 2017;8(1):71–84.

[43] Li X, Zhao J, Zhang H, Cai J. Silencing of lncRNA metas¬tasis-associated lung adenocarcinoma transcript 1 inhibits the proliferation and promotes the apoptosis of gastric cancer cells through regulating microRNA-22-3p-mediated ErbB3. OncoTargets and Therapy 2020;13:559–71.

[44] Li Z, Li J, Tang N. Long noncoding RNA Malat1 is a potent autophagy inducer protecting brain microvascular endothelial cells against oxygen-glucose deprivation/reoxygenation-induced injury by sponging miR-26b and upregulating ULK2 expres¬sion. Neuroscience 2017;354(1–10).

[45] Guo D, Ma J, Yan L, Li T, Li Z, Han X, et al. Down-regulation of lncrna MALAT1 attenuates neuronal cell death through suppressing Beclin1-dependent autophagy by Regulat¬ing Mir-30a in cerebral ischemic stroke. Cellular Physiology and Biochemistry 2017;43(1):182−94.
[46] Chang W, Wang J. Exosomes and their noncoding RNA cargo are emerging as new modulators for diabetes mellitus. Cells 2019;8(8):853.
[47] Cianciaruso C, Phelps EA, Pasquier M, Hamelin R, Demurtas D, Alibashe Ahmed M, et al. Primary human and rat β-cells release the intracellular autoantigens GAD65, IA-2, and proinsulin in exosomes together with cytokine-induced enhancers of immunity. Diabetes 2017;66(2):460−73.
[48] Lei LM, Lin X, Xu F, Shan SK, Guo B, Li FX, Zheng MH, Wang Y, Xu QS, Yuan LQ. Exosomes and obesity-related insulin resistance. Frontiers in Cell and Developmental Biology 2021;9:651996.
[49] Li CJ, Fang QH, Liu ML, Lin JN. Current understanding of the role of adipose-derived extracellular vesicles in metabolic homeostasis and diseases: communication from the distance between cells/tissues. Theranostics 2020;10(16):7422−35.
[50] Castaño C, Kalko S, Novials A, Párrizas M. Obesity-associated exosomal miRNAs modulate glucose and lipid metabolism in mice. Proceedings of the National Academy of Sciences of the United States of America 2018;115(48):12158−63.
[51] Jin Y, Mazza C, Christie JR, Giliani S, Fiorini M, Mella P, et al. Mutations of the Wiskott-Aldrich Syndrome Protein (WASP): hotspots, effect on transcription, and translation and phenotype/genotype correlation. Blood 2004;104(13):4010−9.
[52] Sauer AV, Di Lorenzo B, Carriglio N, Aiuti A. Progress in gene therapy for primary immunodeficiencies using lentiviral vectors. Current Opinion in Allergy and Clinical Immunology 2014;14(6):527−34.
[53] Sheldon H, Andre M, Legg JA, Heal P, Herbert JM, Sainson R, et al. Active involvement of Robo1 and Robo4 in filopodia formation and endothelial cell motility mediated via WASP and other actin nucleation-promoting factors. The FASEB Journal: Official Publication of the Federation of American Societies for Experimental Biology 2009;23(2):513−22.
[54] Brigida I, Sauer AV, Ferrua F, Giannelli S, Scaramuzza S, Pistoia V, et al. B-cell development and functions and therapeutic options in adenosine deaminase-deficient patients. The Journal of Allergy and Clinical Immunology 2014;133(3):799−806.
[55] Pala F, Morbach H, Castiello MC, Schickel JN, Scaramuzza S, Chamberlain N, et al. Lentiviral-mediated gene therapy restores B cell tolerance in Wiskott-Aldrich syndrome patients. Journal of Clinical Investigation 2015;125(10):3941−51.
[56] Ouyang Q, Wu J, Jiang Z, Zhao J, Wang R, Lou A, et al. Microarray expression profile of circular RNAs in peripheral blood mononuclear cells from rheumatoid arthritis patients. Cellular Physiology and Biochemistry: International Journal of Experimental Cellular Physiology, Biochemistry, and Pharmacology 2017;42(2):651−9.
[57] Fairhurst AM, Wandstrat AE, Wakeland EK. Systemic lupus erythematosus: multiple immunological phenotypes in a complex genetic disease. Advances in Immunology 2006;92(1−69).
[58] Liu W, Bai X, Zhang A, Huang J, Xu S, Zhang J. Role of exosomes in central nervous system diseases. Frontiers in Molecular Neuroscience 2019;12:240.
[59] Mondello S, Guedes VA, Lai C, Czeiter E, Amrein K, Kobeissy F, et al. Circulating brain injury exosomal proteins following moderate-to-severe traumatic brain injury: temporal profile, outcome prediction and therapy implications. Cells 2020;9(4):977.
[60] Xia X, Wang Y, Huang Y, Zhang H, Lu H, Zheng JC. Exosomal miRNAs in central nervous system diseases: biomarkers, pathological mediators, protective factors and therapeutic agents. Progress in Neurobiology 2019;183:101694.
[61] D'Anca M, Fenoglio C, Serpente M, Arosio B, Cesari M, Scarpini EA, et al. Exosome determinants of physiological aging and age-related neurodegenerative diseases. Frontiers in Aging Neuroscience 2019;11:232.
[62] Wu X, Zheng T, Zhang B. Exosomes in Parkinson's disease. Neuroscience Bulletin 2017;33(3):331−8.
[63] Tomlinson PR, Zheng Y, Fischer R, Heidasch R, Gardiner C, Evetts S, et al. Identification of distinct circulating exosomes in Parkinson's disease. Annals of Clinical and Translational Neurology 2015;2(4):353−61.

[64] Zagrean AM, Hermann DM, Opris I, Zagrean L, Popa-Wagner A. Multicellular crosstalk between exosomes and the neurovascular unit after cerebral ischemia therapeutic implications. Frontiers in Neuroscience 2018;12:811.
[65] Thome AD, Harms AS, Volpicelli-Daley LA, Standaert DG. microRNA-155 regulates alpha-synuclein-induced inflammatory responses in models of Parkinson disease. Journal of Neuroscience: The Official Journal of the Society for Neuroscience 2016;36(8):2383–90.
[66] Leggio L, Vivarelli S, L'Episcopo F, Tirolo C, Caniglia S, Testa N, et al. microRNAs in Parkinson's disease: from pathogenesis to novel diagnostic and therapeutic approaches. International Journal of Molecular Sciences 2017;18(12):2698.
[67] Yang J, Zhang X, Chen X, Wang L, Yang G. Exosome mediated delivery of miR-124 promotes neurogenesis after ischemia. Molecular Therapy - Nucleic Acids 2017;7:278–87.
[68] Titze-de-Almeida R, Titze-de-Almeida SS. miR-7 replacement therapy in Parkinson's disease. Current Gene Therapy 2018;18(3):143–53.
[69] Finkbeiner S. Huntington's disease. Cold Spring Harbor Perspectives in Biology 2011;3(6):a007476.
[70] Tang BL. Unconventional secretion and intercellular transfer of mutant Huntingtin. Cells 2018;7(6):59.
[71] Li H, Zheng Q, Xie X, Wang J, Zhu H, Hu H, et al. Role of exosomal non-coding RNAs in bone-related diseases. Frontiers in Cell and Developmental Biology 2021;9:811666.
[72] Isola AL, Chen S. Exosomes: the messengers of health and disease. Current Neuropharmacology 2017;15(1):157–65.
[73] Yu H, Wang Z. Cardiomyocyte-derived exosomes: biological functions and potential therapeutic implications. Frontiers in Physiology 2019;10:1049.
[74] Ranjan P, Kumari R, Verma SK. Cardiac fibroblasts and cardiac fibrosis: precise role of exosomes. Frontiers in Cell and Developmental Biology 2019;7:318.
[75] Bellin G, Gardin C, Ferroni L, Chachques JC, Rogante M, Mitrečić D, et al. Exosome in cardiovascular diseases: a complex world full of hope. Cells 2019;8(2):166.
[76] Halushka PV, Goodwin AJ, Halushka MK. Opportunities for microRNAs in the crowded field of cardiovascular biomarkers. Annual Review of Pathology 2019;14:211–38.
[77] Babuin L, Jaffe AS. Troponin: the biomarker of choice for the detection of cardiac injury. Canadian Medical Association Journal: Canadian Medical Association Journal 2005;173(10):1191–202.
[78] Li M, Zhang J. Circulating MicroRNAs: potential and emerging biomarkers for diagnosis of cardiovascular and cerebrovascular diseases. BioMed Research International 2015;730535:2015.
[79] Colpaert RMW, Calore M. MicroRNAs in cardiac diseases. Cells 2019;8(7):737.
[80] Evans S, Mann DL. Myocardial infarction: the long and arduous road from scientific discovery to clinical utility. Circulation Research 2013;113(3):242–4.
[81] Jin J, Sun H, Shi C, Yang H, Wu Y, Li W, et al. Circular RNA in renal diseases. Journal of Cellular and Molecular Medicine 2020;24(12):6523–33.
[82] Mirzaei R, Babakhani S, Ajorloo P, et al. The emerging role of exosomal miRNAs as a diagnostic and therapeutic biomarker in *Mycobacterium tuberculosis* infection. Molecular Medicine 2021;27(34).
[83] Ti D, Hao H, Fu X, et al. Mesenchymal stem cells-derived exosomal microRNAs contribute to wound inflammation. Science China Life Sciences 2016;59:1305–12.

CHAPTER TWO

Biogenesis of exosome and exosomal RNA and its isolation and physical characterizations

Soumyadipta Saha, Ananya Sarkar, Saptarshi Ray, Moumita Mandal, Debalina Bhattacharya
Department of Microbiology, Maulana Azad College, Kolkata, West Bengal, India

1. Introduction

Extracellular vesicles are spherical particles with phospholipid bilayers that are released by different cell types in body fluids in vivo and by cultured cells in vitro [1,2]. These body fluids include serum, urine, cerebrospinal fluid, breast milk, aqueous humor, and amniotic fluid [1,2]. According to the cellular genesis and synthesis, extracellular vesicles are often separated into exosomes, microvesicles, and apoptotic bodies [3]. Exosomes have a size range of 40—150 nm and come from the cell's endosomal compartment [4]. Exosomes have a prospective diagnostic/prognostic value when they are secreted into biological fluids such as saliva, sputum, plasma, and bronchoalveolar lavage fluid [4,5]. Exosomes are not only responsible for cellular waste disposal but also have a significant role in intercellular communications [2]. They are capable of manipulating the characteristics of target cells since they act as bioeffector units [6]. Exosomes consist of numerous constituents of the originating cell, such as cellular surface proteins, lipids, chemicals, and genetic materials such as DNA, mRNA, noncoding RNAs etc. [5]. Over the years, all these exosomal RNAs gained specific interest as they play a crucial role in disease progression and can stimulate angiogenesis and facilitate metastasis in cancers [7,8]. As multivesicular bodies (MVBs), joined to the cytoplasmic membrane and exocytosis is triggered, exosomes are sent out into the extracellular matrix [9]. Biogenesis of exosomes takes place by various steps via different methods: endosomal sorting complex required for transport (ESCRT)-dependent and ESCRT-independent mechanisms [9]. Despite the fact that isolation and characterisation of exosomes are very difficult still finding the optimal method to isolate exosomes is crucial for discovering new biomarkers. However, to assure clinical development of exosome-based therapy and diagnosis, a thorough understanding of exosome biology is required.

Exosomal RNA
ISBN 978-0-443-14008-2
https://doi.org/10.1016/B978-0-443-14008-2.00008-5

Therefore, in this chapter, we have focused about exosomal biogenesis and various types of exosomal RNAs in detail. Additionally, different methods for isolation, purification, and analysis of exosomes have also been discussed.

1.1 Biogenesis and release of exosomes

The biogenesis of exosomes includes multiple steps although the exact mechanism is still not known properly. But the sorting mechanisms have been studied in this case [7,10]. The following steps are described in the following:

- Fusion of different primary endocytic vesicles and the coated vesicles from trans-Golgi network (TGN) with early endosome (EE), located near the cell periphery.
- Early endosomes (EEs) either return the cargos to the cell membrane, termed as "recycling endosomes" [11] or subsequently convert it into late endosomes (LEs), also known as multivesicular bodies (MVBs) in a process called, endosomal maturation.
- In course of maturation, various structural changes occur in early endosomes. Specific cargos are delivered through trans-Golgi network followed by acidification of lumen, then endosomal membrane invaginates and pinches off to form many intraluminal vesicles (ILVs). The formation of MVBs with ILVs is the main event of the biogenesis [12].
- The mechanism for protein sorting and generation of ILVs are classified into two pathways:
  (i) ESCRT-dependent pathways. This pathway is mediated by ESCRT-0, ESCRT-I, ESCRT-II, ESCRT-III complex with accessory proteins such as ALIX, Vsp4, etc.
  (ii) ESCRT-independent pathways. It involves proteins such as tetraspanins, lipids such as ceramide and phosphatidic acid.
- MVBs can fuse with lysosome followed by degradation of contents or it can migrate to the plasma membrane, get fused with it, and release ILVs as exosomes. If MVBs choose secretory route, then they will follow the exocytosis pathway for migration, fusion, and releasing the exosomes [10].
- After releasing to extracellular environment, exosomes are taken up by target recipient cells (located near or far) by direct fusion or by any method of endocytosis. Fig. 2.1 represents how exosomes are synthesized and released by the cells.

Now the question arises, how can the fate of MVBs be determined? Different studies have established the fact that cell can control exosome production and secretion on the basis of its need [13,14]. There is a balanced regulation between them, but the molecular mechanism of determination of fate is not defined [15]. Recently, it has been proposed that ISGylation (ubiquitin-like modification) of MVB protein, can regulate exosome secretion by inducing lysosomal degradetion [16]. MVBs can also follow degradation

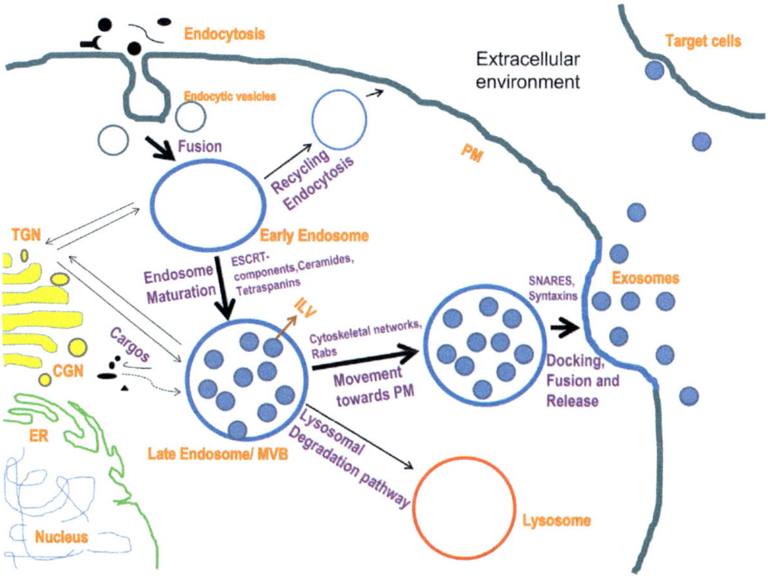

Figure 2.1 Schematic representation of exosome biogenesis.

pathway if the cargos are not deubiquitinated by deubiquitylating enzymes (DUBs) [11]. Studies have shown that cellular stress, heat, chemical exposure, oxidative stress, high or low glucose content, impaired lysosomal function, hypoxic state of tumor cells, increased rate of glycolysis, hypoxia, intracellular Ca^{2+} concentration, pH fluctuations, and energy starvation all can stimulate the exosomal biogenesis and release [17,18].

1.2 Endosomal sorting complex required for transport and exosomal biogenesis

ESCRT proteins are necessary in exosome biogenesis from various cell types or bodily fluids [19]. ESCRT machinery is composed of approximately 30 proteins, and it is arranged into four protein complexes, which are denoted as ESCRT-0, -I, -II, and -III. Numerous proteins such as VPS4, VTA1, and ALIX are also associated with this machinery [19]. The ESCRT complex frequently relocates to the cytosolic side of the endosomal membrane so that a certain protein can be put into ILVs. STAM1/2 (signal transducing adaptor molecule 1/2) and HRS (hepatocyte growth factor–regulated tyrosine kinase substrate) are the two parts that make up ESCRT-0. They interact with each other and can find ubiquitin-rich and PI3P-rich regions. In early endosomes, ESCRT-0 recognizes ubiquitinated proteins and relocates them to the regions that are rich in phosphatidyl inositol monophosphate (PI3P) [3]. The complex aids in the clustering, transportation, and further recruitment of additional protein complexes of ESCRT-I (Alice et al., 2019). ESCRT-I is a heteromeric complex made up of TSG101 (tumor susceptibility gene 101), Vps28, Vps37, and Mvb12 (multivesicular body 12). The TSG101 part of ESCRT-I replaces ESCRT-0 and brings in the ESCRT-II complex [4]. The four components of the ESCRT-II complex are Vps22-EAP30, Vps36-EAP45, and two subunits of Vps25-EAP20. By permitting the internalization of various molecules and cargoes, such as proteins or nucleic acids, ESCRTI-II can start the process that leads to oligomerization and the development of the ESCRT-III complex [20]. ESCRT-III is another heterotetrametric complex containing four subunits, i.e., Vps20-CHMP6, Snf7-CHMP4, Vps24-CHMP3, and Vps2-CHMP2. The joining of Vps25 (a subunit of ESCRT-II) and Vps20 (a subunit of ESCRT-III) is what makes this complex active and brings it to the endosome. This complex participates in the sequestration of MVB proteins. Additionally, auxiliary subunits recruited by ESCRT-III, such as Bro1/Alix, mediate cargo deubiquitination. Before sorting Snf7 proteins into ILVs, the complex will bind to them and then activate the enzyme Doa4 (degradation of alpha 4), and completing the deubiquitination process [20]. The complete invagination of membrane is necessary for generation of ILVs. Accessory proteins VPS4 ATPase dissociate and facilitate in recycling the ESCRT machinery for the next round of exosomal biogenesis [20]. According to a number of research, Syntenin is a 32-kDa protein that has two adjacent PDZ domains: postsynaptic density protein 95/discs large protein/zonula occludens 1 [21]. Both the binding of syntenin to membranes and the high-affinity association with syndecans depend on the two PDZ domains [21]. ILV, MVBs, and exosomes are all formed by the syntenin protein [21]. This cytoskeletal adapter-like molecules also have a pivotal role in modification and maintenance of vesicle morphology [22]. A link between the ESCRT machinery and syndecans is created by the multivalent soluble protein syntenins by binding to ALIX [22]. It is speculated that ALIX (also known as PDCD6IP), a protein that binds with a number of ESCRT proteins (TSG101 and CHMP4), has a role in the expanding and breakdown processes [23,24]. It is shown that exosomal syndecans (membrane proteins carrying heparan sulfate chains in syntenin), ALIX, and syntenin coaccumulate and support the exosomal biogenesis [23]. Here, ESCRT-dependent and independent pathways is represented by a pictorial diagram in Fig. 2.2.

1.3 ESCRT-independent mechanism of exosome biogenesis

Self-organizing lipids and cargo domains facilitate the loading of RNA into exosomes, as opposed to protein sorting by ESCRT. This extensively described exosomal biogenesis pathway is an ESCRT-independent process. Microdomains designed around lipid rafts are necessary for the sideways distribution of cargo along with the endosomal membrane [25,26]. Lipid rafts are subdomains of plasma membrane. They have more sphingolipids, proteins that are attached to glycosylphosphatidylinositol (GPI), and

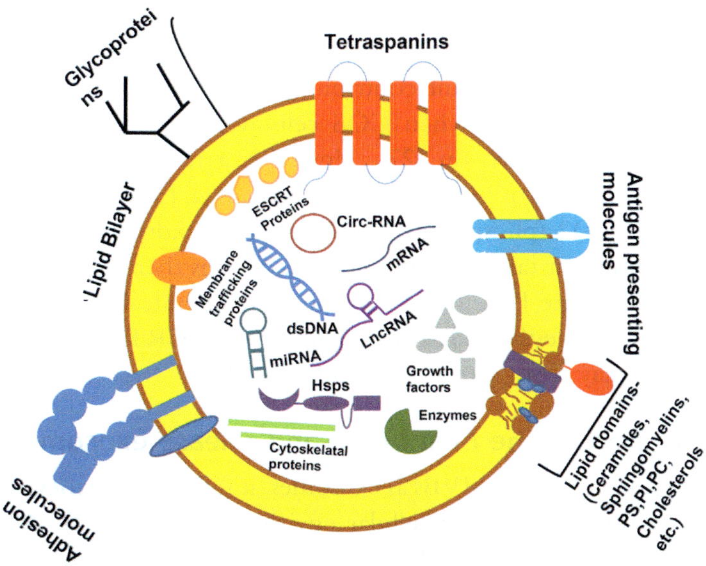

Figure 2.2 ESCRT dependent and independent pathways of exosome biogenesis and release.

cholesterol. This will make it easier for them to connect with other proteins or chemicals and secrete them through exosomes [25]. These microdomains contain significant amounts of sphingomyelinases, which are able to hydrolyze the phosphocholine moiety to create ceramides [26]. Ceramides have been found to promote microdomain coalescence and lateral phase differentiation in the model membranes [26]. The outline of individual membrane lipids, which is influenced by the dimensions of the head group and the length and saturation of the acyl chains, has a significant impact on membrane curvature [27]. The endosomal membrane may naturally curve in a negative direction due to the ceramide's cone-shaped structure [27]. The occurrence of ceramide, lysophospholipid, and glycosphingolipid molecules on the limiting membrane induces the spontaneous budding-in procedure to form ILVs. The ceramide is converted into sphingosine and sphingosine 1 phosphate (S1P) in the presence of ceramidase and sphingosine kinase. Constant stimulation of sphingosine 1 phosphate receptors on the limiting membrane essentially causes tetraspanin to be sorted into ILVs [27]. Tetraspanin is a member of the protein superfamily of membrane proteins that are connected to cell surfaces [28]. It has four transmembrane domains that define it. Tetraspanin-enriched microdomains (TEMs) are widely distributed specialized membrane platforms for compartmentalizing receptors and cytosolic signaling proteins in the plasma membrane, which are organized by tetraspanins [28]. It was reported that the expression of the tetraspanins CD9 and CD82 promotes the exosomal release of beta-catenin from

HEK293 cells [28]. Another protein SIMPLE (small integral membrane protein of the lysosome/late endosome), also known as LITAF (lipopolysaccharide induced TNF factor), is known to have important contribution in exosome formation [29]. In an investigation, it was found that when COS cells were transfected with this LITAF protein, exosomal release went up [27]. However, a mutation of SIMPLE/LITAF prevents the right formation of MVB [27]. Bebelman et al. [30] also looked at the lipid-modifying enzyme PLD2 (phospholipase D2) in the backdrop of how exosomes are made. It is an enzyme that takes phospholipids and turns them into phosphatidic acid (PA). Hessvik and Llorente [29] found that bioactive chemicals will be put into exosomes in a specific way by either ESCRT-dependent or ESCRT-independent pathways, and these pathways may work on various kinds of cells.

1.4 MVBs transport to the membrane and release exosomes

When the cell decides to secrete MVBs as exosomes, it follows the exocytosis pathway to release ILVs as exosomes into extracellular environment (Fig. 2.1). All these events involve several components, cytoskeletal elements, molecular motors, Rab GTPases with effectors, RAL GTPases, SNARES, and tethering factors [31]. Rab GTPases are known to regulate many steps in membrane trafficking, vesicle budding, transport, and fusion [31]. Rab27 (two isoforms, 27A and 27B), Rab35, Rab7, Rab11, Rab5A, Rab9A, and RAL are specifically associated with transportation and release of exosomes [32]. It has been reported that Rab27A involves in this transport toward cell membrane and both Rab27A and Rab27B play their role in docking [32]. In primary oligodendrocytes and brain cells, docking and tethering are carried out with Rab35 [33]. Rab7 effects the lysosomal pathway as well as the exosomal release [32]. Additionally, Rho, Rac, and Cdc42 family of GTPases are involved in this process [33]. The transport of MVBs to cell membrane is mediated by interaction of Rab proteins with corresponding effectors and also with the molecular motors (dyenin, myosin, kinesin) along with actin networks or microtubules (from "−" to "+" end). Dense actin network at the cell periphery, invadopodia (F-actin-rich membrane protrusions) on tumor cells, serves as docking site of MVBs through the binding of exosomal actin—binding protein (e.g., cortactin) with actin [33,34]. The final step, fusion of MVBs carrying ILVs with the cell membrane, is carried out by SNARE protein family [35]. SNAREs are located on both vesicular membrane (v-SNARE/R-SNARE) and cell membrane (t-SNARE/Q-SNARE) [35]. In general, specific Rab GTPases bind with tethering factor (e.g., exocyst). Then, the v-SNARE and t-SNARE pair with each other to form "Trans-SNARE complex," which leads to docking and facilitates the fusion of two membranes, resulting into the release of exosomes to extracellular space [35]. In this case, the SNAREs are also diversified with cellular types; v-SNAREs such as VAMP7, VAMP3, Ykt6, and t-SNAREs such as syntaxin 1A, syntaxin 5, SNAP-23, and SNAP-29, etc. are reported in relation to exosome release [10,29,33]. Additionally, synaptogamin-7 (a calcium sensor)-enhancing exosome secretion upon treatment with Ca^{2+} ionophore indicates the Ca^{2+}-dependent regulation of exosome release [29]. Moreover, tetraspanins also have a role in this event, and it has been observed that the addition of lipid precursor increases the release of exosomes [28].

1.5 Basal composition and cargo

Exosomes are made up of lipid bilayer, which encapsulate several protein complexes, lipid molecules, and different types of nucleic acids. The basal composition and the cargos of exosomes represent the cellular nature and specific physiological states [10,36]. ExoCarta, Vesiclepedia, and EVpedia are the databases for exosomal contents [10,11]. As per the latest update on ExoCarta, 9769 proteins, 3408 mRNAs, 2838 miRNAs, and 1116 lipids have been reported [10]. Fig. 2.3 represents the basal composition and cargo of exosomes.

1.5.1 Nucleic acids

The presence of DNA, mRNAs, and noncoding RNAs in exosomes has been reported [37]. They are described in the following:

- Exosomal DNA, also known as exoDNA, can have mitochondrial or genomic origins. The genotype of recipient cells may change as a result of exosomal DNA transfer (HGT). Additionally, even though exosomes may cause apoptosis, a cell can

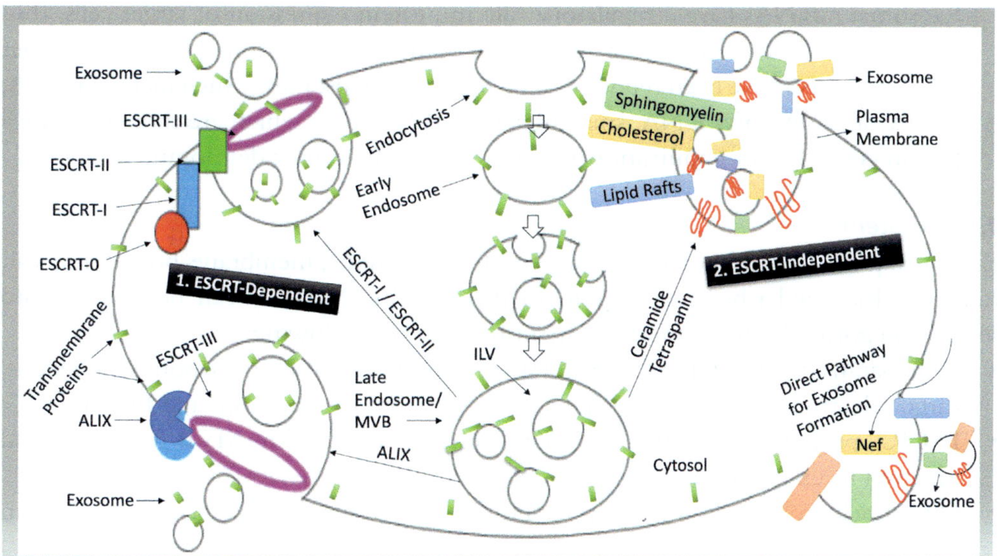

Figure 2.3 The basal composition and cargo of exosomes.

maintain homeostasis by releasing DNA fragments [37,38]. It has been found that exosomes secreted from myeloid-derived regulatory cells contain mitochondrial DNA (mt DNA) along with mitochondria. Also, they may show importance in cell-to-cell communication during inflammation [39].

- MicroRNA (miRNA) is the most abundant RNA species found in exosomal contents. miR-320 and miR-150 are mostly present in exosomes; and miR-214, miR-1, miR-451, miR-155 etc. are some other examples [39]. Exosomal miRNAs (miR-21, miR-29a) are generally responsible for cell-to-cell communication, causing phenotypic changes in target cells by negative regulation, and also some can act as ligand to Toll-like receptors (TLRs) to activate immune cells [40]. Exosomal miRNAs have been found to involve in progression of various diseases such as cancers, myocardial infarction, tumor drug resistance in colorectal cancer, pancreatic cancer, hepatocellular carcinoma, etc. and gain importance as biomarker and therapeutic approaches in recent times [41].
- Long noncoding RNAs (lncRNAs) such as lncRNA TUC339, lncRNA ARSR, and lncRNA HOTAIR have also been identified in exosomes from different cellular origins. They are involved in tumor development, regulation of tumor microenvironment, promotion of cell proliferation, migration by influencing cellular pathways, gene expression as well as affect the drug resistance in tumor cells [26,41,42].
- Circular RNAs (circRNAs) are also a part of exosome's cargo such as circRNA 100284, circ-KLDHC10, and circ-DLEU2 [12]. They are found to regulate the local cellular microenvironment, promote malignancy and metastasis [26], and involve in angiogenesis and epithelial mesenchymal transition [43].
- Exosomal mRNAs have a regulatory role in protein expression of target cells, thus affecting the cell's behavior in response to different conditions [44]. Some other RNA types also have been reported to be present in exosomes, which include rRNA, tRNA, snRNA, piwi RNA, small nucleolar RNA, etc., which are thought to be a part in cell—cell communication and influencing the tumor development [26,36].

1.5.2 Proteins

Exosomal proteins consist of a set of proteins from cytosolic, membrane-bound, nuclear, mitochondrial, and ribosomal regions [3,10,11,26,36,44]. They can be divided into several categories. Some of them are mentioned in the following:

- Tetraspanins: CD63, CD9, CD81, Tspan8, etc. have a role in exosome biogenesis, cargo selection, targeting of recipient cells, penetration, invasion, and fusion.
- ESCRT proteins: TSG-101, Alix, Vps4, etc. are involved in MVB formation and involved in exosome biogenesis (ESCRT-dependent pathway).
- Heat shock proteins: Hsp 20, Hsp 70, Hsp 90, etc. respond to stress and have a role in exosome release.

- Major histocompatibility complex (MHC): MHC class I and MHC class II stimulate immunological responses in recipient cells.
- Cytoskeletal proteins: Actin, coflin-1, cortactin, etc. play an important role in exosomes' migration and secretion.
- Adhesion proteins: Integrin-α, ICAM-1, lactadherin, etc. facilitate the uptake of exosomes.
- Glycoproteins: β-Galactosidase, O-linked and N-linked glycans, etc. are associated with targeting and uptake of exosomes.
- Enzymes: Glyceraldehyde-3-phosphate dehydrogenase (GAPDH), fatty acid synthase, phosphoglycerate kinase 1, ATPase, glucose-6-phosphate isomerase, etc. have metabolic functions.
- Growth factors and cytokines: Tumor necrosis factor (TNF)-α, transforming growth factor-β (TGF-β), etc. involve in targeting, uptake of exosomes, and signaling in recipient cells.
- Trafficking and membrane fusion proteins: Rab27a, Rab35, annexins I, annexins II, syntaxin 1A, etc. help in trafficking and exosomal membrane fusion.

Additionally, death receptors such as FasL, TNF-related apoptosis inducing ligand, iron transport protein such as transferrin receptor, and signal transducing protein such as ARF-1, G proteins, and β-catenin are also reported [3,36].

1.5.3 Lipids

Lipids are the constituent of exosomal membrane bilayer. The lipid contents of exosomes are also of specific or nonspecific types. They are asymmetrically distributed and also form lipid raft-like domain in the membrane [45]. Exosomal lipid contents are enriched of ceramide, phosphatidic acid, Lyobisphosphatidic acid (LBPA), sphingomyelins, cholesterols, phosphatidyl choline, phosphatidyl serine, prostaglandins, leukotrienes, etc. [45]. It has been reported that concentration of sphingomyelin is comparatively higher in exosomes [45]. Lipid molecules of exosomes are known to maintain the structure and rigidity of exosomes and also play an important role in exosome biogenesis (ESCRT-independent pathway), exosome cargo sorting, and regulation of recipient cell's hemostasis [11,26,36,44].

2. Exosome isolation

To study and facilitate the application of exosomes, they need to be isolated from common and easily available biological samples such as blood, urine, saliva, plasma, serum, cerebrospinal fluid (CSF), and breast milk [46]. Different starting materials require different processing approaches. With the recent advancements in technology, many isolation techniques have been discovered to isolate exosomal vesicles, which come with their own sets of pros and cons when applied in the laboratory. Each specific technique focuses on a few particular traits of exosome such as density, shape, size, and availability of surface proteins [47]. Five common exosome isolation techniques are enlisted and briefly described in the following: The exosomal isolation techniques are represented by pictorial diagram in Fig. 2.4.

2.1 Ultrafiltration

It is one of the most widely used size-based separation methods for isolating exosomes. It entails a process where fluids are filtered via a specific filter with a set of pore sizes. Exosome isolation procedures involve a series of filtering processes with pore sizes of 0.8, 0.45, 0.22 μm, and occasionally 0.1 μm [48]. This often eliminates particles that are larger than the pore size of the filtered fluid. The most widely used commercial filters, such the Millipore VVLP poly-vinylidene fluoride filter (pore size of 0.1 mm), are made of hydrophilic polymers that typically have little affinity to proteins [48]. Ultrafiltration is a quicker isolation method that can be used to complete the entire process without the need for any additional specialized equipment. However, the use of shear force may cause large vesicles to rupture or disrupt, which could significantly skew the outcomes of downstream analysis.

2.2 Ultracentrifugation

The principle behind this process is that when a suspension is subjected to centrifugal force, the solid particulate constituents in the suspension will be sedimented below the solvent according to their density, size, and shape. Ultracentrifugation is simply a form of

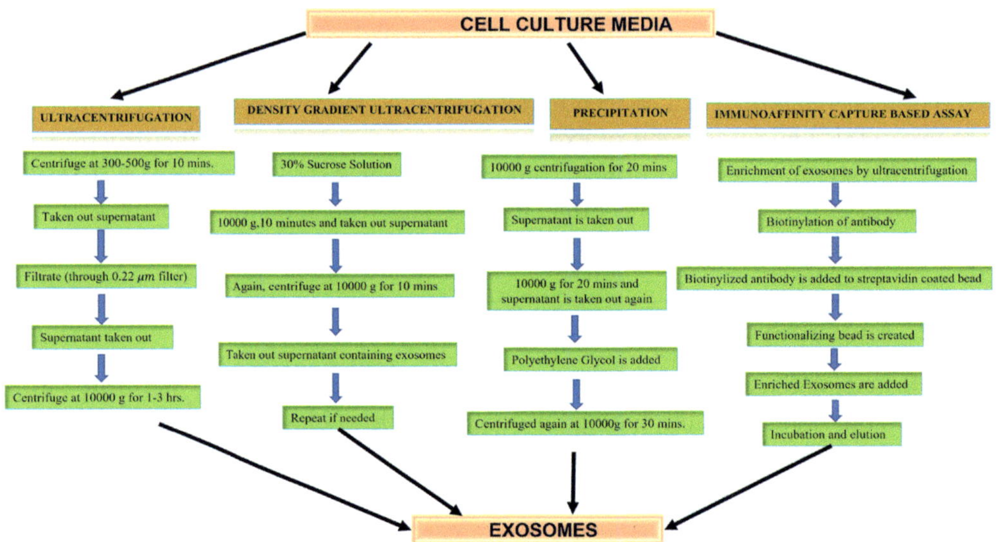

Figure 2.4 Isolation of exosomes by various techniques.

centrifugation that is employed specifically to produce extremely high centrifugal forces up to 1,000,000 × g [49]. Differential and density gradient ultracentrifugation are the two methods of preparative ultracentrifugation that are typically employed in laboratories to isolate exosomes [50].

2.2.1 Differential ultracentrifugation

Currently, differential ultracentrifugation is regarded as the "gold standard" for exosome isolation, with which the efficiency of alternative exosome isolation techniques is evaluated [48]. Differential centrifugation includes several subsequent rounds of centrifugation with constant increase of centrifugal force, which ultimately helps in separating exosomes from the exosomal matrix based on parameters such as density, size, and shape [51]. The main principle behind this procedure is "more dense particles pellet out first" [13]. The supernatant is collected and subjected to additional centrifugation steps after each cycle of ultracentrifugation. Though procedures may differ, some of the common steps are enlisted in the following:

- At first, the cell culture medium is subjected to centrifugation at 300−500 g for approximately 10 min, which pellets out cell debris and larger particles from the matrix.
- Next the supernatant from the cell culture medium is passed through a 0.22 μm filter to clarify the matrix, and it is again subjected to centrifugation at 10,000 g for approximately 30 min, which further pellets out larger vesicles from the matrix.
- Exosomes are extracted from nonexosomal vesicles, which are typically much bigger than 100−150 nm, and will be separated by centrifugation at 10,000 g for a period of 1−3 h. This step is repeated if required [13].
- Finally, the exosome pellet is resuspended in the buffer [13].

This method is relatively inexpensive and needs little to no sample pretreatment making it easier to carry out [48]. Absence of additional chemical reagents throughout the procedure reduces the expense of the experiment as well as the chances of cross-contamination [52]. Needless to say, differential ultracentrifugation requires large amounts of biomaterials as samples for starting volume of experiment and hence cannot be performed on small amounts of sample. This approach requires a lot of manual labor, and its main drawback is the inclusion of nonexosomal contaminants in the exosome fraction.

2.2.2 Density gradient ultracentrifugation

This type of ultracentrifugation separates particles according to size and density, but a prebuilt density gradient (primarily made of sucrose or iodixanol) is always present at the time of centrifugation throughout the process [13]. Exosomes are typically purified by centrifuging them at 210,000 g for 16 h by using a sucrose gradient (2.0−0.25 M) [50]. Following centrifugation, exosome isolation can be accomplished by fractionation collection, typically at densities between 1.1 and 1.2 g/mL. Isopycnic ultracentrifugation and moving-zone ultracentrifugation are the two types of density gradient ultracentrifugation that are currently recognized [49]. Exosomes can be separated with this technique from lipoproteins such as chylomicrons, which resemble them but have a significantly lower flotation density (1.063 g/mL) [50]. However, the limitations of this technique are low productivity and laborious.

2.3 Precipitation

The principle of this technique is instituted on the accumulation of exosomes in the occurrence of precipitating components. In this process, a sample is mixed with a precipitant, and after precipitation, the resultant vesicles are subjected to a low-speed centrifugation [53]. A very common precipitation method is by using PEG (polyethylene glycol). PEG basically binds up the water molecules, causing exosomes to precipitate out of the solution [53]. This method does come with its drawbacks as it precipitates out other vesicles and proteins too during this procedure. So, another different type of precipitation, i.e., lectin-induced agglutination is carried out [14,52]. Lectins bind more specifically to carbohydrates present on the surface of the exosomes (specifically from urinary samples) reducing the probabilities of binding to nonexosomal particles [52]. This method is relatively fast and requires little to no expertise. However, removal of nonexosomal particles during exosomal isolation is an issue in this process.

2.4 Immunoaffinity capture

This technique depends on the employment of an antibody to capture exosomal marker proteins that come into contact on the surface of exosomal membranes, such as major histocompatibility complex antigens (HLA-antigens), tetraspanins CD9, CD63, CD81, and CD82, heat shock proteins Hsp70 and Hsp90, and others [48,54]. The solid-phase carriers (latex, silicone), which are employed in immunoprecipitation, are typically conjugated to the antibodies [54]. These solid-phase carriers can be modified pipette tips, magnetic particles with polymer coatings, or centrifugation columns. Since the bond between the immobilized antibody on the matrix surface and the exosomal wall is fairly strong, the binding exosomes can be cleaned of various elements of biological fluid [48,54]. The isolation of exosome by immunoaffinity was represented in Fig. 2.5. It was shown that this approach produced a larger yield of RNA from the exosomes compared with that in ultracentrifugation. However, this procedure has a relatively high number of steps making it more difficult to complete.

2.5 Size-exclusion chromatography

Size-exclusion chromatography (SEC) is the easiest method for the separation and purification of proteins and carbohydrates such as macromolecular substances. Because of

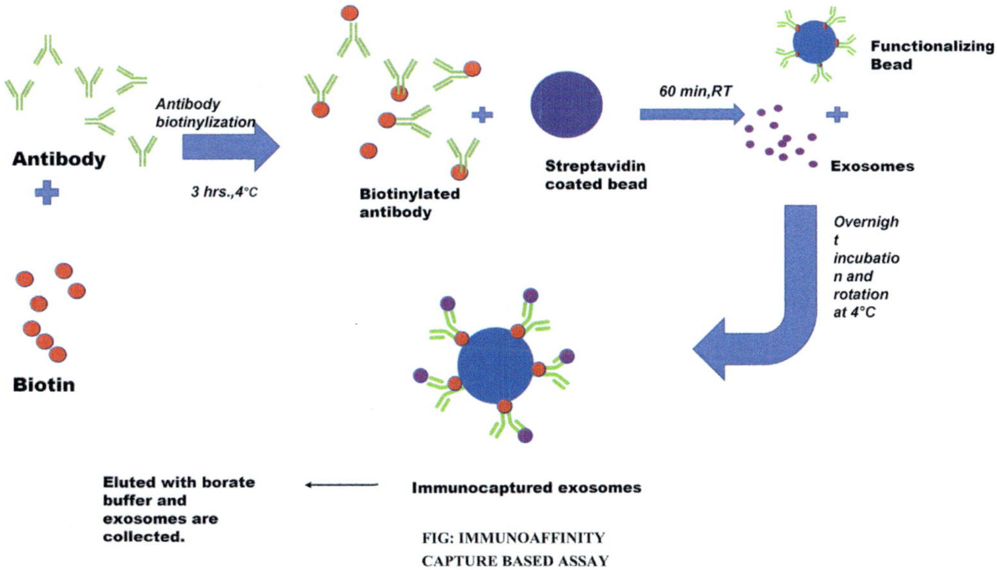

Figure 2.5 Immunoaffinity Capture Based Isolation Technique.

this, SEC is now widely used to isolate exosomes as it retains the vesicles' structure and biological activity and possesses a significant yield [55,56]. SEC sorts biomolecules according to their hydrodynamic radius (Hd) as they move through an inert, low-adsorption resin-based matrix that is porous and packed in a column [55]. Based on their size and how long it takes for them to elute, the bigger particles come out first. This technique can be used to sort items with a wide range of viscosities, from low to high viscosity such as urine or cell culture media to plasma. The drawback in this approach is that samples have to be treated first by ultracentrifugation or ultrafiltration to get rid of any protein contaminations. To make exosomal isolation by SEC simpler, commercially prepacked columns such as qEV (Izon Science) and HiLoad Superdex (GE Healthcare) are now available [56].

3. Characterization of exosomes

Due to a diverse range of applications of exosomes throughout various fields including disease pathway, drug delivery, biomarkers, and diagnostics, proper characterization of exosomes is absolutely necessary. This characterization includes quantitative, qualitative, single vesicle characterization, and topological characterization by which the distinction of one exosome from another one and receptor site identification are performed. For biomolecules such as proteins, lipids, and nucleic acids, quantifiable characterization techniques are employed to assess the yield and purity [55]. This comprises electron microscopy (EM), flow cytometry, fluorescence correlation spectroscopy (FCS), resistive pulse sensing (RPS), dynamic light scattering (DLS), and nanoparticle tracking analysis (NTA). NTA and FCS procedures are the most often employed ones. Western blotting is a technique for qualitative characterization. Exosome features such as size, shape, and chemical content are the focus of single vesicle characterization techniques. This comprises Raman spectroscopy and atomic force microscopy (AFM). The topology of exosomes and their interactions with other biomolecules are the main topics of topological characterization. This encompasses scanning electron microscopy (SEM) and AFM. Some of the methods are discussed in the following.

3.1 Ultrastructural analysis

For the specific study of the exosomal structure and utilization of the exosomal drug delivery model fullest, ultrastructure analysis of different exosome is very important. The most commonly used method for analysis of the ultrastructure of any molecular object is transmission electron microscopy (TEM). Ultrastructure analysis of exosomes can be performed by TEM [57]. Different types of exosomes, with different structures, can be visualized under TEM. For example, cup-shaped exosomes appear as small vesicles packed inside the multivesicular bodies. In the extracellular matrix, exosomes present as a cluster of spherical particles [58]. In the medial layer of smooth muscle cells, there are two types of exosomes that can be identified: The first type is 40–100 nm diameter double-membrane vesicles, enclosed within cytoplasmic multivesicular bodies, and the second type of exosome shows a similar structure as the exosome located in the extracellular space of neointima. Moreover, by comparing the samples from both the diseased individuals and normal individuals, there a clearly identifiable difference can be visualized: (1) an increased number of exosomes are present in the case of diseased cells; (2) a marked heterogeneity was observed in the size and morphology of the diseased exosomes; and (3) in a healthy and atherosclerotic human aorta, a double-layered membrane surrounding exosomes were found [59]. From these ultrastructural analyses, it was stated that exosomes have major potential to be used as a biomarker in cancers and a carrier of antiinflammatory agents [52,60].

3.2 Nanoparticle tracking analysis

The massive challenge to study the structure of the exosome is its very small size. Nanoparticle tracking analysis (NTA) enlightens the way of quantifying and size determination of particles between 30 nm and 1 μm diameter [61]. We know that exosomes along with other concentrations of vesicles are continuously roaming throughout the cytoplasm due to Brownian motion. The process of NTA is to track the scattered light for the absolute size distribution of the exosomes, and for each of the single vesicles, the mean squared velocity is calculated [62]. This process is based on the mechanism of dynamic light scattering (DLS) and the Stokes–Einstein equation to estimate the concentration and size of the particle [63]. Several proteins including CD9, CD81, the tetraspanin protein family especially CD63 are located on the outer surface of exosomes [64]. By designing fluorescent-tagged antibodies against those surface proteins of the exosome, the size of a subgroup of the vesicle can be determined by NTA [65].

3.2.1 Advantages of NTA

1. For the measurement of concentration and size of exosomes, NTA has better reproducibility, compared with flow TEM and flow cytometry.
2. Due to its high resolution, NTA has the ability to detect 30–1000 nm diameter vesicles [66].

3.2.2 Disadvantages of NTA

1. NTA is not fully automated, so user system interference is important for finding experimental data.
2. Detection threshold, viscosity, camera level, and temperature have to be optimized at the time of data obtaining.
3. Sometimes NTA lacks the ability to distinguish between protein aggregates and exosomes [67].

3.3 Atomic force microscopy

AFM is a largely used technique for the topographic study of exosomes. In AFM a suspension of exosome is prepared followed by the spreading of the suspension throughout the mica substrate [68]. Then the mica substrate was allowed to be air-dried. Then ultrapure water is used to wash the sample, followed by drying with nitrogen gas. Silicon probes are used to visualize exosome using AFM [68]. It is a high-resolution process, which involves analysis of the topography, lateral displacement, dissipation, and adhesion properties of exosomes [69]. For better resolution and removing of artifacts, it is best to use the tapping mode of AFM, in which morphological variations of exosome are observed. By AFM, immunogold imaging CD63 receptor sites of exosome are also being identified [70].

3.4 Flow cytometry

This method is used to characterize various surface markers expressed on particular exosomes, sized between 300 and 500 nm [71,72]. As it is based on optical signals, it lacks proper resolution and accuracy [73]. Besides these mentioned processes, there are also various methods available including the ExoView characterization platform, in vivo exosome tracking analysis, etc.

3.5 Western blotting

The process of western blotting is used to detect the protein located on the surface of the exosome. The exosomal surface marker proteins include Alix, TSG101, EpCAM, CD63, CD81, etc., which can be detected by Western blotting [7,74,75]. At first, ultracentrifugation is performed to isolate the exosomal protein. Then the sample is loaded into SDS-PAGE gels, followed by Western blotting with appropriate antibodies against the exosomal marker proteins [40].

3.6 Other methods

3.6.1 Fluid flow fraction

It is a well-known technique for biomolecule characterization, by which size-based exosome separation can be performed [55,76].

3.6.2 Raman spectroscopy

The principle of Raman spectroscopy is the wavelength shift based on scattering of light. It is efficiently used to determine the chemical composition of exosome [65,77].

4. Conclusion

Exosome biology is a fascinating area that requires fervent research to be fully understood. Exosomal research may, however, be useful for treating conditions such as carcinogenesis, metastasis, chemoresistance, Alzheimer's, cardiomyopathy, etc. It is still unknown from an evolutionary perspective whether exosomes may multiply and operate under certain cellular conditions. Exosome isolation procedures are labor-intensive and require thorough screening at every stage. In spite of having immense possible applications, the procedures of isolation of exosomes from biological fluids remain to be a challenging task, which needs further development and expertise.

References

[1] Colombo M, Raposo G, Théry C. Biogenesis, secretion, and intercellular interactions of exosomes and other extracellular vesicles. Annual Review of Cell and Developmental Biology 2014;30:255–89. https://doi.org/10.1146/annurev-cellbio-101512-122326.

[2] Helwa I, Cai J, Drewry MD, Zimmerman A, Dinkins MB, Khaled ML, Seremwe M, Dismuke WM, Bieberich E, Stamer WD, Hamrick MW, Liu Y. A comparative study of serum exosome isolation using differential ultracentrifugation and three commercial reagents. PLoS One 2017;12(1):e0170628. https://doi.org/10.1371/journal.pone.0170628.

[3] Kalluri R, LeBleu VS. The biology, function, and biomedical applications of exosomes. Science 2020;367(6478):eaau6977. https://doi.org/10.1126/science.aau6977.

[4] Chowdhury SG, Bhattacharya D, Karmakar P. Exosomal long noncoding RNAs – the lead thespian behind the regulation, cause and cure of autophagy-related diseases. Molecular Biology Reports 2022;49:7013–24. https://doi.org/10.1007/s11033-022-07514-x.

[5] Théry C, Witwer KW, Aikawa E, Alcaraz MJ, Anderson JD, Andriantsitohaina R, et al. Minimal information for studies of extracellular vesicles 2018 (MISEV2018): A position statement of the International Society for Extracellular Vesicles and update of the MISEV2014 guidelines. Journal of Extracellular Vesicles 2018;7(1):1535750. https://doi.org/10.1080/20013078.2018.1535750.
[6] Mathieu M, Martin-Jaular L, Lavieu G, Théry C. Specificities of secretion and uptake of exosomes and other extracellular vesicles for cell-to-cell communication. Nature Cell Biology 2019;21:9—17. https://doi.org/10.1038/s41556-018-0250-9.
[7] Chowdhury SG, Ray R, Bhattacharya D, Karmakar P. DNA damage induced cellular senescence and it's PTEN-armed exosomes—the warriors against prostate carcinoma cells. Medical Oncology 2022;39:34. https://doi.org/10.1007/s12032-021-01614-7.
[8] Zhang L, Yu D. Exosomes in cancer development, metastasis, and immunity. Biochim Biophys Acta Reviews on Cancer 2019;1871(2):455—68. https://doi.org/10.1016/j.bbcan.2019.04.004.
[9] Conigliaro A, Corrado C, Fontana S, Alessandro R. Exosome basic mechanisms. Academic Press; 2020. p. 1—21. https://doi.org/10.1016/B978-0-12-816053-4.00001-8.
[10] Jadli AS, Ballasy N, Edalat P, Patel VB. Inside(sight) of tiny communicator: exosome biogenesis, secretion, and uptake. Molecular and Cellular Biochemistry 2020;467(1—2):77—94. https://doi.org/10.1007/s11010-020-03703-z.
[11] Mashouri L, Yousefi H, Aref AR, Ahadi AM, Molaei F, Alahari SK. Exosomes: composition, biogenesis, and mechanisms in cancer metastasis and drug resistance. Molecular Cancer 2019;18(1):75. https://doi.org/10.1186/s12943-019-0991-5.
[12] Han QF, Li WJ, Hu KS, Gao J, Zhai WL, Yang JH, Zhang SJ. Exosome biogenesis: machinery, regulation, and therapeutic implications in cancer. Molecular Cancer 2022;21(1):207. https://doi.org/10.1186/s12943-022-01671-0.
[13] Doyle LM, Wang MZ. Overview of extracellular vesicles, their origin, composition, purpose, and methods for exosome isolation and analysis. Cells 2019;8(7):727. https://doi.org/10.3390/cells8070727.
[14] Pegtel DM, Gould SJ. Exosomes Annual Review of Biochemistry 2019;88:487—514. https://doi.org/10.1146/annurev-biochem-013118-111902.
[15] Buratta S, Tancini B, Sagini K, Delo F, Chiaradia E, Urbanelli L, Emiliani C. Lysosomal exocytosis, exosome release and secretory autophagy: the autophagic- and endo-lysosomal systems go extracellular. International Journal of Molecular Sciences 2020;21(7):2576. https://doi.org/10.3390/ijms21072576.
[16] Villarroya-Beltri C, Baixauli F, Mittelbrunn M, Fernández-Delgado I, Torralba D, Moreno-Gonzalo O, et al. ISGylation controls exosome secretion by promoting lysosomal degradation of MVB proteins. Nat Commun 2016;7:13588. https://doi.org/10.1038/ncomms13588.
[17] Ludwig N, Azambuja JH, Rao A, Gillespie DG, Jackson EK, Whiteside TL. Adenosine receptors regulate exosome production. Purinergic Signal 2020;16(2):231—40. https://doi.org/10.1007/s11302-020-09700-7.
[18] Xu L, Faruqu FN, Liam-Or R, Abu Abed O, Li D, Venner K, Errington RJ, Summers H, Wang JT, Al-Jamal KT. Design of experiment (DoE)-driven in vitro and in vivo uptake studies of exosomes for pancreatic cancer delivery enabled by copper-free click chemistry-based labelling. Journal of Extracellular Vesicles 2020;9(1):1779458. https://doi.org/10.1080/20013078.2020.1779458.
[19] Vietri M, Radulovic M, Stenmark H. The many functions of ESCRTs. Nature Reviews Molecular Cell Biology 2020;21(1):25—42. https://doi.org/10.1038/s41580-019-0177-4.
[20] Gurunathan S, Kang MH, Kim JH. A comprehensive review on factors influences biogenesis, functions, therapeutic and clinical implications of exosomes. International Journal of Nanomedicine 2021;16:1281—312. https://doi.org/10.2147/IJN.S291956.
[21] Friand V, David G, Zimmermann P. Syntenin and syndecan in the biogenesis of exosomes. Biology of the Cell 2015;107(10):331—41. https://doi.org/10.1111/boc.201500010.
[22] Shimada T, Yasuda S, Sugiura H, Yamagata K. Syntenin: PDZ protein regulating signaling pathways and cellular functions. International Journal of Molecular Sciences 2019;20(17):4171. https://doi.org/10.3390/ijms20174171.

[23] Baietti MF, Zhang Z, Mortier E, Melchior A, Degeest G, Geeraerts A, Ivarsson Y, Depoortere F, Coomans C, Vermeiren E, Zimmermann P, David G. Syndecan-syntenin-ALIX regulates the biogenesis of exosomes. Nature Cell Biology 2012;14(7):677–85. https://doi.org/10.1038/ncb2502.
[24] Odorizzi G. The multiple personalities of Alix. Journal of Cell Science 2006;119(Pt 15):3025–32. https://doi.org/10.1242/jcs.03072.
[25] Wei D, Zhan W, Gao Y, et al. RAB31 marks and controls an ESCRT-independent exosome pathway. Cell Research 2021;31:157–77. https://doi.org/10.1038/s41422-020-00409-1.
[26] Zhang Y, Liu Y, Liu H, Tang WH. Exosomes: biogenesis, biologic function and clinical potential. Cell Bioscience 2019b;9:19. https://doi.org/10.1186/s13578-019-0282-2.
[27] Hessvik NP, Llorente A. Current knowledge on exosome biogenesis and release. Cellular and Molecular Life Sciences 2018;75:193–208. https://doi.org/10.1007/s00018-017-2595-9.
[28] Andreu, Maria. Tetraspanins in extracellular vesicle formation and function. Frontiers in Immunology, Section Vaccines and Molecular Therapeutics 2014 Sept 16:5. https://doi.org/10.3389/fimmu.2014.00442.
[29] Hessvik NP, Llorente A. Current knowledge on exosome biogenesis and release. Cellular and Molecular Life Sciences 2018;75(2):193–208. https://doi.org/10.1007/s00018-017-2595-9.
[30] Bebelman MP, Smit MJ, Pegtel DM, Baglio SR. Biogenesis and function of extracellular vesicles in cancer. Pharmacology and Therapeutics 2018 Aug;188:1–11. https://doi.org/10.1016/j.pharmthera.2018.02.013.
[31] Li SP, Lin ZX, Jiang XY, Yu XY. Exosomal cargo-loading and synthetic exosome-mimics as potential therapeutic tools. Acta Pharmacol Sinica 2018 Apr;39(4):542–51. https://doi.org/10.1038/aps.2017.178.
[32] van Niel G, D'Angelo G, Raposo G. Shedding light on the cell biology of extracellular vesicles. Nature Reviews of Molecular Cell Biology 2018 Apr;19(4):213–28. https://doi.org/10.1038/nrm.2017.125.
[33] Xu M, Ji J, Jin D, Wu Y, Wu T, Lin R, Zhu S, Jiang F, Ji Y, Bao B, Li M, Xu W, Xiao M. The biogenesis and secretion of exosomes and multivesicular bodies (MVBs): intercellular shuttles and implications in human diseases. Genes and Diseases 2022. https://doi.org/10.1016/j.gendis.2022.03.021.
[34] Han Z, Chen H, Guo Z, Shen J, Luo W, Xie F, Wan Y, Wang S, Li J, He J. Circular RNAs and their role in exosomes. Frontyers in Oncology 2022;12:848341. https://doi.org/10.3389/fonc.2022.848341.
[35] Yu Z, Shi M, Stewart T, Fernagut PO, Huang Y, Tian C, et al. Reduced oligodendrocyte exosome secretion in multiple system atrophy involves SNARE dysfunction. Brain 2020;143(6):1780–97. https://doi.org/10.1093/brain/awaa110.
[36] Gurung S, Perocheau D, Touramanidou L, Baruteau J. The exosome journey: from biogenesis to uptake and intracellular signalling. Cell Communication Signal 2021 Apr 23;19(1):47. https://doi.org/10.1186/s12964-021-00730-1.
[37] Takahashi A, Okada R, Nagao K, et al. Exosomes maintain cellular homeostasis by excreting harmful DNA from cells. Nature Communication 2017;8:15287. https://doi.org/10.1038/ncomms15287.
[38] Cai J, Wu G, Jose PA, Zeng C. Functional transferred DNA within extracellular vesicles. Experimental Cell Research 2016;349(1):179–83. https://doi.org/10.1016/j.yexcr.2016.10.012.
[39] Hough KP, Trevor JL, Strenkowski JG, Wang Y, Chacko BK, Tousif S, Chanda D, Steele C, Antony VB, Dokland T, Ouyang X, Zhang J, Duncan SR, Thannickal VJ, Darley-Usmar VM, Deshane JS. Exosomal transfer of mitochondria from airway myeloid-derived regulatory cells to T cells. Redox Biology 2018;18:54–64. https://doi.org/10.1016/j.redox.2018.06.009.
[40] Zhang J, Li S, Li L, Li M, Guo C, Yao J, Mi S. Exosome and exosomal microRNA: trafficking, sorting, and function. Genomics Proteomics Bioinformatics 2015 Feb;13(1):17–24. https://doi.org/10.1016/j.gpb.2015.02.001.
[41] Guo C, Liu J, Zhou Q, Song J, Zhang Z, Li Z, Wang G, Yuan W, Sun Z. Exosomal noncoding RNAs and tumor drug resistance. Cancer Research 2020;80(20):4307–13. https://doi.org/10.1158/0008-5472.CAN-20-0032.

[42] Poulet C, Njock MS, Moermans C, Louis E, Louis R, Malaise M, Guiot J. Exosomal long non-coding RNAs in lung diseases. International Journal of Molecular Sciences 2020;21(10):3580. https://doi.org/10.3390/ijms21103580.

[43] Wang Y, Liu J, Ma J, et al. Exosomal circRNAs: biogenesis, effect and application in human diseases. Molecular Cancer 2019;18(1):116. https://doi.org/10.1186/s12943-019-1041-z. Published 2019 Jul 5.

[44] Yue B, Yang H, Wang J, Ru W, Wu J, Huang Y, Lan X, Lei C, Chen H. Exosome biogenesis, secretion and function of exosomal miRNAs in skeletal muscle myogenesis. Cell Proliferation 2020;53(7):e12857. https://doi.org/10.1111/cpr.12857.

[45] Skotland T, Hessvik NP, Sandvig K, Llorente A. Exosomal lipid composition and the role of ether lipids and phosphoinositides in exosome biology. Journal of Lipid Research 2019 Jan;60(1):9–18. https://doi.org/10.1194/jlr.R084343.

[46] Patel GK, Khan MA, Zubair H, Srivastava SK, Khushman M, Singh S, et al. Comparative analysis of exosome isolation methods using culture supernatant for optimum yield, purity and downstream applications. Scientific Reports 2019;9(1):5335. https://doi.org/10.1038/s41598-019-41800-2. PMID: 30926864; PMCID: PMC6441044.

[47] Zhang M, Jin K, Zhang Z, Gao L, Zhang Z, Li F, et al. Methods and technologies for exosome isolation and characterization. Small Methods 2018;2(9):180021. https://doi.org/10.1002/smtd.201800021.

[48] Yakubovich EI, Polischouk AG, Evtushenko VI. Principles and problems of exosome isolation from biological fluids. Biochemistry (Moscow) Supplementary Sertes A: Membrane and Cell Biology 2022;16(2):115–26. https://doi.org/10.1134/S1990747822030096.

[49] Li P, Kaslan M, Lee SH, Yao J, Gao Z. Progress in exosome isolation techniques. Theranostics 2017;7(3):789–804. https://doi.org/10.7150/thno.18133.

[50] Coughlan C, Bruce KD, Burgy O, Boyd TD, Michel CR, Garcia-Perez JE, Adame V, Anton P, Bettcher BM, Chial HJ, Königshoff M, Hsieh EWY, Graner M, Potter H. Exosome isolation by ultracentrifugation and precipitation and techniques for downstream analyses. Current Protocols in Cell Biology 2020 Sep;88(1):e110. https://doi.org/10.1002/cpcb.110.

[51] Livshits MA, Khomyakova E, Evtushenko EG, Lazarev VN, Kulemin NA, Semina SE, Generozov EV, Govorun VM. Isolation of exosomes by differential centrifugation: theoretical analysis of a commonly used protocol. Science Reports 2015;5:17319. https://doi.org/10.1038/srep17319.

[52] Ludwig N, Whiteside TL, Reichert TE. Challenges in exosome isolation and analysis in health and disease. International Journal of Molecular Sciences 2019;20(19):4684. https://doi.org/10.3390/ijms20194684.

[53] Macias M, Rebmann V, Mateos B, Varo N, Perez-Gracia JL, Alegre E, Gonzalez A. Comparison of six commercial serum exosome isolation methods suitable for clinical laboratories. Effect in cytokine analysis. Clinical Chemistry and Laboratory Medicine 2019;57(10):1539–45. https://doi.org/10.1515/cclm-2018-1297.

[54] Greening DW, Xu R, Ji H, Tauro BJ, Simpson RJ. A protocol for exosome isolation and characterization: evaluation of ultracentrifugation, density-gradient separation, and immunoaffinity capture methods. Methods in Molecular Biology 2015;1295:179–209. https://doi.org/10.1007/978-1-4939-2550-6_15.

[55] Lai JJ, Chau ZL, Chen SY, Hill JJ, Korpany KV, Liang NW, Lin LH, Lin YH, Liu JK, Liu YC, Lunde R, Shen WT. Exosome processing and characterization approaches for research and technology development. Advancement of Science 2022 May;9(15):e2103222. https://doi.org/10.1002/advs.202103222.

[56] Tian Y, Gong M, Hu Y, Liu H, Zhang W, Zhang M, Hu X, Aubert D, Zhu S, Wu L, Yan XJ. Extracell Vesicles 2020;9(1):1697028. https://doi.org/10.1080/20013078.2019.1697028.

[57] Perrotta I, Aquila S. Exosomes in human atherosclerosis: an ultrastructural analysis study. Ultrastructural Pathology 2016;40(2):101–6. https://doi.org/10.3109/01913123.2016.1154912.

[58] Chen W, Zheng P, Hong T, Wang Y, Liu N, He B, Zou S, Ren D, Duan J, Zhao L, Feng J. Astrocytes-derived exosomes induce neuronal recovery after traumatic brain injury via delivering gap junction alpha 1–20 k. Journal of Tissue Engineering Regenerative Medicine 2020 Mar;14(3):412–23. https://doi.org/10.1002/term.3002.

[59] Li WJ, Chen H, Tong ML, Niu JJ, Zhu XZ, Lin LR. Comparison of the yield and purity of plasma exosomes extracted by ultracentrifugation, precipitation, and membrane-based approaches. Open Chemistry 2022;20(1):182–91. https://doi.org/10.1515/chem-2022-0139.

[60] Giglio A, Perrotta ID, Brandmayr P. Exosomes: ultrastructural evidence in epithelial cells of malpighian tubules. Micron 2017;100:34–7. https://doi.org/10.1016/j.micron.2017.04.008.

[61] He G, Peng X, Wei S, Yang S, Li X, Huang M, Tang S, Jin H, Liu J, Zhang S, Zheng H, Fan Q, Liu J, Yang L, Li H. Exosomes in the hypoxic TME: from release, uptake and biofunctions to clinical applications. Molecular Cancer 2022;21(1):19. https://doi.org/10.1186/s12943-021-01440-5.

[62] Khatun Z, Sharma A. Exosome signalling in stem cells. Stem Cell Fundamentals and Practice 2016;2(2):1. https://doi.org/10.5455/sfp.20160602115050.

[63] Zhang W, Peng P, Kuang Y, et al. Characterization of exosomes derived from ovarian cancer cells and normal ovarian epithelial cells by nanoparticle tracking analysis. Tumour Biology 2016;37(3):4213–21. https://doi.org/10.1007/s13277-015-4105-8.

[64] Dai J, Su Y, Zhong S, Cong L, Liu B, Yang J, Tao Y, He Z, Chen C, Jiang Y. Exosomes: key players in cancer and potential therapeutic strategy. Signal Transduction and Targeted Therapy 2020;5(1):145. https://doi.org/10.1038/s41392-020-00261-0.

[65] Dragovic RA, Gardiner C, Brooks AS, et al. Sizing and phenotyping of cellular vesicles using nanoparticle tracking analysis. Nanomedicine 2011;7(6):780–8.

[66] Zhou M, Weber SR, Zhao Y, Chen H, Sundstrom JM. Methods for exosome isolation and characterization. In: Exosomes: A Clinical Compendium, 23–38. Elsevier; 2019. https://doi.org/10.1016/B978-0-12-816053-4.00002-X.

[67] Franquesa M, Hoogduijn MJ, Ripoll E, Luk F, Salih M, et al. Update on controls for isolation and quantification methodology of extracellular vesicles derived from adipose tissue mesenchymal stem cells. Frontiers in Immunology Section Vaccines and Molecular Therapeutics 2014 Oct 21;5:525. https://doi.org/10.3389/fimmu.2014.00525.

[68] Sharma S, LeClaire M, Gimzewski JK. Ascent of atomic force microscopy as a nanoanalytical tool for exosomes and other extracellular vesicles. Nanotechnology 2018;29(13):132001. https://doi.org/10.1088/1361-6528/aaab06.

[69] Zhao Y, Weber SR, Lease J, Russo M, Siedlecki CA, Xu LC, Chen H, Wang W, Ford M, Simó R, Sundstrom JM. Liquid Biopsy of vitreous reveals an abundant vesicle population consistent with the size and morphology of exosomes. Translation Vision Science Technology 2018;7(3):6. https://doi.org/10.1167/tvst.7.3.6.

[70] Sharma S, Rasool HI, Palanisamy V, Mathisen C, Schmidt M, Wong DT, Gimzewski JK. Structural-mechanical characterization of nanoparticle exosomes in human saliva, using correlative AFM, FESEM, and force spectroscopy. ACS Nano 2010;4:1921–6.

[71] Isola AL, Chen S. Exosomes: the messengers of health and disease. Current Neuropharmacology 2017;15(1):157–65. https://doi.org/10.2174/1570159x14666160825160421.

[72] van der Pol E, Hoekstra AG, Sturk A, Otto C, van Leeuwen TG, Nieuwland R. Optical and non-optical methods for detection and characterization of microparticles and exosomes. Journal of Thrombosis and Haemostasis 2010;8(12):2596–607. https://doi.org/10.1111/j.1538-7836.2010.04074.x.

[73] Jin M, Zhang S, Wang M, Li Q, Ren J, Luo Y, Sun X. Exosomes in pathogenesis, diagnosis, and therapy of ischemic stroke. Frontiers in Bioengineering and Biotechnology 2022;10:980548. https://doi.org/10.3389/fbioe.2022.980548.

[74] Liang B, Peng P, Chen S, Li L, Zhang M, Cao D, Yang J, Li H, Gui T, Li X, Shen K. Characterization and proteomic analysis of ovarian cancer-derived exosomes. Journal of Proteomics 2013;80:171–82. https://doi.org/10.1016/j.jprot.2012.12.029.

[75] Riches A, Campbell E, Borger E, Powis S. Regulation of exosome release from mammary epithelial and breast cancer cells - a new regulatory pathway. European Journal of Cancer 2014;50(5):1025–34. https://doi.org/10.1016/j.ejca.2013.12.019.
[76] Oh S, Kang D, Ahn SM, Simpson RJ, Lee BH, Moon MH. Miniaturized asymmetrical flow field-flow fractionation: application to biological vesicles. Journal of Separation Science 2007 May;30(7):1082–7. https://doi.org/10.1002/jssc.200600394.
[77] Smith ZJ, Lee C, Rojalin T, Carney RP, Hazari S, Knudson A, Lam K, Saari H, Ibañez EL, Viitala T, Laaksonen T, Yliperttula M, Wachsmann-Hogiu S. Single exosome study reveals subpopulations distributed among cell lines with variability related to membrane content. Journal of Extracellular Vesicles 2015;4:28533. https://doi.org/10.3402/jev.v4.28533.

CHAPTER THREE

Exosomal RNA in cancer progression

Anusha Chaudhuri[1] and Shrena Chakraborty[2]
[1]Zentrum für Molekulare Medizin, Institut für kardiovaskuläre Regeneration, Goethe-Universität, Universitätsklinikum, Frankfurt, Germany
[2]TEAM "DNA Recombination, Replication, and Genome Stability", UMR3348, Institut Curie, Orsay, France

Abbreviations

ABCB1 ATP-binding cassette subfamily B member 1
ADIPOQ Adiponectin, C1Q, and collagen domain containing
AGAP2-AS1 ArfGAP with GTPase domain, ankyrin repeat, and PH domain 2 antisense 1
AHIF Antisense transcript of hypoxia-inducible factor 1
AML Acute myeloid leukemia
ASOs Antisense oligonucleotides
CAFs Cancer-associated fibroblast
CC Cervical cancer
CCAT Colon cancer-associated transcript 2
CDKN2B-AS1 Cyclin-dependent kinase inhibitors 2B antisense RNA 1
circRNAs Circular RNAs
CLL Chronic lymphocytic leukemia
COL1A1 Collagen type I alpha 1 chain
CRC Colorectal cancer
CRNDE Colorectal neoplasia differentially expressed
DR5 Death receptor 5
DUSP14 Dual specificity phosphatase 14
ECM Extracellular matrix
EFNA3 gene Ephrin A3
EMT Epithelial—mesenchymal transition
ESCC Esophageal squamous cell carcinoma
EVs Extracellular vesicles
ExoRNAs Exosomal RNAs
EZH2 Enhancer of zeste homolog 2
FUT4 Fucosyltransferase 4 alpha 1,3: fucosyltransferase, myeloid-specific
GAS5 Growth arrest-specific transcript 5
GBM Glioblastoma
GDEs Glioma-derived exosomes
GI Cancer Gastrointestinal cancer
HCC Hepatocellular carcinoma
hnRNPs Heterogenous nuclear ribonucleoproteins
HOTAIR HOX transcript antisense RNA
HOX HomebOX
HPV Human papilloma virus
IARS Isoleucyl-tRNA synthetase

Exosomal RNA
ISBN 978-0-443-14008-2
https://doi.org/10.1016/B978-0-443-14008-2.00010-3

IDH3α Isocitrate dehydrogenase 3α
IFN-γ Interferon gamma
IGF2BP3 Insulin-like growth factor 2 mRNA-binding protein 3
IL Interleukin
KLF4 Krüppel-like factor 4
LCCDEs Lung cancer cell—derived exosomes
LMO7 Lim domain 7
lncRNA SBF2-AS1 lncRNA SBF2 antisense RNA 1
lncRNA-ATB Long noncoding RNA-activated by transforming growth factor β
lncRNAs Long noncoding RNAs
LUZP1 Leucine zipper protein 1
M2-Ex M2 macrophage-derived exosomes non—small-cell lung cancer NSCLC
MALAT1 Metastasis-associated lung adenocarcinoma transcript 1
MEG3 Maternally expressed gene 3
MET Mesenchymal—epithelial transition
miRNAs MicroRNAs
MMP9 Matrix metalloproteinase 9
mRNAs Messenger RNA
NEIL3 Nei-like DNA glycosylase 3
NF Normal fibroblast
NFE2L3 Nuclear factor erythroid 2—like factor 3
PaCa Pancreatic cancer
PCa Prostate cancer
PDCD4 Programmed cell death 4
PDE8A Phosphodiesterase 8A
PDGF Platelet-derived growth factor
piRNAs PIWI-interacting RNAs
PTC Papillary thyroid carcinoma
PTEN Phosphatase and TENsin homolog
PVT1 Plasmacytoma variant translocation 1
RISC-associated RNA-induced silencing complex-associated
RNAi RNA interference
RNF213 Ring finger protein 213
ROR1-AS1 ROR1 antisense RNA 1
SBF2-AS1 SBF2 antisense RNA1
SERPINE2 Serpin family E member 2
shRNAs Short hairpin RNAs
siRNAs Small interfering RNAs
SNHG14 Small nucleolar RNA host gene 14
SOCS5 Suppressor of cytokine signaling-5
Sox2ot SOX2 overlapping transcript
SOX4 SRY-box transcription factor 4
STAT3 Signal transducer and activator of transcription 3
TAMs Tumor-associated macrophages
TC Thyroid cancer
TGF-β Transforming growth factor-B
TME Tumor microenvironment
TMZ Temozolomide
TP53INP1 Tumor p53-induced nuclear protein 1
tsRNAs Transfer RNA-derived small RNAs

TUFT1 Tuftelin 1
UCA1 Urothelial carcinoma-associated 1
VDR Vitamin D receptor
VEGF Vascular endothelial growth factor
ZEB1 Zinc-finger E-box-binding homeobox 1

1. Introduction

Cancer is a convoluted and catastrophic disease that continues to pose a major public health issue across the globe [1]. In spite of the remarkable progress in the field of cancer research and therapy, there is a requirement to better understand the underlying mechanisms that could contribute to cancer development [2]. In recent years, the involvement of exoRNAs in cancer has emerged as an encouraging area for further investigation [3]. Exosomes are compact-structured, membrane-bound vesicles ranging from 25 to around 250 nm in diameter that are secreted from a myriad of cell types, including cancer cells (Fig. 3.1) [4–9]. They are capable of transferring their contents to a recipient cell, thereby modulating the recipient's behavior [10], and hence play a key role in intercellular communications. These vesicles can carry different biomolecules, such as lipids [11], proteins [12], and nucleic acids [12–14] such as DNA, RNA, and different pseudogenes [15–18] (Fig. 3.1). These exoRNAs play a key role in cancer progression by affecting the behavioral pattern of cancer cells in their microenvironment [19,20]. They participate in numerous regulatory processes and thereby contribute to cell proliferation and metastasis [21] (Table 3.1, Fig. 3.1).

In this chapter, we dive deeper into the current background of knowledge regarding the function of exoRNAs surrounding the tumor microenvironment (TME) in the progression and development of cancer, which includes the mechanisms to promote extracellular matrix (ECM) remodeling, epithelial–mesenchymal transition (EMT), endothelial tubulogenesis, tumor metastatic invasion, angiogenesis, chemoresistance, and programmed evasion of the immune system [21] (Fig. 3.3). Additionally, we will also briefly discuss the potential clinical applications of exoRNAs as a diagnostic and prognostic biomarker as well as a therapeutic target for cancer treatment. In cancer, the role of exosomes has been established to promote tumor growth, invasion, and metastasis by transferring oncogenic molecules between cancer and stromal cells [21–25]. Among these molecules, exoRNAs have received considerable attention in recent years, as they can alter gene expression and signaling pathways in recipient cells [8,9]. ExoRNAs can be categorized into messenger RNA (mRNAs) [26], microRNA (miRNAs) [26–28], PIWI-interacting RNAs (piRNAs), transfer RNA-derived small RNAs (tsRNAs) [29], long noncoding RNA (lncRNAs) [30], and circular RNAs (circRNAs) [21,31–34] Fig. 3.1, Table 3.1). All these different types of RNAs are somehow linked to cancer progression, and it has been shown that their expression profiles differ among various cancer types (Fig. 3.2) and stages [35] (Fig. 3.3, Table 3.1). In particular, miRNAs carried

Table 3.1 List of exosomal RNAs in various types of cancers and their role in cancer progression.

Types of exosomal RNA	Examples of exosomal RNA	Types of cancers	Mechanistic role in cancer
MicroRNAs (miRNA)	miR-425-3p	Esophageal squamous cell carcinoma	Enhances cancer progression and tumor cell proliferation
	miR-125b, miR-1290	Bladder cancer	Associated with tumor progression and poor prognosis
	miR-34a	GI cancer	Inhibits tumor growth and metastasis
	miR-21, miR-25-3p, miR-122-5p, miR-155, miR-221-3p	Gastric cancer	Resistance to cisplatin and promotes lymph node metastasis
	miR-21, miR-221, miR-155, miRNA-21-5p and miRNA-451a	Pancreatic cancer	Regulates chemoresistance genes, enhances tumor cell invasion and migration, and regulates immune response
	miRNA-10a-5p, miRNA-125b-5p, miRNA-301a		Promotes migration and invasion, contribute to EMT
	miRNA-21, miRNA-3607-3p, miRNA-1231, miRNA-126-3p, miRNA-210, miRNA-10b, miRNA-451a, and miRNA-1246		Inhibits the migration as well as invasion of PaCa cells
	miR-21, miR-122, MiR-1224-5p	Hepatocellular carcinoma	Promotes tumor cell proliferation, migration, invasion, and metastasis
	miR-21, miR-29a, miR-155, miR146a, miR-223, miR-143	Colorectal cancer	Induces immune tolerance in CRC by suppressing the activity of natural killer (NK) cells and contributes to CRC development and progression, also promotes the M2 polarization of macrophages, which are involved in tumor progression and immune suppression
	miR-320d, miR-34a, miR-29a, miR-92a		Promotes angiogenesis, enhances tumor cell invasion and migration, and regulates immune response
	miR-155, miR-21, miR-126	Blood cancer	Promotes tumor growth and suppresses the immune system by targeting various immune-related genes
	miRNA-96,miR-3917, miR-193a-3p, miR-210-3p, miR-5100,miR-21-5p, miR-34a-5p, miR-1246, miR-17, miR-31	Lung cancer	Regulates apoptosis, enhances cell proliferation, and promotes invasion and migration
	miR-21, miR-146a-5p, miR-485-5p, miR-145		Promotes invasiveness and metastatic qualities

miR-105, miR-10b, miR-7641, miRNA-205, miR-21, miR-141, miR-451, miR-9,miR-210,miR-4269, miR-503-3p, miR-30e-3p, miR-155, miR-7641, miR-221, miR-222, miR-100, miR-30a, miR-9-5p,miR-16, miR-23a, miR-24, miR-26a, miR-27a,miR-16, miR-23a, miR-24, miR-26a, miR-363-5p, miR-7641, and miR-130a-3p	Breast cancer	Enhances tumor cell proliferation, promotes invasion and migration, and regulates immune response
miR-21, miR-221, miR-1246, miR-9, miR-10b, miR-5096, miR-222, miR-26a, and miR-1246	Glioblastoma	Influences cell proliferation, invasion, angiogenesis, and metastasis
miR-124, miR-21, miR-221, miR-451, miR-128-3p		Downregulates and inhibits cell proliferation, invasion and chemoresistance
miR-21-5p, miR-31-5p, miR-221-3p, miR-222-3p, miR-222-3p, miR-221-3p, miR-146a-5p, miR-222-3p, miR-21-5p, let-7i-3p, let-7a, let-7c, let-7d, let-7f, and let-7i	Thyroid cancer	Deregulation in exosomes promotes proliferation, migration, and radioresistance
miR-21, miR-196a, miR-221-3p	Cervical cancer	Activates fibroblast and enhances secretion of proinflammatory cytokines, thereby promoting tumor growth
miR-203, miR-143		Downregulates and suppresses tumor formation
miR-21-5p,miR-27a-3p, miR-141-3p, miR-210-3p, miR-1290	Prostate cancer	Associated with PCa cell proliferation, migration, and invasion
miR-375		Downregulates and inhibits tumor growth
miR-21, miR-10b, miR-1246, miR-146a, miR-29c, miR-155, miR-21, miR-125b, miR-222, miR-1246,miR-200b, miR-205, miR-182	Ovarian cancer	Regulates gene expression, enhances tumor cell proliferation, and promotes metastasis

(Continued)

Table 3.1 List of exosomal RNAs in various types of cancers and their role in cancer progression.—cont'd

Types of exosomal RNA	Examples of exosomal RNA	Types of cancers	Mechanistic role in cancer
	circ-001971, circ-ASAP1, circ-SLC39A6 circRNA PDE8A and IARS	Pancreatic cancer	Regulates chemoresistance genes, enhances tumor cell migration, invasion, and metastasis, and regulates immune response
Circular RNAs (circRNAs)	circRNA-100338	Hepatocellular carcinoma	Inhibits HCC cell proliferation and migration by regulating oncogene expression
	circRNA_100269, circRNA_001569, circRNA-000284, circRNA-ABCC1, circRNA-0000977, circ-PRKCI, circ-SLC8A1, circ-SMAD7	Colorectal cancer	Promotes angiogenesis, enhances tumor cell invasion and migration, and regulates immune response
	circRNA-ITCH, circRNA-100290, circRNA_100284		Associated with advanced tumor stage and poor prognosis as well as inhibits CRC cell proliferation and migration
	circ-LDLRAD3, circFGFR1,circ0013958, circFARSA, hsa_circRNA_0056616, circPVT1, circ-FAM120A, circ-ITCH, circ-ZNF91	Lung cancer	Regulates apoptosis, enhances cell proliferation, and promotes invasion and migration
	hsa_circ_0020123, circPTK2, circNOL10, circSATB2, circFARSA		Inhibits LC apoptosis and cancer development
	circRNA_100876, circBMPR2, circ-1073, hsa_circ_0068033, circ-VRK1, circRNA_000911, circRNA_0001283, circRNA_000554, circDDX17, circRNA cSERPINE2, circ_0009910, circ_0000284, circ_0000190, circ_0001721, circ_0005276, circ-ABCB10, circ-FAM114A2, circ-PRMT5	Breast cancer	Enhances tumor cell proliferation, promotes invasion and migration, and regulates immune response
	circ_0000745, circ_0001946, and circ_0000517		Inhibits breast cancer progression and metastasis by sponging other miRNAs
	circRNA 0001445, ROR1-AS1, circRNA HNRNPA2B1, circATP8B4, circNEIL3	GLIOBLASTOMA	Promotes glioma progression, radioresistance and TAM infiltration

	hsa_circ_007293, hsa_circ_031752, hsa_circ_020135, circRASSF2, circFNDC3B, hsa_circ_0124055, hsa_circ_0101622	Thyroid cancer	Stimulates proliferation, invasion, and metastasis
	circRNA_0000075, circRNA_100290, circMTO1	Prostate cancer	Promotes therapy resistance
	circMYO10, circRNA_100876, circCCDC66, circ-ABCB10, circ-FAM120A, circ-ZNF652	Ovarian cancer	Regulates gene expression, enhances tumor cell proliferation, and promotes metastasis
Long non-coding RNAs (lncRNAs)	lncRNA CCAT2, lncRNA SNHG14, lncRNA H19	Gastric cancer	Associated with tumor stage and lymph node metastasis
	lncRNA HULC, NONHSAT105177, and SBF2-AS1 HULC, SOX2	Pancreatic cancer	Contribute in EMT and enhance the migration, invasion, and metastasis
	HOTAIR, MEG3,LINC00675, LINC00152, SNHG12	Pancreatic cancer	Regulates chemoresistance genes, enhances tumor cell invasion and migration, and regulates immune response
	lncRNA HULC	Hepatocellular carcinoma	Regulates expression of matrix metalloproteinases
	lncRNA PVT1, lncRNA H19, lncRNA CCAT2, lncRNA POU3F3, lncRNA CRNDE, lncRNA MALAT1, XIST	Colorectal cancer	Promotes angiogenesis, enhances tumor cell invasion and migration, and regulates immune response
	lncRNA H19, lncRNA UCA1, lncRNA GAS5, lncRNA NEAT1	Blood cancer	Induces cell proliferation and drug resistance and regulates expression of multiple oncogenes
	lncRNA UFC1, H19, MALAT1, MEG3, FENDRR, CDKN2B-AS1, TUG1, HOTAIR, GAS5, PVT1, ANRIL	Lung cancer	Regulates apoptosis, enhances cell proliferation, and promotes invasion and migration
	lncRNA GAS5, RP11-77G23.5 and PHEX-AS1		Involved in lung cancer progression
	HOTAIR, MALAT1,LINC00963, LINC00152, lncRNAs UCA1, AGAP2-AS1, AFAP1-AS1, SNHG3, lncRNA H19	Breast cancer	Aids in breast cancer metastasis, chemoresistance and drug resistance
	lncRNA GAS5, lncRNA HOTAIR, lncRNA-SNHG14		Enhances tumor cell proliferation, promotes invasion and migration, and regulates immune response

(Continued)

Table 3.1 List of exosomal RNAs in various types of cancers and their role in cancer progression.—cont'd

Types of exosomal RNA	Examples of exosomal RNA	Types of cancers	Mechanistic role in cancer
	lncRNA H19, MALAT1, lncRNA-ATB, lncRNA-AHIF, Linc01060, lncRNA SBF2-AS1, HOTAIR, lnc-CCAT2	GLIOBLASTOMA	Promotes tumor growth and glioma cell invasion
	lncRNA EGFRvIII, lncRNA HOTAIR		Promotes tumor survival and invasiveness
	lncRNA HNF1A-AS1, lncRNA EVC2, lncRNA LUZP1, lncRNA ANKS1B, lncRNA LINC01305	Cervical cancer	Enhances cisplatin resistance, cellular migration, and invasion
	lncRNA PCAT-1, lncRNA H19, lncRNA HOTAIR	Prostate cancer	Activates cell proliferation and invasion by regulating signaling pathways
	LINC00152, lncRNA GAS5, lncRNA MEG3	Ovarian cancer	Regulates gene expression, enhances tumor cell proliferation, and promotes metastasis
	lncRNA HOTAIR, MALAT1, lncRNASNHG16, CCAT1, H19		Upregulates cancer cell proliferation, progression and metastasis
piRNA	piR-651, piR-823, piR-932	Lung cancer	Enhances tumor cell proliferation, promotes invasion and migration, and regulates immune response
	piR-651, piR-823, piR-1414	Colorectal cancer	Enhances tumor cell proliferation, regulates apoptosis, and promotes invasion and migration
	piR-932, piR-34390, piR-39979	Pancreatic cancer	Regulates chemoresistance genes, enhances tumor cell invasion and migration, and regulates immune response
	piR-54265, piR-32499, piR-4280	Ovarian cancer	Regulates gene expression, enhances tumor cell proliferation, and promotes metastasis
	piR-823, piR-4987, piR-57125	Prostate cancer	Enhances tumor cell proliferation, promotes invasion and migration, and regulates immune response
	piR-651, piR-823, piR-4987	Ovarian cancer	Regulates gene expression, enhances tumor cell proliferation, and promotes metastasis

snoRNA	SNORD78, SNORA42, SNORA71B	Breast cancer	Enhances tumor cell proliferation, promotes invasion and migration, and regulates immune response
	SNORD47, SNORD50A, SNORD78	Lung cancer	Regulates apoptosis, enhances cell proliferation, and promotes invasion and migration
	SNORD50A, SNORD56, SNORA42	Colorectal cancer	Promotes angiogenesis, enhances tumor cell invasion and migration, and regulates immune response
		Ovarian cancer	Regulates gene expression, enhances tumor cell proliferation, and promotes metastasis
	SNORD33, SNORD35A, SNORD47	Pancreatic cancer	Regulates chemoresistance genes, enhances tumor cell invasion and migration, and regulates immune response
tRNA	tRNA-Lys, tRNA-Val, tRNA-Leu	Breast cancer	Enhances tumor cell proliferation, promotes invasion and migration, and regulates immune response
	tRNA-Gln, tRNA-Tyr, tRNA-Ser	Colorectal cancer	Promotes angiogenesis, enhances tumor cell invasion and migration, and regulates immune response
	tRNA-Lys, tRNA-Asn, tRNA-Cys	Pancreatic cancer	Regulates chemoresistance genes, enhances tumor cell invasion and migration, and regulates immune response
	tRNA-Asp, tRNA-Val, tRNA-Ser	Ovarian cancer	Regulates gene expression, enhances tumor cell proliferation, and promotes metastasis
	tRF-1001, tRF-03357	Prostate cancer	Inhibits PCa cell proliferation and induces apoptosis

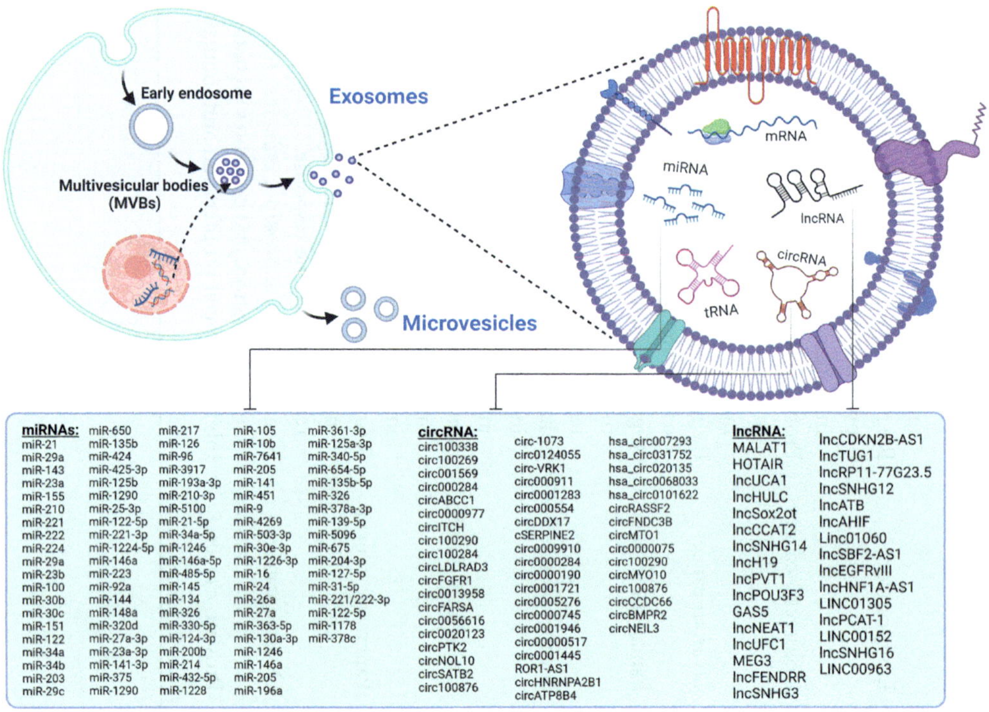

Figure 3.1 Schematic presentation of secretary exosome and different exosomal RNAs, which has roles in cancer progression.

by exosomes have been shown to regulate gene expression in cancer cells by targeting specific mRNA transcripts, leading to their degradation or translational inhibition [36–40]. This can result either in the downregulation of tumor suppressor genes or the upregulation of oncogenes, thereby promoting tumor growth and metastasis [38,41,42]. Moreover, miRNA profiles in circulating exosomes are suggested to be promising biomarkers in cancer prognosis, diagnosis, and treatment [43,44]. Exosome-derived lncRNAs have also been demonstrated to impact cancer progression by influencing gene expression and signaling pathways [45–48]. For example, exosomal lncRNA HOTAIR has implications for breast cancer metastasis through the promotion of EMT and inhibition of apoptosis [49–51]. Similarly, the exosomal lncRNA MALAT1 (metastasis-associated lung adenocarcinoma transcript 1) has been shown to promote the invasion and metastasis of lung cancer by influencing the Wnt/β-catenin pathway [52–54]. Studies suggest that exosomal tsRNAs from plasma such as tRNA-GlyTCC-5, tRNA-GluCTC-5, tRNA-ValTAC-3, and tRNA-ValAAC-5 in hepatic carcinoma can serve as diagnostic biomarkers [55–58], whereas piRNAs (piR-004800 [18], piR-20548188, and piR-10506469 [59]) have been shown to exhibit crucial roles in progression and tumorigenesis in multiple myeloma [60]. Overall, in this chapter, we have

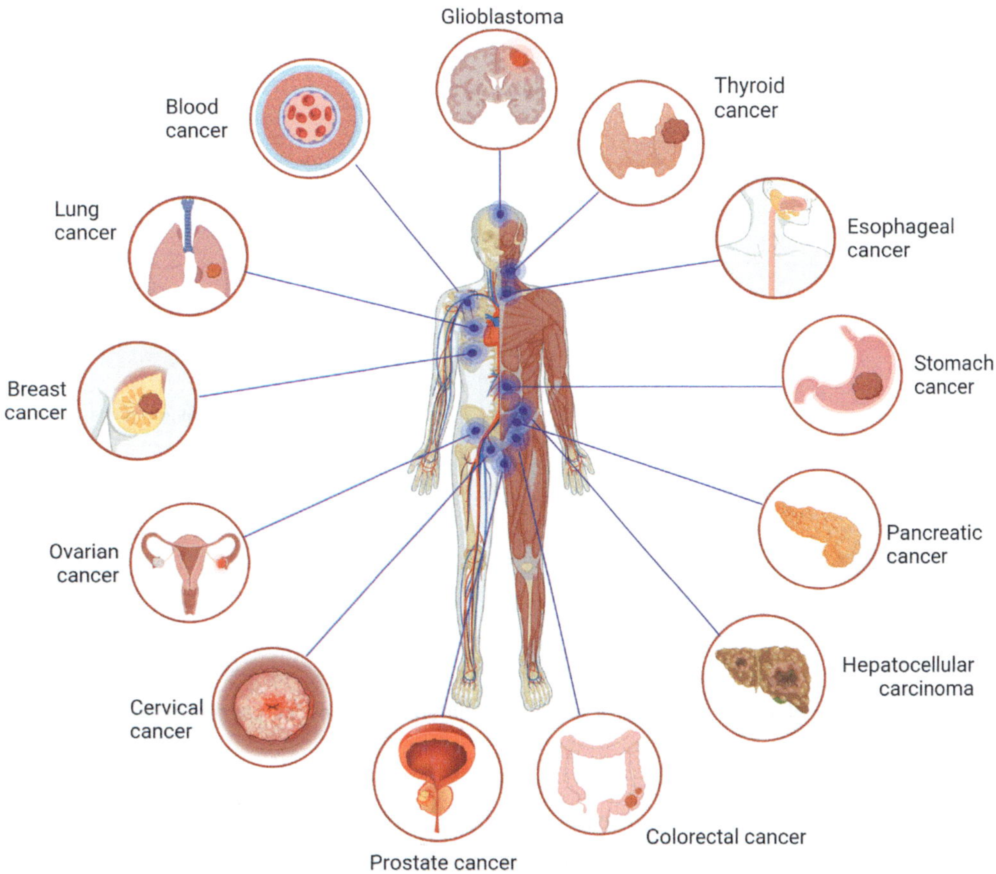

Figure 3.2 Outline of the chapters showing cancer progression in different body parts.

discussed the pivotal roles of cancer cell–derived exoRNAs in TME modulation and focused on the key roles of exosomal ncRNAs derived from various stromal cells in the mediation of different cancer phenotypes and tumor progression (Fig. 3.2). However, more research work is important to fully understand the complex mechanisms underlying exoRNA-mediated communication between stromal cells and cancer cells, and to open up new directions for cancer diagnosis, prognosis, and effective therapeutic strategies (Fig. 3.3).

2. Exosomal RNAs in cancer progression

2.1 Gastrointestinal cancer progression

Gastrointestinal (GI) cancer is one of the major causes of global cancer-related fatalities. The progression and metastasis of GI cancer are complex and involve various genetic and molecular alterations [61]. Over the past decade, researchers have increasingly focused on

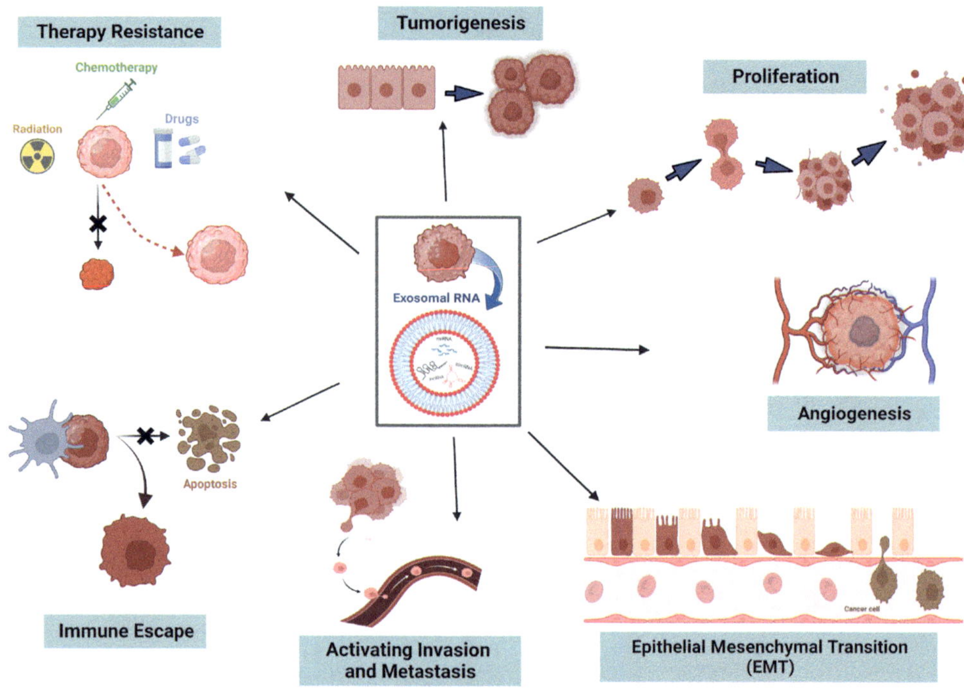

Figure 3.3 Exosomal RNAs play a crucial role in different stages of cancer development and progression.

the role of exoRNAs in various diseases, including cancer. GI cancer is a broad term that refers to a group of cancers that affect the digestive system. In this chapter, we will primarily discuss the role of exoRNAs in GI cancer, focusing on their involvement in cancer progression, metastasis, immune regulation, chemo resistance, therapeutic agents, and biomarkers [62].

It is intriguing to learn that various exoRNAs in GI cancer have very distinct yet discreet roles. One such study discusses that exosomal miR-425-3p has been found to be upregulated in esophageal squamous cell carcinoma cells [63], whereas another study reported that exosomal miR-125b was significantly downregulated in bladder cancer patients and could distinguish between bladder cancer patients and healthy individuals with high sensitivity and specificity [64]. Another study found that exosomal miR-1290 is closely linked with tumor progression and could predict poor prognosis in bladder cancer patients [37,65,66]. In addition to the exosomal miRNAs, there are other types of exoRNA, such as lncRNAs [62], circRNAs [67], snoRNAs [68], tRNAs [69], and small interfering RNAs (siRNAs) [70], which have been found to play robust roles in different kinds of GI cancer progression and metastasis, which we will discuss in the following.

2.1.1 Gastric cancer progression

Gastric cancer (GC) is a common malignancy of the digestive system that poses a major public health concern worldwide. The estimated cases over a year are more than a million, and this disease is responsible for around 700,000 deaths annually. It is a complicated and heterogenous disease with multiple environmental, genetic, and behavioral factors that contribute to its development and progression. In spite of the advances made in diagnosis and treatment, the prognosis for patients suffering from advanced stages of gastric cancer remains below par. Recently, there has been growing interest in the functions of exoRNAs in gastric cancer, as they play a crucial role in intercellular communication and cancer progression [3,71,72].

Several studies have shown that exoRNAs, including both miRNAs and lncRNAs, play a crucial role in gastric cancer [37]. For example, one study found that the levels of exosomal miR-21 were elevated in gastric cancer patients and has been depicted to promote cancer cell proliferation, followed by migration, host tissue invasion, and metastasis of gastric cancer by targeting the tumor suppressor gene, PTEN. This was also associated with poorer survival outcomes [73]. Zeng et al. (2018) discovered that exosomal miR-25-3p was upregulated in the sera of patients with gastric cancer and exhibited lymph node metastasis along with a poor prognosis [74]. Similarly, Jiao et al. (2021) identified exosomal miR-122-5p as a potential biomarker for gastric cancer, which was downregulated in the sera of patients with gastric cancer and was associated with the stage of the tumor and metastasis of lymph nodes [75]. In 2020, another team showed how exosomal lncRNA CCAT2 was upregulated in the sera of patients with gastric cancer and similarly the association with tumor stage and lymph node metastasis [76]. Furthermore, in vitro studies showed that exosomal CCAT2 promoted gastric tumor cell proliferation and migration, suggesting its role in gastric cancer progression. A study also identified exosomal lncRNA SNHG14 as a potential biomarker. It was discovered that exosomal SNHG14 was upregulated in the serum of gastric cancer patients and was associated with tumor stage and lymph node metastasis. Additionally, lncRNA H19 is greatly expressed in gastric cancer exosome and promotes cell proliferation, invasion, and metastasis [77,78].

ExoRNAs even play a major role in gastric cancer drug resistance. One study found that exosomal miR-155 was upregulated in gastric cancer cells resistant to cisplatin, a commonly used chemotherapy drug. The researchers showed that this miRNA could be passed on from resistant cancer cells to sensitive ones via exosomes and that it could confer drug resistance to the sensitive cells [3,37,79]. Another study investigated the function of exosomal miRNAs in gastric cancer metastasis. Here, they found that the exosomal miR-221-3p was highly upregulated in the sera of gastric cancer patients with lymph node metastasis compared with those without metastasis. They showed that this miRNA could promote cancer cell migration and invasion by targeting a specific gene involved in the regulation of the actin cytoskeleton [3,37].

Despite the advances in research leading to the massive evidence of exoRNAs' activity in gastric cancer, very little in the context of a solution is available. In addition, there are a list of technical challenges related to studying exoRNAs. Nevertheless, the study of exoRNAs in gastric cancer constitutes a promising area of research with the potential to uncover new biomarkers and therapeutic targets.

2.1.2 Pancreatic cancer progression

Pancreatic cancer (PaCa) is responsible for almost 7% of cancer-related deaths annually. ExoRNAs' involvement in pancreatic cancer is in fact an area of active research that carries promise for bettering early detection followed by treatment of the ghastly disease [80].

ExoRNAs play key functions for aggressive tumor growth, invasion, and metastasis, as well as misregulating gene expression and contributing to chemotherapy resistance [37,81].

Researchers have explored the possibility that PaCa cancer cells move and accumulate in secondary tissues and organ, which might be a leading issue in treatment failure [81]. Exosomal miRNA-10a-5p released by cancer-associated fibroblasts (CAFs) increases migration and invasion in PaCa cells, but stimulation of vitamin D receptor (VDR) signaling inhibits these effects in PaCa cells. Equivalently, CAF-associated miRNA-221, miRNA-21, and PSC-associated miRNA-451a and miRNA-21-5p have evidence of conferring aggression in PaCa cells [81]. However, in terms of inhibition, NK cell-acquired miRNA-3607-3p inhibits PaCa cell migration and invasion by acting directly on IL-26. Additionally, BM-MSCs-originating miRNA-126-3p and miRNA-1231 also indicate inhibitory roles in migration and invasion of PaCa cells. Surprisingly, rather than proliferation, exosomal lncRNAs and circRNAs appear to play a much stronger role in migration, invasion, and metastasis [81]. This is noted in the instance of tumor-derived exosomal lncRNA SBF2 antisense RNA1 (SBF2-AS1), which is significantly expressed in liver cancer (HULC), NONHSAT105177, SOX2 overlapping transcript (Sox2ot), circRNA phosphodiesterase 8A (PDE8A), and isoleucyl-tRNA synthetase (IARS) [81]. All of these have been linked to increased cancer cell migration, invasion, and metastasis. Spontaneously, exoRNAs also serve as suppressive factors, as mentioned before. Therefore, exoRNAs operate as critical regulators of migration, invasion, and metastasis, supporting PaCa development [81]. One of the other features that requires attention is the EMT, since recently, various exoRNAs have been known to promote EMT in PaCa [81,82]. For example, miRNA-125b-5p is upregulated in very invasive PaCa cells that facilitate migration and invasion but most importantly assist EMT through MEK2/ERK2 signaling activation [81]. In parallel, exosomal lncRNAs HULC, NONHSAT105177, and miRNA-301a have published reports that claim to contribute to EMT in PaCa. Scientists claim that ideas regarding these exoRNAs will provide substantial information to use them for therapeutic intervention [80,83].

Among all exosomal loads, miRNAs are presently described as the best biomarkers due to their abundance and accessibility [39,80,82,84]. This would imply that they can be exploited as possible diagnostic tools. Furthermore, they are inextricably tied to the fact that recurrence occurs in PaCa. According to many research and studies, highly expressed exosomal miRNA-21 not only distinguishes PaCa patients from healthy controls but also individuals suffering from other pancreatic disorders such as intraductal papillary mucinous neoplasm or chronic pancreatitis [41,81—83]. It could be an aid in early diagnosis and evaluation of tumor stage. Furthermore, other exosomal miRNAs such as miRNA-10b, miRNA-210, miRNA-1246, and miRNA-451a can be used for accurately diagnosing PaCa in the early stages [80,82]. Some of the miRNAs, in particular, have been associated with the recurrence of tumors or might be used as independent prognostic markers for PaCa. The details of these biomarkers will be focused on and discussed in greater detail in the following chapter [41,81—83].

Overall, the study of exoRNAs in pancreatic cancer is an exciting field of study with the potential to improve our understanding of this disease and develop new diagnostic and therapeutic strategies [83].

2.1.3 Hepatocellular carcinoma progression

Like any other type of cancer, hepatocellular carcinoma (HCC) is a kind of liver cancer that is influenced by a diverse collection of genetic and environmental variables. In the recent past, exoRNAs have emerged as important players in HCC development and progression and may have potential as diagnostic and therapeutic tools [85].

A few studies have shown that exoRNAs can modulate different cellular processes that contribute to HCC development and progression [10,86—88]. For example, one study identified a group of miRNAs that were expressed differentially in exosomes isolated from HCC patients compared with healthy individuals. One of these miRNAs, miR-21, was highly enriched in HCC-derived exosomes and promoted HCC cell proliferation, followed by invasion, where the main target is the tumor suppressor PTEN [10,86—88]. Similarly, another study found that exosomal miR-122, a liver-specific miRNA, was downregulated in HCC-derived exosomes and correlated with poor prognosis in HCC patients. The researchers engineered mesenchymal stem cells to produce exosomes that contained miR-122, a tumor-suppressive miRNA that is frequently downregulated in HCC. It is shown that exosomal miR-122 inhibits HCC cell proliferation, migration, and invasion by targeting the oncogene ADAM10 [10,86—88]. Various studies even identified a collection of miRNAs that were differentially expressed in serum exosomes from HCC patients compared with healthy controls. They proposed that a combination of these miRNAs could accurately distinguish HCC patients from healthy individuals [10,86—88]. It has been shown that the exosomes are taken up by HCC cells and inhibit their growth as well as invasion in vitro and in vivo. Another study shows that exosomal MiR-1224-5p is downregulated in the exosomes that are derived from HCC cells, and hence, overexpressing it helps inhibit cancer cell proliferation, migration, and invasion by regulating the expression of multiple target genes [8,85–87].

In addition to miRNAs, overexpression of a lncRNA called HULC in HCC-derived exosomes helps promote HCC cell migration and invasion by regulating the expression of the matrix metalloproteinase MMP9. Recently, several lncRNAs, with differential expression in serum exosomes from HCC, were identified, and one of these lncRNAs called HOTAIR was associated with poor prognosis in HCC patients [85,88,89].

Additionally, a paper from 2020 has shown that circRNA-100338 from the serum exosomes helps in promoting HCC metastasis by enhancing invasiveness and angiogenesis. They have therefore been investigated as therapeutic tools for HCC as well. Exosomal circRNA-100338 is also known to inhibit HCC cell proliferation and migration by regulating the expression of the oncogene MDM4. Different scientists show that circulating levels of exosomal circRNA-100338 were decreased in HCC patients [90], suggesting that exosomal circRNAs may also act as potential biomarkers and therapeutic targets for HCC [91]. Moreover, it has been seen that circ-0051443 derived from exosomes of HCC patients suppressed the malignancy of the HCC cells by reducing their proliferation and thereby migration. Therefore, exosomal circ-0051443 can definitely act as a promising biomarker for early HCC detection [85].

2.1.4 Colorectal cancer progression

Colorectal cancer (CRC) is another cause of fatality, with a high proportion of patients presenting with advanced-stage illness at the time of diagnosis, compounding its influence on mortality [92]. Despite varied progress in diagnosis and treatment, CRC prognosis remains a major clinical challenge, and there is a need for better understanding of the molecular mechanisms that drive CRC progression and for developing effective therapeutic strategies [93–95].

It is known that exosomal miRNAs can be transferred between proliferating tumor cells and other cells present in the TME, resulting in changes in gene expression and cellular behavior. Exosomal miRNAs in CRC have been extensively studied and identified as regulators of CRC development and progression. For example, miR-21, an extensively studied oncogenic miRNA that is typically elevated in CRC tissues as well as significantly concentrated in CRC cell exosomes, can be promptly taken up by surrounding cells and contribute to CRC progression. Exosomal miR-21 has also been demonstrated to establish tolerance to immune cells in CRC by reducing the levels of natural killer or NK cell activity. A study showed that miR-21 derived from CRC cells could be taken up by normal colonic fibroblasts and induces the secretion of the proinflammatory cytokine IL-6, leading to enhanced CRC cell proliferation and invasion. In addition to miR-21, other miRNAs, such as miR-29a, miR-155, miR-146a, and miR-223, are also enriched in exosomes derived from CRC cells and contribute to CRC development. MiR-155, for instance, is shown to promote the M2 polarization of macrophages, which are involved in tumor progression and immune suppression. Other examples are miR-155, which promotes CRC cell invasion, and miR-92a, which is associated with poor overall survival [95—99].

Additionally, various tRNA-derived small RNAs (tsRNAs) and the lncRNA PVT1 are highly expressed in CRC exosomes and promote cell proliferation and invasion by targeting the TP53 tumor suppressor gene. One such study shows that lncRNA H19 could promote CRC cell proliferation and invasion by triggering the Wnt/β-catenin signaling pathway. Another exosomal lncRNA CCAT2 does the same by regulating the expression of the EMT-related transcription factor ZEB1 [96,98,100,101]. Other reports suggest that exosomal lncRNA POU3F3 associates with poor prognosis in CRC patients, along with proliferation and invasion. It is also shown in some studies that serum exosomal lncRNA CRNDE (colorectal neoplasia differentially expressed) levels were remarkably higher in CRC patients compared with healthy people and were linked with lymph node metastasis and poor prognosis. It is interesting to learn that plasma exosomal lncRNA MALAT1 levels are elevated in CRC patients and strongly correlate with advanced tumor stage and poor overall survival [95,97,99,100].

Besides, the circular RNAs circRNA_100269 and circRNA_001569 are highly expressed in CRC exosomes. CircRNA-000284 was shown to be upregulated in CRC tissues and could be taken up by various recipient cells to promote CRC cell proliferation and migration. Similarly, exosomal circRNA-ABCC1 could promote CRC cell proliferation and invasion by sponging miR-144 and upregulating the expression of the oncogene MYC [95,97,99—102].

Functionally, exosomal miRNAs also play a role in CAF activation in the TME. In CRC, exosomal miR-92a is found to be highly expressed in CAFs and to promote cancer cell proliferation and invasion by controlling the expression of different genes implicated in EMT and tumor angiogenesis [95—97,99].

ExoRNAs have also been investigated as potential diagnostic and therapeutic tools in CRC. For example, work on exosomal miR-148a inhibiting CRC cell proliferation and migration by targeting the oncogene Wnt1 has been published. Correspondingly, it is also demonstrated that exosomal miR-29a could sensitize CRC cells to chemotherapy by targeting the antiapoptotic protein Mcl-1 [96—98,100].

It was found that exosomal miRNAs from CRC cells could be detected in the serum of CRC patients and that the levels of these miRNAs were correlated with CRC stage and prognosis [103]. For instance, exosomal miR-320d could inhibit CRC cell proliferation and migration, while another exoRNAs, miR-34a, has been shown to suppress tumor growth and metastasis in CRC by targeting the SIRT1/P53 signaling pathway. This provides evidence that exoRNAs can function as a double-edged sword in cancer progression and inhibition [95—97,99].

2.2 Blood cancer progression

Blood cancer, also called hematological malignancy, is a variegated group of cancers that influence the blood, bone marrow, and lymphatic systems. These include leukemia, lymphoma, and myeloma, which can be classified according to the type of blood cell affected and their molecular characteristics [104]. The diagnosis, prognosis, and treatment of blood cancers are often challenging due to the heterogeneity and complexity of the disease. In recent years, exoRNAs have emerged as a promising tool for understanding the pathogenesis of blood cancers, identifying biomarkers for diagnosis and prognosis, and developing advanced therapies [105].

In blood cancers, exoRNAs have an impactful role in disease progression, immune evasion, and drug resistance. Exo-miRNAs such as miR-155, which are upregulated in various blood cancers, including leukemia and lymphoma, are shown to promote the growth and survival of blood cancer cells by suppressing the immune system by targeting various immune-related genes. Exosomal miR-21, which is upregulated in multiple myeloma, initiates tumor growth and drug resistance by targeting various DNA repair pathways [106–110]. Exosomal lncRNAs have also been implicated in blood cancer progression and drug resistance. For example, H19 is shown to stimulate leukemia cell proliferation and drug resistance by modulating various signaling pathways [111,112]. UCA1 has evidence to promote multiple myeloma cell proliferation and drug resistance by affecting the TP53INP1 pathway [113,114], while GAS5 regulates expression of various oncogenes and tumor suppressor genes in chronic lymphocytic leukemia [115].

ExoRNAs have also been investigated as therapeutic targets for blood cancers. Antisense oligonucleotides (ASOs) targeting exosomal miR-155 and miR-21 have been shown to inhibit blood cancer growth and drug resistance in preclinical studies. Exosomal lncRNAs such as H19 and UCA1 have also been targeted using ASOs or small molecule inhibitors, with promising results in preclinical studies [116]. They have also emerged as an optimistic biomarker for diagnosis, prognosis, and therapeutic response to blood cancer [112–114]. As exosomes can be easily isolated from the blood circulation, they represent a minimally invasive and noninvasive approach for blood cancer diagnosis and monitoring. These lncRNAs and miRNAs are reported as potential biomarkers of blood cancer diagnosis and prognosis.

Several studies have furthered our learning regarding the function of exoRNAs in blood cancers and have underlined the prospective of exoRNAs as biomarkers and therapeutic targets that we discuss better in the next chapter.

A study published in 2022 investigated the role of exosomal miRNAs in acute myeloid leukemia (AML). They found that AML cells secrete exosomes containing miR-126, which promotes their survival and proliferation of AML cells. They identified a potential therapeutic target in the exosome-mediated communication between AML cells and the bone marrow microenvironment [117,118].

Similarly, another study investigated the role of exosomal lncRNAs in chronic lymphocytic leukemia (CLL). Exosomes harboring the lncRNA NEAT1 were discovered to boost CLL cell survival and proliferation in this investigation [119—123], thereby identifying a potential therapeutic target in the exosome-mediated communication between CLL cells and the TME.

Overall, exoRNAs have emanated as a favorable tool for understanding the pathogenesis of blood cancers, identifying biomarkers of cancer diagnosis and prognosis, and developing novel therapies. While we await more responses from science, the potential of exoRNAs in blood cancer biology is enormous for improving patient outcomes in the future.

2.3 Lung cancer progression

One other critical source of human death is lung cancer, which occurs every year [124]. The major contributing factor is the communication between tumorigenic cells and the cells in their surrounding environment [125,126]. Moreover, various evidence reveal that exoRNAs discharged by cancerous cells have a crucial role in the recruitment and rewiring of tumor environment reconstitution. Lung cancer cell—derived exosomes (LCCDEs) play a critical role in cell-to-cell communication and have recently gained attention for their role in carcinogenesis, diagnosis, prognosis, and cancer therapy [127]. LCCDEs can stimulate cell proliferation, angiogenesis and metastasis, along with antitumor immune responses during carcinogenesis and therapeutic drug resistance, and are currently considered a crucial factor in liquid biopsy analysis for analyzing lung cancer [127].

An investigation claims that miRNA-96 nurtures lung cancer progression with the help of repression of Lim domain 7 (LMO7) protein activities that participate in the formation and maintenance of the actin cytoskeleton [128]. Similarly, a study on M2 macrophage-derived exosomal miR-3917 exhibited lung cancer progression by targeting GRK6. This is manifested through molecular miRNA transport by M2 macrophage-derived exosomes (M2-Ex) [129]. Other examples are miR-103a-3p, and miR-5100, which aid cancer cell proliferation and invasion by targeting PTEN and FOXO3 respectively [130,131].

In 2021, Cancer Cell International published the story of exosomal miRNA expression in the serum of affected individuals with non—small-cell lung cancer (NSCLC) versus healthy ones. They discovered that several exosomal miRNAs, including miR-21-5p, miR-34a-5p, and miR-1246, were significantly upregulated in NSCLC patients in contrast to controls (Table 3.1) [36]. They observed that miR-21-5p was particularly related to poor survival in NSCLC patients [132].

Another study published in 2021 found that exosomal miR-1246 migrated to fibroblasts from lung cancer cells, thereby promoting their activation and secretion of proinflammatory cytokines, followed by the promotion of lung cancer cells' migration and invasion capacity by downregulating the death receptor 5 (DR5) gene expression [133]. This in turn promotes metastasis and drug resistance by targeting nuclear factor (erythroid 2)—like factor 3 [NFE2L3] [134].

Serum miR-146a-5p, miR1244, and miR-485-5p could distinguish early-stage NSCLC from benign lung lesions in healthy individuals [135]. It is surprising to learning of the evidence about miR-145 sensitizing lung cancer cells to chemotherapy and improving survival outcomes in mice with lung cancer [136]. Additionally, miR-21 and miR-29a bind to TLRs to activate the NF-κB pathway and promote tumor growth and metastasis in lung cancer cells [137,138]. A study from 2016 found that members of the miR200 family, miR-200b and miR-205-5p, seem to get upregulated in the exosomes derived from the serum of lung cancer patients. Knocking out these miRNAs inhibits the proliferation and invasion of lung cancer cells [139].

Exosomal circRNAs are also involved in various malignancies by operating through multiple mechanisms. One such mechanism involves their function as a "miRNA sponge," in which they can bind and regulate miRNA availability. This allows oncogene expression at much higher levels, thereby contributing to cancer progression [140—142]. In lung cancer, there is an elevated level of circRNAs, some of which have oncogenic activities that promote EMT in vitro. For example, suppressing microRNA-137 with circ-LDLRAD3 causes an upregulation in the glutamine transporter, which is one of the SLC1A5 members in NSCLC cells, promoting proliferation and EMT [143]. circFGFR1 is overexpressed in NSCLC and promotes cancer progression. Circ0013958 and circ-FARSA have been found to accentuate the progression of lung cancer by regulating miR-134 and miR-326/miR-330-5p, respectively. By sponging up these miRNAs, these circRNAs allow oncogenes to be expressed at higher levels, which can promote the growth and metastasis of cancer cells. Similarly, hsa_circRNA_0056616 was found to be highly expressed in lung adenocarcinomas with lymph node metastases. In contrast, exosomal circPVT1 from lung cancer cells was found to activate the miR-124-3p/EZH2 pathway and polarize macrophages, increasing cell invasion and migration [140,144]. In this case, the exosomal circRNA is playing a role in shaping the immune response to the cancer cells and promoting their ability to spread. Studies suggest that hsa_circ_0020123 inhibits apoptosis in lung cancer via a reduction of miRNA-144 and an increase in ZEB1 and EZH2 expression. They discovered that silencing hsa_circ_0020123 inhibits cancer cell proliferation and spread. A few studies have also provided evidence that circPTK2 and circNOL10 are downregulated in NSCLC and inhibit lung cancer development. It is relevant that circSATB2 has the ability to provide indices of blood detection for lung cancer metastasis diagnosis with higher sensitivity and specificity [140,141,145].

Further, reports suggest that some lncRNAs are differentially enhanced in exosomes, which promotes tumor progression by aiding tumor premetastatic niche formation and EMT, which in turn promotes the escape of tumor cells from the site of the original tumor to form new metastases. The PI3K/Akt signaling pathway is activated by the exosome-transmitted lncRNA UFC1, which promotes NSCLC development by suppressing PTEN expression. Several other lncRNAs, including MALAT1, H19, MEG3, CDKN2B-AS1, TUG1, HOTAIR, and GAS5, have been linked to lung cancer. By blocking angiogenesis, the lncRNA GAS5 has been shown to be a potential therapeutic target for lung cancer [146–148]. Few lncRNAs, such as RP11-77G23.5 and PHEX-AS1, have been discovered to be highly enhanced in EpCAM-specific exosomes from lung cancer patients and, like a marker, can identify malignant from benign lung tumors [149].

Several studies are underway to develop novel diagnostic and prognostic tools based on LCCDEs, as well as methods for using exosomes as efficient drug delivery vehicles [150,151].

2.4 Breast cancer progression

Breast cancer, as we all know, is the biggest cause of cancer-related death in women aged 25–60, but it also affects men [152,153]. According to GLOBOCAN 2018, around 12% of all cancer cases are diagnosed in 2 million women. ExoRNAs have been implicated in numerous aspects of tumor growth and metastasis in breast cancer, including cell proliferation, invasion, angiogenesis, immune evasion, drug resistance, and TME modulation [152–154].

Exosomal miRNAs influence breast cancer growth through intercellular communication [140,155,156]. Exosomal miRNAs have also been shown in studies to increase breast cancer cell motility by downregulating E-cadherin in normal fibroblasts or to produce drug resistance in cancer cells by targeting apoptosis-related genes. MiR-105, for example, which is abundant in breast cancer exosomes, promotes metastasis by encouraging the establishment of premetastatic niches in distant organs. Similarly, miR-10b, another highly abundant exoRNA in breast cancer, promotes cancer invasion and metastasis by targeting the tumor suppressor gene HOXD10 [157,158]. Exosomal miR-7641 derived from cancer has also been shown to promote breast cancer growth and metastasis [159]. Exosomes, together with RNA-induced silencing complex-associated (RISC-associated) miRNAs, can cause cancer. For example, miRNA-205 targets E2F1, causing breast cancer cells to proliferate. Other exosomal miRNAs involved in breast cancer progression include miR-21, miR-141, and miR-451, which are involved in gene regulation and intercellular communication. Research on exosomal miR-9 shows that it has the potential to change the state of normal human breast fibroblasts, making it easier for NFs to turn into CAFs and help form tumors. MiR-210 was shown to be transported via exosomes in the TME and was associated with the regulation of vascular remodeling-related genes, such as Ephrin A3 and PTP1B, encouraging angiogenesis. Additionally, lymphatic endothelial cells promote breast cancer metastasis through the exosomal miRNAs miR-30e-3p, miR-503-3p, and miR-4269. MiRNAs, such as miR-155 and miR-7641, which target DUSP14 and PTEN, and miR-1226-3p, which target aquaporin-5, are reported to affect breast cancer invasiveness and migration too [155].

New research suggests that miRNA up- and downregulation can cause drug resistance in breast cancer cells via many signal channels. Exosomal miR-221 and miR-222 have been demonstrated to promote breast cancer cell survival as well as medication resistance by targeting PTEN in conjunction with the cell cycle regulator p27. Exosomal miR-100 and miR-30a have been linked to cancer cell resistance to the chemotherapeutic treatments: adriamycin and docetaxel. Exosome-derived miR-9-5p increases tamoxifen resistance in breast cancer cells by decreasing ADIPOQ [155]. Exosomal miRNAs linked to drug resistance could be a target for therapeutic intervention. Other miRNAs found in docetaxel-resistant exosomes, such as miR-27a, miR-16, miR-26a, miR-23a, and miR-24, have been found to reduce breast cancer chemoresistance by influencing resistance transmission via exosomes. Exosomal miRNAs could be valuable biomarkers for detecting resistance to drugs in breast cancer treatment, and their control could help reduce resistant breast cancer cells. Exosomal miR-363-5p, which modulates platelet-derived growth factor (PDGF) signaling and thus inhibits cancer cell proliferation and migration to lymph nodes, was found in different breast cancer patients plasma and was significantly reduced in lymph node—positive patients, according to Wang et al. Furthermore, cancer-associated fibroblast exosomes with low miR-7641 levels promote stemness in breast cancer cells via HIF-1 alpha [155].

CircRNAs are also dysregulated in breast cancer, thereby promoting progression and metastasis [160]. For instance, exosomal circRNA_100876 is found to do the same by sponging miR-361-3p and upregulating its target gene, CDKN1A. circBMPR2, circ-1073, hsa_circ_0068033, circ-VRK1, circRNA_000911, circRNA_0001283, circRNA_000554, and circDDX17 are known to induce apoptosis. circRNA cSERPINE2 can mediate a positive feedback loop among tumor cells and TAMs [161] that promote cancer progression and metastasis, whereas, circ_0009910 can initiate breast cancer progression and chemoresistance by sponging miR-145 and upregulating MTH1. circ_0000284 [162], circ_0000190, circ_0001721, and circ_0005276 have a significant function in enhancing breast cancer progression and metastasis by sponging miR-125a-3p, miR-130a-3p, miR-34a, and miR-340-5p and upregulating FUT4, SOX4, MET, and RNF213, respectively. Some evidence depicts that circ_0000745, circ_0001946, and circ_0000517 can inhibit breast cancer progression and metastasis by sponging miR-654-5p, miR-135b-5p, and miR-326 while downregulating COL1A1 while at the same time upregulating KLF4 and PTEN, respectively [160,163,164].

Additionally, some lncRNAs are involved in breast cancer metastasis. For example, one study asserts that HOTAIR, signal lncRNA can promote metastasis by regulating the expression of HOX genes and EZH2 and by targeting the tumor suppressor gene BRCA1. Another report suggests that MALAT1, a scaffold lncRNA, can enhance metastasis by interacting with members of the miR-200 family and ZEB1/2. Interestingly, exosomal LINC00963 from breast cancer cells can promote tumor growth, progression, and chemoresistance by sponging miR-378a-3p and also upregulating VIM/

ABCB1, whereas, exosomal LINC00152 can induce tumor growth, progression, and drug resistance by targeting miR-139-5p and activating the Wnt/-catenin signaling pathway. Xu et al. found that tamoxifen-resistant breast cancer cells had higher levels of the LncRNA urothelial carcinoma-associated 1 (UCA1) than sensitive cells. Another study published by Zheng et al. stated that hnRNPA2B1 is essential for the exosomal packing of AGAP2-AS1, which is responsible for trastuzumab resistance [165—167].

Also, cancer-associated fibroblast-derived exosomal SNHG3 makes breast tumor cells grow faster by reprogramming their metabolism. Another study found that exosomal transfer of the long noncoding RNA H19 increased doxorubicin resistance in breast cells. GAS5, which is a decoy lncRNA, is an example of an exosomal lncRNA that can stop breast cancer from spreading by binding to miR-21 and stopping it from interacting with PTEN. Furthermore, it has been revealed that lncRNA HOTAIR in circulatory exosomes is positively linked with ErbB2 (HER2/neu) expression, implying its potential utility as a liquid biomarker [166].

2.5 Glioma progression

Despite advancements in surgery, radiation, and chemotherapy, glioma continues to be the most prevalent and lethal central nervous system tumor, with a poor prognosis and a significant treatment challenge due to its high complexity and infiltrative nature. Glioma-derived exosomes (GDEs) were recently discovered in local tissue and circulatory blood specimens, and they have the ability to travel into target cells and promote genetic transfer. GDE ncRNAs are taken in a nonrandom manner by heterologous and homologous cells, influencing posttranscriptional genetic regulation and resulting in tumor growth and metastasis. GDE ncRNAs can also be used in differential diagnosis, postsurgery glioma progression monitoring, and patient response prediction to customized therapy [168].

A vast number of ncRNAs contained in GDEs are thought to play regulatory roles in many glioma pathophysiological processes. MiRNAs, for example, have been linked to glioma development via the activation and regulation of several signaling systems. Targeting the PTEN/PI3K/AKT pathway and the tumor suppressor gene TP53INP1 with exosomal miR-21, miR-221, and miR-1246 has been shown to increase glioma cell growth and spread. Exosomal miR-9 was found to increase angiogenesis by targeting the suppressor gene SOCS5. Exosomal miR-10b was found to increase glioma cell migration and invasion by targeting the tumor suppressor gene HOXD10. Thuringer and colleagues discovered that exosomal miR-5096 isolated from glioblastoma (GBM) patients increased glioma cell invasion via modulating the K+ channel Kir4.1. Exosomal miR-222 and miR-26a expression in glioma-derived exosomes was much higher than in normal brain tissue—derived exosomes, and multiple signal transduction pathways were activated to accelerate tumor growth [168—171].

Exosomal lncRNA H19, in addition to miRNAs, is known to be elevated in gliomas and can enhance tumor growth and invasion by influencing downstream targets such as miR-675 and STAT3. MALAT1, an exosomal lncRNA, has also been found to be elevated in gliomas and has been proven to enhance tumor growth and invasion via altering the PI3K/AKT signaling pathway. Exosomes generated from glioma cells containing lncRNA-ATB were reported to activate astrocytes via Ago-dependent inhibition of miR-204-3p in another study. In turn, this enhances glioma cell invasion. Similarly, the antisense transcript of hypoxia-inducible factor-1 (AHIF) lncRNA is present in exosomes produced by AHIF-knockdown GBM cells and aids in cell growth suppression and invasiveness via modulation of angiogenesis and migratory factors. Similarly, Linc01060-containing exosomes derived from hypoxia GSCs (H-GSCs) accelerate the advancement of glioma illness. Exosomes with high levels of lncRNA SBF2 antisense RNA 1 (lncRNA SBF2-AS1) produced from TMZ-resistant GBM cells may lead to TMZ resistance in chemoresponsive GBM cells. Furthermore, Ma et al. discovered that glioma cells release HOTAIR-containing exosomes into endothelial cells, where the lncRNA stimulates angiogenesis by upregulating VEGF-A levels. Similarly, when linc-CCAT2 is delivered to endothelial cells, it induces angiogenesis by simultaneously activating VEGF-A and TGF [168,172].

Circular RNAs (circRNAs), such as circRNA 0001445, impact glioma growth via the miRNA-127-5p/SNX5 signaling pathway [173]. Exosomal ROR1-AS1 promotes the growth of glioma cell lines SHG-44 and U251, and circRNA HNRNPA2B1 has been shown to increase the expression of oncogenic isoforms of multiple tumor suppressor genes (e.g., RON, BIN1, WWOX, and c-FLIP) in glioma [173]. It is intriguing to find that overexpression of circATP8B4 in normal glioma cells can cause radioresistance by acting as a miRNA sponge. CircNEIL3, on the other hand, causes TAM penetration into the TME when overexpressed. It is then packed into exosomes and distributed to TAMs in glioma cells. This causes IGF2BP3 protein stability, which promotes glioma progression [168,172,174,175].

Exo-RNAs responsible for glioma progression, invasion, and resistance to therapy can be targeted as a therapeutic strategy. Exosomal miR-124, for example, has been demonstrated to suppress glioma cell proliferation and invasion by targeting the oncogene CAV1. In addition, some studies have indicated that inhibiting certain exosomal miRNAs, such as miR-21 and miR-221, can reduce glioma cell growth, invasion, and chemoresistance in vitro/vivo. The oncogene c-Myc is targeted by exosomal miR-451, which inhibits glioma cell proliferation and invasion. Exosomal miR-128-3p has been demonstrated to be downregulated in gliomas and to be capable of suppressing tumor development and invasion by targeting the oncogene Bmi1 and the transcription factor E2F3. In addition to miRNAs, exosomal lncRNAs EGFRvIII and HOTAIR have been shown in preclinical investigations to reduce glioma cell growth and invasion in vitro and in vivo, as well as to improve chemotherapy efficacy [176,177].

One promising approach for targeting exoRNAs in glioma is the use of RNA interference (RNAi) technology, which involves the delivery of siRNAs or short hairpin RNAs (shRNAs) to target and silence specific exoRNAs [178], whereas antisense oligonucleotides (ASOs) [168,179,180] are another potential approach. In preclinical experiments, these techniques were demonstrated to limit glioma cell proliferation and invasion while also improving chemotherapeutic efficacy.

2.6 Thyroid cancer progression

Thyroid cancer (TC) is one of the fastest-growing tumors in the world and has thus become the most prevalent endocrine system malignancy in recent years. Traditional treatments are ineffective for people with poorly differentiated thyroid cancer; the diagnostic accuracy and dynamics of TC must be improved [181].

Similarly, five microRNAs, miR-222-3p, miR-31-5p, miR-21-5p, miR-221-3p, and let-7i-3p, are altered in tumor cell exosomes versus nontumor cell exosomes, according to a 2022 study [121]. These findings led to the identification of exosomal microRNAs produced by aggressive PTC cells as well as the discovery that their visceral targets are TME regulators [121]. An in-depth analysis of these exosomal miRNA targets' functional roles will add additional details on prospective targets of molecular treatments for these neoplasms [121]. As discussed in Chapter 4 of this book, circulating miRNAs miR-146a-5p, miR-221-3p, and miR-222-3p have been reported to be utilized as biomarkers for PTC for distinguishing between patients and healthy controls. MiR-222-3p and miR-21-5p have also been demonstrated to promote PTC proliferation, migration, and invasion by decreasing PTEN expression and activating the AKT signaling pathway. Furthermore, radioresistance may be caused by the upregulation of the miRNAs 221-3p and 222-3p, which activate the STAT3 signaling pathway. Other investigations have discovered that let-7a, let-7c, let-7d, let-7f, and let-7i have elevated levels in PTC patients' plasma [182].

In another study, it was found that CDKN2B-AS1 is an exosomal lncRNA that is made by CSCs in thyroid cancer and sent to other thyroid cancer cells via exosomes to make cancer grow. In the mechanism, CDKN2B-AS1 functions as the ceRNA of miR-122-5p, increasing the expression of P4HA1 and thereby influencing the EMT process [183].

Yang et al. reported in 2019 that various exosomal circRNAs [124], such as hsa_circ_007293 [184], hsa_circ_031752, and hsa_circ_020135, play diverse roles in PTC proliferation, invasion, and metastasis. Furthermore, circRASSF2 promotes PTC advancement via the miR-1178/TLR4 pathway. A few other studies have found that elevated circFNDC3B expression can be seen in isolated serum exosomes from PTC patients [124]. Two further exosomal circRNAs, hsa_circ_0124055 and hsa_circ_0101622, were shown to be highly elevated in the plasma of TC patients [185].

2.7 Cervical cancer progression

Cervical cancer (CC) is a further prevalent malignancy in women, accounting for a large number of deaths due to cancer. The disease is caused by the human papillomavirus (HPV), which can cause changes in the cellular pattern of the cervix, thereby manifesting cancer when left untreated [186].

A growing interest is observed in exoRNA's role in CC progression and therapy. Scientists have identified miRNAs (microRNAs) that are present in CC cell—derived exosomes and discovered that they can promote tumor growth and metastasis. Exosomal miR-21 has been shown to move from CC cells to fibroblasts, where it turns on fibroblasts and makes them release a variety of proinflammatory cytokines that help the tumor grow and make new blood vessels. Another study revealed that exosomal miR-196a was highly upregulated in CC tissues and was associated with lymph node metastasis followed by poor survival. Similarly, miR-221-3p obtained from exosomes appeared to be abundant in cervical squamous cell carcinoma too [37].

In addition to miRNAs, exosomal lncRNAs also have implications for CC progression. Results from publishers in 2019 suggest that exosomal lncRNA HNF1A-AS1 serves as a ceRNA of miR-34b to influence TUFT1 expression and cisplatin resistance in CC cells. Simultaneously, a study discovered that 3099 transcripts were differentially exported in exosomes containing HPV versus exosomes containing no HPV. Among the top 20 overexpressed transcripts, EVC2, LUZP1, and ANKS1B stood out because of their roles in Hh signaling, cellular migration, and invasion, respectively [187]. The exosomal lncRNA LINC01305 is also known to enhance CC growth via KHSRP and exosome-mediated transfer [188,189].

Besides their role in CC progression, exoRNAs can also be exploited as noninvasive biomarkers for CC detection and monitoring. For instance, the exosomal miR-203 that is significantly downregulated in CC tissues compared with healthy controls could be a diagnostic biomarker for CC with high sensitivity. Similar functions can be served by engineered mesenchymal stem cells that produce exosomes containing miR-143, a tumor-suppressive miRNA. Conversely, exosomal siRNAs as a therapeutic approach for CC are an option where the oncogene c-Myc targeting is required [188].

2.8 Prostate cancer progression

Prostate cancer (PCa) is a prominent tumor that affects men worldwide, causing significant morbidity and mortality. PCa prognosis is closely related to tumor stage and grade at diagnosis, with advanced stages and higher grades associated with poorer outcomes. Although localized PCa can be treated successfully, metastatic PCa remains a significant therapeutic challenge with a high fatality rate [190,191].

ExoRNAs are implicated in various aspects of PCa progression, which include tumor growth, invasion, and metastasis. In a study, it was described that miR-21-5p could initiate PCa cell migration and invasion by targeting PDCD4. Others show that miR-27a-3p promotes PCa cell proliferation and migration, while miR-23a-3p promotes angiogenesis by targeting another tumor suppressor, PTEN. Reports suggest that exosomal miR-141-3p promotes PCa cell invasion by targeting transcription factor SOX9 and tumor suppressor DLC-1. A recent article even showed that exosomal miR-210-3p could promote angiogenesis by downregulating the antiangiogenic factor EFNA3, and again, miR-1290 was linked with lymph node metastasis and poor prognosis for PCa patients. Other than miRNAs, other variants of exoRNAs also have implications in PCa progression. For example, it has been demonstrated that exosomal lncRNA PCAT-1 could promote PCa cell proliferation and invasion; plus exosomal lncRNA H19 promotes angiogenesis by overexpression of VEGF and MMP-9. Additional studies describe that exosomal lncRNA HOTAIR enhanced angiogenesis and metastasis by upregulating the Notch signaling pathway in PCa [192]. Furthermore, exosomal tRNA-derived fragments (tRFs) have also been investigated in PCa, revealing the role of exosomal tRF-1001 in the PCa cell proliferation and migration. In contradiction, another study demonstrated that exosomal tRF-03357 and circMTO1 could inhibit PCa cell proliferation and induce apoptosis. It is fascinating to know that a few exoRNAs such as miR-375 were downregulated in PCa and hindered tumor growth by targeting the oncogene SP1 [193,194].

Further studies explain the functions of exosomal circRNAs in PCa therapy resistance, which is a major clinical challenge in the treatment of the disease. It has been shown that circRNA-XIAP helps in promoting docetaxel resistance in PCa patients by regulating of miR-1182/TPD52 axis [195]. A report on exosomal circRNA_0000075 displays upregulation in docetaxel-resistant PCa cells alongside promotion of therapy resistance by sponging miR-145. However, exosomal circRNA_100290 was reportedly overexpressed in enzalutamide-resistant PCa cells and promoted therapy resistance by regulating androgen receptor signaling [192].

Different exoRNAs have been characterized in PCa biology, and their roles in the disease are still undisclosed. Overall, the diverse range of exoRNAs involved in PCa progression and therapy highlights the complex and multifaceted nature of this disease. More research is needed to understand the roles of these molecules in PCa as well as their potential as therapeutic targets in clinical settings [192,196].

2.9 Ovarian cancer progression

Ovarian cancer is also a very lethal gynecological malignancy arising due to shortage of effective treatments for advanced-stage patients [197]. Exosomal miRNAs are investigated as key factors in ovarian cancer progression and metastasis. For instance, exosomal miR-21, miR-10b, and miR-1246 boost ovarian cancer cell proliferation, migration, and invasion by influencing downstream targets, including PTEN, PDCD4, and SOX9, whereas by targeting E2F3, miR2103p regulates cell proliferation and cisplatin sensitivity in ovarian cancer cells [198]. Conversely, exosomal miR-146a and miR-29c inhibit ovarian cancer cell proliferation, migration, and invasion by affecting IRAK1 and CDK6, respectively. ExoRNAs are also important in ovarian cancer angiogenesis and immune response control. Exosomal miR-155 and miR-21 have been reported to initiate angiogenesis by targeting both VHL and PTEN. A recent study published in the journal *Cancer Letters* in 2021 looked into the role of exosomal miRNAs, miR-21 and miR-125b, obtained from ovarian cancer cells, which could be taken up by dendritic cells and stop their maturation, leading to suppression of antitumor immune response. Moreover, exosomal miR-21 and miR-222 enhance cancer-associated fibroblast (CAF) activation and suppress the antitumor immune response by targeting factors STAT3 and SOCS3. Besides, exosomal miR-1246 and miR-21 have been reported to promote tumor-associated macrophage (TAM) polarization toward the M2 phenotype, which has tumor-promoting functions. Additionally, various reports have also revealed that exosomal miRNAs, such as miR-21, miR-200b, and miR-205, can be utilized as biomarkers for early detection and monitoring of ovarian cancer. They demonstrate that targeting these exosomal miRNAs could amplify the antitumor immune response and aid in bettering the efficiency of immunotherapy in ovarian cancer [199—201].

Exosomal lncRNAs, such as MALAT1 and LINC00152, also increase ovarian cancer cell proliferation, migration, and invasion by upregulating SFPQ and activating the PI3K/AKT/mTOR pathway. Exosomal GAS5 and MEG3, on the other hand, have been found to decrease ovarian cancer cell proliferation, migration, and invasion by targeting EZH2 and ZEB1, respectively.

Different exoRNAs have been associated with ovarian cancer therapy resistance as well. MiR-21, for example, has been shown to improve cisplatin resistance by targeting PTEN and activating the PI3K/AKT signaling pathway. Exosomal miR-214, on the other hand, increases paclitaxel resistance through PTEN targeting and PI3K/AKT/mTOR signaling cascade activation. Furthermore, there is evidence that exosomal miR-222 promotes resistance to PARP inhibitors by targeting PTEN and activating the aforementioned pathway. The lncRNAs HOTAIR and SNHG16 are demonstrated to increase cisplatin and paclitaxel resistance by upregulating the Wnt/-catenin signaling pathway. As a result, targeting these exoRNAs could be a viable method for overcoming treatment resistance in ovarian cancer [199,201].

Exosomal circMYO10, on the other hand, has been shown to increase ovarian cancer cell proliferation, migration, and invasion via targeting the miR-432-5p/SIRT7 axis. Exosomal circRNA_100876, on the other hand, inhibits ovarian cancer cell proliferation, migration, and invasion by targeting the miR-1228/STC1 axis. This is especially exciting in terms of finding and describing the many activities of circRNAs in ovarian cancer, stressing the prospect of harnessing and using them as diagnostic biomarkers for ovarian cancer [46,202].

3. Conclusion

ExoRNAs serve a fascinating role in tumor—TME intercellular communication, regulating tumor formation and progression. They amend the function of recipient cells and produce a favorable microenvironment for tumor growth, invasion, and metastasis by regulating energy metabolism, modulating the EMT process, and promoting angiogenesis. This in turn remodels the surrounding microenvironment, creating a tumor-safe space.

The dysregulation of exoRNAs has significant clinical implications, and the manipulation of exoRNAs represents a promising therapeutic method for cancer treatment. The unique properties of exoRNA, including its stability in blood, serum, urine, and other biological fluids and its ability to transfer genetic information between cells, make it an attractive diagnostic and therapeutic target for cancer. Furthermore, exoRNA profiling has been demonstrated to be useful in predicting responses to chemotherapy and other treatments. More study is needed to completely understand the mechanisms underpinning exoRNA's function in cancer progression and to develop new therapeutic techniques targeting exoRNA. However, it appears that further clinical trials focusing on exo-RNAs may be presented in the future. Exo-RNA research has revealed new insights into how circulating exo-RNAs can operate as liquid biopsies and noninvasive biomarkers for cancer detection and treatments. Although clinical trials are still in their early stages, we believe that RNA-based treatment for cancer will be available in the next few years. These will open doors for not just understanding the fundamental foundation of cancer progression but also imparting knowledge in treatment and prognosis, thereby bringing us closer to the far-fetched dream of a systemic cancer-free world.

References

[1] Bray F, Ferlay J, Soerjomataram I, Siegel RL, Torre LA, Jemal A. Global cancer statistics 2018: GLOBOCAN estimates of incidence and mortality worldwide for 36 cancers in 185 countries. CA: A Cancer Journal for Clinicians 2018;68(6):394—424. https://doi.org/10.3322/caac.21492.

[2] Siegel RL, Miller KD, Jemal A. Cancer statistics. CA: A Cancer Journal for Clinicians 2016;66(1):7—30. https://doi.org/10.3322/caac.21332.

[3] Zhao G, Zhou A, Li X, Zhu S, Wang Y, Zhang S, et al. The significance of exosomal RNAs in the development, diagnosis, and treatment of gastric cancer. Genes 2021;12(1):73. https://doi.org/10.3390/genes12010073.

[4] Hade MD, Suire CN, Suo Z. Mesenchymal stem cell-derived exosomes: applications in regenerative medicine. Cells 2021;10(8):1959. https://doi.org/10.3390/cells10081959.

[5] Kalra H, Drummen GPC, Mathivanan S. Focus on extracellular vesicles: introducing the next small big thing. International Journal of Molecular Sciences 2016;17(2):170. https://doi.org/10.3390/ijms17020170.

[6] Stremersch S, Vandenbroucke RE, Van Wonterghem E, Hendrix A, De Smedt SC, Raemdonck K. Comparing exosome-like vesicles with liposomes for the functional cellular delivery of small RNAs. Journal of Controlled Release 2016;232:51—61. https://doi.org/10.1016/j.jconrel.2016.04.005.

[7] Golubovskaya V, Bukrinsky MI, Grizzi F. Editorial on "exosomes, their biogenesis and role in intercellular communication, tumor microenvironment and cancer immunotherapy. Vaccines 2020;8(3):421. https://doi.org/10.3390/vaccines8030421.
[8] Gurung S, Perocheau D, Touramanidou L, Baruteau J. The exosome journey: from biogenesis to uptake and intracellular signalling. Cell Communication and Signaling 2021;19(1):47. https://doi.org/10.1186/s12964-021-00730-1.
[9] Banerjee M, Rajeswari VD. Critical review on the different roles of exosomes in TNBC and exosomal-mediated delivery of MicroRNA/SiRNA/LncRNA and drug targeting signalling pathways in triple-negative breast cancer. Molecules 2023;28(4):1802. https://doi.org/10.3390/molecules28041802.
[10] Margolis L, Sadovsky Y. The biology of extracellular vesicles: the known unknowns. PLoS Biology 2019;17(7):e3000363. https://doi.org/10.1371/journal.pbio.3000363.
[11] Vidal M, Sainte-Marie J, Philippot JR, Bienvenue A. Asymmetric distribution of phospholipids in the membrane of vesicles released during in vitro maturation of Guinea pig reticulocytes: evidence precluding a role for "aminophospholipid translocase.". Journal of Cellular Physiology 1989;140(3):455–62. https://doi.org/10.1002/jcp.1041400308.
[12] Simpson RJ, Lim JW, Moritz RL, Mathivanan S. Exosomes: proteomic insights and diagnostic potential. Expert Review of Proteomics 2009;6(3):267–83. https://doi.org/10.1586/epr.09.17.
[13] Valadi H, Ekström K, Bossios A, Sjöstrand M, Lee JJ, Lötvall JO. Exosome-mediated transfer of MRNAs and MicroRNAs is a novel mechanism of genetic exchange between cells. Nature Cell Biology 2007;9(6):654–9. https://doi.org/10.1038/ncb1596.
[14] Waldenström A, Gennebäck N, Hellman U, Ronquist G. Cardiomyocyte microvesicles contain DNA/RNA and convey biological messages to target cells. PLoS One 2012;7(4):e34653. https://doi.org/10.1371/journal.pone.0034653.
[15] Carpintero-Fernández P, Fafián-Labora J, O'Loghlen A. Technical advances to study extracellular vesicles. Frontiers in Molecular Biosciences 2017;4:79. https://doi.org/10.3389/fmolb.2017.00079.
[16] Felicetti F, De Feo A, Coscia C, Puglisi R, Pedini F, Pasquini L, et al. Exosome-mediated transfer of MiR-222 is sufficient to increase tumor malignancy in melanoma. Journal of Translational Medicine 2016;14(1):56. https://doi.org/10.1186/s12967-016-0811-2.
[17] Li B-L, Lu W, Qu J-J, Ye L, Du G-Q, Wan X-P. Loss of exosomal MiR-148b from cancer-associated fibroblasts promotes endometrial cancer cell invasion and cancer metastasis. Journal of Cellular Physiology 2019;234(3):2943–53. https://doi.org/10.1002/jcp.27111.
[18] Ma H, Wang H, Tian F, Zhong Y, Liu Z, Liao A. PIWI-interacting RNA-004800 is regulated by S1P receptor signaling pathway to keep myeloma cell survival. Frontiers in Oncology 2020;10:438. https://doi.org/10.3389/fonc.2020.00438.
[19] Kahlert C, Kalluri R. Exosomes in tumor microenvironment influence cancer progression and metastasis. Journal of Molecular Medicine 2013;91(4):431–7. https://doi.org/10.1007/s00109-013-1020-6.
[20] Wei F, Li Y. The emerging roles of exosome-derived noncoding RNAs in the tumor immune microenvironment and their future applications. Biomedicine & Pharmacotherapy 2022;156:113863. https://doi.org/10.1016/j.biopha.2022.113863.
[21] Chen Q, Li Y, Liu Y, Xu W, Zhu X. Exosomal non-coding RNAs-mediated crosstalk in the tumor microenvironment. Frontiers in Cell and Developmental Biology 2021;9:646864. https://doi.org/10.3389/fcell.2021.646864.
[22] Dai J, Su Y, Zhong S, Cong L, Liu B, Yang J, et al. Exosomes: key players in cancer and potential therapeutic strategy. Signal Transduction and Targeted Therapy 2020;5(1):1–10. https://doi.org/10.1038/s41392-020-00261-0.
[23] Li I, Nabet BY. Exosomes in the tumor microenvironment as mediators of cancer therapy resistance. Molecular Cancer 2019;18(1):32. https://doi.org/10.1186/s12943-019-0975-5.
[24] Tai Y, Chen K, Hsieh J, Shen T. Exosomes in cancer development and clinical applications. Cancer Science 2018;109(8):2364–74. https://doi.org/10.1111/cas.13697.
[25] Whiteside TL. Tumor-derived exosomes and their role in cancer progression. Advances in Clinical Chemistry 2016;74:103–41. https://doi.org/10.1016/bs.acc.2015.12.005.

[26] Zhang J, Li S, Li L, Li M, Guo C, Yao J, et al. Exosome and exosomal MicroRNA: trafficking, sorting, and function. Genomics, Proteomics & Bioinformatics 2015;13(1):17–24. https://doi.org/10.1016/j.gpb.2015.02.001.
[27] Fong MY, Zhou W, Liu L, Alontaga AY, Chandra M, Ashby J, et al. Breast-cancer-secreted MiR-122 reprograms glucose metabolism in premetastatic niche to promote metastasis. Nature Cell Biology 2015;17(2):183–94. https://doi.org/10.1038/ncb3094.
[28] Zheng D, Huo M, Li B, Wang W, Piao H, Wang Y, et al. The role of exosomes and exosomal MicroRNA in cardiovascular disease. Frontiers in Cell and Developmental Biology 2021:8.
[29] Kim HK, Fuchs G, Wang S, Wei W, Zhang Y, Park H, et al. A transfer-RNA-derived small RNA regulates ribosome biogenesis. Nature 2017;552(7683):57–62. https://doi.org/10.1038/nature25005.
[30] Zheng R, Du M, Wang X, Xu W, Liang J, Wang W, et al. Exosome–transmitted long non-coding RNA PTENP1 suppresses bladder cancer progression. Molecular Cancer 2018;17(1):143. https://doi.org/10.1186/s12943-018-0880-3.
[31] Wang X, Zhang H, Yang H, Bai M, Ning T, Deng T. Exosome-delivered circRNA promotes glycolysis to induce chemoresistance through the miR-122-PKM2 axis in colorectal cancer. Molecular Oncology 2020. https://febs.onlinelibrary.wiley.com/doi/10.1002/1878-0261.12629.
[32] Rynkeviciene R, Simiene J, Strainiene E, Stankevicius V, Usinskiene J, Miseikyte Kaubriene E, et al. Non-coding RNAs in glioma. Cancers 2019;11(1):17. https://doi.org/10.3390/cancers11010017.
[33] Shi X, Wang B, Feng X, Xu Y, Lu K, Sun M. CircRNAs and exosomes: a mysterious frontier for human cancer. Molecular Therapy - Nucleic Acids 2019;19:384–92. https://doi.org/10.1016/j.omtn.2019.11.023.
[34] Kim KM, Abdelmohsen K, Mustapic M, Kapogiannis D, Gorospe M. RNA in extracellular vesicles. WIREs RNA 2017. https://wires.onlinelibrary.wiley.com/doi/10.1002/wrna.1413.
[35] Lee YS, Dutta A. MicroRNAs in cancer. Annual Review of Pathology: Mechanisms of Disease 2009;4:199–227. https://doi.org/10.1146/annurev.pathol.4.110807.092222.
[36] Li B, Cao Y, Sun M, Feng H. Expression, regulation, and function of exosome-derived MiRNAs in cancer progression and therapy. FASEB Journal 2021;35(10):e21916. https://doi.org/10.1096/fj.202100294RR.
[37] Li C, Zhou T, Chen J, Li R, Chen H, Luo S, et al. The role of exosomal MiRNAs in cancer. Journal of Translational Medicine 2022;20:6. https://doi.org/10.1186/s12967-021-03215-4.
[38] Dilsiz N. Role of exosomes and exosomal microRNAs in cancer. Future Science OA 2020;6(4):FSO465. https://doi.org/10.2144/fsoa-2019-0116.
[39] Daoud AZ, Mulholland EJ, Cole G, McCarthy HO. MicroRNAs in pancreatic cancer: biomarkers, prognostic, and therapeutic modulators. BMC Cancer 2019;19(1):1130. https://doi.org/10.1186/s12885-019-6284-y.
[40] Semina EV, Rysenkova KD, Troyanovskiy KE, Shmakova AA, Rubina KA. MicroRNAs in cancer: from gene expression regulation to the metastatic niche reprogramming. Biochemistry (Moscow) 2021;86(7):785–99. https://doi.org/10.1134/S0006297921070014.
[41] Dakubo GD. Cancer Biomarkers in Body Fluids. Cham: Springer International Publishing; 2017. https://doi.org/10.1007/978-3-319-48360-3.
[42] Padda J, Khalid K, Khedr A, Patel V, Al-Ewaidat OA, Tasnim F, Padda S, Cooper AC, Jean-Charles G. Exosome-derived microRNA: efficacy in cancer. Cureus 2021;13(8):e17441. https://doi.org/10.7759/cureus.17441.
[43] Alves dos Santos K, Clemente dos Santos IC, Santos Silva C, Gomes Ribeiro H, de Farias Domingos I, Nogueira Silbiger V. Circulating exosomal miRNAs as biomarkers for the diagnosis and prognosis of colorectal cancer. International Journal of Molecular Sciences 2020;22(1):346. https://doi.org/10.3390/ijms22010346.
[44] Kosaka N, Iguchi H, Ochiya T. Circulating microRNA in body fluid: a new potential biomarker for cancer diagnosis and prognosis. Cancer Science 2010;101(10):2087–92. https://doi.org/10.1111/j.1349-7006.2010.01650.x.
[45] Nie H, Liao Z, Wang Y, Zhou J, He X, Ou C. Exosomal long non-coding RNAs: emerging players in cancer metastasis and potential diagnostic biomarkers for personalized oncology. Genes & Diseases 2020;8(6):769–80. https://doi.org/10.1016/j.gendis.2020.12.004.

[46] Paskeh MDA, Entezari M, Mirzaei S, Zabolian A, Saleki H, Naghdi MJ, et al. Emerging role of exosomes in cancer progression and tumor microenvironment remodeling. Journal of Hematology & Oncology 2022;15(1):83. https://doi.org/10.1186/s13045-022-01305-4.
[47] Han S, Qi Y, Luo Y, Chen X, Liang H. Exosomal long non-coding RNA: interaction between cancer cells and non-cancer cells. Frontiers in Oncology 2021;10:617837. https://doi.org/10.3389/fonc.2020.617837.
[48] Xu Z, Chen Y, Ma L, Chen Y, Liu J, Guo Y, et al. Role of exosomal non-coding RNAs from tumor cells and tumor-associated macrophages in the tumor microenvironment. Molecular Therapy 2022;30(10):3133−54. https://doi.org/10.1016/j.ymthe.2022.01.046.
[49] Shi Y, Huang Q, Kong X, Zhao R, Chen X, Zhai Y, et al. Current knowledge of long non-coding RNA HOTAIR in breast cancer progression and its application. Life 2021;11(6):483. https://doi.org/10.3390/life11060483.
[50] Jin H, Du W, Huang W, Yan J, Tang Q, Chen Y, et al. LncRNA and breast cancer: progress from identifying mechanisms to challenges and opportunities of clinical treatment. Molecular Therapy - Nucleic Acids 2021;25:613−37. https://doi.org/10.1016/j.omtn.2021.08.005.
[51] Chen L, Qian X, Wang Z, Zhou X. The HOTAIR LncRNA: a remarkable oncogenic promoter in human cancer metastasis (review). Oncology Letters 2021;21(4):1−8. https://doi.org/10.3892/ol.2021.12563.
[52] Entezari M, Ghanbarirad M, Taheriazam A, Sadrkhanloo M, Zabolian A, Goharrizi MASB, et al. Long non-coding RNAs and exosomal LncRNAs: potential functions in lung cancer progression, drug resistance and tumor microenvironment remodeling. Biomedicine & Pharmacotherapy 2022;150:112963. https://doi.org/10.1016/j.biopha.2022.112963.
[53] Hu Q, Ma H, Chen H, Zhang Z, Xue Q. LncRNA in tumorigenesis of non-small-cell lung cancer: from bench to bedside. Cell Death Discovery 2022;8(1):1−9. https://doi.org/10.1038/s41420-022-01157-4.
[54] Fu S, Wang Y, Li H, Chen L, Liu Q. Regulatory networks of LncRNA MALAT-1 in cancer. Cancer Management and Research 2020;12:10181−98. https://doi.org/10.2147/CMAR.S276022.
[55] Zhu L, Li J, Gong Y, Wu Q, Tan S, Sun D, et al. Exosomal TRNA-derived small RNA as a promising biomarker for cancer diagnosis. Molecular Cancer 2019;18(1):74. https://doi.org/10.1186/s12943-019-1000-8.
[56] Li X, Liu X, Zhao D, Cui W, Wu Y, Zhang C, et al. TRNA-derived small RNAs: novel regulators of cancer hallmarks and targets of clinical application. Cell Death Discovery 2021;7:249. https://doi.org/10.1038/s41420-021-00647-1.
[57] Zheng B, Song X, Wang L, Zhang Y, Tang Y, Wang S, et al. Plasma exosomal TRNA-derived fragments as diagnostic biomarkers in non-small cell lung cancer. Frontiers in Oncology 2022;12.
[58] Zhu P, Yu J, Zhou P. Role of TRNA-derived fragments in cancer: novel diagnostic and therapeutic targets TRFs in cancer. American Journal of Cancer Research 2020;10(2):393−402.
[59] Gu X, Wang C, Deng H, Qing C, Liu R, Liu S, et al. Exosomal PiRNA profiling revealed unique circulating PiRNA signatures of cholangiocarcinoma and Gallbladder carcinoma. Acta Biochimica et Biophysica Sinica 2020;52(5):475−84. https://doi.org/10.1093/abbs/gmaa028.
[60] Street W. Cancer Facts & Figures 2018. 1930.
[61] Global Burden of Disease Cancer Collaboration; Fitzmaurice C, Abate DL. Global regional, and national cancer incidence, mortality, years of life lost, years lived with disability, and disability-adjusted life-years for 29 cancer groups, 1990 to 2017: a systematic analysis for the global burden of disease study. JAMA Oncology 2019;5(12):1749−68. https://doi.org/10.1001/jamaoncol.2019.2996.
[62] Kang F, Jiang F, Ouyang L, Wu S, Fu C, Liu Y, et al. Potential biological roles of exosomal long non-coding RNAs in gastrointestinal cancer. Frontiers in Cell and Developmental Biology 2022;10:886191. https://doi.org/10.3389/fcell.2022.886191.
[63] Liu L, Zhao Z, Zhou W, Fan X, Zhan Q, Song Y. Enhanced expression of MiR-425 promotes esophageal squamous cell carcinoma tumorigenesis by targeting SMAD2. Journal of Genetics and Genomics 2015;42(11):601−11. https://doi.org/10.1016/j.jgg.2015.09.010.
[64] Zhao X, He W, Li J, Huang S, Wan X, Luo H, et al. MiRNA-125b inhibits proliferation and migration by targeting SphK1 in bladder cancer. American Journal of Translational Research 2015;7(11):2346−54.

[65] Guz M, Jeleniewicz W, Cybulski M. An insight into MiR-1290: an oncogenic MiRNA with diagnostic potential. International Journal of Molecular Sciences 2022;23(3):1234. https://doi.org/10.3390/ijms23031234.
[66] Huang H, Pi X, Xin C, Gong C, Guo F, Wang Y, et al. Differential expression and functions of MiRNAs in bladder cancer. Oncologie 2023;25(1):1–15. https://doi.org/10.1515/oncologie-2022-1030.
[67] Wang H, Zeng X, Zheng Y, Wang Y, Zhou Y. Exosomal circRNA in digestive system tumors: the main player or coadjuvants? Frontiers in Oncology 2021:11.
[68] Wu F, Zhang L, Wu P, Wu Y, Zhang T, Zhang D, et al. The potential role of small nucleolar RNAs in cancers – an evidence map. International Journal of General Medicine 2022;15:3851–64. https://doi.org/10.2147/IJGM.S352333.
[69] Lin C, Zheng L, Huang R, Yang G, Chen J, Li H. TRFs as potential exosome TRNA-derived fragment biomarkers for gastric carcinoma. Clinical Laboratory 2020;66(6). https://doi.org/10.7754/clin.lab.2019.190811.
[70] Losurdo P, de Manzini N, Palmisano S, Grassi M, Parisi S, Rizzolio F, et al. Potential application of small interfering RNA in gastro-intestinal tumors. Pharmaceuticals 2022;15(10):1295. https://doi.org/10.3390/ph15101295.
[71] Thrift AP, El-Serag HB. Burden of gastric cancer. Clinical Gastroenterology and Hepatology 2020;18(3):534–42. https://doi.org/10.1016/j.cgh.2019.07.045.
[72] Wang X, Zhang J, Cao G, Hua J, Shan G, Lin W. Emerging roles of circular RNAs in gastric cancer metastasis and drug resistance. Journal of Experimental & Clinical Cancer Research 2022;41(1):218. https://doi.org/10.1186/s13046-022-02432-z.
[73] Zhang BG, Li JF, Yu BQ, Zhu ZG, Liu BY, Yan M. MicroRNA-21 promotes tumor proliferation and invasion in gastric cancer by targeting PTEN. Oncology Reports 2012;27(4):1019–26. https://doi.org/10.3892/or.2012.1645.
[74] Zeng Z, Li Y, Pan Y, Lan X, Song F, Sun J, et al. Cancer-derived exosomal MiR-25-3p promotes pre-metastatic niche formation by inducing vascular permeability and angiogenesis. Nature Communications 2018;9(1):5395. https://doi.org/10.1038/s41467-018-07810-w.
[75] Exosomal miR-122-5p Inhibits Tumorigenicity of Gastric Cancer by Downregulating GIT1. https://doi.org/10.1177/1724600821990677.
[76] Gomes C, Almeida A, Ferreira JA, Silva L, Santos-Sousa H, Pinto-de-Sousa J, et al. Glycoproteomic analysis of serum from patients with gastric precancerous lesions. Journal of Proteome Research 2013;12(3):1454–66. https://doi.org/10.1021/pr301112x.
[77] Zhang W, Wang Q, Yang Y, Zhou S, Zhang P, Feng T. The role of exosomal LncRNAs in cancer biology and clinical management. Experimental & Molecular Medicine 2021;53(11):1669–73. https://doi.org/10.1038/s12276-021-00699-4.
[78] Pathania AS, Challagundla KB. Exosomal long non-coding RNAs: emerging players in the tumor microenvironment. Molecular Therapy - Nucleic Acids 2020;23:1371–83. https://doi.org/10.1016/j.omtn.2020.09.039.
[79] Guo Q, Wang H, Yan Y, Liu Y, Su C, Chen H, et al. The role of exosomal MicroRNA in cancer drug resistance. Frontiers in Oncology 2020:10.
[80] Zhao Z, Zhao G, Yang S, Zhu S, Zhang S, Li P. The significance of exosomal RNAs in the development, diagnosis, and treatment of pancreatic cancer. Cancer Cell International 2021;21(1):364. https://doi.org/10.1186/s12935-021-02059-8.
[81] Pan Y, Tang H, Li Q, Chen G, Li D. Exosomes and their roles in the chemoresistance of pancreatic cancer. Cancer Medicine 2022;11(24):4979–88. https://doi.org/10.1002/cam4.4830.
[82] Uddin MH, Al-Hallak MN, Philip PA, Mohammad RM, Viola N, Wagner K-U, et al. Exosomal MicroRNA in pancreatic cancer diagnosis, prognosis, and treatment: from bench to bedside. Cancers 2021;13(11):2777. https://doi.org/10.3390/cancers13112777.
[83] Liu H, Qiao S, Fan X, Gu Y, Zhang Y, Huang S. Role of exosomes in pancreatic cancer (review). Oncology Letters 2021;21(4):1–13. https://doi.org/10.3892/ol.2021.12559.
[84] Wu Y, Zeng H, Yu Q, Huang H, Fervers B, Chen Z-S, et al. A circulating exosome RNA signature is a potential diagnostic marker for pancreatic cancer, a systematic study. Cancers 2021;13(11):2565. https://doi.org/10.3390/cancers13112565.

[85] Chen W, Mao Y, Liu C, Wu H, Chen S. Exosome in hepatocellular carcinoma: an update. Journal of Cancer 2021;12(9):2526–36. https://doi.org/10.7150/jca.54566.
[86] Hosseini K, Ranjbar M, Pirpour Tazehkand A, Asgharian P, Montazersaheb S, Tarhriz V, et al. Evaluation of exosomal non-coding RNAs in cancer using high-throughput sequencing. Journal of Translational Medicine 2022;20(1):30. https://doi.org/10.1186/s12967-022-03231-y.
[87] Liu C, Wu H, Mao Y, Chen W, Chen S. Exosomal MicroRNAs in Hepatocellular carcinoma. Cancer Cell International 2021;21(1):254. https://doi.org/10.1186/s12935-021-01941-9.
[88] Yang L, Peng X, Li Y, Zhang X, Ma Y, Wu C, et al. Long non-coding RNA HOTAIR promotes exosome secretion by regulating RAB35 and SNAP23 in Hepatocellular carcinoma. Molecular Cancer 2019;18(1):78. https://doi.org/10.1186/s12943-019-0990-6.
[89] Li L, Bi Y, Diao S, Li X, Yuan T, Xu T, et al. Exosomal LncRNAs and Hepatocellular carcinoma: from basic research to clinical practice. Biochemical Pharmacology 2022;200:115032. https://doi.org/10.1016/j.bcp.2022.115032.
[90] Sukowati CHC, Cabral LKD, Tiribelli C, Pascut D. Circulating long and circular noncoding RNA as non-invasive diagnostic tools of Hepatocellular carcinoma. Biomedicines 2021;9(1):90. https://doi.org/10.3390/biomedicines9010090.
[91] Choi J, Kim G-A, Han S, Lee W, Chun S, Lim Y-S. Longitudinal assessment of three serum biomarkers to detect very early-stage Hepatocellular carcinoma. Hepatology 2019;69(5):1983. https://doi.org/10.1002/hep.30233.
[92] Street, W. Colorectal Cancer Facts & Figures 2020-2022.
[93] Hon KW, Ab-Mutalib NS, Abdullah NMA, Jamal R, Abu N. Extracellular vesicle-derived circular RNAs confers chemoresistance in colorectal cancer. Scientific Reports 2019;9(1):16497. https://doi.org/10.1038/s41598-019-53063-y.
[94] Jin G, Liu Y, Zhang J, Bian Z, Yao S, Fei B, et al. A panel of serum exosomal microRNAs as predictive markers for chemoresistance in advanced colorectal cancer. Cancer Chemotherapy and Pharmacology 2019;84(2):315–25. https://doi.org/10.1007/s00280-019-03867-6.
[95] Titu S, Gata VA, Decea RM, Mocan T, Dina C, Irimie A, et al. Exosomes in colorectal cancer: from physiology to clinical applications. International Journal of Molecular Sciences 2023;24(5):4382. https://doi.org/10.3390/ijms24054382.
[96] Bakhsh T, Alhazmi S, Alburae NA, Farsi A, Alzahrani F, Choudhry H, et al. Exosomal MiRNAs as a promising source of biomarkers in colorectal cancer progression. International Journal of Molecular Sciences 2022;23(9):4855. https://doi.org/10.3390/ijms23094855.
[97] Chen X, Jia M, Ji J, Zhao Z, Zhao Y. Exosome-derived non-coding RNAs in the tumor microenvironment of colorectal cancer: possible functions, mechanisms and clinical applications. Frontiers in Oncology 2022:12.
[98] Wadhonkar K, Singh N, Heralde FM, Parihar SP, Hirani N, Baig MS. Exosome-derived MiRNAs regulate macrophage-colorectal cancer cell cross-talk during aggressive tumor development. Colorectal Cancer 2023;12(1):CRC40. https://doi.org/10.2217/crc-2022-0012.
[99] Amelimojarad M, AmeliMojarad M, Nazemalhosseini-Mojarad E. Exosomal noncoding RNAs in colorectal cancer: an overview of functions, challenges, opportunities, and clinical applications. Pathology, Research & Practice 2022;238:154133. https://doi.org/10.1016/j.prp.2022.154133.
[100] Sun R, He X-Y, Mei C, Ou C-L. Role of exosomal long non-coding RNAs in colorectal cancer. World Journal of Gastrointestinal Oncology 2021;13(8):867–78. https://doi.org/10.4251/wjgo.v13.i8.867.
[101] Zheng R, Zhang K, Tan S, Gao F, Zhang Y, Xu W, et al. Exosomal CircLPAR1 functions in colorectal cancer diagnosis and tumorigenesis through suppressing BRD4 via METTL3–EIF3h interaction. Molecular Cancer 2022;21(1):49. https://doi.org/10.1186/s12943-021-01471-y.
[102] Gao L, Tang X, He Q, Sun G, Wang C, Qu H. Exosome-transmitted CircCOG2 promotes colorectal cancer progression via MiR-1305/TGF-B2/SMAD3 pathway. Cell Death Discovery 2021;7(1):1–13. https://doi.org/10.1038/s41420-021-00680-0.
[103] Matsumura T, Sugimachi K, Iinuma H, Takahashi Y, Kurashige J, Sawada G, et al. Exosomal MicroRNA in serum is a novel biomarker of recurrence in human colorectal cancer. British Journal of Cancer 2015;113(2):275–81. https://doi.org/10.1038/bjc.2015.201.

[104] Blood Cancer UK | Blood Cancer Types. Blood Cancer UK. https://bloodcancer.org.uk/understanding-blood-cancer/blood-cancer-types/ (Accessed 15 May 2023).

[105] Liu Q-W, He Y, Xu WW. Molecular functions and therapeutic applications of exosomal noncoding RNAs in cancer. Experimental & Molecular Medicine 2022;54(3):216–25. https://doi.org/10.1038/s12276-022-00744-w.

[106] Wang D, Zhang W, Zhang C, Wang L, Chen H, Xu J. Exosomal non-coding RNAs have a significant effect on tumor metastasis. Molecular Therapy - Nucleic Acids 2022;29:16–35. https://doi.org/10.1016/j.omtn.2022.05.034.

[107] Liu L, Liu L, Liu R, Liu J, Cheng Q. Exosomal MiR-21-5p derived from multiple myeloma cells promote renal epithelial-mesenchymal transition through targeting TGF-β/SMAD7 signaling pathway. Clinical and Experimental Pharmacology and Physiology 2023. https://doi.org/10.1111/1440-1681.13768.

[108] Calin GA, Ferracin M, Cimmino A, Di Leva G, Shimizu M, Wojcik SE, et al. A MicroRNA signature associated with prognosis and progression in chronic lymphocytic leukemia. New England Journal of Medicine 2005;353(17):1793–801. https://doi.org/10.1056/NEJMoa050995.

[109] Umezu T, Ohyashiki K, Kuroda M, Ohyashiki JH. Leukemia cell to endothelial cell communication via exosomal MiRNAs. Oncogene 2013;32(22):2747–55. https://doi.org/10.1038/onc.2012.295.

[110] Jahromi FNA, Dowran R, Jafari R. Recent advances in the roles of exosomal MicroRNAs (Exo-miRs) in hematologic neoplasms: pathogenesis, diagnosis, and treatment. Cell Communication and Signaling 2023;21:88. https://doi.org/10.1186/s12964-023-01102-7.

[111] Zhang X, Luo M, Zhang J, Guo B, Singh S, Lin X, et al. The role of LncRNA H19 in tumorigenesis and drug resistance of human cancers. Frontiers in Genetics 2022;13:1005522. https://doi.org/10.3389/fgene.2022.1005522.

[112] Asadi M, Gholampour MA, Kompani F, Alizadeh S. Expression of long non-coding RNA H19 in acute lymphoblastic leukemia. Cell Journal (Yakhteh) 2023;25(1):1–10. https://doi.org/10.22074/CELLJ.2022.8315.

[113] Liu Z, Wang Y, Yuan S, Wen F, Liu J, Zou L. Regulatory role of long non-coding RNA UCA1 in signaling pathways and its clinical applications. Oncology Letters 2021;21(5):404. https://doi.org/10.3892/ol.2021.12665.

[114] Li JL, Liu XL, Guo SF, Yang Y, Zhu YL, Li JZ. Long noncoding RNA UCA1 regulates proliferation and apoptosis in multiple myeloma by targeting miR-331-3p/IL6R axis for the activation of JAK2/STAT3 pathway. European Review for Medical and Pharmacological Sciences 2019;23(21):9238–50. https://doi.org/10.26355/eurrev_201911_19416.

[115] Lambrou GI, Hatziagapiou K, Zaravinos A. The non-coding RNA GAS5 and its role in tumor therapy-induced resistance. International Journal of Molecular Sciences 2020;21(20):7633. https://doi.org/10.3390/ijms21207633.

[116] Giudice V, Mensitieri F, Izzo V, Filippelli A, Selleri C. Aptamers and antisense oligonucleotides for diagnosis and treatment of hematological diseases. International Journal of Molecular Sciences 2020;21:3252. https://doi.org/10.3390/ijms21093252.

[117] de Araújo HV, Sakamoto LHT, Bacal NS, Epelman S, Real JM. MicroRNAs and Exosomes: promising new biomarkers in acute myeloid leukemias? Einstein 2022;20:eRB5954. https://doi.org/10.31744/einstein_journal/2022RB5954.

[118] Taverna S, Amodeo V, Saieva L, Russo A, Giallombardo M, De Leo G, et al. Exosomal shuttling of MiR-126 in endothelial cells modulates adhesive and migratory abilities of chronic myelogenous leukemia cells. Molecular Cancer 2014;13:169. https://doi.org/10.1186/1476-4598-13-169.

[119] Lingua MF, Carrà G, Maffeo B, Morotti A. Non-coding RNAs: the "dark side matter" of the CLL universe. Pharmaceuticals 2021;14(2). https://doi.org/10.3390/ph14020168.

[120] Zeng C, Xu Y, Xu L, Yu X, Cheng J, Yang L, et al. Inhibition of long non-coding RNA NEAT1 impairs myeloid differentiation in acute promyelocytic leukemia cells. BMC Cancer 2014;14(1):693. https://doi.org/10.1186/1471-2407-14-693.

[121] Ronchetti D, Favasuli V, Monti P, Cutrona G, Fabris S, Silvestris I, et al. NEAT1 long isoform is highly expressed in chronic lymphocytic leukemia irrespectively of cytogenetic groups or clinical outcome. Non-Coding RNA 2020;6(1):11. https://doi.org/10.3390/ncrna6010011.

[122] Taiana E, Favasuli V, Ronchetti D, Todoerti K, Pelizzoni F, Manzoni M, et al. Long non-coding RNA NEAT1 targeting impairs the DNA repair machinery and triggers anti-tumor activity in multiple myeloma. Leukemia 2020;34(1):234–44. https://doi.org/10.1038/s41375-019-0542-5.

[123] Fabris L, Juracek J, Calin G. Non-coding RNAs as cancer hallmarks in chronic lymphocytic leukemia. International Journal of Molecular Sciences 2020;21(18). https://doi.org/10.3390/ijms21186720.

[124] Brambilla E, Gazdar A. Pathogenesis of lung cancer signaling pathways: roadmap for therapies. European Respiratory Journal 2009;33(6):1485–97. https://doi.org/10.1183/09031936.00014009.

[125] Ungefroren H, Sebens S, Seidl D, Lehnert H, Hass R. Interaction of tumor cells with the microenvironment. Cell Communication and Signaling 2011;9(1):18. https://doi.org/10.1186/1478-811X-9-18.

[126] Dominiak A, Chełstowska B, Olejarz W, Nowicka G. Communication in the cancer microenvironment as a target for therapeutic interventions. Cancers 2020;12(5):1232. https://doi.org/10.3390/cancers12051232.

[127] Li M-Y, Liu L-Z, Dong M. Progress on pivotal role and application of exosome in lung cancer carcinogenesis, diagnosis, therapy and prognosis. Molecular Cancer 2021;20(1):22. https://doi.org/10.1186/s12943-021-01312-y.

[128] Wu H, Zhou J, Mei S, Wu D, Mu Z, Chen B, et al. Circulating exosomal MicroRNA-96 promotes cell proliferation, migration and drug resistance by targeting LMO7. Journal of Cellular and Molecular Medicine 2017;21(6):1228–36. https://doi.org/10.1111/jcmm.13056.

[129] Song S, Zhao Y, Wang X, Tong X, Chen X, Xiong Q. M2 macrophages-derived exosomal MiR-3917 promotes the progression of lung cancer via targeting GRK6. Biological Chemistry 2023;404(1):41–57. https://doi.org/10.1515/hsz-2022-0162.

[130] Li H, Huhe M, Lou J. MicroRNA-103a-3p promotes cell proliferation and invasion in non-small-cell lung cancer cells through Akt pathway by targeting PTEN. BioMed Research International 2021;2021:7590976. https://doi.org/10.1155/2021/7590976.

[131] Liu Y, Ao X, Ding W, Ponnusamy M, Wu W, Hao X, et al. Critical role of FOXO3a in carcinogenesis. Molecular Cancer 2018;17:104. https://doi.org/10.1186/s12943-018-0856-3.

[132] Du J, Qian J, Zheng B, Xu G, Chen H, Chen C. MiR-21-5p is a biomarker for predicting prognosis of lung adenocarcinoma by regulating PIK3R1 expression. International Journal of General Medicine 2021;14:8873–80. https://doi.org/10.2147/IJGM.S337149.

[133] Yuan D, Xu J, Wang J, Pan Y, Fu J, Bai Y, et al. Extracellular MiR-1246 promotes lung cancer cell proliferation and enhances radioresistance by directly targeting DR5. Oncotarget 2016;7(22):32707–22. https://doi.org/10.18632/oncotarget.9017.

[134] Dai Y, Pan Y, Quan M, Chen Q, Pan Y, Ruan Y, et al. MicroRNA-1246 mediates drug resistance and metastasis in breast cancer by targeting NFE2L3. Frontiers in Oncology 2021;11:677168. https://doi.org/10.3389/fonc.2021.677168.

[135] Bottani M, Banfi G, Lombardi G. Circulating MiRNAs as diagnostic and prognostic biomarkers in common solid tumors: focus on lung, breast, prostate cancers, and osteosarcoma. Journal of Clinical Medicine 2019;8(10):1661. https://doi.org/10.3390/jcm8101661.

[136] Xu W, Hua Y, Deng F, Wang D, Wu Y, Zhang W, et al. MiR-145 in cancer therapy resistance and sensitivity: a comprehensive review. Cancer Science 2020;111(9):3122. https://doi.org/10.1111/cas.14517.

[137] Fabbri M, Paone A, Calore F, Galli R, Gaudio E, Santhanam R, et al. MicroRNAs bind to toll-like receptors to induce Prometastatic inflammatory response. Proceedings of the National Academy of Sciences of the United States of America 2012;109(31):E2110–6. https://doi.org/10.1073/pnas.1209414109.

[138] Zhao L, Ma X, Yu J. Exosomes and organ-specific metastasis. Molecular Therapy - Methods & Clinical Development 2021;22:133–47. https://doi.org/10.1016/j.omtm.2021.05.016.

[139] Lin W-T, Hsieh C-F, Chao C-M. Subcutaneous gas of the perineal area: radiographic signs of Fournier's Gangrene. Journal of Korean Medical Science 2016;31(6):1007–8. https://doi.org/10.3346/jkms.2016.31.6.1007.

[140] Hussen BM, Abdullah SR, Hama Faraj GS, Rasul MF, Salihi A, Ghafouri-Fard S, et al. Exosomal circular RNA: a signature for lung cancer progression. Cancer Cell International 2022;22(1):378. https://doi.org/10.1186/s12935-022-02793-7.

[141] Li S, Liu Y, Qiu G, Luo Y, Li X, Meng F, et al. Emerging roles of circular RNAs in non-small cell lung cancer. Oncology Reports 2021;45(4):17. https://doi.org/10.3892/or.2021.7968.

[142] Rong D, Sun H, Li Z, Liu S, Dong C, Fu K, et al. An emerging function of CircRNA-MiRNAs-MRNA Axis in human diseases. Oncotarget 2017;8(42):73271–81. https://doi.org/10.18632/oncotarget.19154.

[143] Xue M, Hong W, Jiang J, Zhao F, Gao X. Circular RNA circ-LDLRAD3 serves as an oncogene to promote non-small cell lung cancer progression by upregulating SLC1A5 through sponging MiR-137. RNA Biology 2020;17(12):1811–22. https://doi.org/10.1080/15476286.2020.1789819.

[144] Liu Y, Li L, Song X. Exosomal CircPVT1 derived from lung cancer promotes the progression of lung cancer by targeting MiR-124-3p/EZH2 Axis and regulating macrophage polarization. Cell cycle (Georgetown, Tex.) 2022;21(5):514–30. https://doi.org/10.1080/15384101.2021.2024997.

[145] Chen H-H, Zhang T-N, Wu Q-J, Huang X-M, Zhao Y-H. Circular RNAs in lung cancer: recent advances and future perspectives. Frontiers in Oncology 2021:11.

[146] Poulet C, Njock M-S, Moermans C, Louis E, Louis R, Malaise M, et al. Exosomal long non-coding RNAs in lung diseases. International Journal of Molecular Sciences 2020;21(10):3580. https://doi.org/10.3390/ijms21103580.

[147] Cao Z, Oyang L, Luo X, Xia L, Hu J, Lin J, et al. The roles of long non-coding RNAs in lung cancer. Journal of Cancer 2022;13(1):174–83. https://doi.org/10.7150/jca.65031.

[148] Bhan A, Soleimani M, Mandal SS. Long noncoding RNA and cancer: a new paradigm. Cancer Research 2017;77(15):3965–81. https://doi.org/10.1158/0008-5472.CAN-16-2634.

[149] Shen X, Yang Y, Chen Y, Zhou C, Zhao X, Li N, et al. Evaluation of EpCAM-specific exosomal LncRNAs as potential diagnostic biomarkers for lung cancer using droplet digital PCR. Journal of Molecular Medicine 2022;100(1):87–100. https://doi.org/10.1007/s00109-021-02145-4.

[150] Wang S, Dong Y, Gong A, Kong H, Gao J, Hao X, et al. Exosomal CircRNAs as novel cancer biomarkers: challenges and opportunities. International Journal of Biological Sciences 2021;17(2):562–73. https://doi.org/10.7150/ijbs.48782.

[151] Li F, Yang Q, He AT, Yang BB. Circular RNAs in cancer: limitations in functional studies and diagnostic potential. Seminars in Cancer Biology 2021;75:49–61. https://doi.org/10.1016/j.semcancer.2020.10.002.

[152] Breast Cancer Statistics | How Common Is Breast Cancer? https://www.cancer.org/cancer/types/breast-cancer/about/how-common-is-breast-cancer.html (Accessed 11 May 2023).

[153] Breast Cancer. https://www.who.int/news-room/fact-sheets/detail/breast-cancer (Accessed 11 May 2023).

[154] Sung H, Ferlay J, Siegel RL, Laversanne M, Soerjomataram I, Jemal A, et al. Global cancer statistics 2020: GLOBOCAN estimates of incidence and mortality worldwide for 36 cancers in 185 countries. CA: A Cancer Journal for Clinicians 2021;71(3):209–49. https://doi.org/10.3322/caac.21660.

[155] Chen W, Li Z, Deng P, Li Z, Xu Y, Li H, et al. Advances of exosomal MiRNAs in breast cancer progression and diagnosis. Diagnostics 2021;11(11):2151. https://doi.org/10.3390/diagnostics11112151.

[156] Zhao Y, Jin L-J, Zhang X-Y. Exosomal MiRNA-205 promotes breast cancer chemoresistance and Tumorigenesis through E2F1. Aging 2021;13(14):18498–514. https://doi.org/10.18632/aging.203298.

[157] Singh R, Pochampally R, Watabe K, Lu Z, Mo Y-Y. Exosome-mediated transfer of MiR-10b promotes cell invasion in breast cancer. Molecular Cancer 2014;13(1):256. https://doi.org/10.1186/1476-4598-13-256.

[158] Wong GL, Abu Jalboush S, Lo H-W. Exosomal MicroRNAs and organotropism in breast cancer metastasis. Cancers 2020;12(7):1827. https://doi.org/10.3390/cancers12071827.

[159] Shen S, Song Y, Zhao B, Xu Y, Ren X, Zhou Y, et al. Cancer-derived exosomal MiR-7641 promotes breast cancer progression and metastasis. Cell Communication and Signaling 2021;19(1):20. https://doi.org/10.1186/s12964-020-00700-z.

[160] Zhou S, Chen W, Yang S, Xu Z, Hu J, Zhang H, et al. The emerging role of circular RNAs in breast cancer. Bioscience Reports 2019;39(6). BSR20190621. https://doi.org/10.1042/BSR20190621.
[161] Zhou B, Mo Z, Lai G, Chen X, Li R, Wu R, et al. Targeting tumor exosomal circular RNA CSERPINE2 suppresses breast cancer progression by modulating MALT1-NF-κB-IL-6 Axis of tumor-associated macrophages. Journal of Experimental & Clinical Cancer Research 2023;42:48. https://doi.org/10.1186/s13046-023-02620-5.
[162] Li L, Zhang Q, Lian K. Circular RNA Circ_0000284 plays an oncogenic role in the progression of non-small cell lung cancer through the MiR-377-3p-mediated PD-L1 promotion. Cancer Cell International 2020;20:247. https://doi.org/10.1186/s12935-020-01310-y.
[163] Wang X, Fang L. Advances in circular RNAs and their roles in breast cancer. Journal of Experimental & Clinical Cancer Research 2018;37(1):206. https://doi.org/10.1186/s13046-018-0870-8.
[164] Hussen BM, Mohamadtahr S, Abdullah SR, Hidayat HJ, Rasul MF, Hama Faraj GS, et al. Exosomal circular RNAs: new player in breast cancer progression and therapeutic targets. Frontiers in Genetics 2023:14.
[165] Zokaei E, Darbeheshti F, Rezaei N. Prospect of exosomal circular RNAs in breast cancer: presents and future. Molecular Biology Reports 2022;49(7). https://doi.org/10.1007/s11033-022-07472-4.
[166] Lakshmi S, Hughes TA, Priya S. Exosomes and exosomal RNAs in breast cancer: a status update. European Journal of Cancer 2021;144:252–68. https://doi.org/10.1016/j.ejca.2020.11.033.
[167] Yi Y, Wu M, Zeng H, Hu W, Zhao C, Xiong M, et al. Tumor-derived exosomal non-coding RNAs: the emerging mechanisms and potential clinical applications in breast cancer. Frontiers in Oncology 2021:11.
[168] Goenka A, Tiek DM, Song X, Iglesia RP, Lu M, Hu B, et al. The role of non-coding RNAs in glioma. Biomedicines 2022;10(8):2031. https://doi.org/10.3390/biomedicines10082031.
[169] Aili Y, Maimaitiming N, Mahemuti Y, Qin H, Wang Y, Wang Z. The role of exosomal MiRNAs in glioma: biological function and clinical application. Frontiers in Oncology 2021;11:686369. https://doi.org/10.3389/fonc.2021.686369.
[170] Ghaemmaghami AB, Mahjoubin-Tehran M, Movahedpour A, Morshedi K, Sheida A, Taghavi SP, et al. Role of exosomes in malignant glioma: MicroRNAs and proteins in pathogenesis and diagnosis. Cell Communication and Signaling 2020;18(1):120. https://doi.org/10.1186/s12964-020-00623-9.
[171] Peng J, Liang Q, Xu Z, Cai Y, Peng B, Li J, et al. Current understanding of exosomal MicroRNAs in glioma immune regulation and therapeutic responses. Frontiers in Immunology 2022;12.
[172] Cheng J, Meng J, Zhu L, Peng Y. Exosomal noncoding RNAs in glioma: biological functions and potential clinical applications. Molecular Cancer 2020;19(1):66. https://doi.org/10.1186/s12943-020-01189-3.
[173] Han Y, Liu Y, Zhang B, Yin G. Exosomal CircRNA 0001445 promotes glioma progression through MiRNA-127-5p/SNX5 pathway. Aging 2021;13(9):13287–99. https://doi.org/10.18632/aging.203013.
[174] Chen X, Chen J, Chen Y, You H, Lin Y, Wu Z, et al. Exosomal circular RNAs in glioma: coexistence of opportunities and challenges for application. Chinese Medical Journal 2022;135(13):1528. https://doi.org/10.1097/CM9.0000000000002088.
[175] Chen M, Yan C, Zhao X. Research progress on circular RNA in glioma. Frontiers in Oncology 2021:11.
[176] Makowska M, Smolarz B, Romanowicz H. MicroRNAs (MiRNAs) in glioblastoma multiforme (GBM)—recent literature review. International Journal of Molecular Sciences 2023;24(4):3521. https://doi.org/10.3390/ijms24043521.
[177] Pancholi S, Tripathi A, Bhan A, Acharya MM, Pillai PP. Emerging concepts on the role of extracellular vesicles and its cargo contents in glioblastoma-microglial crosstalk. Molecular Neurobiology 2022;59(5):2822–37. https://doi.org/10.1007/s12035-022-02752-3.
[178] Teng X-Q, Qu J, Li G-H, Zhuang H-H, Qu Q. Small interfering RNA for gliomas treatment: overcoming hurdles in delivery. Frontiers in Cell and Developmental Biology 2022;10:824299. https://doi.org/10.3389/fcell.2022.824299.

[179] Barranco C. Antisense oligonucleotide slows glioma growth. Nature Reviews Drug Discovery 2023. https://doi.org/10.1038/d41573-023-00077-x.

[180] Adamus T, Hung C-Y, Yu C, Kang E, Hammad M, Flores L, et al. Glioma-targeted delivery of exosome-encapsulated antisense oligonucleotides using neural stem cells. Molecular Therapy - Nucleic Acids 2022;27:611–20. https://doi.org/10.1016/j.omtn.2021.12.029.

[181] Cui Y, Mubarik S, Li R, Nawsherwan, Yu C. Trend dynamics of thyroid cancer incidence among China and the U.S. adult population from 1990 to 2017: a joinpoint and age-period-cohort analysis. BMC Public Health 2021;21(1):624. https://doi.org/10.1186/s12889-021-10635-w.

[182] Maggisano V, Capriglione F, Verrienti A, Celano M, Gagliardi A, Bulotta S, et al. Identification of exosomal MicroRNAs and their targets in papillary thyroid cancer cells. Biomedicines 2022;10(5):961. https://doi.org/10.3390/biomedicines10050961.

[183] Wu Q, He Y, Liu X, Luo F, Jiang Y, Xiang M, et al. Cancer stem cell-like cells-derived exosomal LncRNA CDKN2B-AS1 promotes biological characteristics in thyroid cancer via MiR-122-5p/P4HA1 Axis. Regenerative Therapy 2022;22:19–29. https://doi.org/10.1016/j.reth.2022.11.005.

[184] Lin Q, Qi Q, Hou S, Chen Z, Jiang N, Zhang L, et al. Exosomal circular RNA Hsa_circ_007293 promotes proliferation, migration, invasion, and epithelial-mesenchymal transition of papillary thyroid carcinoma cells through regulation of the MicroRNA-653-5p/paired box 6 Axis. Bioengineered 2021;12(2):10136–49. https://doi.org/10.1080/21655979.2021.2000745.

[185] Xia F, Zhang Z, Li X. Emerging roles of circular RNAs in thyroid cancer. Frontiers in Cell and Developmental Biology 2021:9.

[186] Okunade KS. Human papillomavirus and cervical cancer. Journal of Obstetrics and Gynaecology, 2020;40(5):602–8. https://doi.org/10.1080/01443615.2019.1634030.

[187] Bhat A, Yadav J, Thakur K, Aggarwal N, Chhokar A, Tripathi T, et al. Transcriptome analysis of cervical cancer exosomes and detection of HPVE6*I transcripts in exosomal RNA. BMC Cancer 2022;22(1):164. https://doi.org/10.1186/s12885-022-09262-4.

[188] Ran Z, Wu S, Ma Z, Chen X, Liu J, Yang J. Advances in exosome biomarkers for cervical cancer. Cancer Medicine 2022;11(24):4966–78. https://doi.org/10.1002/cam4.4828.

[189] Huang X, Liu X, Du B, Liu X, Xue M, Yan Q, et al. LncRNA LINC01305 promotes cervical cancer progression through KHSRP and exosome-mediated transfer. Aging 2021;13(15):19230–42. https://doi.org/10.18632/aging.202565.

[190] Taitt HE. Global trends and prostate cancer: a review of incidence, detection, and mortality as influenced by race, ethnicity, and geographic location. American Journal of Men's Health 2018;12(6):1807–23. https://doi.org/10.1177/1557988318798279.

[191] EAU Guidelines on Prostate Cancer - EPIDEMIOLOGY AND AETIOLOGY - Uroweb. Uroweb - European Association of Urology. https://uroweb.org/guidelines/prostate-cancer/chapter/epidemiology-and-aetiology (Accessed 14 May 2023).

[192] Movahedpour A, Khatami SH, Karami N, Vakili O, Naeli P, Jamali Z, et al. Exosomal noncoding RNAs in prostate cancer. Clinica Chimica Acta 2022;537:127–32. https://doi.org/10.1016/j.cca.2022.10.018.

[193] Yu M, Lu B, Zhang J, Ding J, Liu P, Lu Y. TRNA-derived RNA fragments in cancer: current status and future perspectives. Journal of Hematology & Oncology 2020;13(1):121. https://doi.org/10.1186/s13045-020-00955-6.

[194] Fu M, Gu J, Wang M, Zhang J, Chen Y, Jiang P, et al. Emerging roles of TRNA-derived fragments in cancer. Molecular Cancer 2023;22(1):30. https://doi.org/10.1186/s12943-023-01739-5.

[195] Zhang H, Li M, Zhang J, Shen Y, Gui Q. Exosomal circ-XIAP promotes docetaxel resistance in prostate cancer by regulating MiR-1182/TPD52 Axis. Drug Design, Development and Therapy 2021;15:1835. https://doi.org/10.2147/DDDT.S300376.

[196] Lorenc T, Klimczyk K, Michalczewska I, Słomka M, Kubiak-Tomaszewska G, Olejarz W. Exosomes in prostate cancer diagnosis, prognosis and therapy. International Journal of Molecular Sciences 2020;21(6):2118. https://doi.org/10.3390/ijms21062118.

[197] Momenimovahed Z, Tiznobaik A, Taheri S, Salehiniya H. Ovarian cancer in the world: epidemiology and risk factors. International Journal of Women's Health 2019;11:287–99. https://doi.org/10.2147/IJWH.S197604.

[198] Jin Y, Wei J, Xu S, Guan F, Yin L, Zhu H. MiR-210-3p regulates cell growth and affects cisplatin sensitivity in human ovarian cancer cells via targeting E2F3. Molecular Medicine Reports 2019;19(6):4946–54. https://doi.org/10.3892/mmr.2019.10129.
[199] Nakamura K, Sawada K, Kobayashi M, Miyamoto M, Shimizu A, Yamamoto M, et al. Role of the exosome in ovarian cancer progression and its potential as a therapeutic target. Cancers 2019;11(8):1147. https://doi.org/10.3390/cancers11081147.
[200] Croft PK, Sharma S, Godbole N, Rice GE, Salomon C. Ovarian-cancer-associated extracellular vesicles: microenvironmental regulation and potential clinical applications. Cells 2021;10(9):2272. https://doi.org/10.3390/cells10092272.
[201] Gong X, Chi H, Strohmer DF, Teichmann AT, Xia Z, Wang Q. Exosomes: a potential tool for immunotherapy of ovarian cancer. Frontiers in Immunology 2023:13.
[202] Tu C, He J, Qi L, Ren X, Zhang C, Duan Z, et al. Emerging landscape of circular RNAs as biomarkers and pivotal regulators in osteosarcoma. Journal of Cellular Physiology 2020;235(12):9037–58. https://doi.org/10.1002/jcp.29754.

Further reading

[1] Yuan D, Guo T, Zhu D, Ge H, Zhao Y, Huang A, et al. Exosomal LncRNA ATB derived from ovarian cancer cells promotes angiogenesis via regulating MiR-204-3p/TGFβR2 Axis. Cancer Management and Research 2022;14:327–37. https://doi.org/10.2147/CMAR.S330368.
[2] Chai Y, Liu J, Zhang Z, Liu L. HuR-regulated LncRNA NEAT1 stability in tumorigenesis and progression of ovarian cancer. Cancer Medicine 2016;5(7):1588–98. https://doi.org/10.1002/cam4.710.
[3] Buttarelli M, De Donato M, Raspaglio G, Babini G, Ciucci A, Martinelli E, et al. Clinical value of LncRNA MEG3 in high-grade serous ovarian cancer. Cancers 2020;12(4):966. https://doi.org/10.3390/cancers12040966.
[4] Tripathi MK, Doxtater K, Keramatnia F, Zacheaus C, Yallapu MM, Jaggi M, et al. Role of LncRNAs in ovarian cancer: defining new biomarkers for therapeutic purposes. Drug Discovery Today 2018;23(9):1635–43. https://doi.org/10.1016/j.drudis.2018.04.010.

CHAPTER FOUR

Exosomal RNA as a biomarker in cancer diagnostics and therapy

Anirban Roychowdhury[1,2]
[1]Section of Hematology and Oncology, Medicine Service, McGuire Cancer Center, Hunter Holmes McGuire VA Medical Center, Richmond, VA, United States
[2]Division of Hematology, Oncology, and Palliative Care, Department of Internal Medicine, Massey Cancer Center, Virginia Commonwealth University, Richmond, VA, United States

1. Introduction

Extracellular vesicles (EVs) are generally small, membrane-bound bodies released by the cells. EVs are readily found in body fluids, including blood, saliva, and urine [1–3]. EVs are known either as exosomes, microvesicles, or apoptotic bodies [1,2]. The exosomes, among them, have gained the most impetus for biological research [4,5].

Exosomes are small vesicles (typically around 30–150 nm in diameter) that are formed inside multivesicular bodies (MVBs) within the cell and transported out upon diverse extracellular stimuli, including but not limited to growth factors, cytokines, and stress signals [2,4]. Exosomes contain a variety of biological molecules, including RNA, DNA, lipids, and proteins, which can be transferred to recipient cells to modify their behavior [1–4]. Various RNA types have been observed as a part of the exosomal cargo, including messenger RNA (mRNA), long noncoding RNA (lncRNA), transfer RNA (t-RNA), ribosomal RNA (rRNA), microRNA (miRNA), and, recently, circular RNA (circRNA) [6].

Exosomes taken from cancer cells can contain full-length or truncated **mRNA** molecules that can be translated into proteins by the receiving cells [3,6]. **LncRNAs** are RNA species that are often found to be longer than 200 nucleotides but incapable of being translated into proteins [5–7]. The exosome is also found to contain **miRNAs**, a class of small noncoding RNAs that affect the expression of a specific target gene and have been implicated in the pathogenesis of all cancer types [6,8]. The **circRNAs** are also a newly discovered type of noncoding RNA. They can influence the expression of target genes by acting as "sponges" for miRNAs or through cross-talk with RNA-binding proteins (RBPs) [6,9]. There are also a few reports where tRNAs and rRNAs were also observed to be loaded into exosomes and taken up by recipient cells, potentially playing a role in translation regulation [6]. Hence, cancer-derived exosomal RNAs have the potential to influence the pathology of cancer and are actively investigated as diagnostic and therapeutic tools [4,10].

Exosomal RNA
ISBN 978-0-443-14008-2
https://doi.org/10.1016/B978-0-443-14008-2.00004-8

2. Exosomal RNA as a biomarker for diagnosis and therapy outcome of common tumor types

A **biomarker** is a biological entity that can serve as a measurable indicator of cellular processes, response to external stimuli or therapy, or disease state in the body of an organism. In oncology, biomarkers are commonly employed to categorize patients for specific treatments, such as targeted therapies and immunotherapies, which may be effective only for certain tumor attributes [11]. **Liquid biopsy** refers to a biomarker derived from clinical samples such as blood, urine, saliva, or other body fluids, without the need for surgical procedures [12]. Liquid biopsy holds the potential to detect cancer at an early premalignant stage [13].

Fascinatingly, over the past 3 years, the field of exosome research has undergone a significant shift, with the scientific community increasingly gravitating toward RNA-based investigations, despite the initial excitement surrounding exosomal DNA (Fig. 4.1). This transformation in research focus can be attributed to several factors. Initially, exosomal DNA was a subject of great interest but as technology and methodologies advanced, researchers began to understand the crucial role of the broad array of RNA species present in exosomes making them attractive targets for diagnostic and therapeutic purposes [4]. High-throughput sequencing techniques, such as RNA-seq,

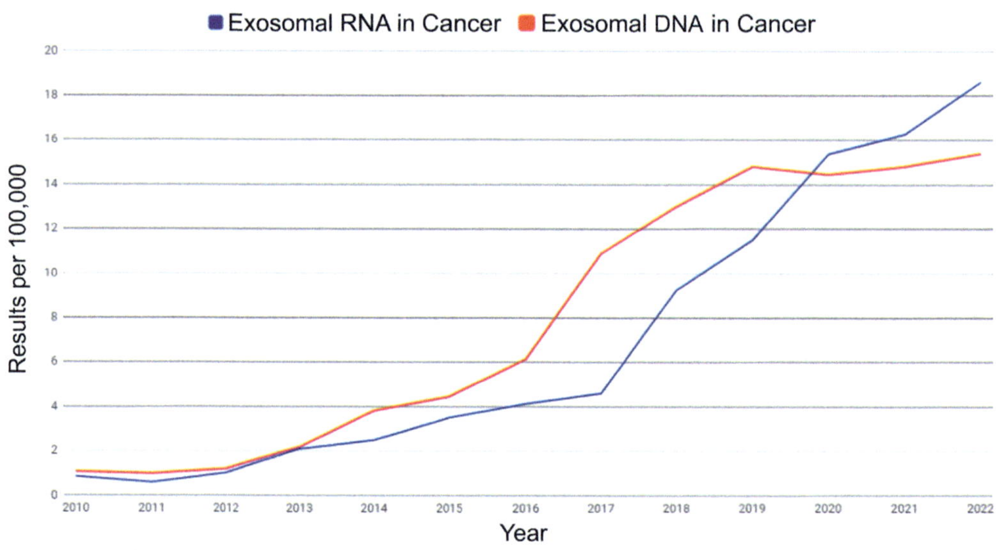

Figure 4.1 The representative plot showing the trend in the number of research published in NCBI Pubmed searched with the terms "exosomal RNA in cancer" and "exosomal DNA in cancer" using the tool esprr.github.io/pubmed-by-year. The data showed that there has been a clear change in the trend in the past 3 years where more research has now been focused on exosomal RNA than exosomal DNA in cancer.

have allowed for identifying and profiling exosomal RNA with greater accuracy and sensitivity, providing deeper insights into their functions and potential clinical applications [14]. Moreover, RNA biomarkers offer dynamic insights into cellular states and regulatory processes compared with DNA or protein biomarkers [9–11]. The shift toward RNA-based research in exosomes has also been driven by the realization that RNA molecules in exosomes are more stable than their free counterparts where the exosomal membrane protects the RNAs from degradation in the body fluids providing a prolonged half-life [4,10]. Therefore, they have immense potential as tumor biomarkers in liquid biopsy.

There has been a plethora of studies on several types of cancer [10]. Here, we focus to provide a comprehensive overview of the exosomal RNA biomarkers that have a role in tumor progression, treatment outcome, and prognosis in cancer patients.

2.1 Brain cancer

Tumors in the central nervous system (CNS) occur rarely in the population but have a high mortality rate. Most CNS tumor cases are gliomas, and the usual treatment approach involves surgically removing the tumor, followed by radiation therapy and oral chemotherapy preferably using temozolomide (TMZ) [15]. Exosomes, which are abundant in cerebrospinal fluid (CSF) and blood serum, have the potential to serve as biomarkers for predicting the prognosis of glioblastoma (GBM) [16–18].

The connection between exosomal lncRNA, **HOXA11-AS**, and the mesenchymal subtype of GBM assists in identifying different subtypes of glioma [16–18]. Furthermore, analysis reveals that the lncRNA, **HOXA-AS3**, is linked to tumor grade and indicates an unfavorable prognosis in glioma, indicating the potential of lncRNAs as biomarkers in the diagnosis of cancer [16–18].

Shi et al. discovered that the exosomal **hsa-miR-21** in the CSF of glioma patients was significantly elevated compared with normal individuals, suggesting its potential as a diagnostic biomarker for glioma [16–18]. Moreover, Yang et al. demonstrated that the expression of **hsa-miR-221** in exosomes from advanced glioma patients was higher in comparison with healthy controls. Lan et al. found a significantly higher expression level of serum exosomal **hsa-miR-301a** in glioma patients than in healthy individuals [16–18]. Additionally, elevated exosomal **hsa-miR-26a** was observed in glioma patients with poor prognosis [16–18]. Vander et al. and other researchers observed a high level of circulating exosomal **hsa-miR-10b** in the CSF of high-grade patients, which was linked to overall survival [16–18]. Therefore, exosomal miRNAs present in bodily fluids have immense potential as biomarkers for diagnosing and predicting the prognosis of glioma [16,17].

Elevated levels of serum exosomal **circMMP1** have been potentially associated with a negative prognosis in glioma patients [16,17]. GBM patients show decreased levels of exosomal **circSMARCA5** and **circHIPK3** in their blood [16,17]. Additionally, GBM patients reportedly display significantly higher levels of plasma exosomal **circ0055202,**

circ0074920, and **circ0043722** [16,17]. A positive correlation has been observed between low exosomal **circFBXW7** expression and overall survival in GBM patients [16,17]. High levels of exosomal **circATP8B4** have been reported to contribute to radioresistance in GBM [16]. Moreover, TMZ-resistant glioma patients exhibit elevated levels of exosomal **circNFIX**, which can promote the progression of oncogenic processes [16]. Hence, the presence of exosomal RNAs in liquid biopsy samples holds tremendous promise as biomarkers for the diagnosis and prognosis of glioma [16,17].

2.2 Oral cancer

Oral cancer refers to the development of malignant tumors encompassing areas such as the mouth, lips, tongue, gums, and throat. They are commonly termed head and neck squamous cell carcinomas (HNSCCs). The standard treatment for oral cancer typically involves surgical removal of the tumor, followed by adjunctive therapies such as chemotherapy and radiation [19]. To explore the potential of exosomal RNAs as biomarkers, several studies have investigated serum and saliva samples from patients with HNSCC [20–22].

According to Sun et al., they noted an increase in the levels of exosomal lncRNA, **LINC00152** in HNSCC [20–23]. This lncRNA shows potential as a diagnostic biomarker for the disease. Similarly, Lei et al. discovered that **HOTTIP**, another lncRNA, was significantly upregulated in exosomes [20–23]. They observed that elevated levels of HOTTIP were associated with advanced tumor stages and poor overall survival among HNSCC patients.

In their respective studies, Wang et al. found elevated levels of **hsa-miR-21** in the serum of oral cancer patients compared with those with benign disease [18–20]. Ye et al. observed an increased presence of **hsa-miR-24-3p** in the plasma of patients with oral tumors [18–20]. Similarly, Momen-Heravi et al. focused on diagnosing HNSCC and examined the presence of **hsa-miR-21, hsa-miR-27, hsa-miR-412-3p,** and **hsa-miR-512-3p** in saliva samples of cancer patients compared with normal controls [20–22,24]. In a separate study, Langevin et al. found an association between the expression of **hsa-miR-486-5p, hsa-miR-31, hsa-miR-486-3p,** and **hsa-miR-10b-5p** in saliva samples of HNSCC patients [20–22,24]. Additionally, Chen et al. identified high expression of **hsa-miR-21** and **hsa-miR-155** in the exosomes from saliva and serum of oral cancer patients, suggesting their potential as distinguishing markers [20–22,24].

In oral cancer, patients with higher levels of exosomal **circ0000199** had a higher rate of tumor recurrence and mortality compared with those with lower levels of the circRNA [20–22,25,26]. Similarly, elevated levels of exosomal **circMYC** in oral carcinoma were closely associated with larger tumor size, lymph node metastasis, and advanced stages, serving as an independent predictor of patient survival [20–22,25].

2.3 Breast cancer

Breast cancer (BC) is a prevalent and heterogenous tumor that requires early detection for effective treatment and reduced mortality rates [27]. There are several subtypes of BC that have been classified for accurate clinical intervention. Broadly, hormone receptor—positive (HR+) human epidermal growth factor receptor 2—positive (HER2+), triple-negative breast cancer (TNBC), luminal A, and luminal B types [27].

The investigation conducted on BC patients revealed that individuals who showed resistance to the commonly used chemotherapy drug doxorubicin exhibited higher expression levels of exosomal **lncRNA-H19** [28—33]. In the case of HER2-positive breast cancer, trastuzumab, a targeted monoclonal antibody, is used to specifically target the receptor tyrosine—protein kinase erbB-2 (HER2) receptor. However, research by Dong et al. found that patients resistant to trastuzumab treatment had higher levels of exosomal **lncRNA-SNHG14** in their serum [28—33] compared with those who responded positively to the treatment. Additionally, patients with TNBC had markedly higher levels of serum exosomal **lncRNA-SUMO1P3** compared with BC patients without TNBC, with benign breast disease, and healthy females [28—33]. Elevated levels of serum exosomal lncRNA-SUMO1P3 were strongly associated with treatment failure and an unfavorable prognosis. Recent studies have reported that **lncRNA-BCRT1** is overexpressed in breast cancer cells and associated with a poor prognosis in patients [28—33].

An exosomal miRNA, **hsa-miR-363-5p**, is reported to be upregulated and displays lower expression in BC patients with sentinel lymph node (LN) metastasis [31,32]. Another example is **hsa-miR-370-3p**, overexpressed in BC patients' tumor tissues, serum, and serum exosomes and associated with larger tumor mass and distant metastasis [28—32,34,35]. Conversely, miRNA, **hsa-miR-148a**, was found to be downregulated and significantly associated with metastasis in the sentinel lymph nodes, tumor differentiation, and 5-year survival [28—32,34,35]. In addition, increased levels of **hsa-miR-3662**, **hsa-miR-146a**, and **hsa-miR-1290** have been correlated with LN metastasis and the progression of BC [31,32]. Studies have also shown differential expression of exosomal miRNAs in BC patients with cerebral and brain metastases. For example, **hsa-miR-576-3p** is significantly upregulated in patients with cerebral metastases, while **hsa-miR-130a-3p** is significantly downregulated [28—32,34,35]. The exosomal miRNA, **hsa-miR-181c**, is significantly observed to be enhanced in the serum of BC patients with metastasis to brain tissues. Higher expression levels of **hsa-miR-150-5p**, **hsa-miR-576-3p**, and **hsa-miR-4665-5p** were observed in TNBC patients with recurrence compared with those without recurrence [31,32]. Notably, a combination of four miRNAs (**hsa-miR-206**, **hsa-miR-24**, **hsa-miR-1246**, and **hsa-miR-373**) demonstrated an impressive accuracy in detecting BC [28—32,34,35]. Similarly, Hirschfeld and colleagues explored the potential of urinary exosomes and analyzed urine samples from BC patients and healthy controls Their findings indicated that a combination of four miRNAs **(hsa-miR-423**, **hsa-miR-424**, **hsa-miR-660**, and **hsa-let-7i**) achieved a high accuracy in detecting BC [28–32,34,35].

Li et al. conducted a study comparing the expression of **circVRK1** in 350 BC-tumor samples and 163 adjacent tissues. The results showed lower expression of circVRK1 in tumor tissues, and patients with low circVRK1 expression exhibited unfavorable clinicopathological features [33,34]. Another study by Yang et al. identified **circPSMA1** as an upregulated circRNA in serum exosomes of TNBC patients [28–32,36,37]. The overexpression of circPSMA1 was associated with tumor migration and metastasis in TNBC. Although these studies provided valuable insights, further research and validation are rolling to understand the potential of exosomal RNA biomarkers in improving the clinical management of BC patients [28–32].

2.4 Lung cancer

Lung cancer is the most prevalent cancer originating in the bronchial region. The most frequent type of lung cancer is non–small cell carcinoma (NSCLC) and has a poor prognosis [38]. Clinically, NSCLC is classified into two main subtypes: lung squamous cell carcinoma (LUSC) and the more prevalent lung adenocarcinoma (LUAD) [38].

Research has demonstrated that certain lncRNAs carried by exosomes play critical roles in the pathogenesis of lung cancer [39–44]. Specifically, studies have indicated that enhanced expression of lncRNA, **MALAT-1**, in serum derived exosomes of NSCLC patients and its levels are positively correlated with the stage of the tumor and the occurrence of LN metastasis [39–44]. Additionally, two other lncRNAs, namely **TBILA** and **AGAP2-AS1**, were found to be elevated in the serum exosomes of NSCLC patients compared with healthy individuals [39–44]. Furthermore, the identification of another exosomal lncRNA called **SOX2-OT** in the plasma holds promise as a potential biomarker for LUSC [39–44]. Conversely, an exosomal lncRNA known as **GAS5** acts as a tumor suppressor, and its levels are significantly reduced in NSCLC patients [39–44] and are indicative of a poor prognosis. Overall, the expression levels of certain exosomal lncRNAs, such as **DLX6-AS1** and **SNHG15**, which promote tumor growth, are elevated in lung cancer patients with unfavorable outcomes [39–44]. For the treatment of advanced NSCLC, three widely used drugs known as gefitinib, erlotinib, and afatinib belong to a class called epidermal growth factor receptor tyrosine kinase inhibitors (EGFR TKIs). However, exosomes carrying a lncRNA called **H19** have been shown to contribute to resistance against both gefitinib and erlotinib in NSCLC patients. Additionally, another exosomal lncRNA, **RP11-838N2.4**, has been found to promote erlotinib resistance in NSCLC [39–44].

Poroyko et al. conducted a study revealing differences in the miRNA content of exosomes between patients with small-cell lung cancer (SCLC) and NSCLC. They identified the miRNAs (**hsa-miR-331-5p, hsa-miR-363-3p,** and **hsa-miR-451a**) that were particularly capable of distinguishing SCLC from NSCLC cases [39–42,45,46]. Other studies have also reported specific exosomal miRNAs (**hsa-miR-15b-5p, hsa-miR-16-2-3p, hsa-miR-25-3p, hsa-miR-363-3p, hsa-miR-451a, hsa-miR-486-5p,** and **hsa-miR-660-5p**) that can differentiate NSCLC patients from healthy individuals [45,46]. Furthermore, a group of three miRNAs (**hsa-miR-21, hsa-miR-155,** and **hsa-miR-205**) has been identified for the early detection of NSCLC [39–42,45,46]. Recent findings also indicate that **hsa-miR-182** and **hsa-miR-210** expression levels are higher in malignant pleural fluid from LUAD compared with benign effusions [45,46]. In advanced NSCLC patients, higher levels of exosomal **hsa-miR-146a** have been associated with better disease progression and a lower recurrence rate [39–42,45,46]. Serum exosomal miRNAs such as **hsa-miR-1, hsa-miR-486,** and **hsa-miR-499** have shown predictive value for the survival rate of NSCLC patients [39–42,45,46]. Elevated levels of **hsa-miR-373** and **hsa-miR-512** in NSCLC patients treated with cisplatin have been linked to better relapse-free survival and potential therapeutic benefits. Additionally, studies have found that higher expression of exosomal **hsa-miR-10b-5p, hsa-miR-21-5p,** and **hsa-miR-23b-3p** is significantly associated with poorer overall survival in NSCLC patients [39–42,45,46]. Correlations have been observed between worse disease-free survival and high expression of **hsa-miR-21** and **hsa-miR-4257** in recurrent NSCLC, suggesting their potential as recurrence-specific predictive biomarkers in patients who underwent curative resection [41,42,45,46]. Low expression of exosomal **hsa-miR-34c-3p** leads to increased cell invasion and motility in NSCLC, highlighting its diagnostic and prognostic value [41,42,45,46]. Recent research by Peng et al. has indicated that plasma exosomal **hsa-miR-320b, hsa-miR-320c,** and **hsa-miR-320d** in advanced-stage NSCLC patients could serve as predictors of immunotherapy efficacy [41,45,46]. Furthermore, increased radiation levels in NSCLC patients resulted in reduced levels of circulating exosomal **hsa-miR29a-3p** and **hsa-miR150-5p**, which may aid in predicting the response to radiotherapy [41,45,46].

According to a recent article, an enhanced expression of a circular RNA called **circPVT1** in blood-derived exosomes of lung cancer patients was reported to be associated with reduced responsiveness to the chemotherapy drugs [41,47,48]. In the context of targeted therapy-resistant lung cancer, another circular RNA called **circ0002130** displayed a high expression level in the serum exosomes of resistant lung cancer patients [41,47,48]. In terms of prognostic markers, **CDR1-AS**, another circular RNA, was identified in exosomes and acts as an independent biomarker for predicting patient outcomes in LUAD [47,48]. Higher expression levels of two exosomal circular RNAs, **circ0008717** and **circ0047921**, were found to promote carcinogenesis in NSCLC with prognostic implications [47,48]. Lastly, **circ-MEMO1** was found to be overexpressed in NSCLC, and such high expression level was predictive of poor prognosis in NSCLC patients [47,48].

2.5 Skin cancer

Melanoma, a malignant form of skin cancer originating from melanocytes responsible for melanin production, poses a significant global health concern [49]. Despite notable advancements in therapeutic approaches, the prognosis for melanoma patients remains unfavorable [49].

The miRNA, **hsa-miR-494**, is frequently encountered in melanoma tumor-derived exosomes. Li et al. demonstrated that **hsa-miR-494** was elevated in the serum exosomes of melanoma patients [50–53]. Alegre et al. compared melanoma patients, disease-free patients with melanoma, and healthy controls and found that **hsa-miR-125b** was among the most deregulated miRNAs. Its expression was significantly reduced in melanoma patients, suggesting its potential as a diagnostic marker [50–53]. Ragusa et al. observed elevated levels of hsa-miR-146a in several biological samples of uveal melanoma patients. These samples included the vitreous humor (gel-like substance in the eye), exosomes derived from the vitreous humor, serum (blood component), and exosomes obtained from the serum of the same patients [50,52]. Pfeffer et al. identified a number of differentially expressed miRNAs, including **hsa-miR-17**, **hsa-miR-19a**, **hsa-miR-21**, **hsa-miR-126**, and **hsa-miR-149**, with higher expression levels in plasma-derived exosomes from patients with metastatic melanoma compared with familial melanoma patients or healthy volunteers [50,52,53]. In a study by Tengda et al., significantly high levels of **hsa-miRNA-532-5p** and **hsa-miR-106b** were detected in serum exosomes of melanoma patients [52,53]. Xiao et al. identified **hsa-miR-191** and **hsa-let-7a** as upregulated in serum exosomes of melanoma patients [50,52,53]. Importantly, these miRNAs in serum exosomes showed potential as identifiers for stratifying stage I melanoma patients from nonmelanoma subjects. Furthermore, Luan et al. observed a significant increase in **hsa-miR-106b-5p** levels in the serum and tissues of melanoma patients and associated with overall survival in melanoma patients [50,52,53]. Additionally, Li et al. confirmed that **hsa-miR-494** was significantly elevated in serum exosomal samples from melanoma patients compared with normal individuals [50,52,53]. Furthermore, **hsa-miR-300** derived from exosomes has been implicated in the immune evasion of melanoma, and its clinical association with poorer prognosis in melanoma patients has been identified [52,53]. A recent study discovered that **hsa-miR-498** and **hsa-miR-3187-3p**, found in exosomes derived from melanoma, played a role in suppressing the cytotoxic effects of CD8+ T cells [53]. Further studies are in progress to unravel the advantages of exosomal RNA in the evaluation of diagnosis and prognosis in melanoma patients [50–53].

2.6 Liver cancer

Hepatocellular carcinoma (HCC) is the most prevalent of all primary liver cancers [54]. HCC is known for its horrid prognosis, higher tendency for metastasis, and, above all, late detection [54]. In the Asian population, chronic hepatitis B virus (HBV) infection is the primary cause of HCC, while hepatitis C virus (HCV) infection is more prevalent in Western countries. Other cofactors that increase the risk of HCC include nonalcoholic fatty liver disease (NAFLD), nonalcoholic steatohepatitis (NASH), alcoholic liver disease, and liver cirrhosis (LC) [54,55]. Extensive research has focused on exosomal RNA markers present in HCC patients [56,57].

The serum exosomal **RAB11A** mRNA showed both high specificity and sensitivity in stratifying healthy individuals, chronic hepatitis C patients (CHC), and HCC patients [56]. Researchers successfully isolated exosomes from the urine of HCC patients and discovered enhanced **LINC00161** levels in the urine samples [55–57]. Another exosomal lncRNA, **TUC339,** was found to be abundant in HCC patients and had the ability to promote cell proliferation. Comparisons between patients with chronic hepatitis B (CHB) and HCC revealed higher expression of the exosomal lncRNA **LINC00635** in HCC patients [55,56]. Elevated levels of exosomal lncRNA **HEIH** in serum were found to differentiate HCV-related HCC patients from those with CHC- or HCV-induced cirrhosis [55,56]. In liver cancer tissues, the exosomal lncRNA **SENP3-EIF4A1** exhibited a significant decrease, allowing it to serve as a distinguishing factor between HCC patients and healthy individuals [55,56]. Similarly, exosomal **LINC00161** was upregulated in HCC patients compared with the control group, making it a potential biomarker for HCC [55,56]. Researchers also discovered that exosomal lncRNAs **ATB**, **CRNDE**, and **FAL1** exhibited increased expression levels in the serum of HCC patients, correlating with HCC progression and indicating a poorer prognosis [55–57]. Furthermore, several exosomal lncRNAs, including lncRNAs, **GPR89B-15**, **FAM72D-3**, **EPC1-4**, and **ZEB2-19**, displayed differential expressions in the serum of HCC patients [55–57]. Collectively, lncRNAs are found to be a promising exosomal biomarker to detect HCC in human patients.

The presence of exosomal **hsa-miR-21-5p** in serum from HCC patients demonstrates a significant association with advanced stage and prognosis in early HCC patients [55,56]. Additionally, exosomal **hsa-miR-210** is upregulated in HCC patients [55,56]. Its presence promotes the proliferation, migration, and tube formation of endothelial cells within HCC [55,56]. The upregulation of **hsa-miR-32-5p** along with low levels of phosphatase and tensin homolog (PTEN) expression is associated with a worse prognosis in HCC patients [55,56]. Furthermore, elevated levels of exosomal **hsa-miR-1247-3p** are positively linked to lung metastasis in individuals with liver cancer [55,56]. Notably, liver cancer patients exhibited significantly higher levels of exosomal **hsa-miR-148a** and **hsa-miR-1246** compared with both healthy individuals and patients with liver cirrhosis. Conversely, the level of exosomal **hsa-miR-125b** in HCC patients is highly downregulated in chronic hepatitis B (CHB) and liver cirrhosis patients, as reported by Liu and colleagues [55,56,58]. Similarly, Min and coauthors found that serum exosomal **hsa-miR-638** levels were remarkably lower in HCC patients compared with healthy controls. This lower level of hsa-miR-638 predicts a poor prognosis for

HCC patients [55,56,58]. Moreover, liver cancer patients demonstrate significant increases in the levels of exosomal **hsa-miR-17-5p**, **hsa-miR-29a**, **hsa-miR-106a**, **hsa-miR-122**, **hsa-miR-192**, and **hsa-miR-194** and can effectively differentiate HCC from healthy individuals [55,56]. Additionally, elevated levels of exosomal **hsa-miR-106a**, **hsa-miR-17-5p**, and **hsa-miR-194** are associated with increased size of the primary tumor tissue. The miRNA, **hsa-miR-23a/b**, displays significant upregulation in exosomes derived from HCC patients with a high body fat ratio compared with those with a low body fat ratio [55–57]. On the other hand, there is a downregulation in the expression of exosomal **hsa-miR-744** in exosomes isolated from the serum of HCC patients [55,56]. Lastly, patients with HCC expressing high levels of serum exosomal **hsa-miR-224** exhibit lower overall survival rates [55,56].

A study showed that exosomal **circ0070396** may prove to be a better biomarker and have the potential to differentiate between liver cancer patients, those with CHB, and liver cirrhosis (LC) [55,56,59]. Higher expression of circulating exosomal **circAKT3** is associated with increased risk of disease relapse in HCC patients who have undergone surgical removal of the tumor and may aid in reducing the recurrence risk and improving prognosis [55,56,59]. Additionally, exosomal **circUHRF1** may act as a biomarker to evaluate resistance against the immunotherapy regime, in HCC patients [59]. Interestingly, the presence of circUHRF1 in exosomes from HCC patient sera is associated with decreased proportions of immune cells. Moreover, the persistent high expression of exosomal **circ100338** in the serum of HCC patients who have undergone curative hepatectomy may serve as a biomarker for subsequent pulmonary metastasis and poor survival [59]. Additionally, exosomal **circTMEM45A** expression levels are markedly increased in HCC patients [59]. Altogether, these studies evidently show the enormous potential of using these biomarkers in the assessment of diagnosis and prognosis for HCC.

2.7 Gastric cancer

Gastric cancer (GC) is a prevalent cancerous tumor worldwide, known for its high rates of occurrence and mortality [60]. One of the most notable risk factors for GC development is the infection by pathogenic bacteria, *Helicobacter pylori* (*H. pylori*) [61]. Regrettably, a significant number of GC patients only seek medical intervention when the disease has already advanced to the advanced stages. As a result, early detection and prompt treatment of GC have become crucial objectives in clinical practice [60,61].

Guo and colleagues conducted a study and successfully detected a specific long noncoding RNA, **GC1,** which showed high sensitivity and specificity in the early detection of GC [62–64]. Similarly, Lin et al. identified two exosomal lncRNAs, **lncUEGC1** and **lncUEGC2**, which have the potential to differentiate early gastric cancer (EGC) through serum analysis [62–64]. Another study by Pan et al. demonstrated that exosomal **HOTTIP**, found in serum, was upregulated in GC compared with healthy controls, and its levels were significantly associated with disease progression, serving as an independent prognostic factor for GC patients [62–64]. Additionally, Pan et al. found a correlation between the expression level of exosomal **ZF-AS1** and lymph node metastasis as well as the advanced stage in GC patients [63,64]. Furthermore, Cao et al. discovered that the lncRNA **SPRY4-IT1** was increased in GC patients and associated with metastasis. Conversely, the lncRNA **PCSK2-2:1** was significantly decreased in GC patients [63,64]. Piao's study revealed that the plasma exosomal lncRNA **CEBPA-AS1** could be a novel diagnostic marker for GC. Xu et al. found reduced levels of exosomal lncRNA **MIAT** in GC patients, while elevated levels of exosomal lncRNA **SLC2A12-10:1** were linked to larger tumors and invasion [63]. Increased expression of exosomal lncRNA **H19** was associated with GC development [63].

Wang et al. conducted a study and discovered that **hsa-miR-19b-3p** and **hsa-miR-106a-5p** are significantly overexpressed in GC patients [63,65]. Another research confirmed that **hsa-miR-15b-3p** is highly expressed in the serum of GC patients and can accurately predict worse outcomes [63,65]. The next study investigated the expressions of **hsa-miR-17-5p**, **hsa-miR-19b-3p**, **hsa-miR-30a-5p**, and **hsa-miR-106a-5p** in the exosomes of GC patients, revealing that **hsa-miR-106a-5p** and **hsa-miR-19b-3p** are overexpressed in GC samples compared with normal controls [63,65]. Researchers observed a substantial elevation in the expression level of **hsa-miR-374a-5p** in the serum of GC patients. Notably, this high expression level of hsa-miR-374a-5p was strongly associated with an unfavorable diagnosis among these GC patients [63,66]. Shi et al. observed elevated levels of **hsa-miR-155-5p** in exosomes derived from tissues of GC patients. The expression of **hsa-miR-1246** in serum exosomes could effectively differentiate early GC patients from both healthy individuals and patients with benign stomach diseases [63]. In addition, Kumata et al. discovered that the level of **hsa-miR-23b** in exosomes derived from GC patients was noticeably lower compared with that in a healthy cohort and had significantly worse outcomes compared with those with high hsa-miR-23b expression [63]. Moreover, research findings indicated that in patients with advanced stages GC, there was a remarkable increase in the levels of serum exosomal **hsa-miR-451**. Additionally, plasma exosomal **hsa-miR-379-5p** and **hsa-miR-410-3p** expressions were significantly elevated in GC patients at advanced stages who had undergone surgery but had reported metastasis. Higher levels of these miRNAs correlated with shorter prognosis in those patients [62,63]. Exosomal **hsa-miR-196a-1** levels were significantly higher in advanced GC patients compared with early-stage GC patients, suggesting its association with advanced GC [62,63]. Various miRNAs were identified as potential biomarkers for specific types of metastases of GC [62,63]. For instance, exosomal **hsa-miR-10b-5p** has demonstrated relevance in identifying GC patients with lymph node metastasis, **hsa-miR-101-3p** in those with ovarian metastasis, **hsa-miR-143-5p** in cases with liver metastasis, and **hsa-miR-301a-3p** in GC patients with peritoneal metastasis. Studies revealed that the levels of serum exosomal **hsa-miR-21** significantly increased in GC patients, especially those with peritoneal recurrence, while **hsa-miR-92a** levels decreased significantly in the same patients [62,63]. Furthermore, high levels of **hsa-miR-423-5p** in exosomes were associated with poor prognosis and lymph node metastasis in GC patients. Finally, Li et al. found that **hsa-miR-25** was elevated in exosomes derived from *H. pylori*-infected GC patients [61,67].

Several exosomal circRNAs derived from the plasma of GC patients play crucial roles in tumor invasion and metastasis [63,67,68]. **CircNRIP1**, specifically, exhibits significantly higher expression levels in exosomes derived from GC plasma compared with normal tissues [63,67,68]. Research by Sang et al. revealed that exosomal **circRELL1** is downregulated in GC [63,67,68]. On the other hand, circSHKBP1 was observed to be significantly overexpressed in the serum of advanced GC patients with poor prognoses. Moreover, **circRanGAP1** also showed a higher expression profile in GC patients when analyzed in parallel with healthy individuals [67,68]. GC patients develop adverse effects upon adriamycin treatment, and enhanced exosomal **circPRRX1** was found to be enhanced in such cases, highlighting its importance as an indicator for therapy response [67]. Shao et al. observed a significant downregulation of **circ0065149** expression in GC patients. Finally, lower expression values of **circ0000419** exhibited a positive association with progression and metastasis in GC patients [67,68].

2.8 Pancreatic cancer

Pancreatic cancer (PC) exhibits a highly malignant and aggressive nature, affecting both men and women alike. The most prevalent form of PC is known as pancreatic ductal adenocarcinoma (PDAC) [69]. Unfortunately, the prognosis for PDAC patients remains discouraging due to challenges associated with early detection, limited treatment options, and unpredictable tumor progression [69]. Nevertheless, emerging research suggests that certain exosomal RNAs hold promise as potential tumor biomarkers for PC [70].

Elevated levels of lncRNA, specifically **SOX2-OT**, have been observed in plasma exosomes, indicating an association with advanced disease and an unfavorable prognosis in individuals with PC [65,71,72]. Additionally, exosomal lncRNAs such as **HULC**, **CRNDE**, and **MALAT-1** have demonstrated their ability to distinguish PDAC from other pancreatic disorders [65,71,72]. Another research study observed that two lncRNAs, namely **HOTAIR** and **PVT1**, as novel noninvasive biomarkers for the detection of PDAC [71,72]. It was discovered that enhanced HOTAIR and PVT1 were observed in the saliva of patients with PDAC compared with healthy individuals, highlighting their potential diagnostic value in identifying the disease [71,72].

The dysregulation of exosomal miRNAs has been established in patients with pancreatic cancer. Remarkably, higher levels of exosomal **hsa-miR-17-5p** and **hsa-miR-21** have been identified in serum samples of PC patients [65,73]. To improve diagnostic capabilities, a group of exosomal miRNAs from serum has been analyzed, consisting of **hsa-miR-1246**, **hsa-miR-3976**, **hsa-miR-4306**, and **hsa-miR-4644**. Another panel includes **hsa-miR-10b**, **hsa-miR-21**, **hsa-miR-30c**, and **hsa-miR-181a**, known to be overexpressed, along with underexpressed **hsa-miR-let7a** in PC [65,73]. These exosomal miRNAs showed immense potential as diagnostic markers for PC. Additionally, significant elevation of exosomal **hsa-miR-191**, **hsa-miR-21**, and **hsa-miR-451a** has been observed in pancreatic cancer [65,73]. Moreover, the levels of exosomal hsa-miR-21 derived from pancreatic juice have shown significant differences between patients with PDAC and those with chronic pancreatitis. In the context of prognosis, hsa-miR-451a present in exosomes isolated from plasma samples of pancreatic cancer patients has been linked to the outcome of this malignancy. The abundance of specific exosomal miRNAs can also be leveraged to assess the response to therapy. For instance, increased levels of **hsa-miR-221** in plasma were observed in PC patients after 3 weeks of treatment [73]. Further examination of the data indicates a correlation with drug resistance [73].

The presence of exosomal circRNAs such as **circPDE8A** and **circIARS** has been found to be correlated with the distant metastasis and prognosis of PC [65,74,75]. Collectively, these exosomal RNAs show promise as potential biomarkers for various aspects of PC, including diagnosis, prognostic prediction, and disease progression.

2.9 Prostate cancer

Prostate cancer (PCa) is a common cancer type encountered in men worldwide with a low survival rate [76]. The nature of PCa is highly diverse, displaying significant variations in clinical progression, ranging from insignificant tumors to life-threatening castration-resistant prostate cancer (CRPC) [76,77]. Consequently, a pressing requirement exists for noninvasive markers to effectively manage PCa [76].

The **ExoDx Prostate IntelliScore (EPI)** is the first commercially available exosomal urine test recommended in the National Comprehensive Cancer Center Network Guidelines (NCCN) for early detection of PCa [77,78]. This test analyzes the expression of three transcripts: prostate cancer antigen 3 (**PCA3**), ETS-related gene (**ERG**), and SAM-pointed domain ETS transcription factor (**SPDEFs**). Two trials have reported its superior sensitivity in predicting high-grade PCa during the initial biopsy, reducing the need for unnecessary biopsies compared with existing risk calculators and standard clinical data [77,78]. The androgen receptor inhibitor (ARi) treatment is given to patients with metastatic CRPC. Enhanced expression of the exosomal androgen receptor splice variant 7 (**AR-V7**) transcript was observed in the plasma of CRPC undergoing ARi and correlates with a worse prognosis [78]. Additionally, urinary exosomes revealed higher expression of full-length androgen receptor (**AR–FL**) in hormone-sensitive prostate cancer (HSPC) patients, while AR-V7 expression levels were elevated in CRPC patients [78]. Moreover, the ratio of AR-V7 to AR-FL was significantly higher in CRPC patients compared with those with HSPC. Another study demonstrated the identification of responders and nonresponders to advanced PCa treatment by detecting the presence of the AR-V7 mRNA variant isolated from the plasma of CRPC patients [77,78]. A clinical trial (NCT03601143) is underway to predict resistance to androgen receptor signaling inhibitors using AR-V7 [77,78].

In their study, Wang et al. successfully isolated exosomes and discovered that **SAP30L-AS1** exhibited diagnostic potential in distinguishing individuals with benign prostatic disease or healthy individuals from those with PCa [79]. Furthermore, serum exosomal **HOXD-AS1** levels were found to be elevated in individuals with metastatic PCa. Another research conducted by Prensner et al. identified the lncRNA **SChLAP1** as being expressed at the highest levels in metastatic tumors [79]. This lncRNA served as a robust and prostate-specific biomarker, demonstrating the ability to predict subsequent tumor metastasis [79].

The miRNA, hsa-miR-1246, has been detected and confirmed as a biomarker in serum exosomes for aggressive PCa. It showed a strong association with unfavorable outcomes and the spread of cancer to other parts of the body [77,80,81]. In another study, Huang et al. identified two plasma exosomal miRNAs, **hsa-miR-1290** and **hsa-miR-375**, which were significantly associated with overall survival (OS) in patients with CRPC [77,80,81]. Guo and his team demonstrated that the expression of hsa-miR-423-3p in plasma exosomes changes during the development of prostate cancer. However, this change is not linked to the response to androgen deprivation therapy (ADT) treatment. Nevertheless, when combined with prostate-specific antigen (PSA) levels, hsa-miR-423-3p improves the CRPC prediction. Conversely, Fredsøe and colleagues developed and validated a biomarker model that includes five urine exosomal miRNAs (**hsa-miR-23b-3p**, **hsa-miR-151a-5p**, **hsa-miR-204-5p**, **hsa-miR-222-3p**, and **hsa-miR-331-3p**) along with the serum PSA test. This model has proven to be valuable in predicting the aggressiveness of the disease and the risk of biochemical recurrence after surgery in multiple groups of PCa patients. It provides crucial guidance for making treatment decisions. After administering curative radiotherapy (RT) to PCa patients with intermediate- and high-risk diseases, the levels of **hsa-let-7a-5p** and **hsa-miR-21** were observed to increase [77,80–82]. Another study revealed that the overexpression of a specific set of miRNAs (**hsa-miR-493-5p**, **hsa-miR-323a-3p**, **hsa-miR-411-5p**, **hsa-miR-494-3p**, **hsa-miR-379-5p**, **hsa-miR-654-3p**, **hsa-miR-409-3p**, **hsa-miR-543**, and **hsa-miR-200c-3p**) significantly predicted the therapeutic benefit of carbon ion radiotherapy (CIRT) [77,80–82]. Additionally, the expression levels of two specific miRNAs, **hsa-miR-654-3p** and **hsa-miR-379-5p**, were associated with the efficacy of CIRT after therapy [80–82]. Exosomes derived from metastatic prostate cancer patients exhibited high levels of **hsa-miR-21** and **hsa-miR-141**, suggesting their potential association with bone metastasis [77,80–82]. Furthermore, exosomal **hsa-miR-21**, **hsa-miR-375**, and **hsa-miR-141** have been observed to aid cancer cells in adapting to low-androgen conditions in distant metastatic organs [80–82].

The expression of four circular RNAs (**circAR-E2**, **circAR-E2In1**, **circARE2In2**, and **circAR-E2**) has been identified in the plasma of patients with CRPC, an advanced form of PCa, and has been associated with disease progression [80,81]. In a study by Li et al., **circ0044516** was found to be upregulated and play a critical role in tumor cell survival and metastasis in PCa [80,81].

2.10 Gynecological cancers

Gynecologic cancers comprise ovarian, cervical, and endometrial tumors and have a significant impact on global mortality rates in women [83]. Among these cancers, ovarian cancer is particularly aggressive, with epithelial ovarian cancer (EOC) accounting for most cases [83]. Cervical squamous cell carcinoma (CSCC) is the primary contributor to cervical cancer (CACX) cases, often caused by persistent infection with high-risk human papillomavirus (HPV) [84,85]. Endometrial cancer is encountered mostly in postmenopausal women, and it can be estrogen-dependent or estrogen-independent [83]. Unfortunately, the lack of reliable biomarkers contributes to the late detection of these cancers, resulting in advanced and recurrent cases with poor prognoses.

Researchers have observed the increased presence of the lncRNA **MALAT1** in serum exosomes in advanced and metastatic forms of EOC, with prognostic implications [86,87]. Similarly, elevated levels of another exosomal lncRNA, **DLX6-AS1,** have been identified in the serum of CACX patients [86,87]. Furthermore, analyses of cervicovaginal lavage fluid revealed heightened expression of exosomal lncRNAs **HOTAIR** and **MALAT1**, along with reduced levels of exosomal lncRNA **MEG3**, in individuals who tested positive for high-risk HPV compared with those who tested negative [86–89].

The levels of exosomal **hsa-miR-940** were found to be increased in the ascites fluid of patients with EOC [86–88]. On the other hand, **hsa-miR-146b-5p** showed higher expression in EOC patients [86–88]. Conversely, the expression of **hsa-miR-101** in serum exosomes was significantly reduced in EOC patients. Zhang and colleagues discovered that in the blood plasma of patients compared with healthy individuals, miRNA, **hsa-miR-106a-5p**, was increased, while **hsa-miR-122-5p** and **hsa-miR-185-5p** were decreased [86–88]. In patients with advanced EOC and chemotherapy resistance, another miRNA called **hsa-miR-214-3p** was found to be highly expressed in serum exosomes [86–88]. Moreover, in serum exosomes from ovarian cancer patients, the levels of **hsa-miR-1307** and **hsa-miR-375** were found to be higher when compared with individuals with noncancerous ovarian tumors as well as healthy individuals [86–88]. In a separate study, researchers identified elevated levels of **hsa-miR-93, hsa-miR-145,** and **hsa-miR-200c**, whereas lower expression of **hsa-miR-34a** and **hsa-miR-484** was found in exosomes derived from the serum of ovarian cancer patients compared with healthy individuals [87,88]. In the plasma of EOC patients, exosomal **hsa-miR-21, hsa-miR-100, hsa-miR-200b**, and **hsa-miR-320** showed high expression, while **hsa-miR-16, hsa-miR-93, hsa-miR-126,** and **hsa-miR-223** exhibited weak expression [87,88]. For cervical cancer (CACX), Ma et al. found increased levels of **hsa-miR-146a-5p**, **hsa-miR-151a-3p**, and **hsa-miR-2110** in plasma exosomes compared with those from healthy individuals [86–89]. Additionally, **hsa-miR-125a-5p** in plasma exosomes showed higher expression in CACX patients than in healthy individuals. Furthermore, Zheng and colleagues demonstrated significant differences in the levels of **hsa-let-7d-3p** and **hsa-miR-30d-5p** between cervical cancer tissues and adjacent normal tissues [86–88]. Finally, in patients with endometrial cancer, elevated expression of **hsa-miR-200c** was discovered in exosomes isolated from the urine [86–88]. Overall, these studies highlight the altered expression of specific miRNAs in exosomes from different biological samples of patients with gynecological cancers, providing potential biomarkers for the detection and monitoring of these diseases.

Next, studies on ovarian cancer patients showed elevated expression of **circ0001068** in serum exosomes [59,86,90]. Xu et al., through a comparison of exosomal circRNA levels in the serum of endometrial cancer patients and healthy individuals, discovered a significant increase in the expression of **circ0109046** and **circ0002577** in the serum exosomes of endometrial cancer patients [59,90]. This finding suggests the potential of these exosomal RNAs as blood-based diagnostic markers for gynecological cancer.

2.11 Colorectal cancer

Colorectal cancer (CRC) is a tumor that arises from the colon or rectum, and it is a highly common cancer globally in both men and women [91]. Conventional treatment involves a comprehensive approach encompassing surgery, chemotherapy, and radiotherapy. However early identification of the disease is critical for optimal treatment and preventing cancer recurrence [91]. Therefore, there is a pressing need to develop exosomal RNA biomarkers that are noninvasive, highly accurate, and capable of detecting CRC at its initial stages.

Low expression of exosomal-derived lncRNA **HOTTIP** is significantly correlated with unfavorable survival in CRC patients [92,93]. Additionally, the expression of the serum exosomal lncRNA **ADAMTS9-AS1** is significantly reduced in CRC patients compared with healthy individuals, indicating its potential as a novel biomarker for diagnosing CRC [92,93].

Exosomal **hsa-miR-671-5p** is highly expressed in individuals with CRC and is linked to metastasis [92,94]. Lower levels of **hsa-miR-671-5p** are associated with the stage of the tumor and the occurrence of metastasis in CRC patients [92,94]. On the other hand, **hsa-miR-654-5p** is abundantly expressed and is correlated with the development of colon cancer, metastasis, and a reduced survival rate in individuals with CRC [94]. The expression of the tumor-suppressive **hsa-miR-193a** significantly increases in the exosomes derived from the plasma of CRC patients with liver metastasis [92,94]. Conversely, **hsa-miR-193b** is depleted in the serum of CRC patients [94]. The reduced expression of **hsa-miR-601** and **hsa-miR-760** is observed in the serum of CRC patients, and these miRNAs have the potential to predict the advancement of CRC [94]. Furthermore, the tumor-suppressive **hsa-miR-150** and **hsa-miR-638** are downregulated in exosomes found in the blood serum of CRC patients. Higher levels of both **hsa-let-7g** and **hsa-miR-181b** strongly correlate with the response of CRC patients to conventional chemotherapy [95]. Additionally, **hsa-miR-128-3p** effectively distinguishes between CRC patients who are resistant to chemotherapy and those who are sensitive [95]. Finally, Jin et al. have confirmed the elevated expression levels of **hsa-miR-21-5p, hsa-miR-96-5p, hsa-miR-1246,** and **hsa-miR-1229-5p** in exosomes present in the serum of chemoresistant CRC patients, highlighting their potential as valuable diagnostic markers [95].

Exosomal circRNAs have been investigated for their diagnostic effectiveness in CRC [92,96]. For instance, the presence of **circLPAR1** in plasma exosomes is noticeably reduced in individuals with CRC, but its levels increase after the surgical removal of the tumor [96]. On the other hand, **circ-PNN** and **circHIPK3** in exosomes derived from CRC serum samples show increased expression, suggesting their potential as reliable diagnostic indicators [96]. Additionally, elevated levels of **circFMN2** and **circ133** in exosomes obtained from serum samples also demonstrate promise as biomarkers for CRC [96].

2.12 Hematological cancers

Hematologic malignancies refer to cancerous conditions that originate in the tissues responsible for producing blood, such as the bone marrow, or in the cells of the immune system [97]. Leukemia initiates in the aberrant bone marrow or blood cells. Lymphoma is a cancer that arises in the lymphatic system of the body [97]. Multiple myeloma (MM) is a cancer that originates in the bone marrow and targets the blood plasma cells [97]. Such are the complexities and clinical challenges posed by these captivating hematologic malignancies, necessitating rigorous scientific exploration and innovative therapeutic interventions.

Higher levels of specific miRNAs (**has-miR-99a-5p**, **hsa-miR-125b-5p**, and **hsa-miR-215**) were detected in exosomes derived from the serum of chemotherapy-resistant patients with diffuse large B cell lymphoma (DLBCL), which is the most common type of lymphoma [98,99]. They found that elevated levels of exosomal hsa-miR-125b were associated with higher rates of disease recurrence and complications in acute myeloid leukemia (AML) subjects [98,99]. Additionally, CML (chronic myeloid leukemia) patients with minimal residual disease were found to have significantly lower levels of exosomal **hsa-miR-215**. The expression levels of **hsa-miR-let7b** and **hsa-miR-18a** were found to be significantly lower in patients with poor outcomes [98,99]. It was demonstrated by Zhang et al. that exosomal miRNAs, including **hsa-miR-15a-5p, hsa-miR-16-5p, hsa-miR-17-5p,** and **hsa-miR-20a-5p**, were notably downregulated in multiple myeloma (MM) patients resistant to chemotherapy, indicating a potential association between these miRNAs and resistance to chemotherapy [98,99]. Furthermore, the survival of MM patients was significantly linked to the expression levels of two specific miRNAs, **hsa-miR-let-7b** and **hsa-miR-18a** [98–101].

In an innovative study, Wu et al. unveiled the presence of a novel circular RNA, **mc-COX2**, derived from the COX2 gene on the mitochondrial genome [100–102]. Notably, this circRNA exhibited high expression levels in the plasma exosomes of patients with chronic lymphocytic leukemia (CLL), suggesting its potential relevance to prognosis [100–102]. Remarkably, MM patients who developed resistance to the chemotherapeutic drug displayed higher levels of exosomal **circMYC** expression following the onset of resistance and unfavorable prognoses [102]. Aside from the primary manifestations of MM, which primarily affect bone marrow plasma cells, myocardial damage stands as an often intractable complication. Sun et al. reported differentially expressed circRNAs, **circG042080** to be closely linked to MM-related myocardial damage [101,102].

3. Conclusion

Over the past decade, there have been enormous research endeavors exploring the possibility of using exosomes as biomarkers for the diagnosis and prognosis of cancer Among the various RNA species present in exosomes, miRNAs have been a major focus of investigation (Fig. 4.2). One of the key reasons for their popularity in exosome research is the unique characteristics that make them promising biomarker candidates [10]. Primarily, exosomal miRNAs are relatively stable in body fluids as compared with other forms of RNA, which can be more fragile and quickly degrade. The miRNAs are more resilient, when protected within exosomes in the extracellular environment [95]. Their stability allows them to circulate in the bloodstream or other bodily fluids for an extended period, making them potentially detectable even in samples that may not be immediately processed. This combination of stability and protection enhances the potential of exosomal miRNAs as reliable biomarkers.

As a result of our comprehensive pan-cancer analysis, two miRNAs, namely hsa-miR-21 and hsa-miR-1246, stood out as having significant potential to serve as biomarkers across various major malignant cancers (Fig. 4.3). The miRNA, has-miR-21, is known to be frequently upregulated in various cancers and has been associated with promoting tumor growth, invasion, and metastasis. Conversely, miRNA, hsa-miR-1246, is also known to be dysregulated in cancer and has been linked to tumor

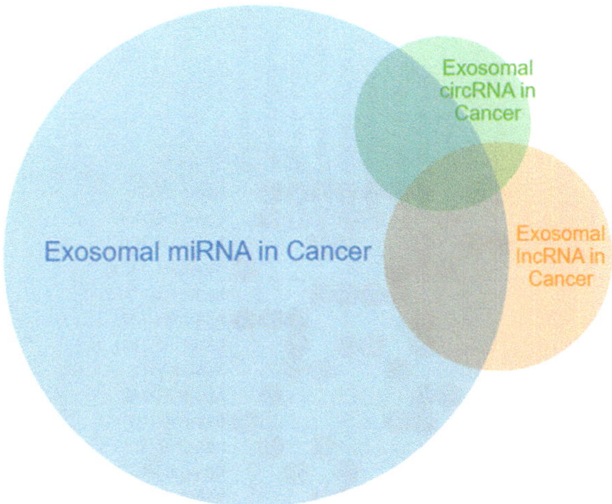

Figure 4.2 The representative Venn diagram showed that among the several RNA species found in exosomes, research on miRNAs has been the most extensive among the other species followed by the lncRNA. The data is derived from NCBI PUBMED using the tool pubvenn.appspot.com.

progression and prognosis. The consistent presence of these miRNAs across different cancer types makes it an attractive candidate for a broad-spectrum cancer biomarker. Its prevalence across multiple cancer types further supports its potential as a valuable biomarker with broader applicability. However, thanks to the emergence of new long-read sequencing technologies, we can now confidently understand the importance of other types of RNA as well [14].

Nonetheless, there are certain drawbacks associated with using exosomes as tumor biomarkers [4,103]. A notable limitation is a rapid and meticulous method to isolate exosomes from a sample before utilizing them as biomarkers to maintain clinical standards [4,103]. Current isolation techniques have their own set of limitations, including being cumbersome, time-consuming, prone to contamination, and expensive. These limitations can lead to the presence of impurities that negatively impact the accuracy of exosome-based diagnosis [4,103]. Hence, it is crucial to develop highly precise separation methods to facilitate the integration of exosome detection into clinical applications [3,14]. Additionally, due to the abundance and heterogeneity of exosome biomarkers, there is a risk of both false positives and false negatives [4]. Consequently, when developing a clinical exosome biomarker [13], it is essential to carefully evaluate sensitivity and specificity using large cohorts of patients.

Overall, exosomes hold promise as a research field for enhancing the diagnosis, treatment, and management of various common types of cancer. However, further research is necessary to gain a comprehensive understanding of the intricate roles played by these RNA types in cancer and to apply this knowledge in clinical settings effectively.

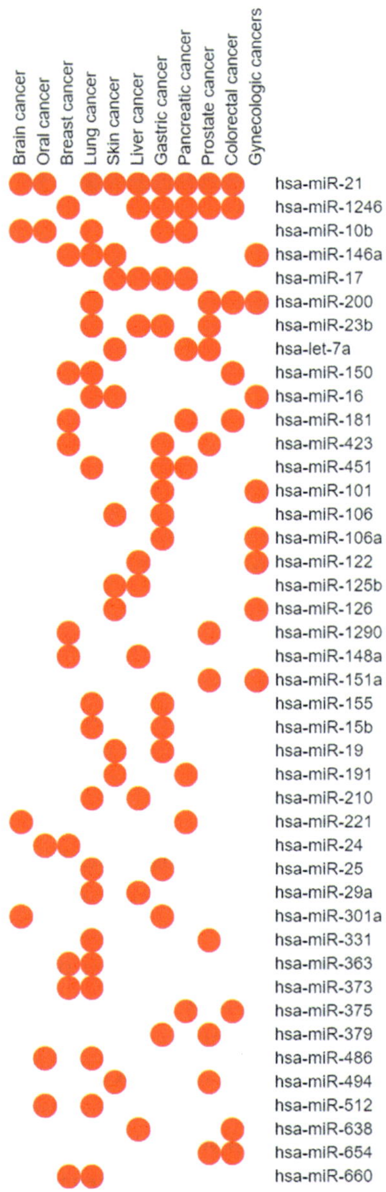

Figure 4.3 The representative heatmap gives an overview of the major miRNAs repeatedly found in exosomes originating from body fluids across at least two major cancer types. Furthermore, hsa-miR-21 and hsa-miR-1246 are one of the most frequently encountered exosomal miRNAs, supporting their potential development for cancer biomarkers.

References

[1] Park SH, Lee EK, Yim J, Lee MH, Lee E, Lee YS, et al. Exosomes: nomenclature, isolation, and biological roles in liver diseases. Biomolecules and Therapeutics (Seoul) 2023;31:253–63.
[2] Doyle LM, Wang MZ. Overview of extracellular vesicles, their origin, composition, purpose, and methods for exosome isolation and analysis. Cells 2019:8.
[3] Shao H, Im H, Castro CM, Breakefield X, Weissleder R, Lee H. New technologies for analysis of extracellular vesicles. Chemistry Review 2018;118:1917–50.
[4] Bamankar S, Londhe VY. The rise of extracellular vesicles as new age biomarkers in cancer diagnosis: promises and pitfalls. Technology in Cancer Research and Treatment 2023;22. 15330338221149266.
[5] Yang G, Waheed S, Wang C, Shekh M, Li Z, Wu J. Exosomes and their bioengineering strategies in the cutaneous wound healing and related complications: current knowledge and future perspectives. International Journal of Biological Sciences 2023;19:1430–54.
[6] Lu Y, Huang W, Li M, Zheng A. Exosome-based carrier for RNA delivery: progress and challenges. Pharmaceutics 2023;15.
[7] Mattick JS, Amaral PP, Carninci P, Carpenter S, Chang HY, Chen LL, et al. Long non-coding RNAs: definitions, functions, challenges and recommendations. Nature Reviews Molecular Cell Biology 2023;24:430–47.
[8] Vishnoi A, Rani S. miRNA biogenesis and regulation of diseases: an updated overview. Methods in Molecular Biology 2023;2595:1–12.
[9] Jagtap U, Anderson ES, Slack FJ. The emerging value of circular noncoding RNA research in cancer diagnosis and treatment. Cancer Research 2023;83:809–13.
[10] Makler A, Asghar W. Exosomal biomarkers for cancer diagnosis and patient monitoring. Expert Review of Molecular Diagnostics 2020;20:387–400.
[11] Califf RM. Biomarker definitions and their applications. Experimental Biology and Medicine 2018;243:213–21.
[12] Michela B. Liquid biopsy: a family of possible diagnostic tools. Diagnostics 2021;11.
[13] Marrugo-Ramirez J, Mir M, Samitier J. Blood-based cancer biomarkers in liquid biopsy: a promising non-invasive alternative to tissue biopsy. International Journal of Molecular Sciences 2018;19.
[14] Elkommos-Zakhary M, Rajesh N, Beljanski V. Exosome RNA sequencing as a tool in the search for cancer biomarkers. Noncoding RNA 2022;8.
[15] Hanif F, Muzaffar K, Perveen K, Malhi SM, Simjee Sh U. Glioblastoma multiforme: a review of its epidemiology and pathogenesis through clinical presentation and treatment. Asian Pacific Journal of Cancer Prevention 2017;18:3–9.
[16] Movahedpour A, Khatami SH, Khorsand M, Salehi M, Savardashtaki A, Mirmajidi SH, et al. Exosomal noncoding RNAs: key players in glioblastoma drug resistance. Molecular and Cellular Biochemistry 2021;476:4081–92.
[17] Russo MN, Whaley LA, Norton ES, Zarco N, Guerrero-Cazares H. Extracellular vesicles in the glioblastoma microenvironment: a diagnostic and therapeutic perspective. Molecular Aspects of Medicine 2023;91:101167.
[18] Goenka A, Tiek DM, Song X, Iglesia RP, Lu M, Hu B, et al. The role of non-coding RNAs in glioma. Biomedicines 2022;10.
[19] Johnson DE, Burtness B, Leemans CR, Lui VWY, Bauman JE, Grandis JR. Head and neck squamous cell carcinoma. Nature Reviews Disease Primers 2020;6:92.
[20] Li T, Li J, Wang H, Zhao J, Yan M, He H, et al. Exosomes: potential biomarkers and functions in head and neck squamous cell carcinoma. Frontiers in Molecular Biosciences 2022;9:881794.
[21] Jangholi A, Muller Bark J, Kenny L, Vasani S, Rao S, Dolcetti R, et al. Exosomes at the crossroad between therapeutic targets and therapy resistance in head and neck squamous cell carcinoma. Biochimica et Biophysica Acta (BBA) - Reviews on Cancer 2022;1877:188784.
[22] Chen B, Leung LL, Qu X, Chan JY. Extracellular vesicles as biomarkers in head and neck squamous cell carcinoma: from diagnosis to disease-free survival. Cancers 2023;15.
[23] Wang Y, Wang S, Ren Y, Zhou X. The role of lncRNA crosstalk in leading cancer metastasis of head and neck squamous cell carcinoma. Frontiers in Oncology 2020;10:561833.

[24] Kang JW, Eun YG, Lee YC. Diagnostic value of salivary miRNA in head and neck squamous cell cancer: systematic review and meta-analysis. International Journal of Molecular Sciences 2021;22.
[25] Nath M, Roy D, Choudhury Y. Circular RNAs are potential prognostic markers of head and neck squamous cell carcinoma: findings of a meta-analysis study. Frontiers in Oncology 2022;12:782439.
[26] Dakubo GD. Cancer biomarkers in body fluids: biomarkers in circulation. Cham: Springer International Publishing: Imprint: Springer; 2017. p. 1. online resource (XIX, 509 pages 520 illustrations in color.
[27] Leong AS, Zhuang Z. The changing role of pathology in breast cancer diagnosis and treatment. Pathobiology 2011;78:99–114.
[28] Famta P, Shah S, Khatri DK, Guru SK, Singh SB, Srivastava S. Enigmatic role of exosomes in breast cancer progression and therapy. Life Sciences 2022;289:120210.
[29] Tang Q, Xiao X, Li R, He H, Li S, Ma C. Recent advances in detection for breast-cancer-derived exosomes. Molecules 2022;27.
[30] Nakase I, Takatani-Nakase T. Exosomes: breast cancer-derived extracellular vesicles; recent key findings and technologies in disease progression, diagnostics, and cancer targeting. Drug Metabolism and Pharmacokinetics 2022;42:100435.
[31] Triantafyllou A, Gazouli M, Theodoropoulos C, Zografos E, Zografos GC, Michalopoulos NV. Exosomes in breast cancer management: where do we stand? A literature review. Biology of the Cell 2022;114:109–22.
[32] Shefer A, Yalovaya A, Tamkovich S. Exosomes in breast cancer: involvement in tumor dissemination and prospects for liquid biopsy. International Journal of Molecular Sciences 2022;23.
[33] Ashekyan O, Abdallah S, Shoukari AA, Chamandi G, Choubassy H, Itani ARS, et al. Spotlight on exosomal non-coding RNAs in breast cancer: an in silico analysis to identify potential lncRNA/circRNA-miRNA-Target Axis. International Journal of Molecular Sciences 2022;23.
[34] Faraji G, Moeini P, Ranjbar MH. Exosomal microRNAs in breast cancer and their potential in diagnosis, prognosis and treatment prediction. Pathology, Research & Practice 2022;238:154081.
[35] Baldasici O, Pileczki V, Cruceriu D, Gavrilas LI, Tudoran O, Balacescu L, et al. Breast cancer-delivered exosomal miRNA as liquid biopsy biomarkers for metastasis prediction: a focus on translational research with clinical applicability. International Journal of Molecular Sciences 2022;23.
[36] Hussen BM, Mohamadtahr S, Abdullah SR, Hidayat HJ, Rasul MF, Hama Faraj GS, et al. Exosomal circular RNAs: new player in breast cancer progression and therapeutic targets. Frontiers in Genetics 2023;14:1126944.
[37] Zokaei E, Darbeheshti F, Rezaei N. Prospect of exosomal circular RNAs in breast cancer: presents and future. Molecular Biology Reports 2022;49:6997–7011.
[38] Sinjab A, Rahal Z, Kadara H. Cell-by-Cell: unlocking lung cancer pathogenesis. Cancers 2022;14.
[39] Chen R, Xu X, Qian Z, Zhang C, Niu Y, Wang Z, et al. The biological functions and clinical applications of exosomes in lung cancer. Cellular and Molecular Life Sciences 2019;76:4613–33.
[40] Qi R, Zhao Y, Guo Q, Mi X, Cheng M, Hou W, et al. Exosomes in the lung cancer microenvironment: biological functions and potential use as clinical biomarkers. Cancer Cell International 2021;21:333.
[41] Khan FH, Reza MJ, Shao YF, Perwez A, Zahra H, Dowlati A, et al. Role of exosomes in lung cancer: a comprehensive insight from immunomodulation to theragnostic applications. Biochimica et Biophysica Acta (BBA) - Reviews on Cancer 2022;1877:188776.
[42] Xu K, Zhang C, Du T, Gabriel ANA, Wang X, Li X, et al. Progress of exosomes in the diagnosis and treatment of lung cancer. Biomedicine & Pharmacotherapy 2021;134:111111.
[43] He L, Lin M, Shen J, Qi H. Emerging role of exosomal long non-coding RNAs in lung cancer. Molecular Biology Reports 2022;49:4989–97.
[44] Entezari M, Ghanbarirad M, Taheriazam A, Sadrkhanloo M, Zabolian A, Goharrizi M, et al. Long non-coding RNAs and exosomal lncRNAs: potential functions in lung cancer progression, drug resistance and tumor microenvironment remodeling. Biomedicine & Pharmacotherapy 2022;150:112963.
[45] Durendez-Saez E, Torres-Martinez S, Calabuig-Farinas S, Meri-Abad M, Ferrero-Gimeno M, Camps C. Exosomal microRNAs in non-small cell lung cancer. Translational Cancer Research 2021;10:3128–39.

[46] Smolarz M, Widlak P. Serum exosomes and their miRNA load-A potential biomarker of lung cancer. Cancers 2021;13.
[47] Hussen BM, Abdullah SR, Hama Faraj GS, Rasul MF, Salihi A, Ghafouri-Fard S, et al. Exosomal circular RNA: a signature for lung cancer progression. Cancer Cell International 2022;22:378.
[48] Chen HH, Zhang TN, Wu QJ, Huang XM, Zhao YH. Circular RNAs in lung cancer: recent advances and future perspectives. Frontiers in Oncology 2021;11:664290.
[49] Strashilov S, Yordanov A. Aetiology and pathogenesis of cutaneous melanoma: current concepts and advances. International Journal of Molecular Sciences 2021;22.
[50] Tan Y, Tang F, Li J, Yu H, Wu M, Wu Y, et al. Tumor-derived exosomes: the emerging orchestrators in melanoma. Biomedicine & Pharmacotherapy 2022;149:112832.
[51] Karami Fath M, Azargoonjahromi A, Jafari N, Mehdi M, Alavi F, Daraei M, et al. Exosome application in tumorigenesis: diagnosis and treatment of melanoma. Medical Oncology 2022;39:19.
[52] Ye Q, Li Z, Li Y, Li Y, Zhang Y, Gui R, et al. Exosome-derived microRNA: implications in melanoma progression. Diagnosis and Treatment, Cancers (Basel) 2022;15.
[53] Dong L, Tian X, Zhao Y, Tu H, Wong A, Yang Y. The roles of MiRNAs (MicroRNAs) in melanoma immunotherapy. International Journal of Molecular Sciences 2022;23.
[54] Ho DW, Lo RC, Chan LK, Ng IO. Molecular pathogenesis of hepatocellular carcinoma. Liver Cancer 2016;5:290–302.
[55] Wang W, Hao LP, Song H, Chu XY, Wang R. The potential roles of exosomal non-coding RNAs in hepatocellular carcinoma. Frontiers in Oncology 2022;12:790916.
[56] Zhang Q, Li H, Liu Y, Li J, Wu C, Tang H. Exosomal non-coding RNAs: new insights into the biology of hepatocellular carcinoma. Current Oncology 2022;29:5383–406.
[57] Li S, Chen L. Exosomes in pathogenesis, diagnosis, and treatment of hepatocellular carcinoma. Frontiers in Oncology 2022;12:793432.
[58] Chen X, Chi H, Zhao X, Pan R, Wei Y, Han Y. Role of exosomes in immune microenvironment of hepatocellular carcinoma. JAMA Oncology 2022;2022:2521025.
[59] Ye D, Gong M, Deng Y, Fang S, Cao Y, Xiang Y, et al. Roles and clinical application of exosomal circRNAs in the diagnosis and treatment of malignant tumors. Journal of Translational Medicine 2022;20:161.
[60] Fu M, Gu J, Jiang P, Qian H, Xu W, Zhang X. Exosomes in gastric cancer: roles, mechanisms, and applications. Molecular Cancer 2019;18:41.
[61] Polakovicova I, Jerez S, Wichmann IA, Sandoval-Borquez A, Carrasco-Veliz N, Corvalan AH. Role of microRNAs and exosomes in *Helicobacter pylori* and epstein-barr virus associated gastric cancers. Frontiers in Microbiology 2018;9:636.
[62] Tang XH, Guo T, Gao XY, Wu XL, Xing XF, Ji JF, et al. Exosome-derived noncoding RNAs in gastric cancer: functions and clinical applications. Molecular Cancer 2021;20:99.
[63] Hu F, Liu J, Liu H, Li F, Wan M, Zhang M, et al. Role of exosomal non-coding RNAs in gastric cancer: biological functions and potential clinical applications. Frontiers in Oncology 2021;11:700168.
[64] Yun BD, Choi YJ, Son SW, Cipolla GA, Berti FCB, Malheiros D, et al. Oncogenic role of exosomal circular and long noncoding RNAs in gastrointestinal cancers. International Journal of Molecular Sciences 2022;23.
[65] Zhao G, Zhou A, Li X, Zhu S, Wang Y, Zhang S, et al. The significance of exosomal RNAs in the development, diagnosis, and treatment of gastric cancer. Genes 2021;12.
[66] Salehi M, Vafadar A, Khatami SH, Taheri-Anganeh M, Vakili O, Savardashtaki A, et al. Gastrointestinal cancer drug resistance: the role of exosomal miRNAs. Molecular Biology Reports 2022;49:2421–32.
[67] Wang X, Zhang J, Cao G, Hua J, Shan G, Lin W. Emerging roles of circular RNAs in gastric cancer metastasis and drug resistance. Journal of Experimental & Clinical Cancer Research 2022;41:218.
[68] Li R, Jiang J, Shi H, Qian H, Zhang X, Xu W. CircRNA: a rising star in gastric cancer. Cellular and Molecular Life Sciences 2020;77:1661–80.
[69] Grant TJ, Hua K, Singh A. Molecular pathogenesis of pancreatic cancer. Progress in Molecular Biology and Translational Science 2016;144:241–75.

[70] Farahmand Y, Tehrany PM, Nazari A, Nava ZH, Alsaffar MF, Yazdani O, et al. A comprehensive survey into the role of exosomes in pancreatic cancer; from the origin of cancer to the progress and possibility of diagnosis and treatment. Pathology, Research & Practice 2023;245:154465.
[71] Robless EE, Howard JA, Casari I, Falasca M. Exosomal long non-coding RNAs in the diagnosis and oncogenesis of pancreatic cancer. Cancer Letters 2021;501:55−65.
[72] Ashrafizadeh M, Rabiee N, Kumar AP, Sethi G, Zarrabi A, Wang Y. Long noncoding RNAs (lncRNAs) in pancreatic cancer progression. Drug Discovery Today 2022;27:2181−98.
[73] Uddin MH, Al-Hallak MN, Philip PA, Mohammad RM, Viola N, Wagner KU, et al. Exosomal microRNA in pancreatic cancer diagnosis, prognosis, and treatment: from bench to bedside. Cancers 2021;13.
[74] Chen K, Wang Q, Kornmann M, Tian X, Yang Y. The role of exosomes in pancreatic cancer from bench to clinical application: an updated review. Frontiers in Oncology 2021;11:644358.
[75] Rong Z, Xu J, Shi S, Tan Z, Meng Q, Hua J, et al. Circular RNA in pancreatic cancer: a novel avenue for the roles of diagnosis and treatment. Theranostics 2021;11:2755−69.
[76] Sekhoacha M, Riet K, Motloung P, Gumenku L, Adegoke A, Mashele S. Prostate cancer review: genetics, diagnosis, treatment options, and alternative approaches. Molecules 2022;27.
[77] Grimaldi AM, Salvatore M, Cavaliere C. Diagnostic and prognostic significance of extracellular vesicles in prostate cancer drug resistance: a systematic review of the literature. Prostate Cancer and Prostatic Diseases 2023;26:228−39.
[78] Ji J, Chen R, Zhao L, Xu Y, Cao Z, Xu H, et al. Circulating exosomal mRNA profiling identifies novel signatures for the detection of prostate cancer. Molecular Cancer 2021;20:58.
[79] Zhou H, Wan H, Feng Y, Zhu L, Mi Y. The diagnostic role and mechanistic functions of exosomal lncRNAs in prostate cancer. Clinical and Translational Oncology 2023;25:592−600.
[80] Liu SY, Liao Y, Hosseinifard H, Imani S, Wen QL. Diagnostic role of extracellular vesicles in cancer: a comprehensive systematic review and meta-analysis. Frontiers in Cell and Developmental Biology 2021;9:705791.
[81] Mugoni V, Ciani Y, Nardella C, Demichelis F. Circulating RNAs in prostate cancer patients. Cancer Letters 2022;524:57−69.
[82] Jin C, Wu P, Li L, Xu W, Qian H. Exosomes: emerging therapy delivery tools and biomarkers for kidney diseases. Stem Cells International 2021;2021:7844455.
[83] Wang Q, Peng H, Qi X, Wu M, Zhao X. Targeted therapies in gynecological cancers: a comprehensive review of clinical evidence. Signal Transduction and Targeted Therapy 2020;5:137.
[84] Roychowdhury A, Samadder S, Islam MS, Chaudhury K, Roy A, Banerjee D, et al. Identification of changes in the human papilloma virus 16 (HPV16) genome during early dissemination of cervical cancer cells may complement histological diagnosis of lymph node metastasis. Pathology and Oncology Research 2017;23:845−52.
[85] Roychowdhury A, Basu M, Pal D, Dutta P, Samadder S, Mondal R, et al. PTPRJ is downregulated in cervical squamous cell carcinoma. Journal of Genetics 2022;101.
[86] Ye M, Wang J, Pan S, Zheng L, Wang ZW, Zhu X. Erratum: nucleic acids and proteins carried by exosomes of different origins as potential biomarkers for gynecologic cancers. Molecular Therapy Oncolytics 2022;25:307.
[87] Wang M, Fu L, Xu Y, Ma S, Zhang X, Zheng L. A comprehensive overview of exosome lncRNAs: emerging biomarkers and potential therapeutics in gynecological cancers. Frontiers in Oncology 2023;13:1138142.
[88] Roy M, Yang YP, Bosquet O, Deo SK, Daunert S. Understanding the role and clinical applications of exosomes in gynecologic malignancies: a review of the current literature. Cancers 2021;14.
[89] Roychowdhury A, Samadder S, Das P, Mazumder DI, Chatterjee A, Addya S, et al. Deregulation of H19 is associated with cervical carcinoma. Genomics 2020;112:961−70.
[90] Najafi S. Circular RNAs as emerging players in cervical cancer tumorigenesis; A review to roles and biomarker potentials. International Journal of Biological Macromolecules 2022;206:939−53.
[91] Kasi A, Handa S, Bhatti S, Umar S, Bansal A, Sun W. Molecular pathogenesis and classification of colorectal carcinoma. Current Colorectal Cancer Reports 2020;16:97−106.

[92] Chen X, Jia M, Ji J, Zhao Z, Zhao Y. Exosome-derived non-coding RNAs in the tumor microenvironment of colorectal cancer: possible functions, mechanisms and clinical applications. Frontiers in Oncology 2022;12:887532.
[93] Kang F, Jiang F, Ouyang L, Wu S, Fu C, Liu Y, et al. Potential biological roles of exosomal long non-coding RNAs in gastrointestinal cancer. Frontiers in Cell and Developmental Biology 2022;10:886191.
[94] Bakhsh T, Alhazmi S, Alburae NA, Farsi A, Alzahrani F, Choudhry H, et al. Exosomal miRNAs as a promising source of biomarkers in colorectal cancer progression. International Journal of Molecular Sciences 2022;23.
[95] Maleki M, Golchin A, Javadi S, Khelghati N, Morovat P, Asemi Z, et al. Role of exosomal miRNA in chemotherapy resistance of Colorectal cancer: a systematic review. Chemical Biology & Drug Design 2023;101:1096−112.
[96] Wang D, Li R, Jiang J, Qian H, Xu W. Exosomal circRNAs: novel biomarkers and therapeutic targets for gastrointestinal tumors. Biomedicine & Pharmacotherapy 2023;157:114053.
[97] Button E, Chan RJ, Chambers S, Butler J, Yates P. A systematic review of prognostic factors at the end of life for people with a hematological malignancy. BMC Cancer 2017;17:213.
[98] Cariello M, Squilla A, Piacente M, Venutolo G, Fasano A. Drug resistance: the role of exosomal miRNA in the microenvironment of hematopoietic tumors. Molecules 2022;28.
[99] Jahromi FNA, Dowran R, Jafari R. Recent advances in the roles of exosomal microRNAs (exomiRs) in hematologic neoplasms: pathogenesis, diagnosis, and treatment. Cell Communication and Signaling 2023;21:88.
[100] Wang H, You Y, Zhu X. The role of exosomes in the progression and therapeutic resistance of hematological malignancies. Frontiers in Oncology 2022;12:887518.
[101] Pandian SRK, Vijayakumar KK, Kunjiappan S, Babkiewicz E, Maszczyk P. Emerging role of exosomes in hematological malignancies. Clinical and Experimental Medicine 2022;23(4):1123−36.
[102] Zhou Q, Xie D, Wang R, Liu L, Yu Y, Tang X, et al. The emerging landscape of exosomal CircRNAs in solid cancers and hematological malignancies. Biomarker Research 2022;10:28.
[103] Logozzi M, Mizzoni D, Di Raimo R, Fais S. Exosomes: a source for new and old biomarkers in cancer. Cancers 2020;12.

CHAPTER FIVE

Exosomal RNA and autophagy

Sougata Ghosh Chowdhury, Parimal Karmakar
Department of Life Science and Biotechnology, Jadavpur University, Kolkata, West Bengal, India

1. Introduction

Autophagy is a cellular mechanism that occurs as a response of the body to various stress conditions. It can respond to various conditions such as nutrient and energy deprivation, hypoxia, and endoplasmic reticulum stress [1]. Under these circumstances, autophagy is able to degrade intracellular macromolecules and allow them to recycle their building blocks. The quality control of organelle tissue is maintained through the basal activation of the autophagy process. It has also been shown that selective autophagy is responsible for degrading different cellular organelles and proteins, which highlights its crucial role in shaping the landscape of the cellular system [2].

Exosomes are single-membrane structures that are generally 30−150 nm in diameter and are formed from endosomes [3,4]. The process of endocytosis occurs when an extracellular substance is acquired by forming a vesicle through the invagination of a plasma membrane. When internalized, the membrane-bound exosomes fuse with the endosome to form a sorter station that carries cargo that mainly consists of lipids and proteins [5]. The early endosomes are capable of forming multivesicular bodies. They are also referred to as late endosomes or recycling endosomes. The early endosomes are involved in the recycling process, which involves transferring cargo to either the Golgi network or the plasma membrane [6]. Most of the time, the recycled proteins are those involved in the cell polarization process. The late endosomes then fuse with the lysosome and form endolysosomes that can be degraded through hydrolases. Once they reach maturity and start to accumulate small endocytic vesicles, they contain a variety of nutrients, proteins, and nucleic acids. Late endosomes that contain intraluminal vesicles (ILVs) typically have a multivesicular structure. Sometimes, these endosomes are multivesicular bodies (MVBs), and they are not likely to get degraded. Once they are integrated with the plasma membrane, they can then enter the extracellular space. Exosomes are differentiated from other types of ILVs by their unique structures and characteristics. They are also different from other extracellular vesicles, such as microvesicles and ectosomes [3]. The presence of markers on the membranes of exosomes determines their origin and structure. In certain circumstances, cells can remove damaged or unwanted substances from their environment by releasing them to the exosomes [7]. Since exosomes can be swapped between different cells, the secretion of

Exosomal RNA
ISBN 978-0-443-14008-2
https://doi.org/10.1016/B978-0-443-14008-2.00003-6

substances from one exosome to the extracellular matrix may have an effect on the neighboring ones. The interactions between exosome biogenesis and autophagy involve different mechanisms. These two play crucial roles in maintaining the homeostasis of the cellular environment. It has been shown that their cellular responses can be triggered by a cross-talk between exosomes and autophagy [8].

Conceptually, the genetic information is sent from one part of the DNA to another in the form of RNA and proteins, which are then utilized by the cell as a functional part. Although the majority of the genome's structure is transcribed, less than 2% is translated into proteins or mRNAs. The rest, also known as noncoding RNAs, play an important role in regulating the activity of genes [9]. The length of the transcript determines the classification of noncoding RNAs. There are two main types of noncoding RNAs: short and long noncoding RNAs. The former is less than 40 nt long, while the latter is more than 200 nt long. The former includes miRNAs, which are composed of 18–25 nt in length, siRNAs, which are 19–23 nt in length, piRNAs (piwi-interacting RNAs), which are 21–36 nt in length [10]. On the other hand, the latter includes tRNA fragments. Due to the role that noncoding RNAs play in various cellular processes, such as differentiation, proliferation, and apoptosis, they have gained increased attention. A mechanism that involves targeting RNA into exosomes allows some of these to be enriched in the exosomes. In next-generation sequencing, it was revealed that the most common types of exosomes are those containing structural and small ribosomal RNAs [11]. In addition to the usual miRNAs, exosomes can also contain other types of RNAs, such as lncRNAs, circRNAs, and mRNAs. The composition of miRNAs is known to be specific. Autophagy-associated proteins and genes are also dependent on the control of miRNAs. A number of studies have shown that a complex miRNA network can orchestrate the process of autophagy [12]. It involves the effects of these miRNAs on various regulatory proteins. The classification of lncRNAs based on their function, location, and repeat elements is generally used. In addition to maintaining and developing biological processes, lncRNAs also play a variety of roles. The downstream effector molecules of lncRNAs can interact with signal pathways to provide the information to the system. In recent years, it has been shown that these small molecules can regulate the cell's autophagy [13]. A large class of noncoding RNAs known as circRNAs are approximately 100 nt long and have a closed-loop structure that does not have polyadenylation or polarity. Unlike the canonical method of back-splicing pre-mRNA, circRNAs are produced through a noncanonical process. Although evidence has shown that circRNAs play a significant role in the development and maintenance of gene expression, their function and complexity remain unclear [14]. For instance, they can act as protein scaffolds and miRNA sponges, and they interact with RNA-binding proteins (RBPs). Over the years, there has been a substantial increase in evidence linking circRNAs to autophagy [15].

2. Exosome biogenesis and composition

Exosomes are produced by late endosomes of living cells, and the development of early endosomes triggers the exosome's biogenesis. When the cell's plasma membrane starts to bud, the early endosomes are transformed into late endosomes, which are characterized by their ILVs. During this process, various proteins are introduced into the membrane, and these are then incorporated into the ILVs. Most of these are then released into extracellular space following the plasma membrane's fusion. Studies have shown that the ILV formation process requires the endosomal sorting complex for transport (ESCRT). ESCRT 0—III complexes blend to form the ESCRT-dependent pathways. The mechanism of ESCRT-dependent pathway commences with the identification and seclusion of ubiquitinated proteins to the specified domains in the membranes of endosomes of ESCRT 0, which interacts with ESCRT I and II complexes [16]. The complex thus formed in turn interacts with ESCRT III to induce the process of budding. As the bud is cleaved off, ILV is formed. VPS4, a sorting protein, separates the ESCRT complex from the MVB membrane. Moreover, it has been reported that various ESCRT proteins are linked with a specific exosomal protein, ALIX, which interacts with SYNDECAN and initiates membrane budding and excision [17]. Rab31 marks and controls an ESCRT-independent exosome pathway. Rab31 is an active component of the epidermal growth factor receptor (EGFR), which is phosphorylated by this and interacts with the flotillin proteins in the microdomain to trigger ILV, an independent pathway of the ESCRT machinery [18]. Exosomes are mainly concocted from lipids, proteins, RNAs, and DNAs. The lipid contents of exosomes are somewhat different in different cells. Besides lipid molecules such as sphingomyelin, cholesterol, ceramides, phosphatidylcholine, and phosphatidylserine were also found in low levels on the surface of platelet-derived exosomes [19]. The function, packaging, and structure of exosomes depend typically on both lipids and proteins. Exosomes are embodied with various kinds of RNAs such as mRNAs, tRNAs, miRNAs, circRNAs, and lncRNAs. Genomic DNA fragments of varying lengths are intriguingly spotted within exosomes, which are often found to be secreted out through this nanocarrier, the level of which heightens during cellular senescence [20]. Although exosomes mainly contain small amounts of proteins derived from nuclear, Golgi, or endoplasmic reticulum origins, they also have numerous ubiquitous cytosolic components [21]. The ubiquitous proteins that can be found inside exosomes include actin-binding and cytoskeleton proteins, exosome biogenesis and signal transduction proteins, metabolic proteins, and heat shock proteins [21]. The members of the tetraspanin group, which include CD9, CD63, CD82, and CD81, are also known to play a role in exosomes' function and biogenesis [22].

3. Overview of autophagy

The term autophagy refers to a process that occurs in every eukaryotic cell. It involves the breakdown of old and worn cellular components to maintain homeostasis. Autophagy is triggered in stressful environments, such as starvation, and recycles macromolecules so that the cells can obtain energy. Autophagy is also involved in regulating the energy balance of the biological system and plays a vital role in regulating the differentiation and cell survival processes [23]. It starts by forming a double-membrane phagophore, which traps the cytosolic materials. After closing, the vesicles are formed and recycled once they are joined with lysosomes. Autophagy can be categorized into three types: macroautophagy, microautophagy, and chaperone-mediated autophagy. These three differ in their ways of selection and transport [24]. For instance, in macroautophagy, the cell's cargos are transported to the cytoplasm for the development of an autophagosome. In microautophagy, the cargos are inactivated through the invagination of a membrane. The chaperones utilized in the CMA process only work in mammalian cells. They are responsible for identifying the cargo proteins with a specific pentapeptide motif. Although it was previously believed that autophagy was not selective, it has been shown that it is capable of removing various cellular components, such as aggregates, mitochondria, and pathogens [25]. The various steps in the process of autophagy are as follows: initiation or autophagosome formation, nucleation of the autophagosome, autophagosome extension, fusion of autophagosome with the lysosome, and degradation.

a. Initiation: The ULK1/2 complex is triggered following the induction of autophagy. It comprises various proteins, such as ATG13 and RB1CC1, as well as ATG101. The ULK1/2 complex is composed of a serine/threonine kinase, which is responsible for the activation of the Class III PI3K complex. This complex produces phagophores on the surface of the membrane, which are usually found at the endoplasmic reticulum membrane [26]. The LC3-ATG13 conjugation adds to the ULK1 activation during the initiation phase. It then triggers the development of the autophagosome.

b. Elongation: The ATG12 and ATG5 complexes are designed to promote the extension of the phagophore around the cargo [27]. ATG12 is covalently linked to ATG5 through a series of biochemical reactions that are triggered by the ATG7 and ATG10 enzymes. The ATG12 and ATG5 complexes seem to have been formed following the translation of these genes. When autophagy is triggered, ATG12-ATG5 complexes bind to ATG16L's molecules, which then form a multiprotein complex and are recruited to the phagophore [28]. Upon its translation, the protein LC3 is cleaved by the ATG4 protease and produces the LC3-I form, which is anchored to the phagophore's membrane [29]. When autophagy is triggered, this protein is bound to phosphatidylethanolamine at the phagophore. ATG7 and ATG3 complexes initiate this reaction. ATG12 and ATG5 complexes already exist on the surface of the phagophore, and they function as E3-like enzymes. Although the lipidated form of LC3-I, known as LC3-II, is anchored to the phagophore, the induction of autophagy triggers a prominent pattern [30]. The recognition of cargo occurs as the phagophore elongates. This process is facilitated by the presence of adaptor proteins, which act as autophagic receptors. They help guide the extension of the phagophore by anchoring the labeled cargo to the surrounding membrane.

c. Closure: The phagophore continues to elongate around its cargo until its membrane fuses. This process leads to the formation of a double-layered structure, known as autophagosomes. Other lipids and proteins are involved in the closure of this body. After this, the LC3II is detached from the membrane, which creates mature autophagosomes.

d. Fusion: Through the use of microtubules, autophagosomes can travel to and fuse with lysosomes. The molecular machinery utilized in this process has recently been reviewed.

e. Degradation: Lysosomes contain hydrolases that break down the inner membrane and cytoplasm of an autolysosomal body. They also digest the various cytoplasmic cargoes and proteins that are involved in the process. The lysosome membranes then elongate as a new one forms before it is separated from the autolysosome. The fate of the rest of the autolysosome remains unclear.

4. Cross-regulation between exosome biogenesis and autophagy

Exosomes are used by cells to communicate with their surroundings and the environment. They are also capable of carrying information such as proteins, nucleic acids, and lipids. The role of autophagy in the degradation and synthesis of these components is vital. The secretion and transport of these components are closely related to the development of autophagosomes. It has been known that the degradation of extracellular components can be performed through the exosome pathway, while the transport of these components can be done through the process of autophagy [31]. This suggests that the role of this process in the transport and secretion of EVs is important. High levels of autophagy inhibited the release of EVs due to the presence of multivesicular bodies and autophagic vacuoles. In contrast, the release of exosomes can be performed through the lysosome pathway, which involves the fusion of lysosomes with autophagosomes [32]. Autophagy and the development and biogenesis of exosomes are linked by the relationship between the two processes (Fig. 5.1). This relationship can be explained by the interactions between the two processes in different cellular contexts. For instance, if the two processes are working together to counter the effects of cellular stress, then the exosome production and the autophagy can compensate each other. The mechanisms by which the selective incorporation of certain proteins is performed during

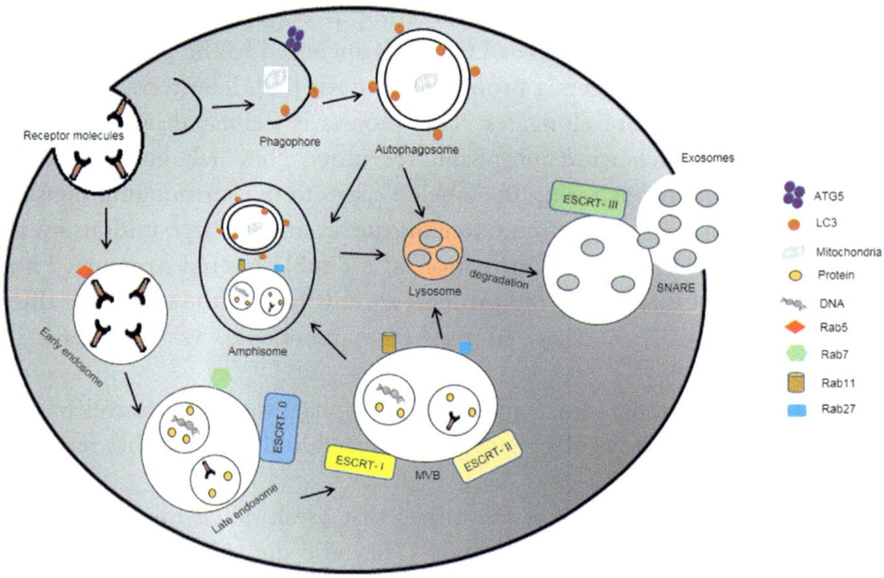

Figure 5.1 The mechanism involves the regulation of exosomes and autophagic pathways. As the endosome develops, it creates MVBs that will eventually fuse with the plasma membrane and secrete exosomes. When the environment triggers stress, the phagophore is formed by extending the protein assembly site. The phagophore grows as it engulfs damaged proteins and organelles. It then forms a bilayered membrane, which can be referred to as an autophagosome. The combination of MVBs and autophagosomes can create an amphisome. Together, the autophagosomes and lysosome form an autolysosome, which is responsible for the degradation of harmful components in the system.

exosome biogenesis and membrane invagination processes have been proposed to be endosomal microautophagy [8]. These endosomal microautophagy process transport cytosolic proteins to the ILVs using either the HSP70 or the ESCRT machinery. During the macroautophagy process, the autophagosomes can fuse with lysosomes or with MVBs. In the case of dysfunctional MVBs, the inhibition of lysosomal function can lead to the degradation of their function and exosome secretion. On the other hand, the presence of EVs can protect the cell from the effects of cellular stress. For instance, if the lysosomal function is inhibited, then the degradation of the exosome secretion can be prevented [32]. For instance, if the presence of EVs is able to stimulate the development and maintenance of exosome secretion, then the degradation of the cells can be prevented. Besides being able to stimulate the development and maintenance of exosome secretion, the presence of EVs can also help to maintain the cytoprotective properties of the cells [33]. Autophagy involves the collection of proteins and molecules that can be affected by various factors. The interaction between the Beclin1 and mTOR signaling pathways is one of the most common factors that can affect this process. This is the point where these two proteins play a role in the regulation of the cell's autophagic response.

The two main pathways that regulate the development and maintenance of autophagy are the AMPK-mTOR and the PI3K-Akt-mTOR pathways [26]. Both of these pathways are known to suppress the cell's autophagic response. AMPK can either inactivate mTOR or stimulate the production of proteins that are related to autophagy. A defect in one of these pathways can prevent the proper digestion of protein aggregates, which can lead to the release of exosomes [8]. The PI3K complex is an important part of the mammalian protein family that is involved in autophagy. It consists of three components: Beclin1, p150, and VPS34. Studies have shown that the presence of the PI3K complex and Beclin1 interacting protein can suppress endocytosis and the autophagic response [34]. However, the effects of exosomes on the activity of the autophagy system are not as apparent. In a study, Nair and colleagues noted that the activity of the three members of the SNARE family, namely syntaxin 7, syntaxin 8, and VAMP7, plays a significant role in the exosome pathway [35,36]. Studies suggest that ATG proteins can regulate the exosome's biogenesis and secretion [37]. For instance, it has been shown that the presence of ATG5 can allow the dissociation of proton pumps from MVBs [38]. This process can then lead to the activation of the fusion of MVBs with the plasma membrane. In another study, it has been shown that the presence of ATG5 can inhibit the acidification of the MVB lumen [39]. ATG5 knockout significantly inhibited the release of exosomes and prevented the enrichment of LC3B [38]. This finding is important because the formation of autophagosomes does not necessitate the release of exosomes. On the other hand, ATG7 knockout did not affect the release of exosomal proteins. It is believed that the absence of LC3B lipidation does not require the release of exosomes. The study revealed that the presence of LC3B plays an important role in the development and maintenance of exosome biogenesis [40]. Murrow and colleagues identified that the inhibition of the ATG12 and ATG3 complex, which is an essential component of the late step of the autophagosome formation process, significantly decreased exosome biogenesis [41]. The interaction between the ATG12-ATG3 complex and ALIX is involved in the membrane fission process and plays a critical role in the development and maintenance of exosome biogenesis. The reduction of ALIX activity, which is also known to reduce basal autophagy flux, suggests that the regulation between exosome biogenesis and the development of autophagosomes is reciprocal [42]. It has been suggested that the interaction between ALIX and exosomal cargo could play a role in the development and maintenance of lysosomal degradation. The reduction of ALIX activity, which is also known to reduce basal autophagy flux, suggests that the regulation between exosome biogenesis and the development of autophagosomes. One of the main markers of the development and maintenance of autophagosomal biogenesis is the presence of LC3B. In the initial step, the activation of the ULK complex leads to the formation of autophagosomes. In the maturation step, the presence of LC3B facilitates the transport and closure of autophagosomes. Although it can be found in the membranes of autophagosomes, it does not need to be formed in the process of phagocytosis initiated by LC3. The presence of LC3B has been found in exosomes in an ATG7-independent manner, which suggests that the LAP-like lipidation pathway could also play a role in the secretion of exosomes [42]. Autophagy modulations are known to regulate the development and release of exosomes. In conditions that are triggered by rapamycin treatment, starvation, or LC3 overexpression, the release of exosomes is inhibited [8]. This suggests that these conditions can lead to the development and release of MVBs in the autophagic pathway. The metabolic state of the cell is known to regulate the development and release of exosomes. It is also believed that the fate of certain molecules that are involved in the degradation of autophagy or the induction of exosomes is regulated by the metabolic sensors [8]. This is the reason why it is important to understand how this regulation affects the autonomy of noncell and autonomous tissues.

5. Regulation of autophagy via exosome vice versa

Autophagy is essential for maintaining cellular homeostasis, but excessive levels lead to cell death. This indicates that it has detrimental implications in pathological settings. There is evidence that the modulation of this process could lead to changes in the exosomes loading, exosome secretion, exosome cargo, and the development of hybrid autophagic vesicles. In a study, Miao and colleagues noted that treatment with rapamycin, an autophagy inducer, significantly decreased exosome secretion but led to an increase in the release of exosomes after the addition of bafilomycin A1, which is an autophagy inhibitor [43]. Various biologically active substances, including carbohydrates, proteins, and nucleic acids, are transported to target cells through exosomes. They play an important role in the induction of the cell's flux by transferring autophagic or other related molecules. Upon exosome internalization, the development of autophagy can be initiated in recipient cells. For instance, exosomal fragments from breast cancer cells can trigger the process. Unfortunately, there is still a lack of information about the cargo of exosomes that can be transferred through it. According to studies, the cargo of exosomes from mesenchymal stem cells (MSCs) contains various mRNAs that are related to the development of autophagy. These include ATG7, Beclin1, and LC3 [44]. It is reasonable to assume that the arrival of materials can facilitate the intracellular autophagy that exosomes can initiate. There is also evidence suggesting that the cargo of exosomes with specific miRNAs can affect the cell's dynamic approach to autophagy in target cells. In an animal model, the researchers found that inhibiting autophagy could improve the injury rate in the heart and ischemia/reperfusion. They also noted that exosomes derived from MSCs were more effective at reducing the formation of autophagosomes and the LC3-II/I ratio [45]. On the other hand, Jin and his colleagues noted that exosomes from adipose-derived stem cells (ADSCs) could improve diabetic nephropathy by reducing the apoptosis rate and inducing autophagy [46]. Studies have shown that exosomes from tumor cells can play a significant role in the progression of different types of cancer. In gastric cancer cells, they can activate the pro-tumor activity of neutrophils by delivering NF-kB, HMGB1, and TLR4 signals [47]. They can promote cancer cell proliferation and migration through the induction of autophagy. In another study, researchers discovered that knockdown of the ATG7 gene could prevent the progression of the hepatitis C virus (HCV) by disrupting its extracellular distribution and progression into exosomes [48]. They also found that the ATG7 deletion led to a significant increase in the level of GAPDH in the exosome's contents. The findings suggest that the link between exosome biogenesis and autophagy could be a mechanism for cells to respond to stress.

6. RNA sorting into exosomes and novel roles in autophagy

Modern sequencing techniques allow for the detection of a wide variety of RNA species and the quantitation of rare and predefined transcript variants. These methods allow the use of RNA as a biomarker, which can provide insight into the activities taking place within biological compartments. Currently, various types of RNA, such as small regulatory and circulating RNAs, are undergoing rigorous analysis as biomarkers. Their investigations are expected to provide insight into the mechanisms by which a disease progresses. As exosomes are an integral component of biological compartments, their studies are likely to provide important insight into the development of the disease. In 2016, Valadi and colleagues reported that there was an abundance of both microRNA and messenger RNA in exosomes. They also discovered that there were no 28S and 18S ribosomal RNAs [49]. More recent studies have shown that there is an abundance of circular and long noncoding RNAs in exosomes. According to the ExoCarta database, over 300 studies identified 3408 mRNAs and 2838 miRNAs, respectively. Besides the usual messenger and microRNAs, exosomes also contain other noncoding RNAs such as rRNAs, snoRNAs, snRNAs, and piRNAs [50]. The database was created by the Shanghai Cancer Center of Fudan University and characterized the RNA species found in human exosomes. Through the analysis of data from other sources, they were able to identify over 30,000 noncoding RNAs and 18,333 mRNAs. In another study, they were able to identify over 1000 microRNAs in exosomes from patients with various diseases. The database, known as EV miRNA, was used to analyze and characterize the expression profile of various noncoding RNAs in exosomes derived from lymphoma cells. It should be noted that the exact nature of circRNAs and lncRNAs in exosomes derived from lymphoma is not known. The total levels of mRNA, miRNA, and lncRNA differ between exosomes and their cells of origin, indicating that the cytoplasmic RNA incorporation into secreted vesicles involves a sorting mechanism [50]. Although the exact mechanism of the process is still unclear, the researchers noted that the discovery of EXO motifs in RNA provides them with an insight into how this process works. These short, specific sequences can bind to certain RNA-binding proteins. These are responsible for the sorting of noncoding and circular RNAs, such as lncRNA and circRNAs, into exosomes. Since exosomes are involved in regulating various functions, the differences in protein and RNA content in exosomes from different sources are not surprising. Noncoding and coding RNAs are known to be one of the most common types of exosome cargo. These are capable of interacting with the recipient cell in a controlled manner. To transport the RNA, exosomes are packed with exosomal-shuttle RNAs. These RNAs are then formed into complexes with the RBPs, which then mediate the exosome's transfer. Due to the increasing number of studies on the expression of noncoding RNAs, such as those focused on the development of new therapeutic targets or the regulation of certain cellular functions, the field has started to narrow its focus. Instead of focusing on the molecular changes that occur in cancer, this approach tends to ignore other RNA species that have the potential to become promising targets.

Other proteins that are involved in autophagy have also been suggested to have the ability to bind to RNA. One example is the MAP1LC3B/LC3B, which can interact with the arginine-rich motif of RNA [51]. In smooth muscles, the cellular LC3B can interact with FN1/fibronectin mRNA after copurifying with ribosomes. The binding of LC3B to the FN1 mRNA's AU-rich region can play a role in its translational regulation. In addition, the triple arginine motif of LC3B can interact with other nuclear proteins, and its nuclear localization is dependent on RNA. Recent studies have shown that the LC3B protein has various novel functions. These include its role in the development of oncogene-induced senescence and chromatin reorganization [52]. It is also believed that there are additional LC3B functions that are not yet known. Some studies suggest that the LC3B may have other potential functions, such as the binding of RNA [53]. One of the most important factors that can be considered when it comes to the development of autophagosomes is their ability to bind to ubiquitin-like proteins. There are two unique reactions that are involved in this process: the conjugation of ATG12 to ATG5 and the conjugation of LC3 to phosphatidylethanolamine. Conjugates of ATG12 and ATG5 are linked to the ATG16L1 complex, which is required for the formation of phagophore elongation (Fig. 5.2). Recently, it has been shown that the presence of RNA can increase the ATG12—TG5 catalyzed reaction in a dose-dependent way.

7. Mechanisms for miRNA sorting into exosomes

MiRNAs are known to suppress the expression of certain protein-coding genes. Moreover, their complex regulatory networks play a vital role in various biological processes. The three types of miRNAs, miR-142-3p, miR-150, and miR-451, were enriched in exosomes as opposed to their parent cells [54]. It was shown that these miRNAs were packaged into exosomes and that the exocytosis process was responsible for their preferential enrichment. There are four potential routes for miRNAs to be

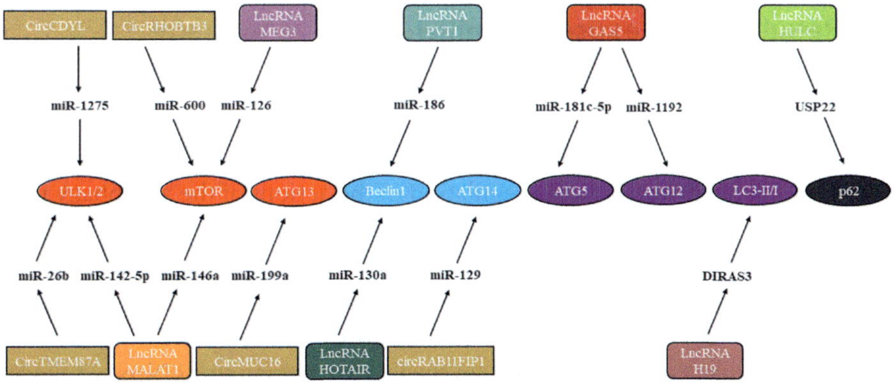

Figure 5.2 The comprehensive mechanism of exosomal lncRNAs and circRNAs in regulation of autophagy. LncRNAs and circRNAs can functions as a sponge for miRNAs to enhance their expression and can act as a competing endogenous RNA. They can also affect the function of autophagy signaling pathways. Red oval represents regulatory initiation proteins; blue oval represents regulatory nucleation proteins; purple oval represents regulatory elongation proteins; black oval represents ubiquitin-binding protein in autophagy pathway.

sorted into exosomes [55]. The nSMase2 pathway was identified as the one that could guide this process. The first molecule to be linked to exosomes' packaging was nSMase2. Overexpression of this molecule increased the number of miRNAs in the exosomes. On the other hand, inhibition of its expression significantly decreased their levels. The second mechanism involved the sumo-catalyzed hnRNP-dependent pathway. The three hnRNP family proteins, hnRNPA2B1, hnRNPA1, and hnRNPC, were bound to exosomes and triggered the loading of miRNAs into them. The sumo-catalyzed hnRNPA2B1 was able to control the exosomal miRNAs' assortment by recognizing the GGAG motif. The third pathway depended on the 3′ end of miRNA. The incorporation of miRNAs into exosomes through the 3′ end of their sequence is dependent on the sorting signal that they provide. MiRNA-induced silencing leads to the final pathway, which is mediated by the miRNA-induced silencing complex (miRISC). The main components of the miRISC family were present in exosomes. The accumulation of these components was caused by the blockade of the MVB turnover. It was also revealed that the suppression of the development of MVBs suppressed the miRNA-mediated silencing. It was believed that sorting process of exosomes depends on the changes in miRNA target abundance [56]. The addition of the Argonaute 2 component to the RISC caused the number of exosomes to increase. Some of the exosomal miRNAs that were known to be affected by this activity included miR-142-3p, miR-15, and miR-451. In addition to these, other sequences in miRNAs that are involved in the exocytosis process can also favor the uptake of these miRNAs into exosomes [57]. Furthermore, the exact mechanisms by which exosomes sort miRNAs remain unclear and warrant further investigation.

8. Regulation of exosomal miRNAs and autophagy

MiRNAs are known to play a role in various genes' modulation at the post-transcriptional level. They can interact with the 3′ UTR of mRNAs to produce different effects. In addition, they can regulate the actions of certain autophagy regulatory proteins. MiRNAs play a dual role in the regulation of the actions of the autophagy pathway, with pro- and antiautophagy functions [58]. They can modify the levels of various proteins, which are involved in the upstream signaling pathways and later stages of the degradation process. The control of the activity of autophagy is essential to maintaining the cell's homeostasis. MiRNA levels are also sensitive to the effects of stress on the organismal and cellular environment. When the levels of certain miRNAs are elevated or decreased in response to stress stimuli, this can trigger the activation of cellular response pathways, such as autophagy [59]. On the contrary, when the miRNAs are lowered in response to stress stimuli, this can cause the development of conditions such as starvation and energy crisis. When packed inside exosomes, cell-free miRNAs can be transported to target specific cells. On the other hand, machinery components and individual miRNA fragments were degraded and denatured by autophagolysosome systems [60]. The identified targets of the compounds were the genes of the proteins that function in the different stages of this pathway. In some cases, the miRNAs were able to target a specific pathway-specific gene. They also identified other genes that play a role in the various stages of the pathway (Table 5.1). Autophagy-associated miRNAs can reduce the expression levels of ATGs, LC3II, and SQSTM1/p62 [12]. In patients with pancreatic cancer, miRNA miR-29a is known to block the progression of the disease by accumulating various autophagosomes and markers of autophagy [61]. It also lowers the levels of TFEB and ATG9A, which are essential for lysosomal function and autophagosome trafficking. In patients with colorectal cancer, miR-214 is known to regulate the activity of ATG12-associated autophagy [62]. Some miRNAs were known to regulate autophagy's induction phase. The mTOR protein, which is the central regulator of this procedure, was identified as a target of miRNAs in the context of autophagy. Most of the pathway's core elements also experienced miRNA-dependent fine-tuning. In melanoma cells, miR-290-295 inhibited the induction of autophagy by targeting the ULK1 gene [63]. In contrast, miR-30A prevented the development of autophagic activity by Beclin1 in patients with alcohol-induced myocardial injury [64]. On the other hand, miR-195 suppressed the activity of autophagy by regulating ATG14 in those with nerve inflammation and pain [65]. Overexpression of miR-181A in leukemia, breast, and liver cancer cells led to a reduction in ATG5's starvation-induced autophagy [66]. In a study, Wang and colleagues found that miR-188-3p prevented the development of autophagic activity by ATG7 in patients with ischemic heart disease [67]. In a study, He et al. discovered that miR-20 inhibited the ATG10 induction in joint chondrocytes [68].

Table 5.1 List of autophagy modulating exosomal miRNAs.

Exosomal miRNAs	Regulation	Target	Outcome	References
miR-29a	Down	TFEB and ATG9A	Pancreatic cancer	[61]
miR-214	Down	ATG12	Colorectal cancer	[62]
miR-290-295	UP	ULK1	Melanoma cancer	[63]
miR-30a	Up	Beclin1	Myocardial injury	[64]
miR-195	Up	ATG14	Neuroinflammation	[65]
miR-181A	Up	ATG5	Breast cancer	[66]
miR-188-3p	Up	ATG7	Ischemic heart disease	[67]
miR-20	Up	ATG10	Osteoarthritis	[68]
miR-4459	Up	p62	Inflammation	[69]
miR-101	Up	Rab5a	Breast cancer	[70]
miR-425-3p	Up	AKT1	Non—small-cell lung cancer	[71]
miR-567	Down	ATG5	Breast cancer	[72]
miR-210	Up	ATG7	COPD	[73]
miR-181-5p	Up	STAT3/Bcl2	Liver fibrosis	[74]
miR-30a	Up	ATG12/Beclin1	Ischemic heart disease	[75]
miR-30d-5p	Down	IL-4/IL-10	Ischemic stroke	[76]
miR-21-5p	Up	SIPA1L2	Atherosclerosis	[77]
miR-155	Up	Rheb-mTOR	Neuronal disease	[78]

Researchers also found that miR-4459 prevented the expression of the LARP1 factor, which is involved in the process of SQSTM1/p62 in endothelial cells [69]. In a study, Frankel and colleagues discovered that miR-101 was able to regulate the induction of autophagy through a number of targets, such as Rab5a [70]. Other studies have shown that exosomes can control the development and maintenance of autophagy in recipient cells through specific pathways. For instance, in non—small-cell lung cancer (NSCLC), exosomal miR-425-3p can reduce the responsiveness of the cells to cisplatin by targeting the AKT/mTOR pathway [71]. This pathway can then lead to an increase in the activity of autophagy. They also found that exosomes that can resist cisplatin can activate autophagic activities and provide resistance to the drug in certain cells. The findings support the idea that miR-425-3p could be a potential biomarker for predicting the response to cisplatin in NSCLC. Patients with trastuzumab-resistant breast cancer cells showed a notable decrease in exosomal miR-567 expression [72]. In vitro and in vivo investigations revealed that the exosomes of these individuals had significantly decreased levels of miR-567. The researchers noted that the activity of miR-567 in reducing trastuzumab's induced autophagy also reverted to the acquired resistance. Their findings suggest that cancer cells can modify the exosomes' control over drug sensitivity in neighboring cells. According to Fujita et al., exosomes from bronchial epithelial cells that were induced by cigarette smoke exposure resulted in airway fibrosis [73]. This condition was triggered by the exosomal miR-210s autophagy regulation. Various exosomes-related miRNAs such as miR-21, miR-30a, miR-214, miR-33, miR-223, miR-125b, etc. are known to regulate the flux of autophagy [79]. They can be used to enhance the levels of cholesterol efflux in the bronchial cells. The researchers noted that these miRNAs can also modify the AMPK-mTOR and PI3K-AKT-mTOR signals. In a study conducted by the researchers, they found that exosomes from genetically modified MSCs transfer miR-181-5p to different cell types, including hepatic stellate cells and mouse liver fibrosis models [74]. Studies have shown that the cargo of exosomes from brain cells and cardiomyocytes that contained miR-30a and miR-30d-5p increased the recipient cells' autophagy [75,76]. In a study, researchers noted that endothelial colony-forming cell (ECFC) exosomes can promote the repair of vascular cells that have been injured due to a balloon injury [77]. They found that these exosomes can protect against vascular injury caused by the ox-LDL-catalyzed damage to the autophagic flux. They discovered that the ECFC exosomes contain miR-21-5p, which binds to the 3′-UTR of signal-induced proliferation-associated 1-like protein 2 (SIPA1L2). It enhances the autophagic flux and prevents its expression. The researchers also found that a knockout of the exosome of miR-21-5p significantly decreased the expression of the ECFC exosome in vascular injury. The researchers concluded that the ECFC exosome repaired vascular injury and promoted autophagy in an atherogenic rat model. Moreover, the ECFC exosomes' effects on lipid balance and vascular endothelial injury were neutralized after the exosome's knockdown. According to the researchers, activated microglia can accelerate the process of neuronal autophagy, which is caused by heat stress, by transferring exosomes to the neurons [78]. They found that the increase in miR-155 expression was detected in the microglia, as well as in the cocultured exosomes and neurons. In addition, the researchers noted that the overexpression of the miR-155 in the activated microglial cells could contribute to the development of neuronal autophagy, which happens when subjected to heat stress. The researchers noted that inhibition of the overexpression of miR-155 in the neurons could help to abolish the neuronal autophagy that occurs when exosomes from the activated microglial cells are combined with cocultured neurons. The presence of miR-155 in exosomes can be seen as a factor that contributes to neuronal pathway of autophagy.

9. Regulation of exosomal lncRNAs and autophagy

LncRNAs are a broad class of biological molecules that can perform various physiological and pathological functions [80]. Due to the advances in the analysis and sequencing of lncRNAs, they can now be further categorized into different classes. The classification of lncRNAs based on their function, location, and presence of repeat elements is generally used. The downstream effectors of lncRNAs interact with the signals generated by different pathways to provide the system with valuable information. These interactions are not coordinated and are not dependent on each other. They are present in a wide variety of species, including fungi, bacteria, plants, animals, and viruses. They are usually categorized according to their size. There have been reports that lncRNAs can regulate the expression of certain ATG genes (Table 5.2). It has been known that long noncoding RNA can also target specific proteins that are involved in autophagy by sponging certain miRNAs. For instance, the lncRNA APF has been shown to enhance the expression of the ATG7. They are also known to sequester miRNAs that are involved in the autophagic process [13]. A study revealed that the lncRNA MEG3 plays an important role in the growth of bladder tumors [81]. It was found that the level of the lncRNA in the tumor tissues was lower than in normal tissue samples. They also found that the level of the autophagy marker LC3-II was negatively correlated with the level of the MEG3. In addition, it can help to prevent cell stress by acting as a synergistic factor. It has also been known that elevated levels of HOTAIR can trigger the development of drug resistance and cancer. Human HCC cells can generate an induction of autophagy after experiencing an increase in HOTAIR expression by ATG7 [82]. Studies have shown that the reduction of lncRNA HOTAIR's activity can prevent the degradation of the PML-RARA protein by ATRA in promyelocytic leukemia protein (PML) cells. It can also increase the cellular level of autophagy by activation of LC3-II/LC3-I protein

Table 5.2 List of autophagy modulating exosomal lncRNAs.

Exosomal lncRNAs	Regulation	Target	Autophagy pathways
MEG3	Up	miR-205-5p	LC3-II (↑), ATG3 (↑), LAMP1 (↑), p62 (↓)
	Down	p53	LC3II/I (↑)
HOTAIR	Up	miR-126-5p	Unknown
	Up	miR-130a	Beclin1 (↑), p62 (↓), LC3II/I (↑)
	Up	LRRK2	ERK/MAPK (↑)
H19	Up	DUSP5	ERK1/2 (↑)
	Up	DIRAS3	ATG7 (↓), LC3-II (↓), Beclin1 (↓)
	Up	ACP5	WNT/β-catenin
MALAT1	Up	miR-26b	ULK2 (↑)
	Down	miR-384	GOLM1 (↓)
	Down	miR-30a	Beclin1 (↓)
	Down	miR-146a	PI3K/AKT/mTOR (↓)
GAS5	Up	miR-222-3p	LC3B (↑), Beclin1 (↑)
	Up	mTOR	LC3II (↓), p62 (↑)
PVT1	Up	miR-186	ATG7 (↑), Beclin1 (↑)
	Up	miR-20a-5p	ULK (↑)
HULC	Up	miR15a	p62 (↑)
	Up	USP22	SIRT1 (↑), p62 (↓), LC3-II (↑)

[13]. Researchers noted that lncRNA H19 plays a role in the initiation of autophagy and high glucose levels can reduce the expression of H19, which can trigger DIRAS3's inactivation by inhibiting the mTOR/PI3K/AKT pathway [36,83]. The researchers noted that the H19 knockout effect increased ATG7 levels and Beclin1 expression, which suggests that H19 is related to major autophagy genes [84]. In another study, the researchers discovered that the lncRNA H19 can activate autophagy by inhibiting the DUSP5. It is known that this enzyme can suppress autophagy by blocking the ERK1/2 phosphorylation [13]. The researchers noted that the lncRNA PTENP1 pseudogene has a 3′ UTR, which is similar to that of the PTEN gene. It can protect the development of miRNAs that contain this gene. The overexpression of PTENP1 leads to a rise in PTEN, which can also trigger the activation of autophagy. It has been shown that the presence of PTENP1 in the cell can enhance the expression of other genes, such as ATG7, ULK1, and P62/SQSTM1 [82]. The researchers noted that vascular inflammation can increase the expression of TGFB2-OT1. This is because it triggers the growth of various other genes, such as ATG7, p62, and ATG3 [69]. According to a study, suppressing the lncRNA MALAT1 can elevate p62 and decrease LAMP2 expression, two important factors in the fusion of autophagosomes and lysosomes [13]. MALAT1 can prevent reoxygenation of endothelial cells. It can also impair their ability to repair damage caused due to the ULK2 upregulation [85]. In a study, Xu and colleagues discovered that lncRNA GAS5 can promote the development of autophagic regulators by upregulating their expression [86]. Researchers noted that PVT1's expression activates ULK1, which is an autophagy-activating factor [87]. In epithelial ovarian cancer cells, higher levels of lncRNA HULC were observed compared with normal samples. They were linked to the development of the disease by blocking the ATG7 pathway and suppressing the process known as autophagy [88]. Studies revealed that the levels of exosomal lncRNA OIP5-AS1 in OS cells inhibited the expression of Beclin1 and LC3-II proteins [89].

10. Regulation of exosomal circRNAs and autophagy

CircRNAs play a role in the development of autophagy by interacting with the proteins that are involved in autophagy (Table 5.3). The lack of a 5' cap and a 3′ tail makes circRNAs undigestible and harmless to RNases. They are commonly expressed in the body fluids and tissues. There are four types of circRNAs, which are classified according to their origin. These include exon circRNAs, intron circRNAs, combination circRNAs, and viral genome-encoding circRNAs. The various ways in which circRNAs play their roles are shown. These include their ability to combine with RNA and form molecular sponges, their ability to adsorb miRNA, and their ability to affect the function of miRNA in the development and occurrence of various diseases. In addition, they can also bind to mRNA. During the initial stage of autophagy, a complex composed of ULK1 and ULK2, as well as ATG13 and FIP200, is activated. In addition to regulating

Table 5.3 List of autophagy modulating exosomal circRNAs.

Exosomal circRNAs	Regulation	Target	Pathway	Outcome	References
circCDYL	Up	miR-1275	ULK1	Breast cancer	[90]
circTMEM87A	Up	miR-142-5p	ULK1	Gastric cancer	[90]
circRHOBTB3	Up	miR-600	AKT/mTOR	Pancreatic ductal adenocarcinoma	[90]
circMUC16	Up	miR-199a	ATG13	Ovarian cancer	[91]
circRAB11FIP1	Up	miR-129	ATG7/ATG14	Ovarian cancer	[92]
circUBE2Q2	Up	miR-370-3p	STAT3	Gastric cancer	[93]
circ-PDE8A	Up	miR338	MACC1/MET	Pancreatic ductal adenocarcinoma	[94]

ULK1's expression, circCDYL and circTMEM87A also promote autophagy by acting as the sponge proteins of miR-1275 and miR-142-5p [90]. CircMUC16 helps ATG13 expression by binding to it. Finally, the complex of ULK1 triggers the activation of the PI3K complex, which contains ATG13, UVRAG, Beclin, and PI3K class III and mediates autophagosome nucleation. CircMUC16 promotes autophagy by interacting with the miR-199a-5p/Beclin1 axis [91]. circRAB11FIP1 also controls the expression of ATG14 by sponging miR-129. The complex of ATG5-ATG12 helps to extend the membrane of the autophagosome [92]. Finally, LC3-II is formed by the combination of LC3 and PE, which is then recruited to the membrane of the autophagosome. The ATG4B-ATG7 complex is responsible for the development of this process. circRAB11 and circCDYL are known to promote the expression of ATG7 by acting as sponges for miR-1275 and miR-129. They also contribute to the fusion of lysosomes and autophagosomes. The multiple pathways that are involved in the regulation of autophagy are known to activate the activity of the stress response pathway, known as STAT3. The various protein structures that are located in the subcellular compartments of the body can affect the process. The phosphorylation of STAT3 protein triggers the expression of certain autophagy-associated genes. These include Beclin1 and Bcl-2 [95]. Unphosphorylated STAT3 can suppress autophagy by sequestering certain proteins, such as FOXO1/3 and EIF2AK2. In a study, circUBE2Q2 was shown to enhance GC development and reduce the activity of the miR-370-3p/STAT3-induced autophagy pathway [93]. The knockdown of this protein significantly decreased the levels of Bcl-2 and p-STAT3. Studies have revealed that circRNAs can be stable and enriched in exosomes, making them an ideal target for detection in urine [96]. For instance, it has been reported that in gastric cancer, exosomes can be used to deliver circRNAs to recipient cells, which suggests that exosomes play a significant role in the disease's invasion and progression [97]. The composition of exosomes circRNAs can be influenced by changes in the levels of miRNA in the donor cells. This information is then transferred to the recipient cells.

Significantly, the development and progression of numerous neoplasias can be initiated and sustained by the interactions between RNA-catalyzed competitive sequences in exosomes. Research has shown that in a KRAS mutation, circRNAs seem to be significantly downregulated, and they can be transferred to colorectal cancer cells via exosomes [98]. In pancreatic ductal adenocarcinoma, exosomal circ-PDE8A has been linked to the development of cancer and lymphatic invasion. It has also been shown that exosome circRNAs can promote white adipose browning by targeting the miR-133 [94]. Moreover, exosomal circRNAs can promote the growth of tumors by targeting the protein USP7 in or from hepatocellular carcinoma and endometrial cancer [99]. Cell proliferation and cancer cell survival were significantly enhanced by the overexpression of circ-DMNT1. Researchers discovered that this circRNAs can interact with AU-binding factor 1 (AUF1) and p53 to promote nuclear translocation [100]. They also noted that the two nuclear translocations can activate cellular autophagy. One particular circRNA that is produced from MUC16 is circMUC16. It is found in epithelial ovarian cancer tissues and has a positive correlation with the tumor stage. Further studies revealed that this gene could enhance cancer cell invasion, proliferation, and metastasis through autophagy. To promote cellular autophagy, circMUC16 was able to regulate the genes that are involved in the process of apoptosis. It also exhibited the ability to bind to the ATG13 homolog. This suggests that circMUC16 could play a role in enhancing the tumors phenotype by directly affecting the ATG13 pathway. A combination of RNA sequencing and molecular analysis revealed that the circTGFBR2 was differentially expressed in exosomes. This study also confirmed that the exosomes that contained enriched circTGFBR2 had an increased resistance to starvation. In the PDAC, Yang and colleagues identified a circRNA known as circRHOBTB3. It functions as a sponge for miRNAs and also facilitates autophagy by suppressing the mTOR/Akt signaling [90]. By analyzing the sequence data of exosomes, Li was able to determine the stability and enrichment of abundant characteristics of the circRNAs in exosomes, which are a promising new tool for cancer diagnosis. Although the field of exosomal circRNA research is starting to gain momentum, the exact traits and functions of these exosomal circRNAs remain unclear.

11. Conclusion

The mechanisms by which exosomes remove harmful proteins are coordinated through the autophagy—lysosomes pathway. This process plays a role in maintaining cellular fitness and protein homeostasis. It is protective because it can remove cellular components that are harmful to the body. Furthermore, exosomes can play a role in facilitating the transmission of signals from one cell to another, which helps coordinate responses from the body's various organs. In this context, it has been proposed that the exosome's function could be linked to the autophagy pathway and that it could mediate the spread of messages to nearby cells. The identity of the different intermediate compartments within exosomes and autophagy is still unclear. But emerging knowledge has provided us with various insights into two systems interconnections. Many types of RNAs are acquired to the phagophore through well-established mechanisms in the lysosome or vacuole, but little is known about the stimuli and sensors that are utilized in this process. Aside from being an autophagic component, RNA can also play a role in the regulation of the pathway. It can either enhance the stability of the protein complexes or mediate the translation of the genes into an improved state. Several ongoing studies in both the autophagy and RNA research arenas will provide valuable insight regarding the various unanswered questions, and their exploration may lead to the discovery of novel therapeutic strategies. Although it is widely believed that exosomes play a role in facilitating communication among cells, there are still a lot of doubt when it comes to the exosomes' contribution to the maintenance and development of cellular health.

Acknowledgments

Authors would like to acknowledge the academic fellowship grant from Department of Science and Technology and Biotechnology, Government of West Bengal.

Author contribution

SGC performed the literature research and drafted the chapter; PK participated in modifying and critical writing of the chapter.

References

[1] He L, Zhang J, Zhao J, Ma N, Kim SW, Qiao S, et al. Autophagy: the last defense against cellular nutritional stress. Advances in Nutrition 2018;9(4):493–504. https://doi.org/10.1093/advances/nmy011.

[2] Li W, He P, Huang Y, Li YF, Lu J, Li M, et al. Selective autophagy of intracellular organelles: recent research advances. Theranostics 2021;11(1):222–56. https://doi.org/10.7150/thno.49860.

[3] Doyle LM, Wang MZ. Overview of extracellular vesicles, their origin, composition, purpose, and methods for exosome isolation and analysis. Cells 2019;8(7):727. https://doi.org/10.3390/cells8070727.

[4] Chowdhury SG, Ray R, Bhattacharya D, Karmakar P. DNA damage induced cellular senescence and it's PTEN-armed exosomes-the warriors against prostate carcinoma cells. Medical Oncology 2022;39(3):34. https://doi.org/10.1007/s12032-021-01614-7.

[5] Zhang Y, Liu Y, Liu H, Tang WH. Exosomes: biogenesis, biologic function and clinical potential. Cell & Bioscience 2019;9:19. https://doi.org/10.1186/s13578-019-0282-2.

[6] O'Sullivan MJ, Lindsay AJ. The endosomal recycling pathway-at the crossroads of the cell. International Journal of Molecular Sciences 2020;21(17):6074. https://doi.org/10.3390/ijms21176074.

[7] H Rashed M, Bayraktar E, K Helal G, Abd-Ellah MF, et al. Exosomes: from garbage bins to promising therapeutic targets. International Journal of Molecular Sciences 2017;18(3):538. https://doi.org/10.3390/ijms18030538.

[8] Baixauli F, López-Otín C, Mittelbrunn M. Exosomes and autophagy: coordinated mechanisms for the maintenance of cellular fitness. Frontiers in Immunology 2014;5:403. https://doi.org/10.3389/fimmu.2014.00403.

[9] Kaikkonen MU, Lam MT, Glass CK. Non-coding RNAs as regulators of gene expression and epigenetics. Cardiovascular Research 2011;90(3):430–40. https://doi.org/10.1093/cvr/cvr097.

[10] Ratti M, Lampis A, Ghidini M, Salati M, et al. MicroRNAs (miRNAs) and long non-coding RNAs (lncRNAs) as new tools for cancer therapy: first steps from bench to bedside. Targeted Oncology 2020;15(3):261–78. https://doi.org/10.1007/s11523-020-00717-x.

[11] Jenjaroenpun P, Kremenska Y, Nair VM, Kremenskoy M, et al. Characterization of RNA in exosomes secreted by human breast cancer cell lines using next-generation sequencing. PeerJ 2013;1:e201. https://doi.org/10.7717/peerj.201.
[12] Gozuacik D, Akkoc Y, Ozturk DG, Kocak M. Autophagy-regulating microRNAs and cancer. Frontiers Oncology 2017;7:65. https://doi.org/10.3389/fonc.2017.00065.
[13] Yang L, Wang H, Shen Q, Feng L, Jin H. Long non-coding RNAs involved in autophagy regulation. Cell Death & Disease 2017;8(10):e3073. https://doi.org/10.1038/cddis.2017.464.
[14] Yu CY, Kuo HC. The emerging roles and functions of circular RNAs and their generation. Journal of Biomedical Science 2019;26(1):29. https://doi.org/10.1186/s12929-019-0523-z.
[15] Cui W, Dang Q, Chen C, Yuan W, Sun Z. Roles of circRNAs on tumor autophagy. Molecular Therapy - Nucleic Acids 2021;23:918–29. https://doi.org/10.1016/j.omtn.2021.01.002.
[16] Frankel EB, Audhya A. ESCRT-dependent cargo sorting at multivesicular endosomes. Seminars in Cell & Developmental Biology 2018;74:4–10. https://doi.org/10.1016/j.semcdb.2017.08.020.
[17] Larios J, Mercier V, Roux A, Gruenberg J. ALIX- and ESCRT-III-dependent sorting of tetraspanins to exosomes. The Journal of Cell Biology 2020;219(3):e201904113. https://doi.org/10.1083/jcb.2019 04113.
[18] Qiu Y, Li P, Zhang Z, Wu M. Insights into exosomal non-coding RNAs sorting mechanism and clinical application. Frontiers Oncology 2021;11:664904. https://doi.org/10.3389/fonc.2021. 664904.
[19] Skotland T, Hessvik NP, Sandvig K, Llorente A. Exosomal lipid composition and the role of ether lipids and phosphoinositides in exosome biology. The Journal of Lipid Research 2019;60(1):9–18. https://doi.org/10.1194/jlr.R084343.
[20] Takasugi M. Emerging roles of extracellular vesicles in cellular senescence and aging. Aging Cell 2018;17(2):e12734. https://doi.org/10.1111/acel.12734.
[21] Jakhar R, Crasta K. Exosomes as emerging pro-tumorigenic mediators of the senescence-associated secretory phenotype. International Journal of Molecular Sciences 2019;20(10):2547. https://doi.org/10.3390/ijms20102547.
[22] Andreu Z, Yáñez-Mó M. Tetraspanins in extracellular vesicle formation and function. Frontiers in Immunology 2014;5:442. https://doi.org/10.3389/fimmu.2014.00442.
[23] Yin Z, Pascual C, Klionsky DJ. Autophagy: machinery and regulation. Microbial Cell 2016;3(12):588–96. https://doi.org/10.15698/mic2016.12.546.
[24] Parzych KR, Klionsky DJ. An overview of autophagy: morphology, mechanism, and regulation. Antioxidants and Redox Signaling 2014;20(3):460–73. https://doi.org/10.1089/ars.2013.5371.
[25] Randow F, Youle RJ. Self and nonself: how autophagy targets mitochondria and bacteria. Cell Host Microbe 2014;15(4):403–11. https://doi.org/10.1016/j.chom.2014.03.012.
[26] Alers S, Löffler AS, Wesselborg S, Stork B. Role of AMPK-mTOR-Ulk1/2 in the regulation of autophagy: cross talk, shortcuts, and feedbacks. Molecular and Cellular Biology 2012;32(1):2–11. https://doi.org/10.1128/MCB.06159-11.
[27] Hollenstein DM, Kraft C. Autophagosomes are formed at a distinct cellular structure. Current Opinion in Cell Biology 2020;65:50–7. https://doi.org/10.1016/j.ceb.2020.02.012.
[28] Walczak M, Martens S. Dissecting the role of the Atg12-Atg5-Atg16 complex during autophagosome formation. Autophagy 2013;9(3):424–5. https://doi.org/10.4161/auto.22931.
[29] Barth S, Glick D, Macleod KF. Autophagy: assays and artifacts. The Journal of Pathology 2010;221(2):117–24. https://doi.org/10.1002/path.2694.
[30] Tanida I, Ueno T, Kominami E. LC3 conjugation system in mammalian autophagy. The International Journal of Biochemistry & Cell Biology 2004;36(12):2503–18. https://doi.org/10.1016/j.biocel.2004.05.009.
[31] Zheng J, Tan J, Miao YY, Zhang Q. Extracellular vesicles degradation pathway based autophagy lysosome pathway. American Journal of Translational Research 2019;11(3):1170–83. PMID: 30972154.
[32] Buratta S, Tancini B, Sagini K, Delo F, Chiaradia E, et al. Lysosomal exocytosis, exosome release and secretory autophagy: the autophagic- and endo-lysosomal systems go extracellular. International Journal of Molecular Sciences 2020;21(7):2576. https://doi.org/10.3390/ijms21072576.

[33] Yáñez-Mó M, Siljander PR, Andreu Z, Zavec AB, Borràs FE, et al. Biological properties of extracellular vesicles and their physiological functions. Journal of Extracellular Vesicles 2015;4:27066. https://doi.org/10.3402/jev.v4.27066.

[34] McKnight NC, Zhenyu Y. Beclin 1, an essential component and master regulator of PI3K-III in health and disease. Current Pathobiology Reports 2013;1(4):231–8. https://doi.org/10.1007/s40139-013-0028-5.

[35] Nair U, Jotwani A, Geng J, Gammoh N, Richerson D, et al. SNARE proteins are required for macroautophagy. Cell 2011;146(2):290–302. https://doi.org/10.1016/j.cell.2011.06.022.

[36] Chowdhury SG, Bhattacharya D, Karmakar P. Exosomal long noncoding RNAs - the lead thespian behind the regulation, cause and cure of autophagy-related diseases. Molecular Biology Reports 2022;49(7):7013–24. https://doi.org/10.1007/s11033-022-07514-x.

[37] Xing H, Tan J, Miao Y, Lv Y, Zhang Q. Crosstalk between exosomes and autophagy: a review of molecular mechanisms and therapies. Journal of Cellular and Molecular Medicine 2021;25(5):2297–308. https://doi.org/10.1111/jcmm.16276.

[38] Papademetrio DL, Garcia MN, Grasso D, Alvarez É. Autophagy-mediated exosomes as immunomodulators of natural killer cells in pancreatic cancer microenvironment. Frontiers Oncology 2021;10:622956. https://doi.org/10.3389/fonc.2020.622956.

[39] Salimi L, Akbari A, Jabbari N, Mojarad B, Vahhabi A, et al. Synergies in exosomes and autophagy pathways for cellular homeostasis and metastasis of tumor cells. Cell & Bioscience 2020;10:64. https://doi.org/10.1186/s13578-020-00426-y.

[40] Gurunathan S, Kang MH, Kim JH. A comprehensive review on factors influences biogenesis, functions, therapeutic and clinical implications of exosomes. International Journal of Nanomedicine 2021;16:1281–312. https://doi.org/10.2147/IJN.S291956.

[41] Murrow L, Malhotra R, Debnath J. ATG12-ATG3 interacts with Alix to promote basal autophagic flux and late endosome function. Nature Cell Biology 2015;17(3):300–10. https://doi.org/10.1038/ncb3112.

[42] Colletti M, Ceglie D, Di Giannatale A, Nazio F. Autophagy and exosomes relationship in cancer: friends or foes? Frontiers in Cell and Developmental Biology January 2021;8:614178. https://doi.org/10.3389/fcell.2020.614178.

[43] Miao Y, Li G, Zhang X, Xu H, Abraham SN. A TRP channel senses lysosome neutralization by pathogens to trigger their expulsion. Cell June 2015;161(6):1306–19. https://doi.org/10.1016/j.cell.2015.05.009.

[44] Hassanpour M, Rezabakhsh A, Rezaie J, Nouri M, Rahbarghazi R. Exosomal cargos modulate autophagy in recipient cells via different signaling pathways. Cell & Bioscience 2020;10:92. https://doi.org/10.1186/s13578-020-00455-7.

[45] Gu X, Li Y, Chen K, Wang X, Wang Z, Lian H, et al. Exosomes derived from umbilical cord mesenchymal stem cells alleviate viral myocarditis through activating AMPK/mTOR-mediated autophagy flux pathway. Journal of Cellular and Molecular Medicine 2020;24(13):7515–30. https://doi.org/10.1111/jcmm.15378.

[46] Jin J, Shi Y, Gong J, Zhao L, Li Y, He Q, et al. Exosome secreted from adipose-derived stem cells attenuates diabetic nephropathy by promoting autophagy flux and inhibiting apoptosis in podocyte. Stem Cell Research & Therapy 2019;10(1):95. https://doi.org/10.1186/s13287-019-1177-1.

[47] Shi Y, Zhang J, Mao Z, Jiang H, Liu W, et al. Extracellular vesicles from gastric cancer cells induce PD-L1 expression on neutrophils to suppress T-cell immunity. Frontiers Oncology 2020;10:629. https://doi.org/10.3389/fonc.2020.00629.

[48] Shrivastava S, Devhare P, Sujijantarat N, Steele R, Kwon YC, Ray R, et al. Knockdown of autophagy inhibits infectious hepatitis C virus release by the exosomal pathway. Journal of Virology 2015;90(3):1387–96. https://doi.org/10.1128/JVI.02383-15.

[49] Valadi H, Ekström K, Bossios A, Sjöstrand M, Lee JJ, Lötvall JO. Exosome-mediated transfer of mRNAs and microRNAs is a novel mechanism of genetic exchange between cells. Nature Cell Biology 2007;9(6):654–9. https://doi.org/10.1038/ncb1596.

[50] Narang P, Shah M, Beljanski V. Exosomal RNAs in diagnosis and therapies. Noncoding RNA Research 2022;7(1):7–15. https://doi.org/10.1016/j.ncrna.2022.01.001.

[51] Frankel LB, Lubas M, Lund AH. Emerging connections between RNA and autophagy. Autophagy 2017;13(1):3–23. https://doi.org/10.1080/15548627.2016.1222992.
[52] Di Micco R, Krizhanovsky V, Baker D, d'Adda di Fagagna F. Cellular senescence in ageing: from mechanisms to therapeutic opportunities. Nature Reviews Molecular Cell Biology 2021;22(2):75–95. https://doi.org/10.1038/s41580-020-00314-w.
[53] Hwang HJ, Ha H, Lee BS, Kim BH, Song HK, Kim YK. LC3B is an RNA-binding protein to trigger rapid mRNA degradation during autophagy. Nature Communications 2022;13(1):1436. https://doi.org/10.1038/s41467-022-29139-1.
[54] Wang M, Yu F, Ding H, Wang Y, Li P, Wang K. Emerging function and clinical values of exosomal MicroRNAs in cancer. Molecular Therapy - Nucleic Acids 2019;16:791–804. https://doi.org/10.1016/j.omtn.2019.04.027.
[55] Zhang J, Li S, Li L, Li M, Guo C, Yao J, et al. Exosome and exosomal microRNA: trafficking, sorting, and function. Genomics, Proteomics & Bioinformatics 2015;13(1):17–24. https://doi.org/10.1016/j.gpb.2015.02.001.
[56] Schwarzenbach H, Gahan PB. MicroRNA shuttle from cell-to-cell by exosomes and its impact in cancer. Noncoding RNA 2019;5(1):28. https://doi.org/10.3390/ncrna5010028.
[57] Penfornis P, Vallabhaneni KC, Whitt J, Pochampally R. Extracellular vesicles as carriers of microRNA, proteins and lipids in tumor microenvironment. International Journal of Cancer 2016;138(1):14–21. https://doi.org/10.1002/ijc.29417.
[58] Zhao Y, Wang Z, Zhang W, Zhang L. MicroRNAs play an essential role in autophagy regulation in various disease phenotypes. BioFactors 2019;45(6):844–56. https://doi.org/10.1002/biof.1555.
[59] Jang JH, Lee TJ. The role of microRNAs in cell death pathways. Yeungnam University Journal of Medicine 2021;38(2):107–17. https://doi.org/10.12701/yujm.2020.00836.
[60] Hale AN, Ledbetter DJ, Gawriluk TR, Rucker 3rd EB. Autophagy: regulation and role in development. Autophagy 2013;9(7):951–72. https://doi.org/10.4161/auto.24273.
[61] Kwon JJ, Willy JA, Quirin KA, Wek RC, Korc M, Yin XM, et al. Novel role of miR-29a in pancreatic cancer autophagy and its therapeutic potential. Oncotarget 2016;7(44):71635–50. https://doi.org/10.18632/oncotarget.11928.
[62] Long J, He Q, Yin Y, Lei X, Li Z, Zhu W. The effect of miRNA and autophagy on colorectal cancer. Cell Proliferation 2020;53(10):e12900. https://doi.org/10.1111/cpr.12900.
[63] Chen Y, Liersch R, Detmar M. The miR-290-295 cluster suppresses autophagic cell death of melanoma cells. Scientific Reports 2012;2:808. https://doi.org/10.1038/srep00808.
[64] Zhu H, Wu H, Liu X, Li B, Chen Y, Ren X, et al. Regulation of autophagy by a beclin 1-targeted microRNA, miR-30a, in cancer cells. Autophagy 2009;5(6):816–23. https://doi.org/10.4161/auto.9064.
[65] Shi G, Shi J, Liu K, Liu N, Wang Y, Fu Z, et al. Increased miR-195 aggravates neuropathic pain by inhibiting autophagy following peripheral nerve injury. Glia 2013;61(4):504–12. https://doi.org/10.1002/glia.22451.
[66] Tekirdag KA, Korkmaz G, Ozturk DG, Agami R, Gozuacik D. MIR181A regulates starvation- and rapamycin-induced autophagy through targeting of ATG5. Autophagy 2013;9(3):374–85. https://doi.org/10.4161/auto.23117.
[67] Wang K, Liu CY, Zhou LY, Wang JX, Wang M, et al. APF lncRNA regulates autophagy and myocardial infarction by targeting miR-188-3p. Nature Communications 2015;6:6779. https://doi.org/10.1038/ncomms7779.
[68] He W, Cheng Y. Inhibition of miR-20 promotes proliferation and autophagy in articular chondrocytes by PI3K/AKT/mTOR signaling pathway. Biomedicine & Pharmacotherapy 2018;97:607–15. https://doi.org/10.1016/j.biopha.2017.10.152.
[69] Huang S, Lu W, Ge D, Meng N, Li Y, Su L, et al. A new microRNA signal pathway regulated by long noncoding RNA TGFB2-OT1 in autophagy and inflammation of vascular endothelial cells. Autophagy 2015;11(12):2172–83. https://doi.org/10.1080/15548627.2015.1106663.
[70] Frankel LB, Wen J, Lees M, Høyer-Hansen M, Farkas T, et al. microRNA-101 is a potent inhibitor of autophagy. EMBO Journal 2011;30(22):4628–41. https://doi.org/10.1038/emboj.2011.331.

[71] Ma Y, Yuwen D, Chen J, Zheng B, Gao J, Fan M, et al. Exosomal transfer of cisplatin-induced miR-425-3p confers cisplatin resistance in NSCLC through activating autophagy. International Journal of Nanomedicine 2019;14:8121–32. https://doi.org/10.2147/IJN.S221383.
[72] Han M, Hu J, Lu P, Cao H, Yu C, Li X, et al. Exosome-transmitted miR-567 reverses trastuzumab resistance by inhibiting ATG5 in breast cancer. Cell Death & Disease 2020;11(1):43. https://doi.org/10.1038/s41419-020-2250-5.
[73] Fujita Y, Araya J, Ito S, Kobayashi K, Kosaka N, et al. Suppression of autophagy by extracellular vesicles promotes myofibroblast differentiation in COPD pathogenesis. Journal of Extracellular Vesicles 2015;4:28388. https://doi.org/10.3402/jev.v4.28388.
[74] Qu Y, Zhang Q, Cai X, Li F, Ma Z, Xu M, et al. Exosomes derived from miR-181-5p-modified adipose-derived mesenchymal stem cells prevent liver fibrosis via autophagy activation. Journal of Cellular and Molecular Medicine 2017;21(10):2491–502. https://doi.org/10.1111/jcmm.13170.
[75] Yang Y, Li Y, Chen X, Cheng X, Liao Y, Yu X. Exosomal transfer of miR-30a between cardiomyocytes regulates autophagy after hypoxia. Journal of Molecular Medicine 2016;94(6):711–24. https://doi.org/10.1007/s00109-016-1387-2.
[76] Jiang M, Wang H, Jin M, Yang X, Ji H, Jiang Y, et al. Exosomes from MiR-30d-5p-ADSCs reverse acute ischemic stroke-induced, autophagy-mediated brain injury by promoting M2 microglial/macrophage polarization. Cellular Physiology and Biochemistry 2018;47(2):864–78. https://doi.org/10.1159/000490078.
[77] Ke X, Liao Z, Luo X, Chen JQ, Deng M, Huang Y, et al. Endothelial colony-forming cell-derived exosomal miR-21-5p regulates autophagic flux to promote vascular endothelial repair by inhibiting SIPL1A2 in atherosclerosis. Cell Communication and Signaling 2022;20(1):30. https://doi.org/10.1186/s12964-022-00828-0.
[78] Li P, Luo X, Luo Z, He GL, Shen TT, Yu XT, et al. Increased miR-155 in microglial exosomes following heat stress accelerates neuronal autophagy via their transfer into neurons. Frontiers in Cellular Neuroscience 2022;16:865568. https://doi.org/10.3389/fncel.2022.865568.
[79] Shah SZA, Zhao D, Hussain T, Sabir N, Yang L. Regulation of MicroRNAs-mediated autophagic flux: a new regulatory avenue for neurodegenerative diseases with focus on prion diseases. Frontiers in Aging Neuroscience 2018;10:139. https://doi.org/10.3389/fnagi.2018.00139.
[80] Chowdhury SG, Ray R, Karmakar P. Exosomal miRNAs — a diagnostic biomarker acting as a guiding light in the diagnosis of prostate cancer. Functional & Integrative Genomics 2022;23(1):23. https://doi.org/10.1007/s10142-022-00951-8.
[81] Li HJ, Gong X, Li ZK, Qin W, He CX, Xing L, et al. Role of long non-coding RNAs on bladder cancer. Frontiers in Cell and Developmental Biology 2021;9:672679. https://doi.org/10.3389/fcell.2021.672679.
[82] Zhang H, Lu B. The roles of ceRNAs-mediated autophagy in cancer chemoresistance and metastasis. Cancers 2020;12(10):2926. https://doi.org/10.3390/cancers12102926.
[83] Zhuo C, Jiang R, Lin X, Shao M. LncRNA H19 inhibits autophagy by epigenetically silencing of DIRAS3 in diabetic cardiomyopathy. Oncotarget 2017;8(1):1429–37. https://doi.org/10.18632/oncotarget.13637.
[84] Wang J, Xie S, Yang J, Xiong H, Jia Y, Zhou Y, et al. The long noncoding RNA H19 promotes tamoxifen resistance in breast cancer via autophagy. Journal of Hematology & Oncology 2019;12(1):81. https://doi.org/10.1186/s13045-019-0747-0.
[85] Wolska M, Jarosz-Popek J, Junger E, Wicik Z, Porshoor T, et al. Long non-coding RNAs as promising therapeutic approach in ischemic stroke: a comprehensive review. Molecular Neurobiology 2021;58(4):1664–82. https://doi.org/10.1007/s12035-020-02206-8.
[86] Xu T, Xu X, Chu Y, Jiang D, Xu G. Long-chain non-coding RNA GAS5 promotes cell autophagy by modulating the miR-181c-5p/ATG5 and miR-1192/ATG12 axes. International Journal of Molecular Medicine 2021;48(6):209. https://doi.org/10.3892/ijmm.2021.5042.
[87] Zou L, Liao M, Zhen Y, Zhu S, Chen X, et al. Autophagy and beyond: unraveling the complexity of UNC-51-like kinase 1 (ULK1) from biological functions to therapeutic implications. Acta Pharmaceutica Sinica B 2022;12(10):3743–82. https://doi.org/10.1016/j.apsb.2022.06.004.

[88] Chen S, Wu DD, Sang XB, Wang LL, et al. The lncRNA HULC functions as an oncogene by targeting ATG7 and ITGB1 in epithelial ovarian carcinoma. Cell Death & Disease 2017;8(10):e3118. https://doi.org/10.1038/cddis.2017.486.

[89] Li Y, Lin S, Xie X, Zhu H, Fan T, Wang S. Highly enriched exosomal lncRNA OIP5-AS1 regulates osteosarcoma tumor angiogenesis and autophagy through miR-153 and ATG5. American Journal of Translational Research 2021;13(5):4211−23.

[90] Yuan Y, Zhang X, Fan X, Peng Y, Jin Z. The emerging roles of circular RNA-mediated autophagy in tumorigenesis and cancer progression. Cell Death & Disease 2022;8(1):385. https://doi.org/10.1038/s41420-022-01172-5.

[91] Gan X, Zhu H, Jiang X, Obiegbusi SC, Yong M, Long X, et al. CircMUC16 promotes autophagy of epithelial ovarian cancer via interaction with ATG13 and miR-199a. Molecular Cancer 2020;19(1):45. https://doi.org/10.1186/s12943-020-01163-z.

[92] Zhang Z, Zhu H, Hu J. CircRAB11FIP1 promoted autophagy flux of ovarian cancer through DSC1 and miR-129. Cell Death & Disease 2021;12(2):219. https://doi.org/10.1038/s41419-021-03486-1.

[93] Yang J, Zhang X, Cao J, Xu P, Chen Z, et al. Circular RNA UBE2Q2 promotes malignant progression of gastric cancer by regulating signal transducer and activator of transcription 3-mediated autophagy and glycolysis. Cell Death & Disease 2021;12(10):910. https://doi.org/10.1038/s41419-021-04216-3.

[94] Zhou Q, Xie D, Wang R, Liu L, Yu Y, et al. The emerging landscape of exosomal CircRNAs in solid cancers and hematological malignancies. Biomarker research 2022;10(1):28. https://doi.org/10.1186/s40364-022-00375-3.

[95] You L, Wang Z, Li H, Shou J, Jing Z, Xie J, et al. The role of STAT3 in autophagy. Autophagy 2015;11(5):729−39. https://doi.org/10.1080/15548627.2015.1017192.

[96] Guo X, Tan W, Wang C. The emerging roles of exosomal circRNAs in diseases. Clinical and Translational Oncology 2021;23(6):1020−33. https://doi.org/10.1007/s12094-020-02485-6.

[97] Li J, Zhang G, Liu CG, Xiang X, Le MTN, et al. The potential role of exosomal circRNAs in the tumor microenvironment: insights into cancer diagnosis and therapy. Theranostics 2022;12(1):87−104. https://doi.org/10.7150/thno.64096.

[98] Dou Y, Cha DJ, Franklin JL, Higginbotham JN, et al. Circular RNAs are down-regulated in KRAS mutant colon cancer cells and can be transferred to exosomes. Scientific Reports 2016;6:37982. https://doi.org/10.1038/srep37982.

[99] Zhang H, Deng T, Ge S, Liu Y, Bai M, et al. Exosome circRNA secreted from adipocytes promotes the growth of hepatocellular carcinoma by targeting deubiquitination-related USP7. Oncogene 2019;38(15):2844−59. https://doi.org/10.1038/s41388-018-0619-z.

[100] Jiang MP, Xu WX, Hou JC, Xu Q, et al. The emerging role of the interactions between circular RNAs and RNA-binding proteins in common human cancers. Journal of Cancer 2021;12(17):5206−19. https://doi.org/10.7150/jca.58182.

CHAPTER SIX

Exosomal RNA in type 2 diabetes mellitus and other metabolic diseases

Kakali Ghoshal
Department of Medicine, Division of Nephrology and Hypertension, Vanderbilt University School of Medicine, Nashville, TN, United States

1. Introduction

Accurate prediction of disease progression and diagnosis are the key components in clinical research. With modern advancements in techniques, precise detection of soluble biomarkers in biofluids is a critical step in the early detection of a disease. Clinical studies investigating the roles of extracellular vesicles (EV) have gained their importance in recent times for their remarkable roles in various pathological and physiological conditions. EVs are lipid bilayer-delimited particles released by almost every type of cell. Exosomes are small EVs of endosomal origin in a size range of ~40–160 nm in diameter (~100 nm on average) [1]. Since exosomes are widely distributed in body fluids and can reflect physiological and pathological conditions, they have received increased attention in recent times.

Exosomes are universally present in all biofluids such as blood, plasma, serum, urine, and breast milk and can be released by almost every cell type, such as mesenchymal stem cells (MSCs), endothelial cells, myoblasts, and adipocytes [2]. Exosomes vary in size, contents, and functions, and their contents include RNAs, DNAs, proteins, and lipids, which participate in physiological processes such as intercellular communication and material transport [3]. Strong evidence suggests that exosomal RNAs can regulate gene expression and functions in recipient cells, and thus, they are emerging as a promising target in various human diseases [3–5]. Exosomal RNAs constitute both mRNA and noncoding RNAs such as miRNAs, lncRNAs, and circRNAs [2].

Metabolic diseases are a group of diseases often intricately linked with each other, raising the chances of serious health hazards and mortality manifold. Insulin resistance and diabetes mellitus are classes of metabolic disorders that are characterized by hyperglycemia, often caused by genetic and/or environmental factors [2]. According to the International Diabetes Federation 2021, about 537 million adults are affected by diabetes, and the prediction of a rise is 643 million by 2030 and 783 million by 2045 [6]. Even after significant progress in therapeutic management, the morbidity and mortality rates have continued to increase. The presence of other metabolic disorders such as obesity, hyperlipidemia, fatty liver diseases, and cardiovascular disorders alone or together

Exosomal RNA
ISBN 978-0-443-14008-2
https://doi.org/10.1016/B978-0-443-14008-2.00009-7

with diabetes can even magnify the threats several times. Therefore, advancements in early detection, monitoring the disease progression, and targeted therapy of metabolic disorders are of utmost importance [3]. In this chapter, we provide information of the recent advancements of exosomal RNA based therapeutic strategies in metabolic disease diagnosis, management, and treatment.

2. Characteristics features and function of exosomes and exosomal RNA

First discovered in sheep reticulocytes, exosomes were later found to be secreted by almost all mammalian cell types [7]. Exosomes are nanosized vesicles with a density of 1.10–1.19 g/mL, consisting of intracellular multivesicular bodies that fuse with the cell membrane to be released into the extracellular matrix [8]. The morphology of exosomes was studied under an electron microscope, revealing their flattened sphere-like shapes wrapped in lipid bilayer [8]. Exosomes are formed by double invagination of the plasma membrane creating intracellular multivesicular bodies (MVBs) containing intraluminal vesicles (ILVs) [1,8]. ILVs are secreted as exosomes by exocytosis through MVB fusion to the plasma membrane (Fig. 6.3) [1,8].

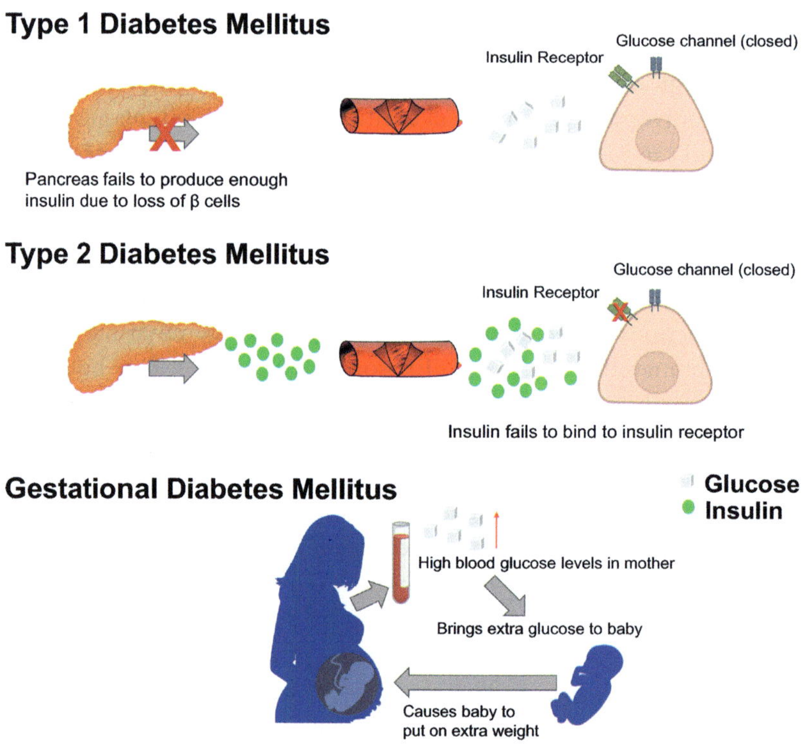

Figure 6.1 Types of diabetes mellitus.

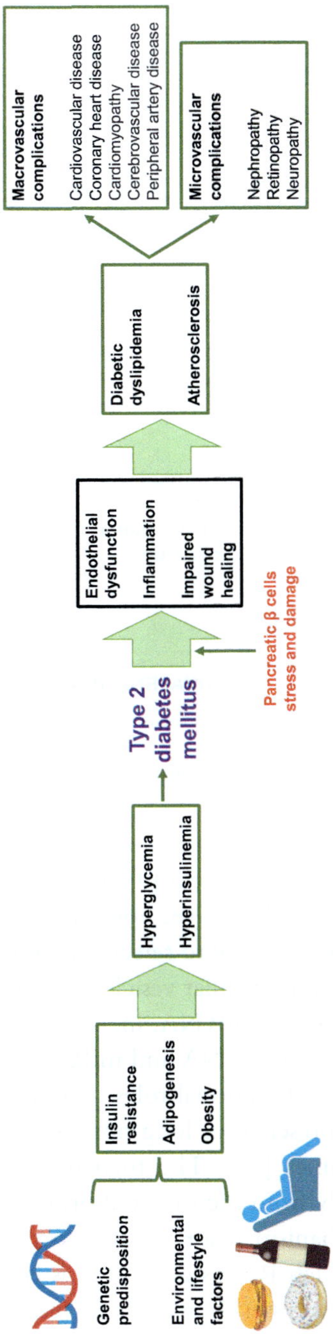

Figure 6.2 Pathophysiology and complications of type 2 diabetes mellitus and related metabolic complications.

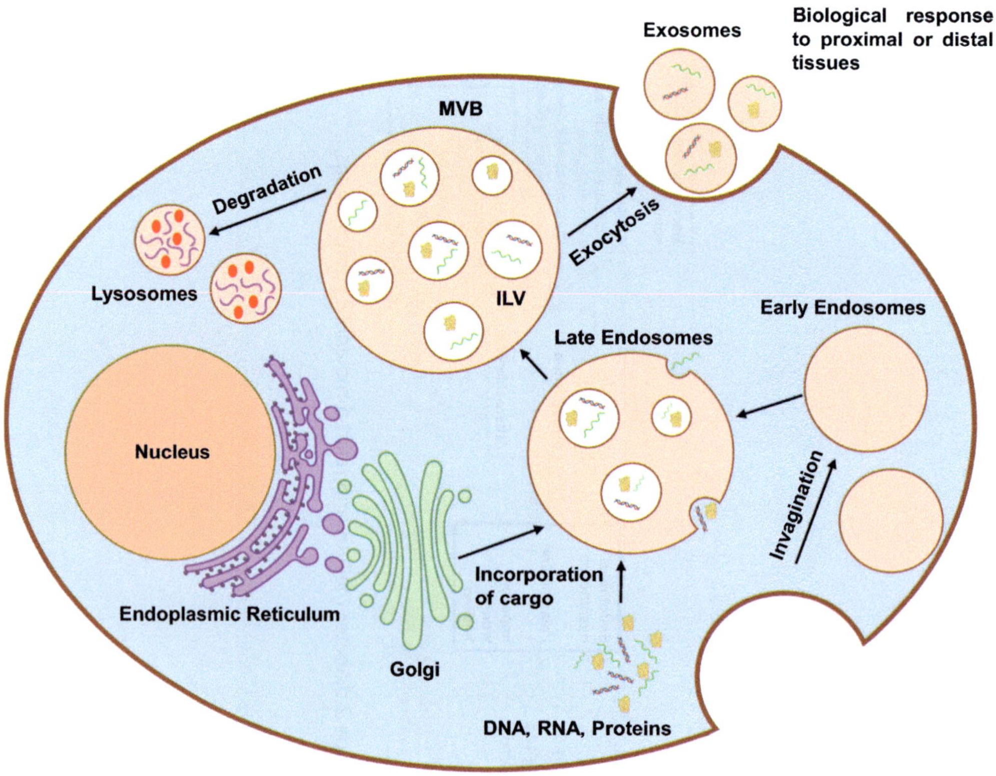

Figure 6.3 Schematic representation of biogenesis of exosomes.

Exosomal RNA is a mixture of coding and noncoding RNA that is interdependent on the microenvironment of the secreting cells [9]. The underlying mechanisms of how RNA is loaded into exosomes are still elusive; however, a few RNA-binding proteins have been identified that possesses capacity to selectively bind to RNA through specific motifs and then deliver them into intracellular vesicles [9]. Exosomes harbor both coding mRNA and noncoding RNA such as miRNA, lncRNAs, and circRNAs. Valadi et al. first reported the abundance of both mRNA and miRNA in exosomes, which can be delivered to other cells with active functional roles in a new location [10]. Various investigations also confirmed the presence of long noncoding RNA (lncRNA) and circular RNA (cirRNA) in exosomes [11]. The total mRNA, miRNA, and lncRNA content in exosomes differ greatly from their cells of origin, indicating that cytoplasmic RNA undergoes a sorting mechanism prior to its incorporation in exosomes [7,10]. However, the mechanism by which RNAs get sorted is largely unknown. It has been postulated that exosomal RNA possesses short, specific sequences known as EXOmotifs that can bind to different RNA-binding proteins such as hnRNPA2B1, hnRNPA1, YBX1, or SYNCRIP to facilitate the sorting procedure in different cells [7,12].

3. Pathophysiology of insulin resistance, diabetes mellitus, and related metabolic complications

Diabetes mellitus, a global noncommunicable metabolic disorder, is characterized by the presence of chronic hyperglycemia accompanied by greater or lesser impairment in metabolism of carbohydrates, lipids, and proteins [13]. Chronic hyperglycemia in diabetes is often linked with long-term system damage and dysfunction, and failure of vital organs such as eyes, kidneys, nerves, heart, and blood vessels [13]. Diabetes mellitus is subdivided in three broad categories (Fig. 6.1). Type 1, or insulin-dependent diabetes mellitus (IDDM) or "juvenile diabetes," results from the failure of pancreas to produce enough insulin. This affects ~10% of the total diabetic population [13]. Type 2, or noninsulin-dependent diabetes mellitus (NIDDM) or "adult-onset diabetes," affects ~90% of the total diabetic population and begins with insulin resistance, a condition in which cells fail to respond to insulin [13]. However, in type 2 diabetes, both impaired insulin secretion through a dysfunction of the pancreatic β-cell and attenuated insulin action through insulin resistance are implicated [14]. Insulin resistance and hyperinsulinemia eventually progress to impaired glucose tolerance [13]. Insulin resistance thus serves as the primary risk factor for the development of type 2 diabetes and related metabolic complications [15]. Gestational diabetes is a condition where pregnant women develop hyperglycemia [13].

Diabetes is often closely linked with other metabolic complications such as obesity, dyslipidemia, and cardiovascular disorders [16,17], together called the metabolic syndrome. Type 2 diabetes mellitus (T2DM) also results in different macro- and microvascular disorders, and the most prevalent diabetic macrovascular complication being cardiovascular disorder [18]. Increased fatty acid release from insulin-resistant adipocytes aggravates the progression of diabetic dyslipidemia [17]. Furthermore, enhanced free fatty acids flux in abundance of glycogens produces triglycerides, prompting apolipoprotein B (ApoB) and VLDL secretion [16,17]. Through several underlined pathways, VLDL can decrease HDL and increase LDL levels. Along with hepatic insulin resistance, it is often linked with the progression of nonalcoholic fatty liver disease (NAFLD) [19]. The pathophysiology and complications of T2DM and related metabolic disorders are described in Fig. 6.2. In this chapter, we will focus on the versatile roles and clinical applications of various tissue-specific exosomal RNAs, especially those particularly contributing to diabetes and metabolic complications.

3.1 Role of exosomal RNA in insulin resistance and metabolic syndrome

3.1.1 *In adipose tissue*

Historically, adipose tissue was identified as an energy storage organ; however, recent studies identify it as a dynamic endocrine organ secreting a variety of adipokines, enzymes, growth factors, and hormones regulating glucose and lipid metabolism [8,20].

Insulin acts on adipose tissue either by stimulating glucose uptake and triglyceride synthesis or by suppressing triglyceride hydrolysis and the release of FFA and glycerol into the circulation [21]. Adipocyte-derived exosomes, also known as adiposome, can contribute to insulin resistance and obesity [22]. Adiposomes contain approximately 7000 mRNAs and 140 miRNAs, most of which are adipocyte specific and work in paracrine or endocrine manner [23]. In addition, adipocyte-derived exosomes are believed to be mediators linking obesity and insulin resistance in surrounding tissues such as the liver, interacting with adjacent sites to promote lipid esterification, signifying the role of exosomes in systemic metabolism [24].

Extensive research was done on adipose tissue/adipocyte-derived exosomes and exosomal RNA, and their regulatory roles in metabolic disorders. Thomou et al. have shown that mice lacking the adipose tissue–specific microRNA (miRNA)-processing enzyme Dicer (ADicerKO) as well as humans with lipodystrophy exhibit a substantially low level of circulating exosomal miRNAs [25]. However, transplantation of both white and brown adipose tissue in the knockout mice successfully restored the circulating miRNA levels with improved glucose tolerance. Similarly, administration of serum exosomes from healthy wild-type mice to the knockouts manifested a similar outcome as the transplanted group [25]. They have also shown that ADicerKO mice have a significantly higher level of circulating fibroblast growth factor 21 (FGF21), which influences metabolism in several tissues. Along with, they hypothesized that how miR-99b could be a possible candidate for regulating FGF21.

Yu et al. proposed that adipocyte-derived exosomal miR-27a induces insulin resistance in skeletal muscle through repression of PPARγ [26]. Adipose tissue inflammation, a signature in metabolic disorders, corresponds to immune cell (particularly macrophage) infiltration in adipose tissues. Song et al. demonstrated that adiposomes carrying sonic hedgehog (Shh) induce the proinflammatory M1 polarization of macrophages, which contributes to insulin resistance in adipocytes through the Ptch/PI3K pathway [27]. Wang et al. have shown that activated macrophages secrete miR-155-enriched exosomes, which act as a paracrine regulator for fibroblast proliferation and inflammation in myocardial infarction [28]. Ying et al. have shown that exosomal miRNAs (particularly miR-155) released by adipose tissue–derived macrophages cause glucose intolerance and insulin resistance [29]. On the contrary, Lin et al. demonstrated a possible negative correlation of miR-155 with insulin resistance and diabetes in humans and mice [30]. They reported that the serum of type 2 diabetic patients has a lower level of miR-155. They have also shown that transgenic overexpression of miR-155 in mice leads to hypoglycemia, improved glucose tolerance, and insulin sensitivity. On the contrary, miR-155 deletion leads to hyperglycemia, insulin resistance, and glucose intolerance in mice [30]. Adiponectin is an adipose tissue–derived hormone that plays significant protective roles against insulin resistance, diabetes, and obesity [20,31]. Pan et al. reported that exosomal miR-34a secreted by adipocytes can inhibit the antiinflammatory M2 polarization of macrophages, eventually repressing Klf4, promoting adipose tissue inflammation [32]. They have also shown that deletion of miR-34a restores inflammation and increases adiponectin.

Ferrante et al. have identified several visceral adipose tissue—derived exosomal miRNAs that are differentially regulated in obese individuals compared with their lean counterparts [33]. Seven such miRNAs (miR-23b, miR-148b, miR-182, miR-3681, miR-4269, miR-4319, and miR-4429) were found to target the mRNA in the TGF-β and Wnt/β-catenin signaling pathways. They confirmed downregulation of miR-148b and miR-4269 and upregulation of miR-23b are linked with obesity [33].

A study suggested that in obese and lean individuals, 10 and 58 exosomal miRNAs were differentially expressed in subcutaneous and visceral adipose tissues, respectively [34]. These mRNAs are responsible for targeting inflammatory genes related to obesity. CD86 is regulated by hsamiR31565p, whereas FGF2, FOSL2, and AMPD3 are regulated by hsamiR5825p, hsamiR566, and miR548, respectively [34].

Castaño et al. identified several miRNA profiles of plasma exosomes that increased in obesity, including miR-122, miR-192, miR-27a-3p, and miR-27b-3p [35]. Most importantly, they have shown that administration of control exosomes transfected with obesity-associated miRNA mimics (MIMIC) strongly induces glucose intolerance in lean mice and results in central obesity and hepatic steatosis. MIMIC-treated mice significantly increased PPARα expression in white adipose tissues but not in livers, resulting in increased circulating free fatty acids and hypertriglyceridemia. Wen et al. have shown that hypertrophic adipocyte-derived exosomal miR-802-5p can induce insulin resistance in cardiac myocytes by targeting HSP60 [36]. Thus, evidence suggests that adipose tissue—derived exosomes and exosomal RNA contribute significantly to glucose and lipid homeostasis in various physiological conditions. In one study, it was found that adipose tissue—derived stem cell—derived exosomes (ADSC-Exo) contain microRNAs with proangiogenic (miR-126, miR-130a, and miR-132) and antifibrotic (miR-let7b and miR-let7c) properties [37]. Thus, transplantation of ADSC-Exo could successfully restore erectile function in diabetic rats [37]. Putting these recent studies together, they implicate that adipocytes/adipose tissue—derived exosomes and their RNA content play critical roles in glucose and lipid homeostasis, thus having tremendous clinical significance.

3.1.2 In livers

Liver is an insulin-sensitive organ, and defective insulin signaling and the development of hepatic insulin resistance can majorly imbalance whole body energy homeostasis and metabolism [38]. However, only a few studies have discussed the role of exosomal RNA in hepatic insulin resistance and related metabolic complications. For instance, Ji et al.

have shown that early-onset obesity induces the secretion of exosomal miR-3075 from hepatocytes, exhibiting protective roles against insulin resistance [39]. They also postulated that miR-3075 targets FA2H, and FA2H depletion increases insulin sensitivity in adipocytes, myocytes, and hepatocytes. Su et al. have shown that bone marrow mesenchymal stem cells (BMSCs)-derived exosomal miR-29b-3p governs aging-associated insulin resistance in hepatocytes, myocytes, and adipocytes by targeting sirtuin 1 [40]. Few studies have implicated the role of exosomal RNAs in fatty liver diseases. For instance, Yamada et al. have depicted that increased serum levels of miR-21, miR-34a, miR-122, and miR-451 are associated with the development and progression of nonalcoholic fatty liver disease [41]. In another study, Becker et al. have demonstrated that differential expression of serum exosomal miR-122, miR-192, and mIR-21 may regulate the progression of nonalcoholic steatohepatitis [42].

3.1.3 *In muscle*

Skeletal muscle, being extremely insulin sensitive, is responsible for over 80% of glucose uptake from oral glucose load, thus having an essential role in glucose clearance [43,44]. Skeletal muscle insulin resistance can appear decades before the onset of pancreatic β-cell failure and other symptoms of type 2 diabetes mellitus [43,44]. However, only a few studies have depicted the role of exosomal RNA in the pathophysiology of skeletal muscle in diabetic complications. Katayama et al. identified several differentially expressed exosomal miRNAs in the serum of type 2 diabetic individuals, of which miR-20b-5p was the most prevalent [45]. Moreover, when miR-20b-5p is transfected into primary human skeletal muscle cells, it alters cellular glucose metabolism and the STAT3 and AKT signaling pathways. They propose that miR-20b-5p may be part of a homeostatic mechanism attempting to improve metabolic control [45]. Wang et al. postulated that exosomes and exosomal RNA derived from pancreatic cancer cells induce insulin resistance in skeletal muscle cells through the PI3K/Akt/FoxO1 pathway [46]. Another study revealed that inhibition of miR-182-3p can alleviate gestational diabetes mellitus by improving insulin resistance in skeletal muscle [47]. Only one study reported that smooth muscle density improved after PDCD4 silencing or treatment with miR-21-5p-containing MSC-derived exosomes in a diabetic rat model with erectile dysfunction [48].

3.2 Role of exosomal RNA in pancreatic β-cells dysfunction

The pancreas is an organ that has both exocrine and endocrine functions. The endocrine role of the pancreas involves secreting insulin from its β cells and a few other hormones from pancreatic islets, which help to maintain glucose homeostasis

[49]. The role of exosomal RNA in pancreatic β cells functions has been elucidated in some studies. For example, inhibition of miR-375 has been implicated in defective pancreatic islet development [50]. At the same time, miR-375 helps to maintain normal pancreatic alpha- and beta-cell masses [51]. On the contrary, in another study, it was found that overexpression of miR-375 suppressed glucose-induced insulin secretion, and its inhibition enhanced insulin secretion [52]. Exosomal RNA released by pancreatic β cells has a profound effect on surrounding cells and can constitute a cell-to-cell communication mode, permitting a concerted adaptation of β cells to environmental cues [53]. Guo et al. have shown that β-cell-derived exosomes promote differentiation of induced pluripotent stem cells into insulin-producing cells through microRNA-dependent (especially miR-706, miR-709, miR-466c-5p, and miR-423-5p) mechanisms [54]. Li et al. have shown that pancreatic β cells control glucose homeostasis via the secretion of exosomal miR-29 [55]. An increased load of free fatty acids also enhanced the expression of miR-29 [55]. Xu et al. demonstrated that serum exosomal miR-26a is reduced in individuals with obesity and T2DM [56]. Mechanistically, they have provided evidence that pancreatic β-cells—derived exosomal miR-26a prevents hyperinsulinemia through targeting several critical regulators of insulin secretion and β-cell proliferation [56]. Kamalden et al. have shown that exosomal miR-15a from pancreatic β-cells can travel through the circulation and worsen diabetic complications via their transfer to different cell types, inducing oxidative stress and cell injury [57]. They have also witnessed an increased concentration of miR-15a in the plasma of diabetic patients, correlating with their disease severity [57]. Despite these studies, the role of exosomal RNA in maintaining the functional aspects of pancreatic β cells is poorly understood, and further investigation may open new clinical and therapeutic ventures.

3.3 Role of exosomal RNA in hyperglycemia-induced inflammation

Diabetes mellitus exerts low-grade systemic inflammation, and often, it governs the diabetic pathophysiology and metabolic abnormalities [58]. Natural killer (NK) cells are believed to be associated with type 2 diabetes by regulating systemic inflammation; however, the underlying mechanism is yet unknown. Wang et al. demonstrated that NK-derived exosomal miR-1249-3p dramatically induces cellular insulin sensitivity and relieves inflammation in hepatocytes and adipocytes [59]. Mechanistically, they have shown that exosomal miR-1249-3p can directly target SKOR1 to govern the formation of the ternary complex SMAD6/MYD88/SMURF1, which mediates glucose homeostasis by suppressing the TLR4/NF-κB signaling pathway [59]. A study by Shyu et al. shows that high glucose can significantly upregulate macrophage-derived exosomal metastasis-associated lung adenocarcinoma transcript 1 (MALAT1) expression to suppress miR-150-5p, counteracting the inhibitory effect of miR-150-5p on resist in expression in macrophages to promote vascular disease [60]. Ying et al. have shown that M2-polarized bone marrow—derived macrophages release exosomes enriched with miR-690 targeting NAD kinase, hence modulating macrophage inflammation and insulin signaling [61].

3.4 Exosomal RNA in diabetic macrovascular complications

Diabetes results in various macrovascular complications, including cardiovascular disease (CVD), coronary heart disease, cardiomyopathy, cerebrovascular disease, and peripheral artery disease [62]. Macrovascular disease starts with atherosclerotic plaque formation in the vasculature and is closely linked with dyslipidemia, obesity, and hypertension [62,63]. Studies indicate that exosomes stimulate vascular growth and maturation by delivering angiogenic miRNA molecules to exocrine bodies, and thus, exosomal miRNA could be a potential clinical and therapeutic target for diabetic macrovascular complications [64].

For instance, Barile et al. have shown that treatment with exosomes released from cardiac progenitor cells (CPCs) improved cardiac function after injury [65]. They have identified that exosomal miR-210 from CPCs inhibits apoptosis in cardiomyocytic cells by downregulating ephrin A3 and PTP1b. They have also shown that exosomal miR-132 downregulates RasGAP-p120, enhancing tube formation in endothelial cells [65]. In another study, a group of researchers showed that BMSCs-derived exosomal miR-214 significantly decreased apoptosis rates and reactive oxygen species (ROS) production in cardiac stem cells (CSCs) after oxidative stress injury [66]. They provided evidence that in CSCs, miR-214/CaMKII axis regulates oxidative stress—related injury, apoptosis, and calcium homeostasis disequilibrium. In another study, the protective role of BMSCs derived from exosomal miR-125b by targeting SIRT7 in myocardial ischemia reperfusion injury was well elucidated [67]. Wang et al. have shown that exosomes secreted by MSCs are enriched with miR-210 [68]. miR-210 is necessary to exert a proangiogenic effect through the miR-210-Efna3 pathway for its therapeutic function against myocardial infarction. Thus, exosomes are gaining importance for treating target cells and thereby exerting therapeutic properties. In another study, MSC-derived exosome cargo containing miRNA-181a was delivered to exert its therapeutic effect on myocardial I/R injury [69]. miRNA-181a has been shown to exert an antiinflammatory effect in the target cells by raising the regulatory T cell ratio through inhibition of the c-Fos protein. Adamiak et al. did an extensive study on murine-induced pluripotent stem cell (iPSC)—derived exosomes and exosomal RNA and their protective role in myocardial tissues [70]. They postulated that exosomal microRNAs such as let-7, miR-145, miR-17-92, and miR-302a-5p play crucial roles in the regulation of cell proliferation, self-renewal, and apoptosis, exerting a beneficial role in myocardial tissues. Their heatmap analysis also predicted that these miRNA contents regulate various pathways, including Wnt, PI3K-Akt, and MAPK, which are involved in the regulation of the actin cytoskeleton, focal adhesion, and ECM—receptor interactions. Additionally, they identified several miRNAs (miR-294, miR-16, miR-34, and miR-20) in iPSC-derived exosomes that enhance angiogenesis by regulating VEGFA [70].

The promising therapeutic benefits of exosomes and their miRNA contents in heart diseases have been extensively studied. For instance, it was shown that exosomes secreted by myocardial cells and cardiomyocytes cultured in ischemic conditions are rich with miR-222 and miR-143. Also, metalloproteases (MMPs) are abundant in exosomes in ischemic conditions, which together with miRNA confer protection against oxidative-induced lesion, promote proliferation, and sprout of endothelial cells, along with promoting angiogenesis [71]. Mouse BMSCs treated in hypoxic conditions released exosomes enriched with miR-210, which exerts a protective effect on myocardial infarction [72]. These include significantly higher survival rates, smaller scar sizes, and better cardiac function recovery. Plasma exosomal miRNA profiling from severe peripheral artery disease patients revealed a significant increase in certain miRNAs (miR-21, miR-92a, miR-126, miR-143, miR-181b, and miR-221) compared with healthy or mild peripheral artery disease patients [73]. In addition, circulating exosomes have been shown to exert in vitro functional effects on vascular smooth muscle cells and endothelial cells in terms of vessel remodeling. A study by Huang et al. revealed a selected panel of 12 circulating miRNAs associated with diabetic heart failure (miR-1-3p, miR-21-5p, miR-29a-5p, miR-30d-5p, miR-34a-5p, miR-126a-5p, miR-143-3p, miR-145-5p, miR-195-5p, miR-206-3p, miR-320-3p, and miR-378-3p) [74]. Maternal diabetes can also contribute to changing the exosomal miRNA profile, leading to congenital defects. For instance, maternal diabetes altered the exosomal miRNA profile (miR-133, miR-30, miR-99, miR-23) impacting the fetal cardiac development [75]. Thus, these studies indicate that the secretory miRNAs carried by exosomes and transported to corresponding target cells play different roles in diabetes-related cardiovascular complications [64].

3.5 Exosomal RNA in diabetic microvascular complications

Diabetic microvascular complications are long-term complications that affect small blood vessels, typically manifesting three pathological conditions known as nephropathy, retinopathy, and neuropathy [76]. The role of exosomal miRNAs in governing renal function and physiology in diabetes is thus gaining importance as exosomes play critical roles in disease occurrence and development. For instance, Zhou et al. reported differential expression profiles of several exosomal RNAs (169 lncRNAs, 885 mRNAs, 3 circRNAs, and 152 miRNAs) secreted by high glucose challenged HK-2 cells compared with a controlled condition [76]. Barutta et al. have shown that urinary exosomes from type 1 diabetic nephropathy patients have enriched miR-130a and miR-145, while miR-155 and miR-424 are reduced [77]. Overexpression of miR-145 has also been also witnessed in urine and glomerulus in animal models of diabetic nephropathy, making it a promising diagnostic biomarker. In another study, Delić et al. found several differentially regulated urinary exosomal miRNAs to be associated with the progression of type 2 diabetes—associated nephropathy [78]. Zang et al. revealed significantly upregulated urinary exosomal miR-21-5p, let-7e-5p, and miR-23b-3p and downregulated miR-30b-5p and miR-125b-5p to be associated with type 2 diabetic nephropathy [79]. On the other hand, Kim et al. demonstrated that seven serum exosomal miRNAs (miR-1246, miR-642a-3p, let-7c-5p, miR-1255b-5p, let-7i-3p, miR-5010-5p, and miR-150-3p) are upregulated in diabetic subjects with nephropathy [80]. BMSC-derived exosomal miR-30e-5p inhibited caspase-1-mediated pyroptosis by targeting ELAVL1 in high-glucose-induced HK-2 cells, providing a new strategy for treating diabetic kidney disease [80]. Podocyte-derived signal transduction factors are good candidates to assess podocyte injuries in diabetic nephropathy progression. Both urinary and blood exosomal mRNA Wilms tumor 1 (WT1) is reflective of damage in diabetic glomeruli and thus an important diagnostic and prognostic biomarker for diabetic nephropathy [81,82]. A study on rats indicated that an increase in urinary exosomal miR-451-5p and miR-16 expression in diabetic nephropathy exerts protective effects against diabetes-induced kidney fibrosis [83]. In one study, Li et al. showed that urine-derived stem cells—derived exosomal miR-146a-5p could protect against renal ischemia/reperfusion (I/R) injury by targeting the 3′UTR of IRAK1 and subsequently inhibiting the activation of NF-κB signaling and infiltration of inflammatory cells to protect renal function [84]. Apart from miRNA, exosomal lncRNAs and circRNAs are also believed to play critical roles in diabetic nephropathy progression. For instance, the transcriptional regulatory network on urine and plasma exosomes showed five lncRNAs (LINC02614, BAALC-AS1, FAM230B, LOC100505824, and LINC01484) and the miR-301a-3p to be associated in network organization in critical pathways regulating filtration barrier integrity and tubule reabsorption [85]. Yang et al. identified 27 macrophage-derived differentially expressed exosomal lncRNAs that are involved in high-glucose-induced epithelial—mesenchymal transition (EMT), a signature of diabetic nephropathy progression [86]. In another study, Sun et al. proposed that urine-derived stem cells—derived exosomal lncRNA TUG1 regulates ASCL4-mediated ferroptosis by interacting with SRSF1, protecting ischemia/reperfusion injury—induced acute kidney injury [87]. Dong et al. demonstrated that circular ribonucleic acid nucleoporin 98 (circNUP98) knockdown alleviated high-glucose-induced proliferation, fibrosis, inflammation, and oxidative stress in human glomerular mesangial cells by regulating the miR-151-3p-HMGA2 axis, providing potential therapeutics for diabetic nephropathy [88]. CircRNA is more stable than lncRNA, and it gets encapsulated in exosomes to participate in the various pathological processes of several diseases. A study by Li et al. demonstrated that high-glucose-treated glomerular epithelial cells (GECs) secrete exosomes enriched with circTAOK1, boosting proliferation, fibrosis, and EMT of glomerular mesangial cells through targeting the miR-520h/SMAD3 axis, providing new insights into the pathogenesis of diabetic nephropathy [89]. Bai et al. proposed that high glucose in diabetic conditions induces the enrichment of exosomal circDLGAP4 from mesangial cells [90]. They have also shown that overexpression of circDLGAP4 markedly promoted diabetic kidney disease progression via modulating the miR-143/ERBB3/NF-κB/MMP-2 axis [90]. Zhu et al. proposed that high glucose induces the expression of exosomal circ0125310 from mesangial cells, promoting cell proliferation and fibrosis in diabetic nephropathy via sponging miR-422a and targeting the IGF1R/p38 axis [91].

Diabetic retinopathy is another common microvascular complication resulting in visual impairment. Cao et al. have shown that MSCs-derived exosomal lncRNA SNHG7 suppresses endothelial—mesenchymal transition and tube formation in human retinal microvascular endothelial cells via the miR-34a-5p/XBP1 axis [92]. Ye et al. proposed that hypoxia can upregulate circEhmt1 expression in pericytes, and exosomal transfer of circEhmt1 to endotheliocytes protects them from high-glucose-induced injury. They demonstrated that overexpression of circEhmt1 in pericytes upregulates the transcriptional factor NFIA to suppress NLRP3-mediated inflammasome formation [93]. In another study, Gu et al. demonstrated that retinal pigment epithelial cells secrete miR-202-5p-containing exosomes to protect against proliferative diabetic retinopathy through the TGF/Smad pathway [94]. Exosomal lncRNAs also exhibit the potential to act as predictive biomarkers to detect the progression of diabetic retinopathy. For example, Ye et al. have identified that in diabetic retinopathy, plasma exosomal lncRNA DLX6-AS1 acts as a risk factor, whereas lncRNAs PRINS and FAM190A-3 have protective roles [95]. Li et al. demonstrated that CircFndc3b derived from high-glucose-induced endothelial cells may play an important role in the angiogenesis of proliferative diabetic retinopathy [96]. Thus, these studies give a lot of proof that exosomal RNA could make a big difference in both treating and diagnosing microvascular complications of diabetes.

3.6 Exosomal RNA in diabetic wound healing

Another major challenge of diabetes is impaired wound healing due to the weakened functionality of angiogenesis caused by persistent hyperglycemia [97]. Chen et al. reported that two types of serum exosomal circRNAs (hsa_circ_0000907 and hsa_circ_0057362) are markedly increased in diabetic foot ulcers, thus having novel diagnostic potential [98]. In another study, Kuang et al. have reported that human keratinocyte-derived exosomal MALAT1 facilitates wound healing by suppressing miR-1914-3p to activate MFGE8, which enhances macrophage phagocytosis and the prohealing phenotype while reducing apoptosis [99]. Also, bone marrow stromal cell—derived exosomal circITCH was reported to inhibit ferroptosis and improve angiogenesis, accelerating the wound healing process in diabetic foot ulcers by recruiting the TAF15 protein to activate the Nrf2 signaling pathway [100]. Fu et al. have demonstrated that human amniotic MSC—derived exosomes promote diabetic wound healing by enhancing angiogenesis through several lncRNAs (PANTR1, H19, OIP5-AS1, and NR2F1-AS1), thus representing a promising strategy for diabetic wound healing [97]. In another study, the role of BMSC-derived exosomal lncRNA KLF3-AS1 in promoting diabetic cutaneous wound healing has been well documented [101]. Li et al. have demonstrated that MSC-derived exosomal lncRNA H19 prevents apoptosis and inflammation of fibroblasts by impairing miR-152-3p-mediated PTEN inhibition, facilitating wound healing in diabetic foot ulcers [102]. Liang et al. have demonstrated that umbilical cord—derived MSC-derived exosomal circHIPK3 can promote diabetes-induced ulcer wound healing through the miR-20b-5p/Nrf2/VEGFA axis [103], thus providing a novel therapeutic strategy.

3.7 Exosomal RNA in type 1 diabetes mellitus

Exosomal RNA has also been implicated in the diagnosis and therapy of type 1 diabetes mellitus (T1DM), which occurs in children. Kong et al. have proposed that urinary upregulation of exosomal miR-424 and miR-218 could be predictive biomarkers for detecting the progression of T1DM [104]. Ibrahim et al. witnessed an increased expression of serum exosomal miR-34a in T1DM patients [105]. Furthermore, they identified a strong association between miR-34a with biomarkers of dyslipidemia and endothelial dysfunction, elucidating the critical role miR-34a plays in glucose and lipid homeostasis. Pang et al. identified several plasma exosomal circRNAs that are differentially expressed in T1DM [106]. They predicted that hsa_circ0005630-miR-1247-5p (downregulation)—ATXN1/ARL6IP1 and hsa_circ0007026-miR-324-5p (upregulation)—NCAPD2/PGAM1 axis might be implicated in the development of T1DM. In another study, the plasma exosomal lncRNA profile revealed several (77 upregulated and 85 downregulated) differentially expressed lncRNAs to be associated with the progression of T1DM [107].

3.8 Exosomal RNA in gestational diabetes

Gestational diabetes mellitus (GDM) can severely affect both the health of the mother and the fetus, and there is a dearth of effective diagnostic biomarkers to detect the development of GDM during early pregnancy. Jiang et al. demonstrated that in humans, plasma exosomal hsa_cirRNA_0039480 is highly expressed at different stages of GDM and thus acts as an effective candidate biomarker for early detection of GDM [108]. Cao et al. found several exosomal circRNAs (9 upregulated circRNAs; circ_0022742, circ_0024650, circ_0046060, circ_0061201, circ_0054905, circ_0091988, circ_0042554, circ_0020530, and circ_0087091 and six downregulated circRNAs; circ_0075548, circ_0092108, circ_0073438, circ_0011012, circ_0075313, and circ_0058166) from umbilical cord blood that are differentially expressed in GDM as compared with healthy subjects [109]. Huang et al. suggested that in GDM, exosomes released from umbilical cord blood are enriched with circ_0074673, which in turn attenuates proliferation, migration, and angiogenesis of umbilical vein endothelial cells via the miR-1200/MEOX2 axis [110]. Alternatively, Cao et al. proposed that in GDM, exosomal circ_0046060 of umbilical cord mesenchymal stromal cells can ameliorate insulin resistance via the miR-338-3p/G6PC2 axis [111]. The roles of exosomal mRNAs and lncRNAs in the progression of GDM are, however, poorly understood. Cao et al. did microarray analysis to identify 84 mRNAs and 256 lncRNAs that are differentially expressed in the umbilical cord blood exosomes of GDM patients compared with healthy controls [112]. Tang et al. proposed that the lncRNA GAS5 in the peripheral blood exosomes interacts with HERC5/TAC1 and can act as a novel therapeutic target to identify the pathogenesis of GDM [113].

4. Discussion and conclusion

Exosomes are gaining importance in recent times for their critical roles in cell-to-cell communication in both endocrine and paracrine manner, thus mediating various vital cellular functions [114]. So far, there is no cure for diabetes and related metabolic complications, and often these disorders persist throughout a person's lifetime. The current clinical manifestations to identify the complexities and severeness of different types of diabetes and related metabolic disorders often fail to precisely address the disease pathophysiology and progression [8]. Since exosomes and their RNA contents are versatile carriers of molecular signals and can precisely reflect the pathophysiological characteristics of disease progression in real time, they are emerging as novel therapeutic targets for metabolic disorders in the near future (Fig. 6.4; Table 6.1). Development of next-generation sequencing technologies identified several exosomal miRNAs implicated in diabetes and related complications [115]. Compared with exosomal proteins, the isolation and characterization of exosomal RNA are relatively easier, inexpensive, sensitive, and selective. Circulatory exosomes have unique advantages as a natural endogenous drug carrier because of their good immune compatibility and low immunogenicity [8]. At the same time, exosomes are not easily targeted by immune cells

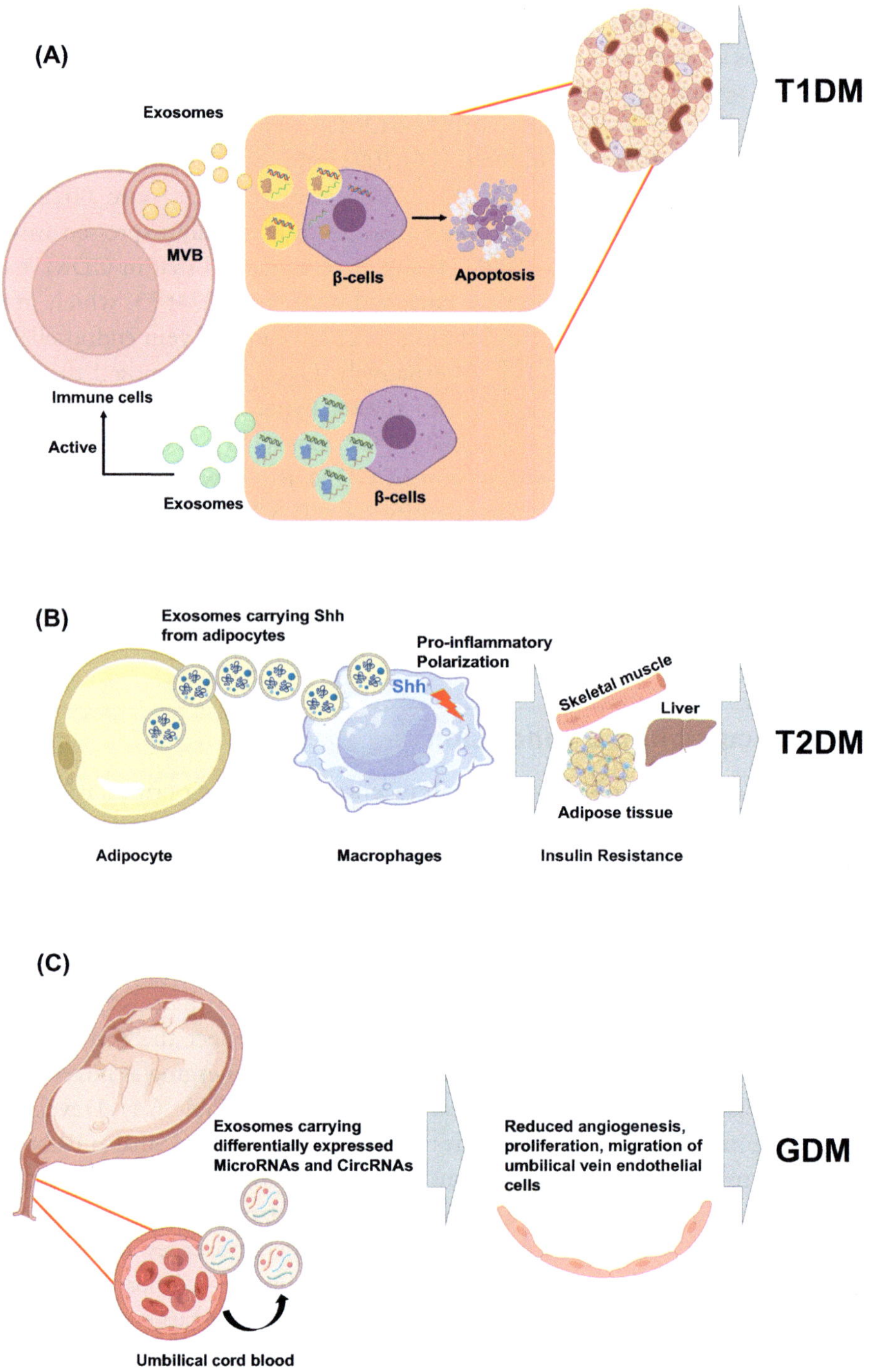

Figure 6.4 Role of exosomal RNA in regulating the pathophysiology and progression of diabetes mellitus. (A) In type 1 diabetes mellitus (T1DM), activated immune cells—derived exosomal RNA induces β-cell apoptosis and death. β-cells of pancreatic islets can also release exosomes carrying RNAs

possessing greater stability; likewise, they can easily cross the blood—brain barrier [8,116]. These attractive attributes of exosomes and exosomal RNAs are advantageous for delivering drugs targeting specific tissues or organs.

Recent studies gave ample evidence of exosomal miRNA's ability to regulate insulin signaling [25,26,39] as well as diabetic macro- [84] and microvascular complications [70,73]. At the same time, serum, plasma, or urine levels of specific exosomal miRNAs are often differentially regulated in diabetes and related metabolic complications [30,77]. These advance the therapeutic and diagnostic applications of exosomal miRNA in predicting disease progression, as well as inventing new therapeutic targets. For instance, Kumar et al. have shown that a high-fat diet caused insulin resistance in mice with increased expression of the aryl hydrocarbon receptor and inhibited expression of miR-375 and VAMP7 [117]. However, when they advanced to increase miR-375 expression by feeding the mice ginger-derived nanoparticles, the mice gained insulin sensitivity [117]. The pivotal role of exosomal miRNA is to act on mRNA in target tissues; thus, manipulating the release of exosomal miRNA can selectively alleviate disease complexities. Despite the recent advances in exosomal RNAs in diabetic pathophysiology, studies elucidating the roles of exosomal lncRNA or circRNA are relatively rare. Only a few studies have depicted how exosomal circRNAs can sponge with miRNAs to mediate various critical signaling pathways involved in metabolic complications [89,90].

Research on exosomal miRNA comes with a few additional challenges. Isolation and purification of exosomes are not yet fully developed and often consist of residual non-vesicular compartments, affecting downstream experiments of exosomal RNAs [115]. The abundance of exosomal RNAs in body fluids is relatively low, which restricts the potential clinical and therapeutic applications. Exosomal miRNA could possibly be linked with one or more diseases, and further investigation is needed to specifically address diabetes and related complications. Lastly, most studies on exosomal RNAs are either on cell lines or in animal models, and there remains a lack of clinical trials to confirm the accuracy and safety of these findings [115]. Despite these challenges, with further investigation and the development of technological advances, exosomal RNAs are expected to provide new therapeutic strategies for the prevention, diagnosis, and treatment of diabetes mellitus and related complications.

to further activate immune cells favoring T1DM progression. (B) Adipocyte-derived exosomes carrying sonic hedgehog (Shh) induce proinflammatory M1 polarization of macrophages inducing insulin resistance in skeletal muscle, liver, and adipose tissues, thus exacerbating the pathophysiology of type 2 diabetes mellitus (T2DM). (C) In gestational diabetes mellitus (GDM), exosomes carrying microRNAs and circRNAs from umbilical cord blood reduce the angiogenesis, proliferation, and migration of umbilical vein endothelial cells, thus facilitating diabetic progression in both the mother and the baby.

Table 6.1 List of exosomal RNA in diabetes mellitus (up—upregulated; down—downregulated).

Exosomal RNA	Regulation	Source of exosomes	Effected cells	Disease model	Application	References
miR-27a (up)	Represses PPARγ	Adipocytes	Skeletal muscle	Insulin resistance	Unveiling the pathophysiology	[26]
miR-155 (up)	Attenuates insulin signaling	Activated macrophages	Myocytes, hepatocytes, adipocytes	Insulin resistance, diabetes, myocardial infarction	Unveiling the pathophysiology	[28,29]
miR-34a (up)	Inhibits the antiinflammatory M2 polarization of macrophages repressing Klf4	Adipocytes	Macrophages	Adipose tissue inflammation, insulin resistance	Unveiling the pathophysiology	[32]
miR-148b (down), miR-4269 (down), miR-23b (up)	Target TGF-β and Wnt/β-catenin signaling pathways	Visceral adipose tissues	A549 cells	Obesity	Unveiling the pathophysiology	[33]
miR-122 (up), miR-192 (up), miR-27a-3p (up), miR-27b-3p (up)	Increase PPARα expression	Plasma	In white adipose tissues	Central obesity and hepatic steatosis	Unveiling the pathophysiology	[35]
miR-802-5p (up)	Targets HSP60	Adipocytes	Cardiac myocytes	Insulin resistance	Unveiling the pathophysiology	[36]
miR-3075 (down)	Increases FA2H	Hepatocytes	Myocytes, hepatocytes, adipocytes	Insulin resistance	Unveiling the pathophysiology	[39]
miR-29b-3p (up)	Targets sirtuin 1	Bone marrow mesenchymal stem cells	Hepatocytes, myocytes, and adipocytes	Aging associated insulin resistance	Unveiling the pathophysiology	[40]
miR-20b-5p (up)	Targets STAT3 and AKT signaling pathway	Serum	Skeletal muscle cells	T2DM	Unveiling the pathophysiology	[45]
miR-21-5p (up)	PDCD4 silencing	Mesenchymal stem cells	Smooth muscle	Diabetic rat model with erectile dysfunction	Treatment	[48]

miR-706 (up), miR-709 (up), miR-466c-5p (up), miR-423-5p (up)	Induce pluripotent stem cells into insulin producing cells	Pancreatic β-Cells	Pluripotent stem cells	T1DM	Unveiling the pathophysiology	[54]
miR-29 (up)	IRS-Akt-mediated hepatic insulin signaling	Pancreatic β-Cells	Pancreas, liver	High-fat-diet-induced insulin resistance	Treatment	[55]
miR-26a (up)	Targets several critical regulators of insulin secretion and β cell proliferation	Pancreatic β-Cells	Pancreas	Obesity and T2DM	Treatment	[56]
miR-15a (up)	Travels through the circulation and worsen diabetic complications	Pancreatic β-Cells	Various cell types	Diabetic complications	Unveiling the pathophysiology	[57]
miR-1249-3p (up)	Targets SKOR1 to induce SMAD6/MYD88/SMURF1 complex formation, suppressing TLR4/NF-κB signaling pathway	Natural killer cells	Hepatocytes and adipocytes	Glucose metabolism	Treatment	[59]
MALAT1 (up)	Suppresses miR-150-5p and hence increased resistin expression	Macrophage	Macrophage	Vascular disease	Treatment	[60]
miR-690 (up)	Targets NAD kinase	Bone marrow−derived macrophages	Myocytes, hepatocytes, adipocytes	Inflammation and insulin signaling	Treatment	[61]
miR-210 (up)	Inhibits apoptosis by downregulating ephrin A3 and PTP1b	Cardiac progenitor cells	Cardiomyocytic cells	Cardiomyopathy	Treatment	[65]
miR-214 (up)	Targets miR-214/CaMKII axis	Bone marrow−derived mesenchymal stem cells	Cardiac stem cells	Oxidative stress related injury and apoptosis	Treatment	[66]
miR-125b (up)	Targets SIRT7	Bone marrow−derived mesenchymal stem cells	Myocardium cells	Myocardial ischemia reperfusion injury	Treatment	[67]

(Continued)

Table 6.1 List of exosomal RNA in diabetes mellitus (up—upregulated; down—downregulated).—cont'd

Exosomal RNA	Regulation	Source of exosomes	Effected cells	Disease model	Application	References
miR-210 (up)	Exerts proangiogenic effect through Efna3	Mesenchymal stem cells HUVEC	Endothelial cells	Myocardial infarction	Treatment	[68]
miRNA-181a (up)	Raises the regulatory T cell ratio through inhibition of the c-Fos protein	Mesenchymal stromal cell	PBMCs	Myocardium I/R injury	Treatment	[69]
let-7 (up), miR-145 (up), miR-17-92 (up), miR-302a-5p (up)	Regulate Wnt, PI3K-Akt, and MAPK pathway	Murine induced pluripotent stem cell (iPSC)	Murine cardiac endothelial cells	Acute myocardial infarction	Treatment	[70]
miR-222 (up), miR-143 (up)	Protect against oxidative-induced lesion, promotes proliferation and angiogenesis	Myocardial cells and cardiomyocytes	Endothelial cells	Ischemia, myocardial infarction	Treatment	[71]
miR-210 (up)	Transfers to nSMase2-dependent way	Bone marrow—derived mesenchymal stem cells	Cardiac progenitor cells	Myocardial infarction	Treatment	[72]
miR-21 (up), miR-92a (up), miR-126 (up), miR-143 (up), miR-181b (up), miR-221 (up)	Influence atherosclerosis progression, vascular repair, and remodeling	Plasma	Vascular smooth muscle cells and endothelial cells	Peripheral artery disease	Diagnosis	[73]
miR-145 (up)	Targets STAT-1	Urine and glomerulus	Mesangial cells	Type 1 diabetic nephropathy	Diagnosis	[77]
miR-21-5p (up), let-7e-5p (up), miR-23b-3p (up), miR-30b-5p (down), miR-125b-5p (down)	Development and differentiation of kidney cells and may be markers of kidney cell damage	Urine	Kidney	Type 2 diabetic nephropathy	Diagnosis	[79]

miR-30e-5p (up)	Inhibits caspase-1-mediated pyroptosis by targeting ELAVL1	Bone marrow mesenchymal stem cell	HK-2 cells	Diabetic kidney disease	Unveiling the pathophysiology	[80]
mRNA Wilms tumor 1 (WT1)	Regulates podocyte injuries in diabetic nephropathy progression	Urine and blood	Glomerulus	Diabetic nephropathy	Unveiling the pathophysiology	[81,82]
miR-146a-5p (up)	Targets the 3′UTR of IRAK1 to inhibit NF-κB signaling	Urine-derived stem cells	Kidney	Renal ischemia/reperfusion (I/R) injury	Treatment	[84]
lncRNA TUG1(up)	Regulates ASCL4-mediated ferroptosis by interacting with SRSF1	Urine-derived stem cells	Kidney	Acute kidney injury	Treatment	[87]
circNUP98 (up)	Regulating the miR-151-3p-HMGA2 axis	Serum	Glomerular mesangial cells	Diabetic nephropathy	Treatment	[88]
circTAOK1 (up)	Targets the miR-520h/SMAD3 axis	Glomerular epithelial cells	Glomerular mesangial cells	Diabetic nephropathy	Treatment	[89]
circDLGAP4 (up)	Modulates miR-143/ERBB3/NF-κB/MMP-2 axis	Mesangial cells	Mesangial cells	Diabetic nephropathy	Unveiling the pathophysiology	[90]
circ0125310 (up)	Sponges miR-422a to target the IGF1R/p38 axis	Mesangial cells	Mesangial cells	Diabetic nephropathy	Unveiling the pathophysiology	[91]
lncRNA SNHG7 (up)	Suppresses endothelial-mesenchymal transition and tube formation via miR-34a-5p/XBP1 axis	Mesenchymal stem cells	Retinal microvascular endothelial cells	Diabetic retinopathy	Unveiling the pathophysiology	[92]
circEhmt1 (up)	Upregulates transcriptional factor NFIA to suppress NLRP3-mediated inflammasome formation	Mesenchymal stem cells	Human retinal microvascular endothelial cells	Diabetic retinopathy	Unveiling the pathophysiology	[93]

(Continued)

Table 6.1 List of exosomal RNA in diabetes mellitus (up—upregulated; down—downregulated).—cont'd

Exosomal RNA	Regulation	Source of exosomes	Effected cells	Disease model	Application	References
miR-202-5p (up)	Protects through the TGF/Smad pathway	Retinal pigment epithelial cells	Endothelial cells	Diabetic retinopathy	Treatment	[94]
MALAT1(up)	Suppresses miR-1914-3p to activate MFGE8	Keratinocyte	Macrophages	Diabetic wound healing	Treatment	[99]
circITCH (up)	Recruits TAF15 protein to activate Nrf2 signaling pathway	Bone marrow stromal cell	Endothelial cells	Diabetic wound healing	Treatment	[100]
lncRNA H19 (up)	Impairs miR-152-3p-mediated PTEN inhibition	Mesenchymal stem cells	Fibroblasts	Diabetic wound healing	Treatment	[102]
circHIPK3 (up)	Promotes wound healing through miR-20b-5p/Nrf2/VEGFA axis	Umbilical cord—derived mesenchymal stem cells	Endothelial cells	Diabetic wound healing	Treatment	[103]
miR-424 (up), miR-218 (up)	Predictive biomarker	Urine	—	T1DM	Diagnosis	[104]
miR-34a (up)	Predictive biomarker	Serum	—	T1DM, dyslipidemia	Diagnosis	[106]
hsa_cirRNA_0039480 (up)	Predictive biomarker	Plasma	—	Gestational diabetes mellitus	Diagnosis	[108]
circ_0046060 (up)	Ameliorates insulin resistance via miR-338-3p/G6PC2 axis	Umbilical cord mesenchymal stromal cell	L-02 cells	Gestational diabetes mellitus	Unveiling the pathophysiology	[111]

Acknowledgments

I would like to thank https://pixabay.com/ for providing free images, which were used to make figures in this chapter. I am indebted to my father late **Bijoy Kumar Ghoshal** for his constant support throughout my scientific career. I lost him in December 2021, and I would dedicate this work as a tribute to his loving memories.

References

[1] Kalluri R, LeBleu VS. The biology, function, and biomedical applications of exosomes. Science February 7, 2020;367(6478):eaau6977. https://doi.org/10.1126/science.aau6977. PMID: 32029601; PMCID: PMC7717626.

[2] Wang J, Yue BL, Huang YZ, Lan XY, Liu WJ, Chen H. Exosomal RNAs: novel potential biomarkers for diseases—a review. International Journal of Molecular Sciences February 23, 2022;23(5):2461. https://doi.org/10.3390/ijms23052461. PMID: 35269604; PMCID: PMC8910301.

[3] Dini L, Tacconi S, Carata E, Tata AM, Vergallo C, Panzarini E. Microvesicles and exosomes in metabolic diseases and inflammation. Cytokine and Growth Factor Reviews February 2020;51:27−39. https://doi.org/10.1016/j.cytogfr.2019.12.008. Epub 2020 Jan 3. PMID: 31917095.

[4] Gehrmann U, Näslund TI, Hiltbrunner S, Larssen P, Gabrielsson S. Harnessing the exosome-induced immune response for cancer immunotherapy. Seminars in Cancer Biology October 2014;28:58−67. https://doi.org/10.1016/j.semcancer.2014.05.003. Epub 2014 May 22. PMID: 24859748.

[5] D'Anca M, Fenoglio C, Serpente M, Arosio B, Cesari M, Scarpini EA, et al. Exosome determinants of physiological aging and age-related neurodegenerative diseases. Frontiers in Aging Neuroscience August 28, 2019;11:232. https://doi.org/10.3389/fnagi.2019.00232. PMID: 31555123; PMCID: PMC6722391.

[6] Magliano DJ, Boyko EJ. IDF Diabetes Atlas 10th edition scientific committee. IDF DIABETES ATLAS [Internet]. 10th ed. Brussels: International Diabetes Federation; 2021. PMID: 35914061.

[7] Narang P, Shah M, Beljanski V. Exosomal RNAs in diagnosis and therapies. Noncoding RNA Research January 14, 2022;7(1):7−15. https://doi.org/10.1016/j.ncrna.2022.01.001. PMID: 35087990; PMCID: PMC8777382.

[8] Sufianov A, Kostin A, Begliarzade S, Kudriashov V, Ilyasova T, Liang Y, et al. Exosomal non coding RNAs as a novel target for diabetes mellitus and its complications. Noncoding RNA Research February 7, 2023;8(2):192−204. https://doi.org/10.1016/j.ncrna.2023.02.001. PMID: 36818396; PMCID: PMC9929646.

[9] Zhang Y, Liu Y, Liu H, Tang WH. Exosomes: biogenesis, biologic function and clinical potential. Cell and Bioscience February 15, 2019;9:19. https://doi.org/10.1186/s13578-019-0282-2. PMID: 30815248; PMCID: PMC6377728.

[10] Valadi H, Ekström K, Bossios A, Sjöstrand M, Lee JJ, Lötvall JO. Exosome-mediated transfer of mRNAs and microRNAs is a novel mechanism of genetic exchange between cells. Nature Cell Biology June 2007;9(6):654−9. https://doi.org/10.1038/ncb1596. Epub 2007 May 7. PMID: 17486113.

[11] Geng X, Lin X, Zhang Y, Li Q, Guo Y, Fang C, et al. Exosomal circular RNA sorting mechanisms and their function in promoting or inhibiting cancer. Oncology Letters May 2020;19(5):3369−80. https://doi.org/10.3892/ol.2020.11449. Epub 2020 Mar 6. PMID: 32269609; PMCID: PMC7114721.

[12] Garcia-Martin R, Wang G, Brandão BB, Zanotto TM, Shah S, Kumar Patel S, et al. MicroRNA sequence codes for small extracellular vesicle release and cellular retention. Nature January 2022;601(7893):446−51. https://doi.org/10.1038/s41586-021-04234-3. Epub 2021 Dec 22. PMID: 34937935; PMCID: PMC9035265.

[13] American Diabetes Association. Standards of medical care in diabetes. Diabetes Care January 2004;27(Suppl. 1):S15−35. https://doi.org/10.2337/diacare.27.2007.s15. PMID: 14693923.

[14] Holt RI. Diagnosis, epidemiology and pathogenesis of diabetes mellitus: an update for psychiatrists. British Journal of Psychiatry — Supplement April 2004;47:S55–63. https://doi.org/10.1192/bjp.184.47.s55. PMID: 15056594.

[15] Ghoshal K, Li X, Peng D, Falck JR, Anugu RR, Chiusa M, et al. EET analog treatment improves insulin signaling in a genetic mouse model of insulin resistance. Diabetes October 21, 2021:db210298. https://doi.org/10.2337/db21-0298. Epub ahead of print. PMID: 34957494.

[16] Ghoshal K, Das S, Aich K, Goswami S, Chowdhury S, Bhattacharyya M. A novel sensor to estimate the prevalence of hypochlorous (HOCl) toxicity in individuals with type 2 diabetes and dyslipidemia. Clinica Chimica Acta July 1, 2016;458:144–53. https://doi.org/10.1016/j.cca.2016.05.006. Epub 2016 May 11. PMID: 27178483.

[17] Ghoshal K, Chakraborty S, Das C, Chattopadhyay S, Chowdhury S, Bhattacharyya M. Dielectric properties of plasma membrane: a signature for dyslipidemia in diabetes mellitus. Archives of Biochemistry and Biophysics December 1, 2017;635:27–36. https://doi.org/10.1016/j.abb.2017.10.002. Epub 2017 Oct 10. PMID: 29029878.

[18] Cade WT. Diabetes-related microvascular and macrovascular diseases in the physical therapy setting. Physical Therapy November 2008;88(11):1322–35. https://doi.org/10.2522/ptj.20080008. Epub 2008 Sep 18. PMID: 18801863; PMCID: PMC2579903.

[19] Kumashiro N, Erion DM, Zhang D, Kahn M, Beddow SA, Chu X, et al. Cellular mechanism of insulin resistance in nonalcoholic fatty liver disease. Proceedings of the National Academy of Sciences of the United States of America September 27, 2011;108(39):16381–5. https://doi.org/10.1073/pnas.1113359108. Epub 2011 Sep 19. PMID: 21930939; PMCID: PMC3182681.

[20] Ghoshal K, Bhattacharyya M. Adiponectin: probe of the molecular paradigm associating diabetes and obesity. World Journal of Diabetes February 15, 2015;6(1):151–66. https://doi.org/10.4239/wjd.v6.i1.151. PMID: 25685286; PMCID: PMC4317307.

[21] Gastaldelli A, Gaggini M, DeFronzo RA. Role of adipose tissue insulin resistance in the natural history of type 2 diabetes: results from the san antonio metabolism study. Diabetes April 2017;66(4):815–22. https://doi.org/10.2337/db16-1167. Epub 2017 Jan 4. PMID: 28052966.

[22] Müller G. Control of lipid storage and cell size between adipocytes by vesicle-associated glycosylphosphatidylinositol-anchored proteins. Archives of Physiology and Biochemistry February 2011;117(1):23–43. https://doi.org/10.3109/13813455.2010.513393. Epub 2010 Sep 30. PMID: 20883086.

[23] Ogawa R, Tanaka C, Sato M, Nagasaki H, Sugimura K, Okumura K, et al. Adipocyte-derived microvesicles contain RNA that is transported into macrophages and might be secreted into blood circulation. Biochemical and Biophysical Research Communications August 6, 2010;398(4):723–9. https://doi.org/10.1016/j.bbrc.2010.07.008. Epub 2010 Jul 17. PMID: 20621060.

[24] Mei R, Qin W, Zheng Y, Wan Z, Liu L. Role of adipose tissue derived exosomes in metabolic disease. Frontiers in Endocrinology May 4, 2022;13:873865. https://doi.org/10.3389/fendo.2022.873865. PMID: 35600580; PMCID: PMC9114355.

[25] Thomou T, Mori MA, Dreyfuss JM, Konishi M, Sakaguchi M, Wolfrum C, et al. Adipose-derived circulating miRNAs regulate gene expression in other tissues. Nature February 23, 2017;542(7642):450–5. https://doi.org/10.1038/nature21365. Epub 2017 Feb 15. Erratum in: Nature. 2017 May 10;545(7653):252. PMID: 28199304; PMCID: PMC5330251.

[26] Yu Y, Du H, Wei S, Feng L, Li J, Yao F, et al. Adipocyte-derived exosomal MiR-27a induces insulin resistance in skeletal muscle through repression of PPARγ. Theranostics March 8, 2018;8(8):2171–88. https://doi.org/10.7150/thno.22565. PMID: 29721071; PMCID: PMC5928879.

[27] Song M, Han L, Chen FF, Wang D, Wang F, Zhang L, et al. Adipocyte-derived exosomes carrying sonic hedgehog mediate M1 macrophage polarization-induced insulin resistance via Ptch and PI3K pathways. Cellular Physiology and Biochemistry 2018;48(4):1416–32. https://doi.org/10.1159/000492252. Epub 2018 Jul 31. PMID: 30064125.

[28] Wang C, Zhang C, Liu L, Xi A, Chen B, Li Y, et al. Macrophage-derived mir-155-containing exosomes suppress fibroblast proliferation and promote fibroblast inflammation during cardiac injury. Molecular Therapy January 4, 2017;25(1):192–204. https://doi.org/10.1016/j.ymthe.2016.09.001. Epub 2017 Jan 4. PMID: 28129114; PMCID: PMC5363311.

[29] Ying W, Riopel M, Bandyopadhyay G, Dong Y, Birmingham A, Seo JB, et al. Adipose tissue macrophage-derived exosomal miRNAs can modulate in vivo and in vitro insulin sensitivity. Cell October 5, 2017;171(2):372–384.e12. https://doi.org/10.1016/j.cell.2017.08.035. Epub 2017 Sep 21. PMID: 28942920.

[30] Lin X, Qin Y, Jia J, Lin T, Lin X, Chen L, et al. MiR-155 enhances insulin sensitivity by coordinated regulation of multiple genes in mice. PLoS Genetics October 6, 2016;12(10):e1006308. https://doi.org/10.1371/journal.pgen.1006308. PMID: 27711113; PMCID: PMC5053416.

[31] Ghoshal K, Chatterjee T, Chowdhury S, Sengupta S, Bhattacharyya M. Adiponectin genetic variant and expression coupled with lipid peroxidation reveal new signatures in diabetic dyslipidemia. Biochemical Genetics June 2021;59(3):781–98. https://doi.org/10.1007/s10528-021-10030-5. Epub 2021 Feb 4. PMID: 33543406.

[32] Pan Y, Hui X, Hoo RLC, Ye D, Chan CYC, Feng T, et al. Adipocyte-secreted exosomal microRNA-34a inhibits M2 macrophage polarization to promote obesity-induced adipose inflammation. Journal of Clinical Investigation February 1, 2019;129(2):834–49. https://doi.org/10.1172/JCI123069. Epub 2019 Jan 22. PMID: 30667374; PMCID: PMC6355214.

[33] Ferrante SC, Nadler EP, Pillai DK, Hubal MJ, Wang Z, Wang JM, et al. Adipocyte-derived exosomal miRNAs: a novel mechanism for obesity-related disease. Pediatric Research March 2015;77(3):447–54. https://doi.org/10.1038/pr.2014.202. Epub 2014 Dec 17. PMID: 25518011; PMCID: PMC4346410.

[34] Yang Z, Wei Z, Wu X, Yang H. Screening of exosomal miRNAs derived from subcutaneous and visceral adipose tissues: determination of targets for the treatment of obesity and associated metabolic disorders. Molecular Medicine Reports September 2018;18(3):3314–24. https://doi.org/10.3892/mmr.2018.9312. Epub 2018 Jul 24. PMID: 30066923; PMCID: PMC6102639.

[35] Castaño C, Kalko S, Novials A, Párrizas M. Obesity-associated exosomal miRNAs modulate glucose and lipid metabolism in mice. Proceedings of the National Academy of Sciences of the United States of America November 27, 2018;115(48):12158–63. https://doi.org/10.1073/pnas.1808855115. Epub 2018 Nov 14. PMID: 30429322; PMCID: PMC6275521.

[36] Wen Z, Li J, Fu Y, Zheng Y, Ma M, Wang C. Hypertrophic adipocyte-derived exosomal miR-802-5p contributes to insulin resistance in cardiac myocytes through targeting HSP60. Obesity October 2020;28(10):1932–40. https://doi.org/10.1002/oby.22932. Epub 2020 Aug 25. PMID: 32844579.

[37] Zhu LL, Huang X, Yu W, Chen H, Chen Y, Dai YT. Transplantation of adipose tissue-derived stem cell-derived exosomes ameliorates erectile function in diabetic rats. Andrologia March 2018;50(2). https://doi.org/10.1111/and.12871. Epub 2017 Oct 23. PMID: 29057541.

[38] Meshkani R, Adeli K. Hepatic insulin resistance, metabolic syndrome and cardiovascular disease. Clinical Biochemistry September 2009;42(13–14):1331–46. https://doi.org/10.1016/j.clinbiochem.2009.05.018. Epub 2009 Jun 6. PMID: 19501581.

[39] Ji Y, Luo Z, Gao H, Dos Reis FCG, Bandyopadhyay G, Jin Z, et al. Hepatocyte-derived exosomes from early onset obese mice promote insulin sensitivity through miR-3075. Nature Metabolism September 2021;3(9):1163–74. https://doi.org/10.1038/s42255-021-00444-1. Epub 2021 Sep 6. PMID: 34489604; PMCID: PMC8460610.

[40] Su T, Xiao Y, Xiao Y, Guo Q, Li C, Huang Y, et al. Bone marrow mesenchymal stem cells-derived exosomal MiR-29b-3p regulates aging-associated insulin resistance. ACS Nano February 26, 2019;13(2):2450–62. https://doi.org/10.1021/acsnano.8b09375. Epub 2019 Feb 11. PMID: 30715852.

[41] Yamada H, Suzuki K, Ichino N, Ando Y, Sawada A, Osakabe K, et al. Associations between circulating microRNAs (miR-21, miR-34a, miR-122 and miR-451) and non-alcoholic fatty liver. Clinica Chimica Acta September 23, 2013;424:99–103. https://doi.org/10.1016/j.cca.2013.05.021. Epub 2013 May 30. PMID: 23727030.

[42] Becker PP, Rau M, Schmitt J, Malsch C, Hammer C, Bantel H, et al. Performance of serum microRNAs -122, -192 and -21 as biomarkers in patients with non-alcoholic steatohepatitis. PLoS One November 13, 2015;10(11):e0142661. https://doi.org/10.1371/journal.pone.0142661. PMID: 26565986; PMCID: PMC4643880.

[43] Merz KE, Thurmond DC. Role of skeletal muscle in insulin resistance and glucose uptake. Comprehensive Physiology July 8, 2020;10(3):785–809. https://doi.org/10.1002/cphy.c190029. PMID: 32940941; PMCID: PMC8074531.
[44] DeFronzo RA, Tripathy D. Skeletal muscle insulin resistance is the primary defect in type 2 diabetes. Diabetes Care November 2009;32(Suppl. 2):S157–63. https://doi.org/10.2337/dc09-S302. PMID: 19875544; PMCID: PMC2811436.
[45] Katayama M, Wiklander OPB, Fritz T, Caidahl K, El-Andaloussi S, Zierath JR, et al. Circulating exosomal miR-20b-5p is elevated in type 2 diabetes and could impair insulin action in human skeletal muscle. Diabetes March 2019;68(3):515–26. https://doi.org/10.2337/db18-0470. Epub 2018 Dec 14. PMID: 30552111.
[46] Wang L, Zhang B, Zheng W, Kang M, Chen Q, Qin W, et al. Exosomes derived from pancreatic cancer cells induce insulin resistance in C2C12 myotube cells through the PI3K/Akt/FoxO1 pathway. Scientific Reports July 14, 2017;7(1):5384. https://doi.org/10.1038/s41598-017-05541-4. PMID: 28710412; PMCID: PMC5511275.
[47] Rao J, Chen Y, Huang J, Wu R, Dong Z, Gao Y, et al. Inhibiting miR-182-3p alleviates gestational diabetes mellitus by improving insulin resistance in skeletal muscle. Balkan Medical Journal March 14, 2022;39(2):121–9. https://doi.org/10.4274/balkanmedj.galenos.2021.2021-8-140. PMID: 35330559; PMCID: PMC8941230.
[48] Huo W, Li Y, Zhang Y, Li H. Mesenchymal stem cells-derived exosomal microRNA-21-5p downregulates PDCD4 and ameliorates erectile dysfunction in a rat model of diabetes mellitus. The FASEB Journal October 2020;34(10):13345–60. https://doi.org/10.1096/fj.202000102RR. Epub 2020 Aug 17. Erratum in: FASEB J. 2023 Jan;37(1):e22721. PMID: 32808325.
[49] Yao ZY, Chen WB, Shao SS, Ma SZ, Yang CB, Li MZ, et al. Role of exosome-associated microRNA in diagnostic and therapeutic applications to metabolic disorders. Journal of Zhejiang University – Science B March 2018;19(3):183–98. https://doi.org/10.1631/jzus.B1600490. PMID: 29504312; PMCID: PMC5854634.
[50] Kloosterman WP, Lagendijk AK, Ketting RF, Moulton JD, Plasterk RH. Targeted inhibition of miRNA maturation with morpholinos reveals a role for miR-375 in pancreatic islet development. PLoS Biology August 2007;5(8):e203. https://doi.org/10.1371/journal.pbio.0050203. PMID: 17676975; PMCID: PMC1925136.
[51] Poy MN, Hausser J, Trajkovski M, Braun M, Collins S, Rorsman P, et al. miR-375 maintains normal pancreatic alpha- and beta-cell mass. Proceedings of the National Academy of Sciences of the United States of America April 7, 2009;106(14):5813–8. https://doi.org/10.1073/pnas.0810550106. Epub 2009 Mar 16. PMID: 19289822; PMCID: PMC2656556.
[52] Poy MN, Eliasson L, Krutzfeldt J, Kuwajima S, Ma X, Macdonald PE, et al. A pancreatic islet-specific microRNA regulates insulin secretion. Nature November 11, 2004;432(7014):226–30. https://doi.org/10.1038/nature03076. PMID: 15538371.
[53] Guay C, Menoud V, Rome S, Regazzi R. Horizontal transfer of exosomal microRNAs transduce apoptotic signals between pancreatic beta-cells. Cell Communication and Signaling March 19, 2015;13:17. https://doi.org/10.1186/s12964-015-0097-7. PMID: 25880779; PMCID: PMC4371845.
[54] Guo Q, Lu Y, Huang Y, Guo Y, Zhu S, Zhang Q, et al. Exosomes from β-cells promote differentiation of induced pluripotent stem cells into insulin-producing cells through microRNA-dependent mechanisms. Diabetes, Metabolic Syndrome and Obesity December 11, 2021;14:4767–82. https://doi.org/10.2147/DMSO.S342647. PMID: 34934332; PMCID: PMC8678630.
[55] Li J, Zhang Y, Ye Y, Li D, Liu Y, Lee E, et al. Pancreatic β cells control glucose homeostasis via the secretion of exosomal miR-29 family. Journal of Extracellular Vesicles January 2021;10(3):e12055. https://doi.org/10.1002/jev2.12055. Epub 2021 Jan 21. PMID: 33520119; PMCID: PMC7820156.
[56] Xu H, Du X, Xu J, Zhang Y, Tian Y, Liu G, et al. Pancreatic β cell microRNA-26a alleviates type 2 diabetes by improving peripheral insulin sensitivity and preserving β cell function. PLoS Biology February 24, 2020;18(2):e3000603. https://doi.org/10.1371/journal.pbio.3000603. PMID: 32092075; PMCID: PMC7058362.

[57] Kamalden TA, Macgregor-Das AM, Kannan SM, Dunkerly-Eyring B, Khaliddin N, Xu Z, et al. Exosomal MicroRNA-15a transfer from the pancreas augments diabetic complications by inducing oxidative stress. Antioxidants and Redox Signaling November 1, 2017;27(13):913–30. https://doi.org/10.1089/ars.2016.6844. Epub 2017 Mar 17. PMID: 28173719; PMCID: PMC5649125.

[58] Tsalamandris S, Antonopoulos AS, Oikonomou E, Papamikroulis GA, Vogiatzi G, Papaioannou S, et al. The role of inflammation in diabetes: current concepts and future perspectives. European Cardiology Review April 2019;14(1):50–9. https://doi.org/10.15420/ecr.2018.33.1. PMID: 31131037; PMCID: PMC6523054.

[59] Wang Y, Li M, Chen L, Bian H, Chen X, Zheng H, et al. Natural killer cell-derived exosomal miR-1249-3p attenuates insulin resistance and inflammation in mouse models of type 2 diabetes. Signal Transduction and Targeted Therapy November 30, 2021;6(1):409. https://doi.org/10.1038/s41392-021-00805-y. PMID: 34848693; PMCID: PMC8632983.

[60] Shyu KG, Wang BW, Fang WJ, Pan CM, Lin CM. Exosomal MALAT1 derived from high glucose-treated macrophages up-regulates resistin expression via miR-150-5p downregulation. International Journal of Molecular Sciences January 20, 2022;23(3):1095. https://doi.org/10.3390/ijms23031095. PMID: 35163020; PMCID: PMC8834900.

[61] Ying W, Gao H, Dos Reis FCG, Bandyopadhyay G, Ofrecio JM, Luo Z, et al. MiR-690, an exosomal-derived miRNA from M2-polarized macrophages, improves insulin sensitivity in obese mice. Cell Metabolism April 6, 2021;33(4):781–790.e5. https://doi.org/10.1016/j.cmet.2020.12.019. Epub 2021 Jan 14. PMID: 33450179; PMCID: PMC8035248.

[62] Viigimaa M, Sachinidis A, Toumpourleka M, Koutsampasopoulos K, Alliksoo S, Titma T. Macrovascular complications of type 2 diabetes mellitus. Current Vascular Pharmacology 2020;18(2):110–6. https://doi.org/10.2174/1570161117666190405165151. PMID: 30961498.

[63] Ghoshal K, Bhattacharyya M. Overview of platelet physiology: its hemostatic and nonhemostatic role in disease pathogenesis. The Scientific World Journal March 3, 2014;2014:781857. https://doi.org/10.1155/2014/781857. PMID: 24729754; PMCID: PMC3960550.

[64] Zhao S, Wang H, Xu H, Tan Y, Zhang C, Zeng Q, et al. Targeting the microRNAs in exosome: a potential therapeutic strategy for alleviation of diabetes-related cardiovascular complication. Pharmacological Research November 2021;173:105868. https://doi.org/10.1016/j.phrs.2021.105868. Epub 2021 Sep 3. PMID: 34481974.

[65] Barile L, Lionetti V, Cervio E, Matteucci M, Gherghiceanu M, Popescu LM, et al. Extracellular vesicles from human cardiac progenitor cells inhibit cardiomyocyte apoptosis and improve cardiac function after myocardial infarction. Cardiovascular Research September 1, 2014;103(4):530–41. https://doi.org/10.1093/cvr/cvu167. Epub 2014 Jul 11. PMID: 25016614.

[66] Wang Y, Zhao R, Liu D, Deng W, Xu G, Liu W, et al. Exosomes derived from miR-214-enriched bone marrow-derived mesenchymal stem cells regulate oxidative damage in cardiac stem cells by targeting CaMKII. Oxidative Medicine and Cellular Longevity August 7, 2018;2018:4971261. https://doi.org/10.1155/2018/4971261. PMID: 30159114; PMCID: PMC6109555.

[67] Chen Q, Liu Y, Ding X, Li Q, Qiu F, Wang M, et al. Bone marrow mesenchymal stem cell-secreted exosomes carrying microRNA-125b protect against myocardial ischemia reperfusion injury via targeting SIRT7. Molecular and Cellular Biochemistry February 2020;465(1–2):103–14. https://doi.org/10.1007/s11010-019-03671-z. Epub 2019 Dec 19. PMID: 31858380; PMCID: PMC6955239.

[68] Wang N, Chen C, Yang D, Liao Q, Luo H, Wang X, et al. Mesenchymal stem cells-derived extracellular vesicles, via miR-210, improve infarcted cardiac function by promotion of angiogenesis. Biochimica et Biophysica Acta, Molecular Basis of Disease August 2017;1863(8):2085–92. https://doi.org/10.1016/j.bbadis.2017.02.023. Epub 2017 Feb 27. PMID: 28249798.

[69] Wei Z, Qiao S, Zhao J, Liu Y, Li Q, Wei Z, et al. miRNA-181a over-expression in mesenchymal stem cell-derived exosomes influenced inflammatory response after myocardial ischemia-reperfusion injury. Life Sciences September 1, 2019;232:116632. https://doi.org/10.1016/j.lfs.2019.116632. Epub 2019 Jul 3. Erratum in: Life Sci. 2020 Sep 1;256:118045. PMID: 31278944.

[70] Adamiak M, Cheng G, Bobis-Wozowicz S, Zhao L, Kedracka-Krok S, Samanta A, et al. Induced pluripotent stem cell (iPSC)-derived extracellular vesicles are safer and more effective for cardiac repair than iPSCs. Circulation Research January 19, 2018;122(2):296–309. https://doi.org/10.1161/CIRCRESAHA.117.311769. Epub 2017 Nov 8. PMID: 29118058; PMCID: PMC5775034.

[71] Ribeiro-Rodrigues TM, Laundos TL, Pereira-Carvalho R, Batista-Almeida D, Pereira R, Coelho-Santos V, et al. Exosomes secreted by cardiomyocytes subjected to ischaemia promote cardiac angiogenesis. Cardiovascular Research September 1, 2017;113(11):1338–50. https://doi.org/10.1093/cvr/cvx118. PMID: 28859292.

[72] Zhu J, Lu K, Zhang N, Zhao Y, Ma Q, Shen J, et al. Myocardial reparative functions of exosomes from mesenchymal stem cells are enhanced by hypoxia treatment of the cells via transferring microRNA-210 in an nSMase2-dependent way. Artificial Cells, Nanomedicine and Biotechnology December 2018;46(8):1659–70. https://doi.org/10.1080/21691401.2017.1388249. Epub 2017 Nov 16. PMID: 29141446; PMCID: PMC5955787.

[73] Sorrentino TA, Duong P, Bouchareychas L, Chen M, Chung A, Schaller MS, et al. Circulating exosomes from patients with peripheral artery disease influence vascular cell migration and contain distinct microRNA cargo. JVS Vascular Science 2020;1:28–41. https://doi.org/10.1016/j.jvssci.2020.02.001. Epub 2020 Feb 28. PMID: 32550603; PMCID: PMC7299234.

[74] Huang JP, Chang CC, Kuo CY, Huang KJ, Sokal EM, Chen KH, et al. Exosomal microRNAs miR-30d-5p and miR-126a-5p are associated with heart failure with preserved ejection fraction in STZ-induced type 1 diabetic rats. International Journal of Molecular Sciences July 6, 2022;23(14):7514. https://doi.org/10.3390/ijms23147514. PMID: 35886860; PMCID: PMC9318774.

[75] Shi R, Zhao L, Cai W, Wei M, Zhou X, Yang G, et al. Maternal exosomes in diabetes contribute to the cardiac development deficiency. Biochemical and Biophysical Research Communications January 29, 2017;483(1):602–8. https://doi.org/10.1016/j.bbrc.2016.12.097. Epub 2016 Dec 18. PMID: 27998767.

[76] Flyvbjerg A. Pathogenesis of microvascular complications, textbook of diabetes. 5th ed. December 9, 2016. https://doi.org/10.1002/9781118924853.ch37 [Chapter 37].

[77] Barutta F, Tricarico M, Corbelli A, Annaratone L, Pinach S, Grimaldi S, et al. Urinary exosomal microRNAs in incipient diabetic nephropathy. PLoS One November 4, 2013;8(11):e73798. https://doi.org/10.1371/journal.pone.0073798. PMID: 24223694; PMCID: PMC3817183.

[78] Delić D, Eisele C, Schmid R, Baum P, Wiech F, Gerl M, et al. Urinary exosomal miRNA signature in type II diabetic nephropathy patients. PLoS One March 1, 2016;11(3):e0150154. https://doi.org/10.1371/journal.pone.0150154. PMID: 26930277; PMCID: PMC4773074.

[79] Zang J, Maxwell AP, Simpson DA, McKay GJ. Differential expression of urinary exosomal MicroRNAs miR-21-5p and miR-30b-5p in individuals with diabetic kidney disease. Scientific Reports July 29, 2019;9(1):10900. https://doi.org/10.1038/s41598-019-47504-x. PMID: 31358876; PMCID: PMC6662907.

[80] Kim H, Bae YU, Jeon JS, Noh H, Park HK, Byun DW, et al. The circulating exosomal microRNAs related to albuminuria in patients with diabetic nephropathy. Journal of Translational Medicine July 22, 2019;17(1):236. https://doi.org/10.1186/s12967-019-1983-3. PMID: 31331349; PMCID: PMC6647278.

[81] Abe H, Sakurai A, Ono H, Hayashi S, Yoshimoto S, Ochi A, et al. Urinary exosomal mRNA of WT1 as diagnostic and prognostic biomarker for diabetic nephropathy. Journal of Medical Investigation 2018;65(3.4):208–15. https://doi.org/10.2152/jmi.65.208. PMID: 30282862.

[82] Hashemi E, Dehghanbanadaki H, Baharanchi AA, Forouzanfar K, Kakaei A, Mohammadi SM, et al. WT1 and ACE mRNAs of blood extracellular vesicle as biomarkers of diabetic nephropathy. Journal of Translational Medicine July 10, 2021;19(1):299. https://doi.org/10.1186/s12967-021-02964-6. PMID: 34246281; PMCID: PMC8272332.

[83] Mohan A, Singh RS, Kumari M, Garg D, Upadhyay A, Ecelbarger CM, et al. Urinary exosomal microRNA-451-5p is a potential early biomarker of diabetic nephropathy in rats. PLoS One April 21, 2016;11(4):e0154055. https://doi.org/10.1371/journal.pone.0154055. PMID: 27101382; PMCID: PMC4839711.

[84] Li X, Liao J, Su X, Li W, Bi Z, Wang J, et al. Human urine-derived stem cells protect against renal ischemia/reperfusion injury in a rat model via exosomal miR-146a-5p which targets IRAK1. Theranostics July 25, 2020;10(21):9561–78. https://doi.org/10.7150/thno.42153. PMID: 32863945; PMCID: PMC7449916.

[85] Riffo-Campos AL, Perez-Hernandez J, Ortega A, Martinez-Arroyo O, Flores-Chova A, Redon J, et al. Exosomal and plasma non-coding RNA signature associated with urinary albumin excretion in hypertension. International Journal of Molecular Sciences January 13, 2022;23(2):823. https://doi.org/10.3390/ijms23020823. PMID: 35055008; PMCID: PMC8775608.
[86] Yang H, Bai Y, Fu C, Liu W, Diao Z. Exosomes from high glucose-treated macrophages promote epithelial-mesenchymal transition of renal tubular epithelial cells via long non-coding RNAs. BMC Nephrology January 30, 2023;24(1):24. https://doi.org/10.1186/s12882-023-03065-w. PMID: 36717805; PMCID: PMC9887774.
[87] Sun Z, Wu J, Bi Q, Wang W. Exosomal lncRNA TUG1 derived from human urine-derived stem cells attenuates renal ischemia/reperfusion injury by interacting with SRSF1 to regulate ASCL4-mediated ferroptosis. Stem Cell Research and Therapy July 15, 2022;13(1):297. https://doi.org/10.1186/s13287-022-02986-x. PMID: 35841017; PMCID: PMC9284726.
[88] Dong Q, Dong L, Zhu Y, Wang X, Li Z, Zhang L. Circular ribonucleic acid nucleoporin 98 knockdown alleviates high glucose-induced proliferation, fibrosis, inflammation and oxidative stress in human glomerular mesangial cells by regulating the microribonucleic acid-151-3p-high mobility group AT-hook 2 axis. Journal of Diabetes Investigation August 2022;13(8):1303–15. https://doi.org/10.1111/jdi.13821. Epub 2022 Jun 6. PMID: 35482475; PMCID: PMC9340880.
[89] Li B, Sun G, Yu H, Meng J, Wei F. Exosomal circTAOK1 contributes to diabetic kidney disease progression through regulating SMAD3 expression by sponging miR-520h. International Urology and Nephrology September 2022;54(9):2343–54. https://doi.org/10.1007/s11255-022-03139-y. Epub 2022 Feb 10. PMID: 35142978.
[90] Bai S, Xiong X, Tang B, Ji T, Li X, Qu X, et al. Exosomal circ_DLGAP4 promotes diabetic kidney disease progression by sponging miR-143 and targeting ERBB3/NF-κB/MMP-2 axis. Cell Death and Disease November 23, 2020;11(11):1008. https://doi.org/10.1038/s41419-020-03169-3. PMID: 33230102; PMCID: PMC7683700.
[91] Zhu Y, Zha F, Tang B, Ji TT, Li XY, Feng L, et al. Exosomal hsa_circ_0125310 promotes cell proliferation and fibrosis in diabetic nephropathy via sponging miR-422a and targeting the IGF1R/p38 axis. Journal of Cellular and Molecular Medicine January 2022;26(1):151–62. https://doi.org/10.1111/jcmm.17065. Epub 2021 Dec 2. PMID: 34854210; PMCID: PMC8742240.
[92] Cao X, Xue LD, Di Y, Li T, Tian YJ, Song Y. MSC-derived exosomal lncRNA SNHG7 suppresses endothelial-mesenchymal transition and tube formation in diabetic retinopathy via miR-34a-5p/XBP1 axis. Life Sciences May 1, 2021;272:119232. https://doi.org/10.1016/j.lfs.2021.119232. Epub 2021 Feb 16. PMID: 33600866.
[93] Ye L, Guo H, Wang Y, Peng Y, Zhang Y, Li S, et al. Exosomal circEhmt1 released from hypoxia-pretreated pericytes regulates high glucose-induced microvascular dysfunction via the NFIA/NLRP3 pathway. Oxidative Medicine and Cellular Longevity March 18, 2021;2021:8833098. https://doi.org/10.1155/2021/8833098. PMID: 33815662; PMCID: PMC7994074.
[94] Gu S, Liu Y, Zou J, Wang W, Wei T, Wang X, et al. Retinal pigment epithelial cells secrete miR-202-5p-containing exosomes to protect against proliferative diabetic retinopathy. Experimental Eye Research December 2020;201:108271. https://doi.org/10.1016/j.exer.2020.108271. Epub 2020 Sep 29. PMID: 33007305.
[95] Ye Q, Li L, Shao Z, Xu M, Li L, Yan Q, et al. Association between lncRNAs in plasma exosomes and diabetic retinopathy. Frontiers in Endocrinology September 14, 2022;13:987488. https://doi.org/10.3389/fendo.2022.987488. PMID: 36187126; PMCID: PMC9519175.
[96] Li X, Wang J, Qian H, Wu Y, Zhang Z, Hu Z, et al. Serum exosomal circular RNA expression profile and regulative role in proliferative diabetic retinopathy. Frontiers in Genetics August 10, 2021;12:719312. https://doi.org/10.3389/fgene.2021.719312. PMID: 34447414; PMCID: PMC8383346.
[97] Fu S, Zhang H, Li X, Zhang Q, Guo C, Qiu K, et al. Exosomes derived from human amniotic mesenchymal stem cells facilitate diabetic wound healing by angiogenesis and enrich multiple lncRNAs. Tissue Engineering and Regenerative Medicine April 2023;20(2):295–308. https://doi.org/10.1007/s13770-022-00513-w. Epub 2023 Jan 25. PMID: 36696086; PMCID: PMC10070558.

[98] Chen ZJ, Shi XJ, Fu LJ, Liu J, Shi K, Zhang WB, et al. Serum and exosomal hsa_circ_0000907 and hsa_circ_0057362 as novel biomarkers in the early diagnosis of diabetic foot ulcer. European Review for Medical and Pharmacological Sciences August 2020;24(15):8117–26. https://doi.org/10.26355/eurrev_202008_22498. PMID: 32767340.

[99] Kuang L, Zhang C, Li B, Deng H, Chen R, Li G. Human keratinocyte-derived exosomal MALAT1 promotes diabetic wound healing by upregulating MFGE8 via microRNA-1914-3p. International Journal of Nanomedicine February 21, 2023;18:949–70. https://doi.org/10.2147/IJN.S399785. PMID: 36852184; PMCID: PMC9961177.

[100] Chen J, Li X, Liu H, Zhong D, Yin K, Li Y, et al. Bone marrow stromal cell-derived exosomal circular RNA improves diabetic foot ulcer wound healing by activating the nuclear factor erythroid 2-related factor 2 pathway and inhibiting ferroptosis. Diabetic Medicine July 2023;40(7):e15031. https://doi.org/10.1111/dme.15031. Epub 2023 Apr 5. PMID: 36537855.

[101] Han ZF, Cao JH, Liu ZY, Yang Z, Qi RX, Xu HL. Exosomal lncRNA KLF3-AS1 derived from bone marrow mesenchymal stem cells stimulates angiogenesis to promote diabetic cutaneous wound healing. Diabetes Research and Clinical Practice January 2022;183:109126. https://doi.org/10.1016/j.diabres.2021.109126. Epub 2021 Nov 3. PMID: 34742784.

[102] Li B, Luan S, Chen J, Zhou Y, Wang T, Li Z, et al. The MSC-derived exosomal lncRNA H19 promotes wound healing in diabetic foot ulcers by upregulating PTEN via MicroRNA-152-3p. Molecular Therapy – Nucleic Acids March 6, 2020;19:814–26. https://doi.org/10.1016/j.omtn.2019.11.034. Epub 2019 Dec 14. PMID: 31958697; PMCID: PMC7005423.

[103] Liang ZH, Lin SS, Pan NF, Zhong GY, Qiu ZY, Kuang SJ, et al. UCMSCs-derived exosomal circHIPK3 promotes ulcer wound angiogenesis of diabetes mellitus via miR-20b-5p/Nrf2/VEGFA axis. Diabetic Medicine February 2023;40(2):e14968. https://doi.org/10.1111/dme.14968. Epub 2022 Oct 31. PMID: 36209373.

[104] Kong Q, Guo X, Guo Z, Su T. Urinary exosome miR-424 and miR-218 as biomarkers for type 1 diabetes in children. Clinical Laboratory June 1, 2019;65(6). https://doi.org/10.7754/Clin.-Lab.2018.180921. PMID: 31232015.

[105] Ibrahim AA, Wahby AA, Ashmawy I, Saleh RM, Soliman H. Association of exosomal miR-34a with markers of dyslipidemia and endothelial dysfunction in children and adolescents with T1DM. Journal of Clinical Research in Pediatric Endocrinology November 25, 2020;12(4):401–9. https://doi.org/10.4274/jcrpe.galenos.2020.2020.0134. Epub 2020 Jul 13. PMID: 32654473; PMCID: PMC7711647.

[106] Pang H, Fan W, Shi X, Luo S, Wang Y, Lin J, et al. Differential expression and bioinformatics analysis of plasma-derived exosomal circRNA in type 1 diabetes mellitus. Journal of Immunology Research October 27, 2022;2022:3625052. https://doi.org/10.1155/2022/3625052. PMID: 36339941; PMCID: PMC9634467.

[107] Pang H, Fan W, Shi X, Li J, Wang Y, Luo S, et al. Characterization of lncRNA profiles of plasma-derived exosomes from type 1 diabetes mellitus. Frontiers in Endocrinology May 12, 2022;13:822221. https://doi.org/10.3389/fendo.2022.822221. PMID: 35634499; PMCID: PMC9135040.

[108] Jiang B, Zhang J, Sun X, Yang C, Cheng G, Xu M, et al. Circulating exosomal hsa_-circRNA_0039480 is highly expressed in gestational diabetes mellitus and may be served as a biomarker for early diagnosis of GDM. Journal of Translational Medicine January 3, 2022;20(1):5. https://doi.org/10.1186/s12967-021-03195-5. PMID: 34980149; PMCID: PMC8722188.

[109] Cao M, Zhang L, Lin Y, Li Z, Xu J, Shi Z, et al. Circular RNA expression profiles in umbilical cord blood exosomes from normal and gestational diabetes mellitus patients. Bioscience Reports November 27, 2020;40(11):BSR20201946. https://doi.org/10.1042/BSR20201946. PMID: 33146699; PMCID: PMC7670577.

[110] Huang Y, Liang B, Chen X. Exosomal circular RNA circ_0074673 regulates the proliferation, migration, and angiogenesis of human umbilical vein endothelial cells via the microRNA-1200/MEOX2 axis. Bioengineered December 2021;12(1):6782–92. https://doi.org/10.1080/21655979.2021.1967077. PMID: 34516311; PMCID: PMC8806572.

[111] Cao M, Bu C, Zhang J, Ren Y, Zhou G, Chen C, et al. Exosomal circular RNA hsa_circ_0046060 of umbilical cord mesenchymal stromal cell ameliorates glucose metabolism and insulin resistance in gestational diabetes mellitus via the miR-338-3p/G6PC2 Axis. The Internet Journal of Endocrinology June 11, 2022;2022:9218113. https://doi.org/10.1155/2022/9218113. PMID: 35726320; PMCID: PMC9206588.

[112] Cao M, Zhang L, Lin Y, Li Z, Xu J, Shi Z, et al. Differential mRNA and long noncoding RNA expression profiles in umbilical cord blood exosomes from gestational diabetes mellitus patients. DNA and Cell Biology November 2020;39(11):2005–16. https://doi.org/10.1089/dna.2020.5783. Epub 2020 Sep 28. PMID: 32986505.

[113] Tang GY, Yu P, Zhang C, Deng HY, Lu MX, Le JH. The neuropeptide-related HERC5/TAC1 interactions may Be associated with the dysregulation of lncRNA GAS5 expression in gestational diabetes mellitus exosomes. Disease Markers February 8, 2022;2022:8075285. https://doi.org/10.1155/2022/8075285. PMID: 35178132; PMCID: PMC8847027.

[114] Nguyen BY, Azam T, Wang X. Cellular signaling cross-talk between different cardiac cell populations: an insight into the role of exosomes in the heart diseases and therapy. American Journal of Physiology – Heart and Circulatory Physiology April 1, 2021;320(4):H1213–34. https://doi.org/10.1152/ajpheart.00718.2020. Epub 2021 Jan 29. PMID: 33513083; PMCID: PMC8260388.

[115] He X, Kuang G, Wu Y, Ou C. Emerging roles of exosomal miRNAs in diabetes mellitus. Clinical and Translational Medicine June 2021;11(6):e468. https://doi.org/10.1002/ctm2.468. PMID: 34185424; PMCID: PMC8236118.

[116] García-Romero N, Carrión-Navarro J, Esteban-Rubio S, Lázaro-Ibáñez E, Peris-Celda M, Alonso MM, et al. DNA sequences within glioma-derived extracellular vesicles can cross the intact blood-brain barrier and be detected in peripheral blood of patients. Oncotarget January 3, 2017;8(1):1416–28. https://doi.org/10.18632/oncotarget.13635. PMID: 27902458; PMCID: PMC5352065.

[117] Kumar A, Ren Y, Sundaram K, Mu J, Sriwastva MK, Dryden GW, et al. miR-375 prevents high-fat diet-induced insulin resistance and obesity by targeting the aryl hydrocarbon receptor and bacterial tryptophanase (tnaA) gene. Theranostics February 19, 2021;11(9):4061–77. https://doi.org/10.7150/thno.52558. PMID: 33754048; PMCID: PMC7977461.

CHAPTER SEVEN

Exosomal RNAs in immunological diseases

Bidisha Saha[1], Supriya Das[1], Debalina Bhattacharya[2], Mainak Mukhopadhyay[1] and Kursad Turksen[3,a]

[1]Department of Biotechnology, JIS University, Kolkata, West Bengal, India
[2]Department of Microbiology, Maulana Azad College, Kolkata, West Bengal, India
[3]Ottawa Hospital Research Institute, Ottawa, ON, Canada

1. Introduction

Protein-coding genes and their accompanying transcripts have been the focus of countless studies for decades, but they make up less than 2% of the human genome [1]. Noncoding RNAs (ncRNAs) are the most abundant type of RNA in eukaryotic cells, making up as much as 90% of the transcriptome [2]. It is possible to broadly categorize ncRNAs into long ncRNAs (lncRNAs), circular RNAs (circRNAs), small interfering RNAs (siRNAs), and microRNAs (miRNAs), all based on their length (in nucleotides; nt) [3].

Inappropriate stimulation of B and T cells contributes to the production of autoantibodies, which in turn generate the persistent inflammation that characterizes autoimmune diseases. The endocrine system, pancreas, lungs, brain, bones, intestines, skin, and kidneys could all experience autotoxic effects from these diseases [4]. Pathogenic inflammatory reactions caused by T cells specific for self-antigens are facilitated by epigenetic, genetic, cellular, and molecular components involved in autoimmune disease mechanisms [5]. Acquired and/or genetic abnormalities of host immune regulatory mechanisms, chemical mimics of bacterial and viral proteins, and defective clearance of apoptotic bodies all contribute to these processes [6].

Maintaining internal homeostasis requires good immunological tolerance and control, and the immune system is in charge of both surveillance and defense against the invasiveness of many foreign infections. Surveillance and defense make the most of an immune system's activities. Various immune cells (immunocytes) work together to create a synthetic immune defense that, depending on the types of exogenous antigens encountered, can be antiviral, antibacterial, or anticancer. Autoimmunity and the development of some immunological illnesses may be triggered by factors such as improper self-antigen exposure, immune system

[a] Retired.

Exosomal RNA
ISBN 978-0-443-14008-2
https://doi.org/10.1016/B978-0-443-14008-2.00013-9

dysregulation, and the stimulation of cross-antigens. Even though proteins have been the primary focus of current research, exosomal RNAs, notably miRNAs and circular RNAs, have been reported to be novel candidates implicated in the control of immune responses and immunological disorders [7]. In this chapter, we will discuss the role of circRNAs in the immune system and the diseases they regulate.

2. Function and mechanism of exosomal RNAs in immune response

Exosomal RNAs facilitate cellular communication and regulate several immunological processes, making them crucial to immune responses. MiRNAs, long non-coding RNAs (lncRNAs), and small nuclear RNAs (snRNAs) are all types of RNA that can be found in the RNA inside of exosomes, which are small extracellular vesicles (EVs) released by many types of cells, including immune cells. Immune responses can be modulated by exosomal RNAs that are delivered to recipient cells. Exosomal RNAs play an essential role in the immune response, and they do so in the following ways [8]:

1. *Regulation of inflammatory responses:* Exosomal RNAs can activate immune cells involved in the inflammatory response, such as macrophages and dendritic cells. These RNAs can carry specific miRNAs or messenger RNAs (mRNAs) that target and regulate gene expression in recipient cells. By affecting the expression of inflammatory mediators, cytokines, or chemokines, exosomal RNAs can help turn immune cells into ones that are more likely to cause inflammation. By regulating the expression of these molecules, exosomal RNAs can fine-tune the intensity and duration of inflammatory responses. Exosomal RNAs can amplify inflammatory signals by transferring proinflammatory miRNAs or mRNAs to recipient cells. These exosomal RNAs can activate specific pathways or enhance the expression of inflammatory molecules, leading to an intensified inflammatory response. They promote antiinflammatory responses or dampen excessive inflammation. These RNAs can facilitate the resolution of inflammation and the restoration of tissue homeostasis. Exosomal RNAs serve as mediators of intercellular communication during inflammation. They can be released by inflamed cells and taken up by neighboring or distant cells, influencing their inflammatory status and responses. Exosomal RNAs can transfer inflammatory signals, genetic material, or regulatory factors, enabling the exchange of information between different cell types and the coordination of the inflammatory response. Exosomal RNAs can participate in the host defense against pathogens by transferring pathogen-derived RNA to immune cells. These RNAs can activate innate immune responses, induce the production of antiviral or antibacterial factors, and enhance immune surveillance against pathogens. Overall, exosomal RNAs play diverse roles in the inflammatory response, including immune cell activation, regulation of inflammatory signaling, amplification or resolution of inflammation, intercellular communication, and pathogen defense. They

contribute to the complex orchestration of the inflammatory response and its subsequent resolution, aiming to maintain immune homeostasis and tissue integrity.

2. *Immune evasion and tolerance:* Exosomal RNAs have the ability to transport regulatory compounds that block immunological reactions. After immune cells or other cell types release exosomal RNAs like these, immune cells such as T cells and antigen-presenting cells (APCs) can take them up. Once within recipient cells, these RNAs can prevent immune activation-related genes from being expressed, suppressing immunological responses and encouraging immune evasion. These may assist in the development and maintenance of immunological tolerance. Regulatory T cells (Tregs) can be produced and function more effectively when specific miRNAs or other noncoding RNAs are present in regulatory exosomes. Tregs, a subgroup of T cells, are essential for maintaining immunological tolerance because they reduce overactive immune reactions and prevent autoimmune disease. Exosomal RNAs may help regulate signal transmission to recipient cells, promoting immunological tolerance. RNAs have the ability to cause immune cells to undergo apoptosis as well as "programmed cell death," inhibiting immunological responses. Apoptotic pathways can be activated, resulting in cell death, by targeting and downregulating antiapoptotic genes in recipient cells. Exosomal RNAs have the ability to alter how well APCs deliver antigens, which can affect immunological reactions and tolerance. Exosomal RNAs generated by APCs may contain particular molecules that change how antigen-presenting machinery parts, such as MHC molecules or costimulatory molecules, are expressed. The activation and development of T cells can be affected by this modification of antigen presentation, which may promote immunological tolerance or prevent immune recognition. Exosomal RNAs have the ability to alter immune cell activity, including that of APCs and effector T cells, to encourage immune evasion or tolerance. These exosomal RNAs may contain genetic material or regulatory components that modify the activity of immune cells or suppress immunological responses by altering signaling pathways or the expression of genes in recipient cells. Exosomal RNAs generally decrease immunological responses, foster immune tolerance, induce immune cell death, modulate antigen presentation, and alter immune cell function to contribute to immune evasion and tolerance. These pathways enable pathogens or malignancies to avoid immune identification and to develop an immune-tolerance state, which facilitates their survival and expansion. To develop techniques to circumvent these systems and improve immune responses against infections or cancer cells, it is critical to understand the precise exosomal RNA-mediated mechanisms of immune evasion and tolerance.
3. *Antigen presentation:* Exosomal RNAs in antigen-presenting cells (APCs) can contribute to antigen presentation to T cells. Exosomes released by APCs can contain mRNA molecules encoding major histocompatibility complex (MHC) molecules, which are crucial for presenting antigens to T cells. The transfer of MHC mRNA via exosomes can enhance the expression of MHC molecules on recipient cells, leading to efficient antigen

presentation and T cell activation. Exosomal RNAs derived from APCs can influence the function of recipient immune cells. For example, exosomal miRNAs transferred from APCs to T cells can regulate gene expression in the recipient T cells. These miRNAs can target specific mRNA transcripts, leading to the suppression or enhancement of immune cell function. By modulating the expression of genes involved in immune cell activation, proliferation, or differentiation, exosomal RNAs can impact the overall immune response. Exosomal RNAs from APCs can contribute to immune regulation and the induction of tolerance. APCs, such as dendritic cells, can release exosomes containing regulatory miRNAs or other noncoding RNAs. These exosomal RNAs can be taken up by immune cells, including other APCs and T cells, and modulate their function. They can suppress immune responses, promote the generation of regulatory T cells, and dampen inflammation, thereby promoting immune tolerance. During infection, APCs can capture and process pathogens, including viruses or bacteria. Exosomes released by infected APCs can contain pathogen-derived RNAs, such as viral or bacterial RNA. These exosomes can transfer the pathogen-derived RNA to other APCs or immune cells, facilitating the recognition of the pathogen and the initiation of appropriate immune responses. Overall, exosomal RNAs in APC cells contribute to antigen presentation, modulation of immune cell function, immune regulation, transfer of pathogen-derived RNA, and intercellular communication. These roles highlight the significance of exosomal RNAs in shaping the immune response and facilitating effective immune surveillance and defense against pathogens.

Through their roles in mediating intercellular communication, regulating gene expression, and impacting a variety of immune cell functions, exosomal RNAs play a crucial role in monitoring immunological responses. Exosomal RNAs help control the immune system in a number of ways, such as by changing how cells work, separating immune cells into groups, letting cells talk to each other, regulating inflammation, allowing tolerance and stopping the immune response, and protecting the body from pathogens. They assist in the control of the immune system, the preservation of immunological balance, and the defense against infections and aberrant immune responses. Developing therapeutic techniques to modify immune responses in various diseases and circumstances requires a thorough understanding of the unique processes of exosomal RNA-mediated immune regulation (Fig. 7.1).

3. Exosomal RNAs in immune regulation

The immune system is a sophisticated network responsible for preserving an organism's interior environment. To prevent damage to healthy tissues caused by either overactive immune responses or inefficient responses against pathogen invasion, proper immune control is essential. This highlights the need to research immune responses and biological mechanisms. Functional biological macromolecules, such as exosomal circRNAs, have recently gained attention as possible immune regulators.

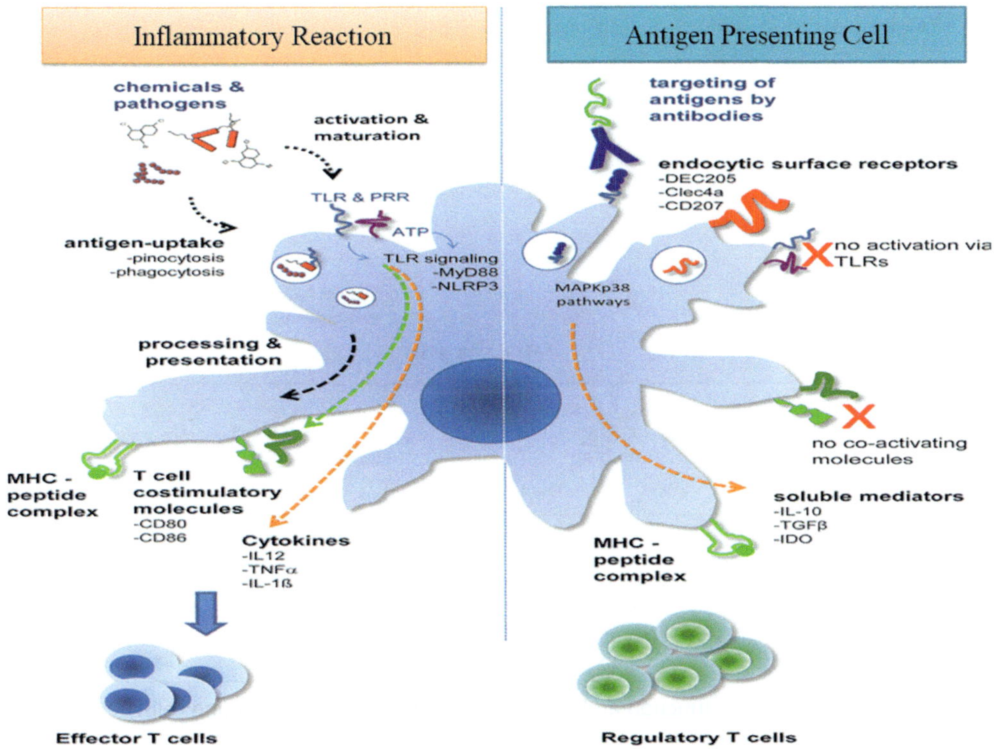

Figure 7.1 Exosomal RNAs in the immunological response: role and mechanism.

One study discovered that circRNAs can regulate cellular immunological signaling pathways by competitively binding to double-stranded RNA-activated protein kinase (PKR) [9]. Many circRNAs, such as dsRNAs, can bind to PKR and limit its activity by forming small defective RNA duplexes. RNase L destroys circRNAs in response to viral infection, freeing PKR and activating downstream antiviral immune systems.

Macrophages are vital to innate immunity and are an essential immune cell. Their roles are established by their ability to become activated and polarized into distinct subtypes in response to microenvironmental cues. M1 and M2 macrophages, for example, have distinct roles in various immunological responses, disease progression, and healing. Exosomal RNAs such as circZC3H4 and Circ-RasGEF1B have been linked to macrophage activation in research. For instance, the HECTD1/ZC3H12A-dependent ubiquitin-proteasome pathway [10] is where circZC3H4 plays a role in SiO_2-induced macrophage activation. Lipopolysaccharide (LPS)-induced macrophage activation requires the presence of Circ-RasGEF1B [11]. Further evidence linking circRNAs with macrophage polarization comes from the observation of differential expression of circRNAs between M1 and M2 macrophages [12]. In addition, stimulation of macrophages can trigger the creation of circRNAs that modulate macrophage immunological

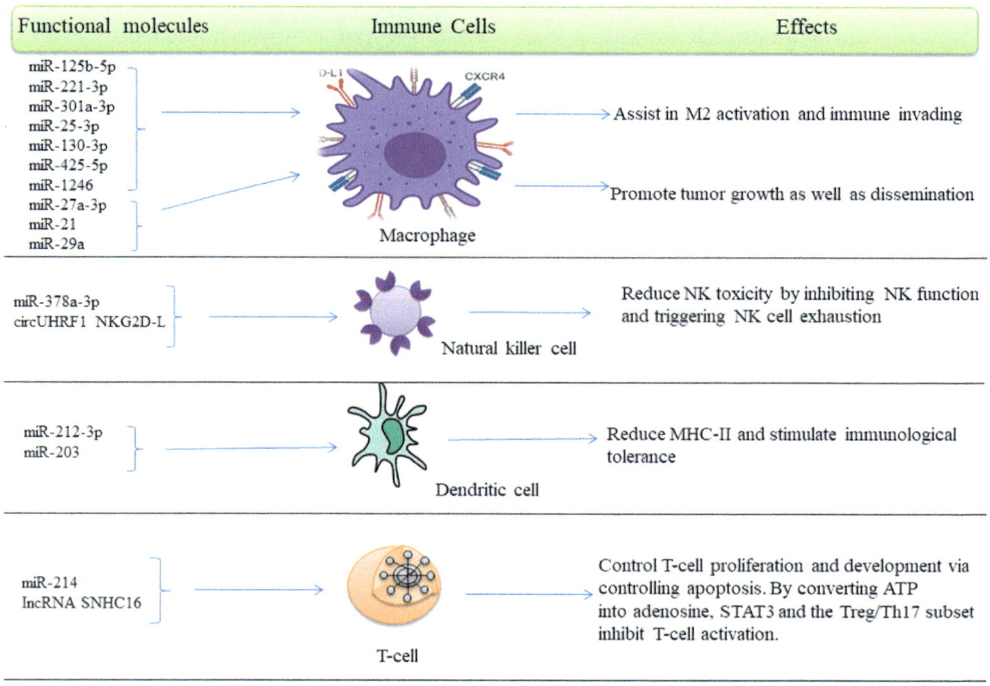

Figure 7.2 The function and impact of exosomal RNAs on immune cells.

activity. For instance, LPS induces the RNA circ-RasGEF1B, which in turn favorably influences the expression of the adhesion protein intercellular adhesion molecule 1 (ICAM-1). Regulation of the stability of mature ICAM-1 mRNA occurs via the lipopolysaccharide (LPS)/Toll-like receptor 4 (TLR4) signaling pathway and facilitates intercellular contacts during antigen presentation [12]. Macrophage apoptosis and the suppression of macrophage growth may also involve circ-ANRIL [13].

In conclusion, circRNAs are crucial for macrophage maturation, activation, and function. They influence macrophage polarization and function, regulate cellular immune signaling pathways, and participate in immune control by competitively binding to PKR. Additional study of circRNAs in the immune system has the potential to shed light on the mechanisms of immune response and identify novel therapeutic targets for disorders associated with the immune system. The function and impact of exosomal RNAs on immune cells are depicted in Fig. 7.2.

4. Pathogenesis of autoimmune diseases

When the immune system assaults healthy tissues and organs, this is known as autoimmunity. It manifests itself either through the activity of autoreactive T cells or the generation of autoantibodies that attack self-antigens. Natural autoimmunity occurs as part of the immune response, and autoimmune illnesses develop when the immune system is stimulated in an abnormal way [6,14]. Systemic autoimmune diseases, including systemic lupus erythematosus (SLE), can affect several organs and tissues across the body. Organ-specific autoimmune diseases affect only one organ or tissue. There are a number of factors at play in the onset of pathogenic autoimmunity. Tolerance mechanisms are protective in healthy contexts because they suppress the activation of self-reactive cells. Clonal ignorance, in which the immune system ignores self-reactive lymphocytes; anergy, in which self-reactive lymphocytes become unresponsive; and activation-induced cell death, in which self-reactive cells are eliminated, are all examples of these methods [15]. But in autoimmune diseases, these tolerance mechanisms fail, allowing self-reactive B and T cells to survive and become activated in response to autoantigens [16–18]. In autoimmune illnesses, the breakdown of tolerance mechanisms is caused by a confluence of variables. Nonhuman leukocyte antigen (HLA) and HLA gene abnormalities, as well as environmental variables and deficiencies in immunoregulatory pathways, can all play a role. Organ damage can occur when self-reactive B and T lymphocyte clones that target autoantigens remain in the body [18,19]. Autoimmune diseases form when the body's immune system becomes dysregulated, leading to an imbalance in the body's inflammatory and immunological systems [15]. Small membrane-bound vesicles generated by cells, known as EVs, can transport bioactive substances such as RNA and proteins, and their dysregulation has been linked to autoimmune processes. Autoimmune diseases can be caused by anomalies in a number of different systems, including those involved in cytokine generation, antigen presentation, and autoantigen scavenging. In conclusion, the immune system becomes misdirected and begins attacking self-antigens in people with autoimmune illnesses. Autoimmune disorders are caused in part by a breakdown in the body's tolerance mechanisms, a lack of equilibrium between immunoregulatory and inflammatory processes, and anomalies in autoantigen scavenging systems, antigen presentation, cytokines, and EVs [6]. Targeted medications and interventions for autoimmune disorders require an understanding of these pathways.

Inflammation that fails to subside and immune system attacks that cause tissue damage are characteristics of autoimmune illnesses. Research is needed to learn more about the causes of these conditions and find new ways to treat them. The acetylcholine receptor in myasthenia gravis (MG), the thyroid-stimulating hormone receptor (TSHR) in Graves' disease (GD), and the myelin basic protein in multiple sclerosis are all examples of autoimmune disorders with recognized autoantigens [6]. These disorders manifest themselves in a variety of ways due to a combination of factors, including chronic inflammation and signaling via the autoantigen receptor. Unfortunately, our knowledge of the pathophysiology of many chronic inflammatory illnesses is currently limited. There is no single explanation that can cover the wide range of autoimmune disorders, despite the fact that some pathogenic features have been established in particular diseases. Since so little is known about the causes of autoimmune diseases and how to effectively diagnose, treat, and manage them, further study is required.

In autoimmunity, mutations only disrupt the host's tolerance checkpoints after a protracted dormant period [20]. Mutations in the autoimmune regulator (AIRE) transcription factor impair negative thymic selection of tissue-specific T cells. Self-antigens expressed by specialized epithelial cells in the thymus medulla, such as proinsulin in the pancreas, are among the autoantigens in the body that AIRE partially controls [21]. The clinical presentation of AIRE abnormalities, however, may lie dormant for years, with autoimmune polyendocrinopathy spreading progressively and randomly to new tissues over time. Usually, one type of tissue, such as the parathyroid or adrenal glands, is afflicted in the first few years of life, and then other types of tissues are impacted later. Autoimmunity against certain antigens and organs is also shown in AIRE-deficient animals. Antigenic integrity relies on people keeping the balance between antiinflammatory and proinflammatory reactions [22]. Suppressing autoimmune and immunological reactions is a group effort that involves many different types of immunocompetent cells working together through cell-to-cell interaction and soluble antiinflammatory chemicals like cytokines. Tregs, or regulatory T cells, can originate in the thymus (CD4+ CD25 FoxP3+ natural Tregs) or in the periphery (type 1 regulatory T cells, generating interleukin 10 [IL-10], or transforming growth factor beta [TGF-β]) [23]. Autoimmune mechanisms can develop in those with a predisposition to the condition when pathological conditions cause either a quantitative or qualitative shift in Tregs and proinflammatory responses. Prolonged dysfunction of regulatory T cells has been reported in autoimmune illnesses such as rheumatoid arthritis (RA), systemic lupus erythematosus (SLE), Sjögren's syndrome (SS), and mixed connective tissue disease (MCTD) [24–27]. This may result in a general decline in Tregs or a reduction in their ability to reduce inflammation. These results emphasize the significance of Tregs and show that long-term dysregulation of Tregs contributes to the onset and progression of autoimmune illnesses.

Th17 lymphocytes are a subset of T helper (Th) cells that generate interleukin-17 (IL-17) and are induced to differentiate from T lymphocytes in inflammatory environments. Interleukin-1 (IL-1), interleukin-6 (IL-6), interleukin-21 (IL-21), and interleukin-23 (IL-23) induce a protein called signal transducer and activator of transcription 3 (STAT3) [15], which in turn mediates T lymphocyte development into Th17 cells. It has been found that Th17 cells aid in the maturation of both Th1 and Th2 cells [28,29]. Persistent IL-17 production by Th17 cells is a factor in the pathophysiology of autoimmune and inflammatory illnesses and also contributes to chronic inflammation more generally. Th17 cells, on the other hand, are essential in the host's defense mechanism against infections. Inducing inflammation by bringing macrophages and neutrophils to the scene, Th17 cells are themselves products of the inflammatory response [15,29]. Rheumatoid arthritis (RA) is distinguished from osteoarthritis by the presence of elevated IL-17 levels in synovial biopsies, sera, and synovial fluids (Fig. 7.3). Increased levels of IL-6 and TGF-β, which encourage the formation of Th17 cells, have been documented in autoimmune diseases

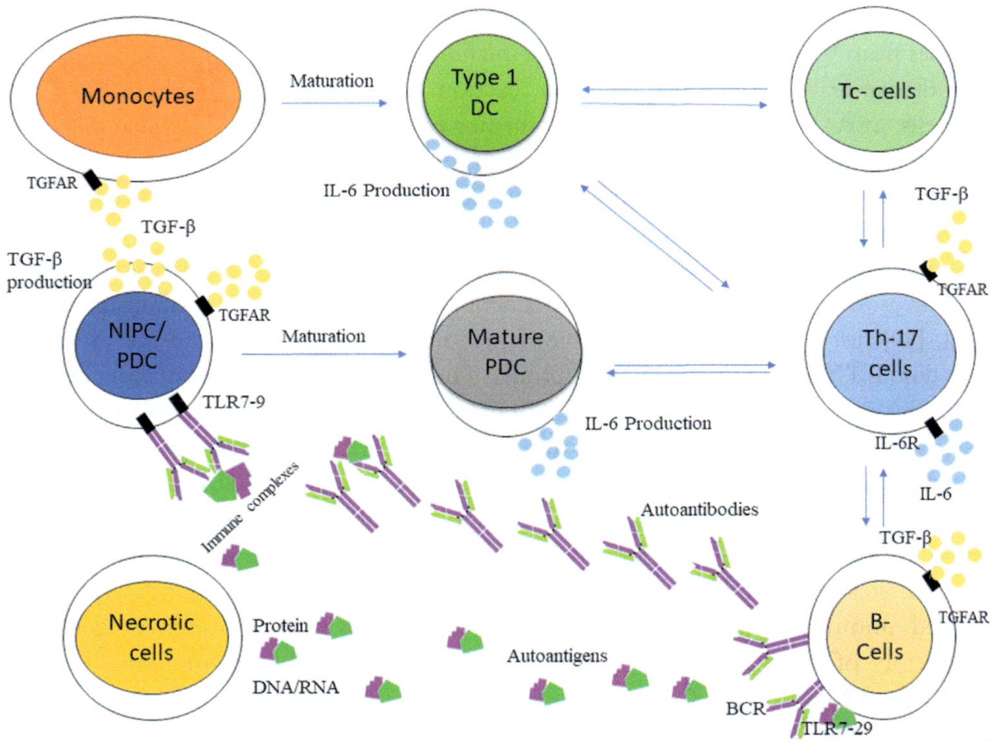

Figure 7.3 Pathogenesis of autoimmune disease.

[30,31]. Another element that contributes to an imbalance between antiinflammatory and proinflammatory responses is high levels of tumor necrosis factor-alpha (TNF), which are seen in various autoimmune illnesses [32,33].

5. Role of exosomal RNAs in autoimmune diseases

5.1 Myasthenia gravis

MG is an autoimmune disorder characterized by muscle weakness and fatigue due to the impairment of neuromuscular transmission. Even though no one knows for sure what causes MG, it is thought that immune dysregulation and the production of autoantibodies against parts of the neuromuscular junction, such as the acetylcholine receptor (AChR) or muscle-specific kinase (MuSK), are to blame [34].

Many studies show that miRNAs are effective at regulating the pathophysiology of MG. For example, patients with MG who have a positive AchR-Ab do not have regulation of hsa-miR21-5p, hsa-miR150-5p, and hsa-R27a-3p [35]. Furthermore, miRNAs have been found to be useful as disease biomarkers in various investigations [36,37]. Derkacz et al. [38] evaluated the role of exosomal miRNAs in MG development by studying their appearance in the plasma of MG patients. They analyzed plasma samples from 92 MG patients and 49 age-matched controls using deep sequencing and quantitative real-time polymerase chain reaction. Given its potential role in MG pathogenesis, miR-106a-5p was selected for further investigation. MG cases, such as QMGS cases, had significantly reduced miR-106a-5p expression levels relative to healthy volunteers. MicroRNA-146a overexpression reduced chronic clinical MG in mouse models, shifted Th1/Th17 to Th2/Treg T cell patterns in serum and spleen, and induced exosome production in a mouse model of autoimmune MG. The therapeutic benefits of these exosomes were partially dose-dependent and Ag specific [39]. In addition, Zhong et al. [40] discovered that the MG biomarker miR-150-5p influenced the efficacy of low-dose RTX in alleviating MG symptoms. These data point to a possible link between miR-150-5p and contacts between CD19+ and CD27+ B cells, which would explain the observed connection. Another study identified that exosomes derived from MG patients' sera contained higher levels of miR-146a compared with healthy controls. MiR-146a is known to regulate immune responses and inflammation, and its upregulation in exosomes suggests a potential role in MG pathogenesis. These miRNAs have been implicated in immune regulation and could potentially contribute to the dysregulated immune responses observed in MG. Moreover, a study investigated the expression of circular RNAs (circRNAs) in peripheral blood mononuclear cells (PBMCs) of MG patients. CircRNAs are a type of noncoding RNA with emerging roles in gene regulation. The study identified differentially expressed circRNAs in MG patients compared with healthy controls, suggesting their potential involvement in MG pathogenesis and immune dysregulation.

5.2 Autoimmune thyroid disease

Autoimmune thyroid disease (AITD) is one of the most common autoimmune diseases, and its two forms, Hashimoto's thyroiditis (HT) and GD, are both rather common [41]. According to Rodrguez-Munoz et al. [42], microvesicles isolated from people with AITD can inhibit Treg cell development while stimulating Th1 and Th17 subsets. Okoye et al. found, however, that exosome miRNAs released by Treg cells may be taken up by other T cells, including Th1 cells. They found that exosomes from Treg cells could send Let-7d, which targets Ptgs2 (Cox-2), to Th1 cells. This would suppress the immune system and keep a systemic illness from happening [43]. Patients with AITD, such as GD and HT, have had their plasma exosome miRNA profile analyzed in a new study. They discovered that miR-146a and miR-155 levels in circulating microvesicles are elevated in AITD patients compared with healthy controls. MiR-146a was upregulated in patients with both HT and GD, whereas miR-155 was upregulated in HT patients relative to controls. In contrast, PBMCs and the IL-8 and SMAD4 genes, as well as their targets miR-146a and miR-155, were down in AITD patients [42]. Moreover, Hiratsuka et al. demonstrated that EVs produced from intractable GD patients' sera can stimulate IL-1 and TNF-mRNA expression in PBMCs and activate immune system cells, in contrast to remission GD patients or healthy controls. They also demonstrated that miRNA levels in the blood of GD patients changed with the severity of their condition. While the miRNA composition of plasma exosomes has not yet been described, four widely distributed miRNAs—let7g-3p, miR-23b-5p, miR-339-5p, and miR-92a-3p—have. MiR-23b-5p and miR-92a-3p levels were found to be elevated in GD patients in remission, while let-7g-3p and miR-339-5p levels were found to be downregulated in participants with uncontrollable GD [44]. As regulators of immune suppression in autoimmune disease and illness mitigation, they argue that the two miRNAs' enhanced production is related to these roles. Susceptibility to radioactive iodine treatment is also associated with the expression of the sodium iodine symporter, which is regulated by miR-339-5p. Therefore, they argue that the serum exosomes of people with intractable GD may have a major impact on the etiology of GD [45]. In a separate study, exosomes extracted from Hashimoto's thyroiditis patients' thyroid tissue showed variable expression of miRNAs. It was discovered that miRNAs such as miR-451a and miR-21 may regulate the immune system and thyroid function by targeting specific genes.

Along with this, exosomal long noncoding RNAs (lncRNAs) were studied to see if they play a part in autoimmune thyroid disorders. Long noncoding RNAs (lncRNAs) are another type of exosomal RNAs with roles in controlling gene expression. Exosomes isolated from the blood of patients with Graves' disease were shown to include lncRNAs that were differently expressed in comparison with those from healthy controls. Immune modulation and inflammatory pathways have both been linked to these lncRNAs.

Insights into the potential significance of exosomal RNAs in autoimmune thyroid illnesses have been provided by these investigations, but more work is needed to determine their precise functions in disease etiology, immunological dysregulation, and intercellular communication. To further understand how exosomal RNAs play a role in the onset and progression of autoimmune thyroid disorders, future research could investigate the functional effects of individual exosomal RNAs on thyroid cells, immune cells, and their interactions. With this information, we may be able to find better ways to diagnose and treat these diseases.

5.3 Systemic lupus erythematosus

Multiple systems and organs are affected by the autoimmune illness known as systemic lupus erythematosus (SLE), which typically affects young women [46]. Various organs, including the urinary, circulatory, pulmonary, and central neurological systems, can be damaged or destroyed due to the formation of autoantibodies against self-antigens, immune complex deposition, complement depletion, inflammation, and organ destruction [47,48]. Despite clinical expertize and test criteria, correctly diagnosing and treating people with SLE remains difficult. The use of circular RNAs (circRNAs), which can act as miRNA "sponges," is one possible solution to this issue. For instance, overexpression of CDR1as decreases miR-7 and increases the expression of its target mRNAs, whereas miR-7 restricts the action of PTEN and leads to B cell hyperreactivity in SLE [49,50]. High-throughput RNA sequencing has shown 207 dysregulated circRNAs in lupus patients, and a number of miRNAs have been linked to their regulation via interactions with these circRNAs. Additional study is required to elucidate the roles and processes of four highly dysregulated circRNAs, including hsa_circ_102584, hsa_circ_400011, hsa_circ_101471, and hsa_circ_100226; however, they show promise as noninvasive biomarkers for SLE [51]. The potential of urine-derived exosomal miR-129 [52] as a biomarker for primary fibrosis in lupus nephritis (LN) has been demonstrated. MicroRNA-146a (miR-146a) has been demonstrated to regulate MSC cell death by targeting TRAF6/NF-κB, and this has led to speculation that it plays a role in the pathophysiology of systemic lupus erythematosus. Exosomes carrying miR-214 are released from endothelial cells, and they may suppress senescence and stimulate angiogenesis in target cells [53]. The adverse effects of TGF-β, including mesangial proliferation and fibrogenesis, can be attenuated by exosomal miR-302d generated from urine in dormant LN [54].

In conclusion, SLE is a multiorgan, multisystem autoimmune illness. Exosomal miRNAs and circRNAs have been identified as promising targets for the study of SLE pathophysiology, the identification of biomarkers, and the development of new treatments. However, more study is required to clarify the mechanisms and activities of these compounds in SLE.

6. Role of exosomal RNAs in other immunological diseases

6.1 Primary biliary cirrhosis

Primary biliary cirrhosis (PBC), also known as primary biliary cholangitis, causes hepatic fibrosis or cirrhosis by destroying the tiny intrahepatic bile channels [55]. Because patients with PBC frequently experience no symptoms until the latter stages of the disease, they are almost always in a very advanced stage of the disease when they are finally diagnosed. The development of sensitive biomarkers for the early detection of PBC is, hence, of paramount importance. Zheng et al. conducted a circRNA microarray analysis in plasma from patients with PBC and healthy individuals, and they identified 18 upregulated and 4 downregulated circRNAs that may be associated with PBC. Then, one of the 18 upregulated circRNAs, hsa_circ_402458, was identified as a biomarker for PBC [56]. There are two possible biotargets for hsa_circ_402458: miR-522 and miR-943 [57]. The significance of miR-522 in chronic inflammatory illnesses and miR-943 [58] in the progression of PBC has been proven by numerous studies. Hsa_circ_402458 contributes to PBC regulation and may antagonize miR-522 and miR-943. Because of its high stability and intrinsic conservation, hsa_circ_402458 is well suited for early PBC detection using patient exosomes and blood plasma [59].

6.2 Fibrosis

Infection, allergies, and tissue injury all contribute to a state of persistent inflammation that eventually leads to fibrosis [60]. Studies show that exosomal circRNAs drive fibrosis development in several tissues. Ingesting excessive amounts of silica over a lengthy period of time can lead to lung fibrosis. SiO_2 causes a high expression of Circ-ZC3H4. Zinc finger CCCH-type containing 4 (ZC3H4) promotes proliferation and migration in alveolar macrophages and fibroblasts after exposure to SiO_2 via restricting miR-212 as an endogenous competitor through its binding sites [10]. Liver fibrosis requires the activation of hepatic stellate cells (HSCs). It is clear that HSCs play a crucial role in promoting fibrosis since they are the primary source of the ECM and can develop into cells that are similar to myofibroblasts. Stimulating HSCs over an extended period of time with inflammatory stimuli is a major trigger for their activation. Studies show that when hsa_circ_0071410 is turned down, miR-9-5p expression goes up, which slows down HSC activity along a number of immune pathways [61]. Myocardial fibrosis is made worse by hsa_circ_010567 because it inhibits miR-141 expression by activating TGF-β1 signaling and increases the influence of circRNAs on fibrosis and the cardiovascular system. There is a need for additional research on the roles of circRNAs in organ fibrosis [62].

6.3 Psoriasis

Psoriasis is an autoimmune skin illness in which keratinocytes and activated immune cells combine to produce silvery scales on the skin [63]. MiRNAs in EVs and exosomes have received a lot of attention recently for their potential role in the pathogenesis of psoriasis. Psoriasis relies heavily on T cells, and the miRNA content of EVs varies between T cell subsets. The exosomes released by Treg cells are rich in miR-146a-5p, miR-150-5p, and miR-21-5p, whereas those released by Th1/Th17 cells are richer in miR-106a-5p, miR-155-5p, and miR-19a-3p. Psoriasis patients had elevated levels of these miRNAs, but after taking Etanercept, an anti-TNF-α medication, their levels returned to normal. This miRNA profile in exosomes is a promising tool for monitoring inflammation in psoriasis patients because it represents communication between T cell subsets and immune response activation. EVs, or exosomes, released by keratinocytes, the key cells implicated in psoriasis lesions, can also affect the epidermal microenvironment [64,65]. The miRNA profiles of EVs from cytokine-treated keratinocytes are changed, with higher levels of 28 miRNAs and lower levels of 114 miRNAs. T cells can uptake these EVs and be polarized to the Th1 or Th17 subsets. Psoriasis disease progression is facilitated by exosomal miR-381-3p, which is present in cytokine-treated keratinocytes and activates Th1 and Th17 cells [66]. Psoriatic arthritis (PsA) is a form of inflammatory arthritis that can affect people with psoriasis [64]. MiRNA expression in plasma exosomes is different in people with PsA compared with those with cutaneous psoriasis (PsC). Diagnostic indicators for PsA include the miRNAs let-7b-5p and miR-30e-5p, both of which are downregulated in PsA patients relative to PsC patients [67]. Psoriasis pathogenesis has also been linked to a number of other miRNAs. Psoriasis is linked to changes in WNT pathways and a lower level of miR-1305 in serum exosomes [68]. Autoimmune disorders and bone destruction are shared characteristics of PsA, psoriasis, rheumatoid arthritis, and gout; these conditions are linked to specific miRNAs such as miR-769-5p, miR-589-5p, miR-370-3p, miR-199a-5p, and miR-151a-3p. Psoriasis pathogenesis may be influenced by miRNAs that are transported by EVs and exosomes. The different miRNA profiles in T cell subsets, keratinocytes, and plasma exosomes [67,69] give information about how the disease works and could be used as diagnostic markers.

6.4 Inflammatory bowel disease

Ulcerative colitis (UC) and Crohn's disease (CD) are two forms of inflammatory bowel disease (IBD) that damage the digestive tract and cause chronic inflammation [38]. Exosomes, which contain RNAs and proteins among other functional elements, play significant roles in IBD by modulating connections between cells and their environment. MiRNA expression patterns are dysregulated in IBD, and these changes are linked to canonical pathways in IBD pathogenesis. Using exosomes, researchers have looked at how mast cells (MCs) communicate with intestinal epithelial cells (IECs) [70]. Exosomes released by MCs were discovered to increase the permeability of IECs and disturb the formation of the intestinal barrier. The downregulation of tight junction-related proteins such as tight junction proteins 1 (TJP1, ZO-1), occludin (OCLN), and claudin 8 (CLDN8) was linked to the transfer of exosomal miRNAs from MCs to IECs. The increased aggregation of MCs in the intestinal mucosa of people with IBD suggests a possible function for these cells in the pathogenesis of this disease.

Exosomal miR-195a-3p from Tregs has been found to target Caspase 12, a proapoptotic enzyme involved in innate immune responses and IBD pathogenesis, thereby promoting the proliferation and suppressing the apoptosis of intestinal epithelial cells. Targeting caspase 12 and encouraging intestinal epithelial barrier restoration, miR-195a-3p, supplied by Treg-derived exosomes, was hypothesized to have therapeutic potential in IBD [71].

Exosomes from intestinal cells are particularly abundant in miR-21, which has been shown to be dysregulated in active IBD [72]. Both the protective and inflammatory effects of miR-21 deletion in colitis models point to its multifaceted role in IBD etiology. Exosomal miR-223 is highly abundant and has been linked to IBD progression through its regulation of intestinal macrophages and dendritic cells and promotion of the IL-32 inflammatory cascade. Exosomal miR-223 interferes with the stability of tight junctions by downregulating claudin 8, which compromises the integrity of the intestinal epithelial barrier. In conclusion, exosomes and their cargo, which includes miRNAs, play important roles in the development of inflammatory bowel disease [73,74]. Exosome miRNA expression dysregulation is associated with intestinal barrier dysfunction, inflammation, and immunological responses in IBD. The development of treatment methods and biomarkers for IBD may benefit from a better understanding of the roles and mechanisms represented by exosomal miRNAs [75,76].

6.5 Sjogren syndrome

Dryness of the mouth and eyes are symptoms of Sjögren's syndrome (SS), an inflammatory illness characterized by inflammation of the exocrine glands, most notably the salivary and lacrimal glands [77]. With implications for diagnosis and disease development, exosomes, small EVs, have been studied as possible carriers of miRNAs in SS [78].

Researchers have discovered that exosomes contain a disproportionately high number of miRNAs found in saliva. Insightful information on the possible use of salivary exosomal miRNAs as a noninvasive diagnostic technique for SS was gleaned from a study in which miRNA-containing exosomes were extracted from the saliva of both healthy individuals and SS patients [79]. Patients with SS often experience inflammation of the salivary glands, and miRNAs have been discovered as possible indicators of this inflammation [80]. This provides support for the idea that exosomal miRNAs can be used as biomarkers of illness. Inflammation of the salivary glands in SS has been linked to viral infections, including Epstein—Barr virus (EBV). Salivary gland exosomes have been identified to transport an EBV-encoded miRNA, namely EBV-miR-BART13-3p, from B cells to epithelial cells, where it inhibits secretion [81]. This demonstrates how exosomal miRNAs mediate interactions across several cell types in SS pathogenesis. In addition, miR-142-3p, which is secreted by T cells, can specifically target proteins involved in calcium signaling and protein secretion in salivary gland epithelial cells when it is delivered to these cells in the form of exosomes [82]. This may contribute to the development of SS by reducing the efficiency of the epithelial cells that line the salivary glands. Exosomal miRNAs are a viable route for molecular studies of the development and diagnosis of SS because of their stability and lack of invasiveness [83]. Exosomal miRNAs in SS need additional investigation since they may shed light on disease processes and treatment strategies.

7. Conclusion

This research looked into the prospect that exosomal RNAs play a causative function in autoimmune disorders. These results emphasized the potential role of exosomal miRNAs and circRNAs in modulating immune responses in a variety of contexts. Exosomal RNAs have been found to exhibit unique expression profiles in different autoimmune disorders, suggesting their potential as biomarkers. Exosomal RNAs have shown promise as indicators of disease activity; however, their sensitivity and specificity in bodily fluids still need to be determined. In addition, methods for enhancing the repeatability of exosomal RNA screening and information gaps concerning the pathophysiology of autoimmune disorders need to be addressed. The clinical application of exosomal RNAs is hindered by the difficulty of isolating disease-specific RNAs. The development of molecular treatments targeting miRNA-induced changes in autoimmune disorders may be facilitated by a better understanding of the molecular pathways involving exosomal RNAs. Though preliminary, more study of exosomal RNAs in autoimmune diseases is required. Overall, the study emphasizes the need for more research into the functions and mechanisms of exosomal RNAs, given their potential as biomarkers and therapeutic agents in autoimmune disorders.

References

[1] Lander ES, Linton LM, Birren B, Nusbaum C, Zody MC, Baldwin J, et al. International human Genome Sequencing Consortium. Initial sequencing and analysis of the human genome. Nature 2001;409(6822):860–921.
[2] Warner JR. The economics of ribosome biosynthesis in yeast. Trends in Biochemical Sciences 1999;24(11):437–40.
[3] St Laurent G, Wahlestedt C, Kapranov P. The Landscape of long noncoding RNA classification. Trends in Genetics: TIG 2015;31(5):239–51.
[4] Lee KH, Ahn BS, Cha D, Jang WW, Choi E, Park S, et al. Understanding the immunopathogenesis of autoimmune diseases by animal studies using gene modulation: a comprehensive review. Autoimmunity Reviews 2020;19(3):102469.
[5] Bluestone JA, Bour-Jordan H. Current and future immunomodulation strategies to restore tolerance in autoimmune diseases. Cold Spring Harbor Perspectives in Biology 2012;4(11):a007542.
[6] Jose J, Muralinaidu R, Sunil P, Varghese SS. Pathogenesis of autoimmune diseases: a short review. Oral and Maxillofacial Pathology Journal 2014;5(1):2014.
[7] Chen YG, Satpathy AT, Chang HY. Gene regulation in the immune system by long noncoding RNAs. Nature Immunology 2017;18(9):962–72.
[8] Li C, Ni YQ, Xu H, Xiang QY, Zhao Y, Zhan JK, et al. Roles and mechanisms of exosomal non-coding RNAs in human health and diseases. Signal Transduction and Targeted Therapy 2021;6(1):383.
[9] Liu CX, Li X, Nan F, Jiang S, Gao X, Guo SK, et al. Structure and degradation of circular RNAs regulate PKR activation in innate immunity. Cell 2019;177(4):865–880.e21.
[10] Yang X, Wang J, Zhou Z, Jiang R, Huang J, Chen L, et al. Silica-induced initiation of circular ZC3H4 RNA/ZC3H4 pathway promotes the pulmonary macrophage activation. The FASEB Journal: Official Publication of the Federation of American Societies for Experimental Biology 2018;32(6):3264–77.
[11] Ng WL, Marinov GK, Liau ES, Lam YL, Lim YY, Ea CK. Inducible RasGEF1B circular RNA is a positive regulator of ICAM-1 in the TLR4/LPS pathway. RNA Biology 2016;13(9):861–71.
[12] Zhang Y, Zhang Y, Li X, Zhang M, Lv K. Microarray analysis of circular RNA expression patterns in polarized macrophages. International Journal of Molecular Medicine 2017;39(2):373–9.
[13] Holdt L, Stahringer A, Sass K, Holdt LM, Stahringer A, Sass K, Pichler G, Kulak NA, Wilfert W. Circular non-coding RNA ANRIL modulates ribosomal RNA maturation and atherosclerosis in humans. Nature Communications 2016;7:12429.
[14] Tan L, Wu H, Liu Y, Zhao M, Li D, Lu Q. Recent advances of exosomes in immune modulation and autoimmune diseases. Autoimmunity 2016;49(6):357–65.
[15] Lleo A, Invernizzi P, Gao B, Podda M, Gershwin ME. Definition of human autoimmunity–autoantibodies versus autoimmune disease. Autoimmunity Reviews 2010;9(5):A259–66.
[16] Abbas AK, Lohr J, Knoechel B, Nagabhushanam V. T cell tolerance and autoimmunity. Autoimmunity Reviews 2004;3(7–8):471–5.

[17] Davies AJ. Immunological tolerance and the autoimmune response. Autoimmunity Reviews 2008;7(7):538–43.
[18] Ermann J, Fathman CG. Autoimmune diseases: genes, bugs and failed regulation. Nature Immunology 2001;2(9):759–61.
[19] Cooper GS, Miller FW, Pandey JP. The role of genetic factors in autoimmune disease: implications for environmental research. Environmental Health Perspectives 1999;107(Suppl. 5):693–700.
[20] Goodnow CC. Multistep pathogenesis of autoimmune disease. Cell 2007;130(1):25–35.
[21] Liston A, Lesage S, Wilson J, Peltonen L, Goodnow CC. Aire regulates negative selection of organ-specific T cells. Nature Immunology 2003;4(4):350–4.
[22] Anderson MS, Venanzi ES, Klein L, Chen Z, Berzins SP, Turley SJ, et al. Projection of an immunological self shadow within the thymus by the aire protein. Science 2002;298(5597):1395–401.
[23] Toda A, Piccirillo CA. Development and function of naturally occurring CD4+ CD25+ regulatory T cells. Journal of Leukocyte Biology 2006;80(3):458–70.
[24] Baráth S, Sipka S, Aleksza M, Szegedi A, Szodoray P, Végh J, et al. Regulatory T cells in peripheral blood of patients with mixed connective tissue disease. Scandinavian Journal of Rheumatology 2006;35(4):300–4.
[25] Dejaco C, Duftner C, Grubeck-Loebenstein B, Schirmer M. Imbalance of regulatory T cells in human autoimmune diseases. Immunology 2006;117(3):289–300.
[26] Liu MF, Wang CR, Fung LL, Wu CR. Decreased CD4+ CD25+ T cells in peripheral blood of patients with systemic lupus erythematosus. Scandinavian Journal of Immunology 2004;59(2):198–202.
[27] Szodoray P, Papp G, Horvath IF, Barath S, Sipka S, Nakken B, et al. Cells with regulatory function of the innate and adaptive immune system in primary Sjögren's syndrome. Clinical and Experimental Immunology 2009;157(3):343–9.
[28] O'Garra A, Arai N. The molecular basis of T helper 1 and T helper 2 cell differentiation. Trends in Cell Biology 2000;10(12):542–50.
[29] Steinman L. A rush to judgment on Th17. The Journal of Experimental Medicine 2008;205(7):1517–22.
[30] Kotake S, Udagawa N, Takahashi N, Matsuzaki K, Itoh K, Ishiyama S, et al. IL-17 in synovial fluids from patients with rheumatoid arthritis is a potent stimulator of osteoclastogenesis. The Journal of Clinical Investigation 1999;103(9):1345–52.
[31] Hwang SY, Kim HY. Expression of IL-17 homologs and their receptors in the synovial cells of rheumatoid arthritis patients. Molecules and Cells 2005;19(2):180–4.
[32] Wu J, Shen Z. Exosomal miRNAs as biomarkers for diagnostic and prognostic in lung cancer. Cancer Medicine 2020;9(19):6909–22.
[33] Valencia X, Lipsky PE. CD4+ CD25+ FoxP3+ regulatory T cells in autoimmune diseases. Nature Clinical Practice Rheumatology 2007;3(11):619–26.
[34] Xu H, Bao Z, Liang D, Li M, Wei M, Ge X, et al. Plasma exosomal miR-106a-5p expression in myasthenia gravis. Muscle and Nerve 2020;61(3):401–7.
[35] Punga T, Le Panse R, Andersson M, Truffault F, Berrih-Aknin S, Punga AR. Circulating miRNAs in myasthenia gravis: miR-150-5p as a new potential biomarker. Annals of Clinical and Translational Neurology 2014;1(1):49–58.
[36] Fricke A, Cimniak AFV, Ullrich PV, Becherer C, Bickert C, Pfeifer D, et al. Whole blood miRNA expression analysis reveals miR-3613-3p as a potential biomarker for dedifferentiated liposarcoma. Cancer Biomarkers: Section A of Disease Markers 2018;22(2):199–207.
[37] Kumar P, Dezso Z, MacKenzie C, Oestreicher J, Agoulnik S, Byrne M, et al. Circulating miRNA biomarkers for Alzheimer's disease. PLoS One 2013;8(7):e69807.
[38] Derkacz A, Olczyk P, Komosinska-Vassev K. Diagnostic markers for nonspecific inflammatory bowel diseases. Disease Markers 2018;7451946:2018.
[39] Yin W, Ouyang S, Luo Z, Zeng Q, Hu B, Xu L, et al. Immature exosomes derived from MicroRNA-146a overexpressing dendritic cells act as antigen-specific therapy for myasthenia gravis. Inflammation 2017;40(4):1460–73.
[40] Zhong H, Lu J, Jing S, Xi J, Yan C, Song J, et al. Low-dose rituximab lowers serum Exosomal miR-150-5p in AChR-positive refractory myasthenia gravis patients. Journal of Neuroimmunology 2020;348:577383.

[41] Ramos-Leví AM, Marazuela M. Pathogenesis of thyroid autoimmune disease: the role of cellular mechanisms. Endocrinología y Nutrición: Organo de la Sociedad Espanola de Endocrinologia y Nutricion 2016;63(8):421–9.

[42] Rodríguez-Muñoz A, Martínez-Hernández R, Ramos-Leví AM, Serrano-Somavilla A, González-Amaro R, Sánchez-Madrid F, et al. Circulating microvesicles regulate Treg and Th17 differentiation in human autoimmune thyroid disorders. The Journal of Clinical Endocrinology and Metabolism 2015;100(12):E1531–9.

[43] Okoye IS, Coomes SM, Pelly VS, Czieso S, Papayannopoulos V, Tolmachova T, et al. MicroRNA-containing T-regulatory-cell-derived exosomes suppress pathogenic T helper 1 cells. Immunity 2014;41(1):89–103.

[44] Huang X, Yuan T, Tschannen M, Huang X, Yuan T, Tschannen M, Sun Z, Jacob H, Du M. Characterization of human plasma-derived exosomal RNAs by deep sequencing. BMC Genomics 2013;14:319.

[45] Hiratsuka I, Yamada H, Munetsuna E, Hashimoto S, Itoh M. Circulating MicroRNAs in Graves' disease in relation to clinical activity. Thyroid: Official Journal of the American Thyroid Association 2016;26(10):1431–40.

[46] Fairhurst AM, Wandstrat AE, Wakeland EK. Systemic lupus erythematosus: multiple immunological phenotypes in a complex genetic disease. Advances in Immunology 2006;92:1–69.

[47] D'Cruz DP. Systemic lupus erythematosus. BMJ (Clinical Research) 2006;332(7546):890–4.

[48] Smith PP, Gordon C. Systemic lupus erythematosus: clinical presentations. Autoimmunity Reviews 2010;10(1):43–5.

[49] Wu XN, Ye YX, Niu JW, Li Y, Li X, You X, et al. Defective PTEN regulation contributes to B cell hyperresponsiveness in systemic lupus erythematosus. Science Translational Medicine 2014;6(246):246ra99.

[50] Zhao Y, Alexandrov PN, Jaber V, Lukiw WJ. Deficiency in the ubiquitin conjugating enzyme UBE2A in Alzheimer's disease (AD) is linked to deficits in a natural circular miRNA-7 sponge (circRNA; ciRS-7). Genes 2016;7(12):116.

[51] Li H, Li K, Lai W, Li X, Wang H, Yang J, et al. Comprehensive circular RNA profiles in plasma reveals that circular RNAs can be used as novel biomarkers for systemic lupus erythematosus. Clinica Chimica Acta: International Journal of Clinical Chemistry 2018;480:17–25.

[52] Wu H, Zeng J, Yin J, Peng Q, Zhao M, Lu Q. Organ-specific biomarkers in lupus. Autoimmunity Reviews 2017;16(4):391–7.

[53] Dong C, Zhou Q, Fu T, Zhao R, Yang J, Kong X, et al. Circulating exosomes derived-miR-146a from systemic lupus erythematosus patients regulates senescence of mesenchymal stem cells. BioMed Research International 2019;2019:6071308.

[54] Perez-Hernandez J, Forner MJ, Pinto C, Chaves FJ, Cortes R, Redon J. Increased urinary exosomal MicroRNAs in patients with systemic lupus erythematosus. PLoS One 2015;10(9):e0138618.

[55] Carey EJ, Ali AH, Lindor KD. Primary biliary cirrhosis. Lancet 2015;386(10003):1565–75.

[56] Zheng J, Li Z, Wang T, Zhao Y, Wang Y. Microarray expression profile of circular RNAs in plasma from primary biliary cholangitis patients. Cellular Physiology and Biochemistry: International Journal of Experimental Cellular Physiology, Biochemistry, and Pharmacology 2017;44(4):1271–81.

[57] Kang GJ, Lee HJ, Byun HJ, Kim EJ, Kim HJ, Park MK, et al. Novel involvement of miR-522-3p in high-mobility group box 1-induced prostaglandin reductase 1 expression and reduction of phagocytosis. Biochimica et Biophysica Acta. Molecular Cell Research 2017;1864(4):625–33.

[58] Danza K, De Summa S, Pinto R, Pilato B, Palumbo O, Carella M, et al. TGFbeta and miRNA regulation in familial and sporadic breast cancer. Oncotarget 2017;8(31):50715–23.

[59] Meng S, Zhou H, Feng Z, Xu Z, Tang Y, Li P, et al. CircRNA: functions and properties of a novel potential biomarker for cancer. Molecular Cancer 2017;16(1):94.

[60] Wynn TA. Cellular and molecular mechanisms of fibrosis. The Journal of Pathology 2008;214(2):199–210.

[61] Chen Y, Yuan B, Wu Z, Dong Y, Zhang L, Zeng Z. Microarray profiling of circular RNAs and the potential regulatory role of hsa_circ_0071410 in the activated human hepatic stellate cell induced by irradiation. Gene 2017;629:35–42.

[62] Zhou B, Yu JW. A novel identified circular RNA, circRNA_010567, promotes myocardial fibrosis via suppressing miR-141 by targeting TGF-β1. Biochemical and Biophysical Research Communications 2017;487(4):769–75.

[63] Hawkes JE, Chan TC, Krueger JG. Psoriasis pathogenesis and the development of novel targeted immune therapies. The Journal of Allergy and Clinical Immunology 2017;140(3):645–53.

[64] Van den Bosch F, Coates L. Clinical management of psoriatic arthritis. Lancet 2018;391(10136):2285–94.

[65] Rendon A, Schäkel K. Psoriasis pathogenesis and treatment. International Journal of Molecular Sciences 2019;20(6):1475.

[66] Jiang M, Fang H, Dang E, Zhang J, Qiao P, Yu C, et al. Small extracellular vesicles containing miR-381-3p from keratinocytes promote T helper type 1 and T helper type 17 polarization in psoriasis. Journal of Investigative Dermatology 2021;141(3):563–74.

[67] Pasquali L, Svedbom A, Srivastava A, Rosén E, Lindqvist U, Ståhle M, et al. Circulating microRNAs in extracellular vesicles as potential biomarkers for psoriatic arthritis in patients with psoriasis. Journal of the European Academy of Dermatology and Venereology: JEADV 2020;34(6):1248–56.

[68] Aira L, Debes G. IL-10 (+) regulatory B cell migration into inflamed skin limits cutaneous inflammation. Journal of Investigative Dermatology 2020;140(7):S19.

[69] Yang C, Ouyang L, Wang W, Chen B, Liu W, Yuan X, et al. Sodium butyrate-modified sulfonated polyetheretherketone modulates macrophage behavior and shows enhanced antibacterial and osteogenic functions during implant-associated infections. Journal of Materials Chemistry B 2019;7(36):5541–53.

[70] Li M, Zhao J, Cao M, Li M, Zhao J, Cao M, Liu R, Chen G, Li S. Mast cells-derived MiR-223 destroys intestinal barrier function by inhibition of CLDN8 expression in intestinal epithelial cells. Biological Research 2020;53(12).

[71] Liao F, Lu X, Dong W. Exosomes derived from T regulatory cells relieve inflammatory bowel disease by transferring miR-195a-3p. IUBMB Life 2020;72:2591–600.

[72] Cordes F, Brückner M, Lenz P, Veltman K, Glauben R, Siegmund B, et al. MicroRNA-320a strengthens intestinal barrier function and follows the course of experimental colitis. Inflammatory Bowel Diseases 2016;22(10):2341–55.

[73] Ocansey DK, Zhang L, Wang Y, Yan Y, Qian H, Zhang X, et al. Exosome-mediated effects and applications in inflammatory bowel disease. Biological Reviews 2020;95(5):1287–307.

[74] Ando Y, Mazzurana L, Forkel M, Okazaki K, Aoi M, Schmidt PT, et al. Downregulation of MicroRNA-21 in colonic CD3+ T cells in UC remission. Inflammatory Bowel Diseases 2016;22(12):2788–93.

[75] Wang H, Chao K, Ng SC, Wang H, Chao K, Ng SC, Bai AH, Yu Q, Yu J. Pro-inflammatory miR-223 mediates the cross-talk between the IL23 pathway and the intestinal barrier in inflammatory bowel disease. Genome Biology 2016;17(58).

[76] Zhou H, Xiao J, Wu N, Liu C, Xu J, Liu F, et al. MicroRNA-223 regulates the differentiation and function of intestinal dendritic cells and macrophages by targeting C/EBPβ. Cell Reports 2015;13(6):1149–60.

[77] Brito-Zerón P, Baldini C, Bootsma H, Bowman SJ, Jonsson R, Mariette X, et al. Sjögren syndrome. Nature Reviews Disease Primers 2016;2:16047.

[78] Gallo A, Tandon M, Alevizos I, Illei GG. The majority of microRNAs detectable in serum and saliva is concentrated in exosomes. PLoS One 2012;7(3):e30679.

[79] Michael A, Bajracharya SD, Yuen PS, Zhou H, Star RA, Illei GG, et al. Exosomes from human saliva as a source of microRNA biomarkers. Oral Diseases 2010;16(1):34–8.

[80] Alevizos I, Alexander S, Turner RJ, Illei GG. MicroRNA expression profiles as biomarkers of minor salivary gland inflammation and dysfunction in Sjögren's syndrome. Arthritis and Rheumatism 2011;63(2):535–44.

[81] Gallo A, Jang SI, Ong HL, Perez P, Tandon M, Ambudkar I, et al. Targeting the Ca(2+) sensor STIM1 by exosomal transfer of Ebv-miR-BART13-3p is associated with Sjögren's syndrome. EBioMedicine 2016;10:216–26.

[82] Cortes-Troncoso J, Jang SI, Perez P, Hidalgo J, Ikeuchi T, Greenwell-Wild T, et al. T cell exosome-derived miR-142-3p impairs glandular cell function in Sjögren's syndrome. JCI Insight 2020;5(9):e133497.
[83] Keller S, Ridinger J, Rupp AK, Janssen JW, Altevogt P. Body fluid derived exosomes as a novel template for clinical diagnostics. Journal of Translational Medicine 2011;9:86.

Further reading

[1] Wu F, Dong F, Arendovich N, Zhang J, Huang Y, Kwon JH. Divergent influence of microRNA-21 deletion on murine colitis phenotypes. Inflammatory Bowel Diseases 2014;20(11):1972–85.

CHAPTER EIGHT

Potential role of exosomal miRNA in disease progression and therapeutics

Abhishek Bhattacherjee[1,2]
[1]Jubilant Biosys, Bengaluru, Karnataka, India
[2]Department of Chemistry, Department of AFNS, University of Alberta, Edmonton, AB, Canada

1. Introduction

Exosomes are extracellular vesicles (EVs) (30–100 nm) that can transmit molecular cargo between cells. Though different nomenclatures exist for these particles based on their origin, morphology, and location in the literature, as per the International Society for Extracellular Vesicles, it is recommended to identify these particles (ranging between 30 and 100 nm) as EVs [1]. It is important to understand that exosomes can be released from different types of cells, ranging from blood cells to immunocytes; hence, the cargo within exosomes varies significantly depending on their origin [2–4]. Fig. 8.1 indicates the versatile cargo present within the exosome [5]. In several cancer cell lines, it has been shown that it influences cellular function; however, how and which molecule is the key factor that influences recipient cells is still a matter of debate [6–8]. Recently, miRNA's presence inside exosomes has drawn much attention due to their functional roles in cancer, neurodegenerative diseases, and obesity. Long ago, it was thought these exosomal miRNAs were generated during a random process and were nonfunctional, but in the era of personalized medicines, their role is very prominent as biomarkers to identify early disease onset [9,10]. Briefly, microRNA (miRNA) is originated from noncoding RNAs and is often noticed to be dysregulated in cancer and other neurodegenerative disease [11–13]. It is not only important as a biomarker but its role as a cell signaling molecule is also prominent [14–16]. Hence, it is important to understand the role of miRNAs in the context of their versatile applications.

2. The cellular interaction and sorting mechanism for exosomal miRNAs

Before understanding the functional role of exosomal miRNA, it is important to understand how it helps in cellular communications. Exosomes, which are found in bodily fluids, are crucial for cell communication. Exosomes and their recipient cells interact with each other through three main processes. First, target cells' signaling

Exosomal RNA
ISBN 978-0-443-14008-2
https://doi.org/10.1016/B978-0-443-14008-2.00012-7

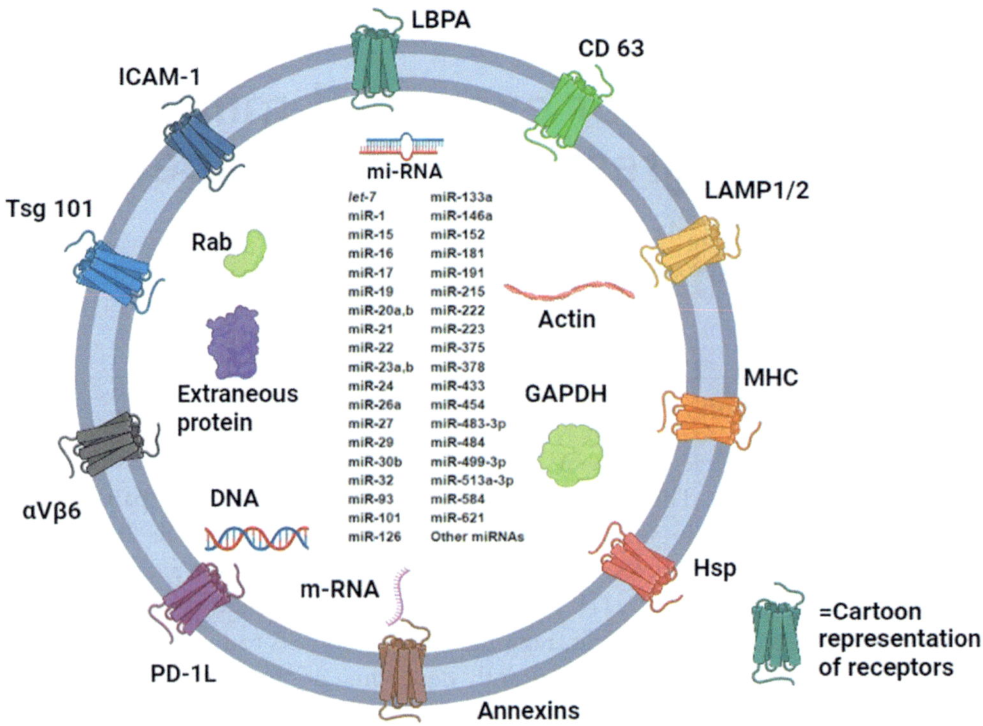

Figure 8.1 Exosome compositions. The diverse origin of exosomal particles accounts for the exosome's diverse payload. The three main components that make up exosomal cargos are protein, DNA, and miRNA. One of these three miRNAs has an intriguing cargo that can actually control how cells behave [5].

receptors are directly contacted by the transmembrane proteins of exosomes [17,18]. The next stage involves exosomes fusing with recipient cells' plasma membranes to release their content into the cytoplasm [17–19]. The exosomes then have two fates after being internalized by the recipient cells. One scenario involves absorbed exosomes fusing with endosomes and going through a process called transcytosis, which moves the exosomes across the recipient cells and releases them into nearby cells. On the other hand, endosomes attached to exosomes that have been absorbed will mature into lysosomes and go through degradation [17–19]. Fig. 8.2 indicates how miRNA is packed inside the exosome and transferred to adjacent cells to modulate cellular function [20].

Exosome internalization in recipient cells is influenced by a number of variables. According to a recent study [21], in the breast cancer cell line BT-549, disruption of exosomal lipid rafts inhibited exosome internalization. It is also reported that annexins, which are involved in cell adhesion and proliferation, were crucial for exosome uptake [21]. After treating the ovarian cancer cell line SKOV3 with protease K, Escrevente et al. reported that

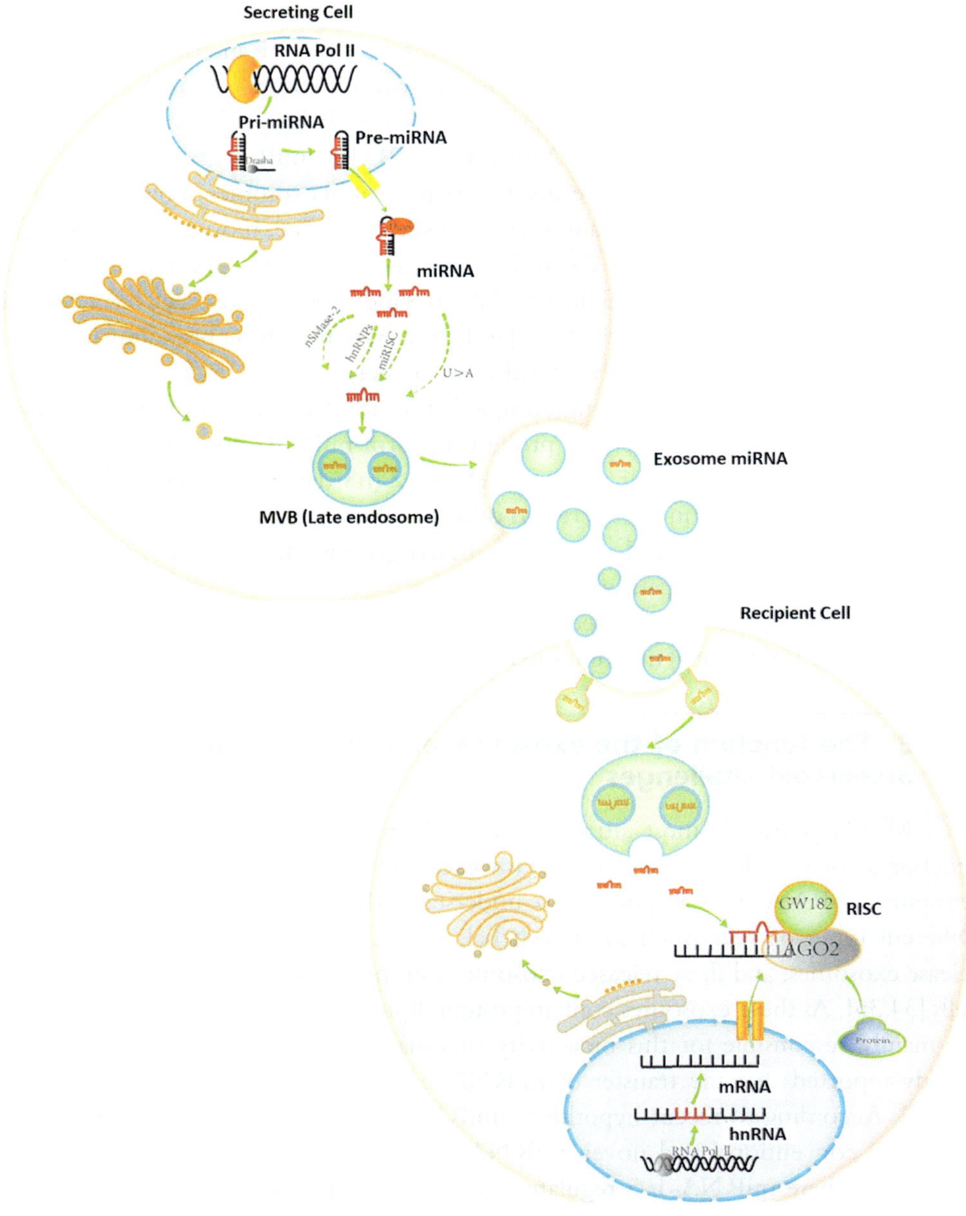

Figure 8.2 Exosome biogenesis and biological function. It shows that the majority of miRNA is encoded by their own gene and contained within exosomes. Additionally, they are transported to other cells and integrated with the DNA to control cellular function [20].

exosome uptake decreased, indicating that the protein interaction promoted exosome internalization [22]. However, the detailed mechanism of exosome internalization is still not well understood. Among all the content present within exosomes, miRNA is the most abundant. Furthermore, miRNA has several regulatory functions, which makes it interesting [23]. Different subsets of miRNA including miR-150, miR-142-3p, miR-451, let-7 miRNA, let-7f, miR-20b, miR-30e-3p, etc. are present in exosomes of different cell lines [24–26]. Interestingly, their level also varies across the cell line; moreover, it is also highlighted by several researchers [26–28]. miRNA levels in exosomes vary significantly under pathophysiological conditions. All these variabilities indicate the parent cell has a definitive shorting mechanism that dictates the fate of miRNA. As of now, most prevailing hypotheses indicate the possibility of five different pathways, namely:

(1) The miRNA-induced silencing complex (miRISC)-related pathway. In this case, miRNAs interact with assembly proteins to form miRISC complex [29].
(2) The route reliant on neural sphingomyelinase 2 (nSMase2). It is widely known that nSMase2 is involved in the secretion of miRNA into exosomes.
(3) The miRNA motif and simulated heterogenous nuclear ribonucleoproteins (hnRNPs)-dependent pathway [30,31].
(4) The 30-end of the miRNA sequence-dependent pathway [32].
(5) Argonaute RISC catalytic component 2 (AGO2)—dependent pathway [23,32,33].

3. The function of the exosome, exosomal miRNAs, and associated challenges

MiRNAs are the most attractive material present inside the exosome and play functional roles in the recipient cells. However, before understanding the function of exosomal miRNA, it is important to understand the major functions of exosomes. Different types of cells (such as endothelial, blood, platelet, immunocytes, etc.) can release exosomes, and these released exosomes can modulate the bioactivity of recipient cells [34,35]. As these exosomes contain protein, lipid, and nucleic acids (miRNA), they are mainly responsible for this bioactivity modulation (Fig. 8.1). Several studies have already reported that the transfer of miRNA between cells via exosomes is possible [36,37]. According to recent hypotheses, miRNAs can perform two different types of functions: conventional and novel. miRNA-105 is a classic example of conventional regulation where miRNA-105 regulates ZO-1 gene expression negatively in different breast cancer and endothelial cell lines [38]. Similarly, miRNA-214, and miRNA-92a are examples of conventional regulation where they significantly regulate expression of the integrin-a5 gene and influence recipient cell migration [39]. On the contrary, when miRNA performs additional functions along with conventional functions, it is denoted as a novel miRNA. miRNA-21 and miRNA-29a, along with gene regulation, perform as ligands of TLRs (Toll-like receptors) and help regulate cellular function

[40–42]. Though exosomes and exosomal miRNA look very attractive, there are several limitations to accessing the function of miRNA [42,43]. One of the major limitations among them is exosomal extraction. Despite several technologies ranging from simple ultracentrifugation to immunoaffinity exosome extractions, it remains challenging. On the other hand, exosomes carry a large number of miRNAs, and sometimes, an abundance of specific types of miRNA is low within exosomes. Hence, identifying the specific type of miRNA and understanding its specific function is quite challenging. However, with existing tools, researchers were able to identify several crucial functions of exosomal miRNAs in different disease models.

4. Importance of exosomal miRNA

(1) Development of new gene therapy/delivery methods:
Virus-based gene therapy is common nowadays. However, the potential of exosomes to transfer functional miRNA to different cells makes researchers interested in using exosomes and exosomal miRNA in the development of active carriers and as a potential new therapy. Along with the advantage of using the exosome-based delivery system in humans, it will not sensitize immune responses and can also cross the blood–brain barrier [44,45]. Some preliminary studies on this aspect have already been reported in animal models; however, a deeper understanding is required in this respect [46].

(2) New ways to tackle intercellular communication:
To date, researchers believed chemical signaling was the only way for intercellular communications; however, the existence of exosomes and their ability to regulate cellular function via miRNA delivery challenge this hypothesis [47]. Growing pieces of evidence suggest exosomes and exosomal miRNA play a huge role in regulating disease pathology, positively or negatively. Similar observations were observed in neuronal regulation, where it has been reported neurons. See the sentence construction.

(3) Early detection of disease (potential biomarkers):
To date, the development of symptoms is considered the main tool for diagnosing early disease onset [48]. However, in neurological disease, these symptoms are observed at later stages, so they are not very helpful for early diagnosis [49,50]. Hence, researchers keep looking for biomarkers that can be useful for the early detection of disease. Recently, it has been reported that the presence of miRNAs in peripheral blood samples of Alzheimer's and Parkinson's patients is positively correlated with disease progression. Similarly, two exosomal miRNA-based commercial kits for the detection of prostate and lung cancer are already on the market [51,52]. This early evidence shows the huge potential of exosomal miRNAs, and soon these can be used as biomarkers for early disease detection.

5. Applications of exosomal miRNAs

5.1 Exosomal miRNAs as novel cancer biomarkers

Exosomal miRNAs have tremendous importance in cancer biology. They can modulate cellular function as well as act as biomarkers to indicate disease progression. Fig. 8.3 indicates the major roles of exosomal miRNA in cancer biology [53]. In a CD9-GFP-tagged macrophage system, Fabbriet al. showed that exosomal miRNAs directly bind to TLR and can change how cells work [54]. Furthermore, in their publication, they have shown colocalization of TLR and exosomal miRNAs. TLR–miRNA interaction–mediated cytokine modulation was also demonstrated in TLR−/− mice models. All this evidence indicates the effective binding of miRNAs with surface TLRs, which can modulate cellular function. These functions are considered noble. miRNA-200 is another type of exosomal miRNA that is abundantly present in epithelial exosomes. These miRNAs can influence the metastasis of cancer cells via mesenchymal–epithelial interaction and modulation of ZEB2 [55]. miRNA-21,143,378 is abundant in cancer-associated fibroblasts. Researchers showed that these miRNAs can migrate inside

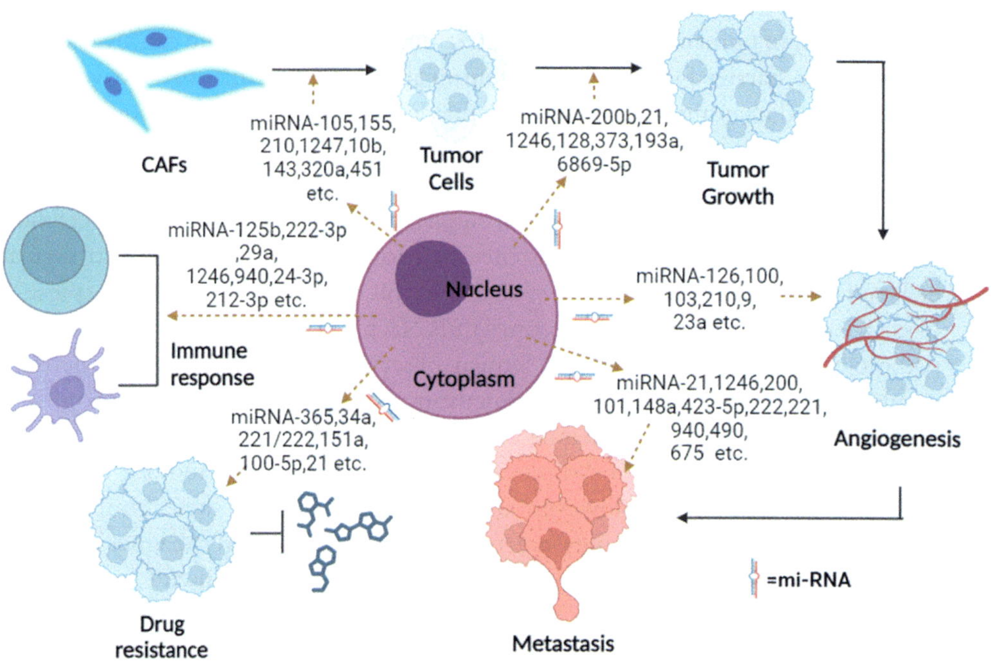

Figure 8.3 Exosomal miRNAs play multiple roles in tumor formation and development. Tumor cells may transmit exosomal miRNAs to surrounding tumor cells. In recipient tumor cells, exosomal miRNAs can regulate cell proliferation, invasion, and metastasis by modulating their downstream targets. Exosomal miRNAs play an important role in tumor angiogenesis by mediating specific communication between tumor cells and endothelial cells. Drug-resistant tumor cells can transfer a resistant phenotype to drug-sensitive tumor cells by lateral transfer of exosomal miRNAs [53].

breast cancer cells and assist in forming the mammosphere. In turn, they also stimulate epithelial—mesenchymal transcription factors [56]. It is reported that miRNA-21 increases liver metastasis, which is direct evidence of how exo-miRNA influences disease progression. Similarly, it is evident that in bone marrow—derived mesenchymal stem cells, miRNA-23b reduces cellular proliferation by targeting cell cycle regulators [57]. In another interesting study, Challagundla et al. showed that exosomal miRNA-21 can transfer between neuroblastoma and monocytes [58,59], and this interaction helps M2 polarization. Interestingly, miRNA-21 also helps monocytes secrete more exosomal miRNA-115, which in turn acts as a telomerase inhibitor [59].

MiRNAs also serve an important function as biomarkers in lung cancer. MiRNA-17-3p, -21, -106a, -146,-155, -191, -192, -203, -205,-210,-212, and -214 profiles were shown to substantially correlate with the progression of lung adenocarcinoma in a clinical investigation [60,61]. Other important panels of miRNA biomarkers for lung cancer and squamous cell cancer are let-7b, let-7e, miRNA-24, miRNA-181b, and miRNA-361b, respectively [62,63], which show significant correlation with disease progression. Expression of miRNA biomarkers, namely, miRNA-141, -200a, -200b, -200c, -203, etc., can clinically distinguish between ovarian papillary adenocarcinoma and benign ovarian cancer [28,64]. It is also reported that breast cancer tumor—derived exosomes are enriched with miRNA-1246 [65]. Patients with luminal tumors show an abundance of miRNA-373 in their serum, which is positively correlated with disease progression [66].

Along with breast and lung cancer, in prostate cancer, exosomal miRNA plays a crucial role as a biomarker. Table 8.1 summarizes the major prognostic exosomal miRNA biomarkers in cancer biology [67]. An elevated level of miRNA-141 is positively correlated with disease progression. This miRNA is mostly present in enriched serum as compared with the whole blood of advanced prostate cancer patients [68]. Furthermore, in prognosis, miRNA-141 shows 80% accuracy. The levels of miRNA 1290 and 375 in patient plasma dictate the overall fate of survival. Similarly, Bryant et al. showed how exo miRNAs such as miRNA-375 and miRNA-141 can be useful as prediction tools for recurrence after radical prostatectomy [69,70].

In colorectal cancer patients, exosomal miRNAs serve as crucial biomarkers. In a recent study [71], researchers showed several miRNAs, namely, miRNA-21, let-7a, miRNA-1229, miRNA-223, miRNA-1246, miRNA-150, miRNA-23a etc., abundantly present in patient serum [72]. It is also observed that the level of these miRNAs significantly decreases in healthy patients or postsurgery patients. It is also evident that miRNA-17-92 is abundant in CRC patients; however, this study was performed in a relatively lower number of patient samples [73]. So the validity of this biomarker is questionable.

All this evidence suggests that exosomal miRNA is directly correlated with cancer progression and hence can be very useful as a biomarker at different stages of disease progression. Furthermore, all these studies nicely compare and contrast the conventional and noble roles of exosomal miRNA in cancer.

Table 8.1 Role of exosomal miRNA as biomarkers in cancer [67].

Cancer	Exosomal mIRNAs	Donor	Recipient	Type of biomarker
Lung cancer	mIR-21	Bronchial epithelial (HBE) cells	Normal HBE cells	Angiogenesis
	mIR-23b-3p, mIR-10b-5p, and mIR-21-5p	Plasmatic exosomes	Non–small-cell lung cancer cells (NSCLC)	Progression, angiogenesis, and metastasis
	mIR-210	Lung adenocarcinoma	Stromal cells	Angiogenesis
	mIR-155/-146a	Immune cells	Immune cells	Inflammation
	mIR-192	A549	Endothelial cells	Bone metastasis
	mIR-494	Lung adenocarcinoma cells	Lymph nodes, lung cells	Premetastasis
Esophageal cancer (ESCC)	mIR 30a	ESCC cells	/	Proliferation
Gastric cancer	mIR-21	Macrophage	BGC-823	Metastasis
	mIR-221	Mesenchymal stem cells	HGC-27	Metastasis
Colorectal cancer (CRC)	mIR-19a	CRC cells	/	Metastasis
	mIR-21,-192, and-221	HCT-15, SW480, and WIDr	HepG2 and A549	/
	mIR-23b-3p	Blood plasma isolated from CRC patients	Colon cancer cells	Inhibitor
Liver cancer	mIR-122	Huh 7 cells	HepG2 cells	Inhibitor
	mIR-223	Macrophages	Hepatocellular carcinoma cells (HuH7 and HepG2)	Inhibitor
Pancreatic cancer	mIR-19b	PCa cells	Urine	Diagnosis
	mIR-23b-3p	PANC-1 cells	PANC-1 cells	Proliferation and metastasis
	mIR-122-Sp and mIR-193b-3p	Plasma samples	Pancreatic cancer cells	Proliferation
	mIR-141	PCa cells	Serum	Metastasis
	mIR-141, mIR-375	PCa cells	Serum	Metastasis
	mIR-145	Urine	PCa cells	Suppressor
	mIR-196a-5p	Urine	PCa cells	Metastasis
	mIR-200c-3p	Urine	PCa cells	Suppressor
	mIR-1246	Serum	PCa cells	Metastasis
	mIR-1290, mIR-375	PCa cells	Plasma	Survival prognosis

(Continued)

Table 8.1 Role of exosomal miRNA as biomarkers in cancer [67].—cont'd

Cancer	Exosomal mIRNAs	Donor	Recipient	Type of biomarker
Bladder cancer	Exosome-derived mIR-29c	mIR-29c	BIU-87 cells	Apoptosis
Prostate cancer	mIR-34a	Docetaxel-resistant PC cells	Docetaxel-resistant	Drug resistance
	mIR-125a	DIAPH3-silenced cells	Macrophages	Proliferation
	mIR-141	Bone metastatic PCa cells	Bone cells	Metastasis
	mIR-375	Serum	PC cells	Metastasis
	mIR-290,-373	PC cells	/	Prognosis
	mIR-1290	Plasma	/	Prognosis
Ovarian cancer	mIR-21-5p	CP70	A2780	Drug resistance
	mIR-127-3p	/	OVCAR-3 and Caov-3 cells	Proliferation
	mIR-200a/b/c/141	SKOV-3 and OVCAR-3	Ovarian cancer cells (OC)	Proliferation
	mIR-339-5p	OCES-2 cells	Endothelial cells	Proliferation and metastasis
	mIR-409-3p	OCES-2 cells	Endothelial cells	Proliferation and metastasis
Breast cancer	mIR-16	EGCG-treated 4 T1 cells	Macrophages	Metastasis
	mIR-21	Bone marrow-derived MSCs	Breast cancer cells	Proliferation
	mIR-23b	Bone marrow mesenchymal stem cells	Breast cancer cells	Dormancy
	mIR-105	MDA-MB-231	Endothelial cells	Metastasis
	mIR-l0b	MDA-MB-231	HMLE (MCF-7)	Metastasis
	mIR-122	Breast cancer patients/MCF10A	Recipient premetastatic niche cells	Metastasis
	mIR-134	Hs578T and Hs578Ts008	Breast cancer cells	Drug resistance
	mIR-200	Metastatic breast cells	Non-metastatic breast cells	Metastasis
	mIR-221/−222	MCF-7 (tamoxifen-resistant)	MCF-7 (tamoxifen-sensitive)	Drug resistance
	mIR-223-3p	IL 4-activated macrophages	MDA-MB-231	Metastasis
	mIR-373	Breast cancer cells	/	Malignant prediction
	mIR-503	Endothelial cells	Breast cancer cells	Metastasis

(Continued)

Table 8.1 Role of exosomal miRNA as biomarkers in cancer [67].—cont'd

Cancer	Exosomal mIRNAs	Donor	Recipient	Type of biomarker
Neuroblastoma	mIR-21	NBL cells	Human monocytes	Drug resistance
Hematological malignancies	mIR-21	CLL cells	MSCs and endothelial cells	Metastasis
	mIR-92a	K562 cells	Umbilical vein endothelial cells	Metastasis
	mIR-126	LAMA84	Endothelial cells	Metastasis
	mIR-135b	Multiple myeloma cells	Endothelial cells	Metastasis, angiogenesis
	mIR-202-3p	Chronic lymphocytic leukemia (MEC1)	Human stromal cells	Proliferation
	mIR-210	K562 under hypoxic conditions	Umbilical vein endothelial cells	Metastasis
Melanoma	mIR-31, -185, and -34b	A375 and SK-MEL-28	Normal melanocytes	Metastasis
Melanoma	mIR-125b-5p	PLX4032-resistant melanoma cell line	Primary melanoma cell lines	Metastasis
	mIR-222	Metastatic melanoma cell lines	Primary melanoma cell lines	Metastasis
Merkel cell carcinoma (MCC)	mIR-30a, mIR-34, mIR-142-3p, mIR-1539	MCV-positive or -negative tumors	/	MCPyV infection

6. Role of exosomal miRNAs in cardiovascular disease

Different types of stroke, such as ischemic stroke, and hemorrhagic stroke, are the most common types of cardiovascular disease. Most of these strokes cause pathophysiological changes and initiate several inflammatory reactions [74—78]. Fig. 8.4 highlights the major role of exosomal miRNA in cardiovascular disease [79]. It is reported that miRNA-30d-5P is abundantly present in adipose-derived stem cells. These exosomal miRNAs inhibit neuronal apoptosis and promote M2 maturation in macrophages. Hence, they are beneficial for cardiovascular disease. Exosomal miRNA-24 is another

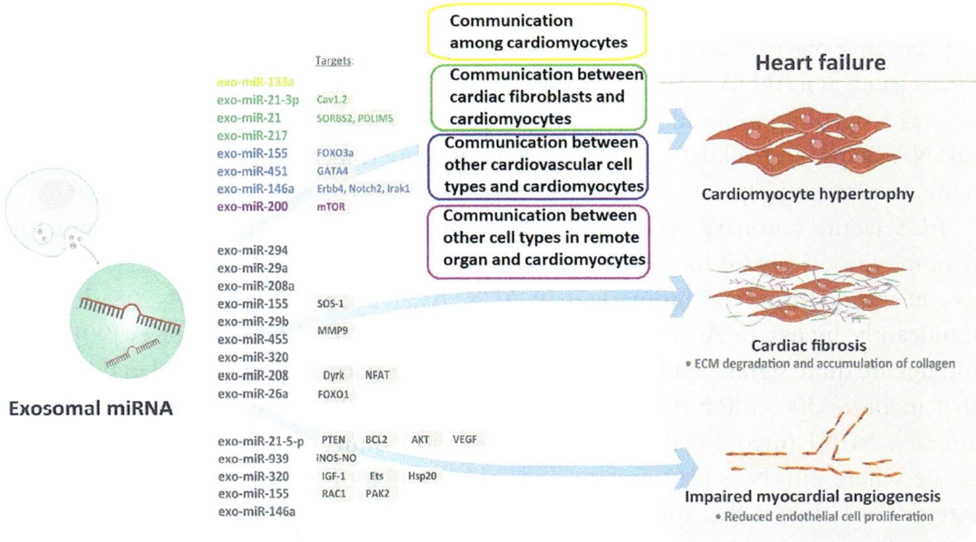

Figure 8.4 Role of exosomal miRNA in cardiovascular disorders [79].

important factor that is secreted by M2 macrophages and minimizes apoptosis. Similarly, therapeutic benefits of miRNA-29b-3p and miRNA-17-92 have also been reported recently [80–82]. In a recent study, it was shown that miRNA-181b-5p can interact with TRPM7 and promote cell migration under stress conditions [83], while on the contrary, miRNA-493 inhibits cellular migration by targeting p-Akt and p-ERK pathways [84]. So exosomal miRNAs can regulate cardiovascular diseases both positively and negatively. Furthermore, miRNAs can be used as biomarkers. MiRNA-21-5p, 30a-5p, miRNA-134, and miRNA-422a are significantly upregulated in different stages of cardiac strokes [85]. MiRNA-126 is also an important biomarker that can differentiate between permanent and transient cardiac injuries [86].

Atherosclerosis is another common heart disease that leads to myocardial stroke. Mainly inflammatory cells and endothelial cells contribute most to disease progression. The deposition of extracellular lipids is another important factor. Endothelial cells and macrophage interactions are mostly mediated by exosomal miRNA, and thus, new treatments mainly focus on these cell types and further try to revert them to their natural function. Several studies suggest miRNAs-143 and 145 show promising results in atherosclerosis treatment [87,88]. It has been further reported that these miRNAs can interact with KLF2 and help reduce aortic injuries [89]. It is also clear that mesenchymal exosomal miRNAs (like miRNA-let7) promote macrophage M2 polarization through the HMGA2/NF-κB pathway and have antiatherosclerotic effects via the IGFZBP1/

PTEN pathway. MiRNA-155 is another important factor that can prevent inflammation by polarizing macrophages to the M2 phenotype [20]. It is also reported that exosomal RNAs (such as miRNA-92a) can be upregulated during an endothelial injury while they serve as a signaling molecule for macrophages. All these studies suggest how exosomal miRNA helps intercellular communication and functionally regulates cellular function in atherosclerosis [20].

ACS (acute coronary syndrome) is a serious heart disease caused by complete or incomplete occlusive thrombosis, where blood flow to the heart decreases. MiRNAs play an important role as biomarkers in ACS. As an example, miRNA-208a levels are significantly higher in ACS patients as compared with healthy volunteers [90]. These findings are quite significant because almost 500 ACS patients participated in this analysis (that includes 300 miRNA-208a high and 200 miR-208a low participants). On the contrary, MIRI (myocardial ischemia—reperfusion injury) is another important heart disease where miRNAs play a vital role in the context of therapy. Preclinical rat model systems have shown that miRNA-126 increases cell survival and cardiac function by decreasing apoptosis in newborn cardiomyocytes [91,92]. Furthermore, it was observed in a different study that miRNA-423 secreted by CFs reduces apoptosis in ex vivo conditions by targeting RAP2C [20]. These examples justify the importance of exosomal miRNA in the context of heart disease.

7. Role of exosomal miRNA in neurodegenerative diseases

Alzheimer's disease (AD), Parkinson's disorder (PD), and multiple sclerosis (MS) are a few of the important pathophysiological disorders that target the brain and CNS. Recent findings interestingly elucidate the role of exosomal miRNA in neurodegenerative disease. A higher amyloid plaque burden is one of the key features of AD. Exosomal miRNA-193b targets directly APP (amyloid precursor protein); hence, it might have a therapeutic benefit to prevent the early onset of AD [93]. Furthermore, miRNA-135a and 384 expressions are higher in AD patients as compared with healthy people, while miRNA-193B and 223 are negatively correlated with AD progression. Hence, these exosomal miRNAs can be used for AD diagnosis [46]. It is also reported that exosomal miRNA-113b and miRNA-124 can promote neurite growth and enhance brain repair, respectively, so these can be beneficial in PD [94,95]. Similarly, it is reported that miRNA-331-5p and miRNA-505 are positively correlated with PD progression and hence can be used as diagnostic markers [96]. Recently, it was reported that some exosomal miRNAs, such as miRNA-34a, act in the neurotoxic pathway via the nuclear factor 2-related factor 2 (NRF2) regulator and hence play an important role in disease pathology [97,98]. However, further investigation is required in this context.

8. Role of exosomal miRNAs in brain injury

Concerningly, 50–60 million people are diagnosed with various sorts of brain injuries every year, making it one of the most prevalent issues around the globe [99]. The main underlying reasons for brain injuries are neurological abnormalities and inflammation [100]. Exosomal miRNA is said to play a significant role in regulating traumatic brain damage. For example, upon brain damage, a substantial increase in miRNA-21 was seen in neurons [101,102]. Since they polarize microglia to the M1 phase during neuroinflammation, they have a negative effect on damage healing. In contrast, miRNA-124 encourages macrophages to undergo the M2 transition, aiding in the improvement of injury-related diseases. It is also reported that miRNA-24-mediated effects are associated with their inhibitory role in the TLR4 pathway [103]. In another study, researchers showed that miRN-124 3p inhibits neuroinflammation and can target microglial autophagy by targeting FIP200. All these results suggest that exosomal miRNAs play promising roles in brain injury–related problems [104–106].

9. Role of exosomal miRNA in spinal cord injury

Spinal cord injuries are very similar to brain injuries, where glial scarring, demyelination, apoptosis, and neuroinflammation are a few of the key contributing factors [107,108]. Due to the abundance and specificity of exosomal miRNAs in body fluid, they play an important role in the prognosis of spinal cord injury. MiRNAs 125b-5p, and 152-3p are a few of the important miRNAs that can be used as diagnostic markers for spinal cord injury. It is reported that miRNA-544 originates from mesenchymal cells and can significantly reduce neuroinflammation; however, the underlying mechanism is not clear [109]. Similarly, miRNA-133b stimulates the STAT3, CREB, and RhoA-ROCK signaling pathways and helps in spinal cord injury recovery [110,111]. On the contrary, miRNA-21 showed the opposite result in a preclinical rat model. Similarly, miRNA-21 expression is negatively correlated with neuroinflammation in spinal cord injury.

10. Role of exosomal miRNA in glioma

Glioma is a brain tumor, and unfortunately, the average survival time for glioblastoma is 1.5 years [112,113]. It is reported that several exosomal miRNAs such as miRNA 29a or 92a activate the immunosuppressive environment in gliomas by activating the mobility box transcription factor or protein kinase cAMP; hence, they escape immune surveillance [114]. On the contrary, a different set of exosomal miRNAs, namely miRNAs-10 and 21, inhibits immune escape by targeting RAR-associated orphan receptors [115]. Exosomal miRNA-26a and 148a are two important miRNAs that promote angiogenesis and proliferation, respectively [116,117]. miRNA-199a can be transferred to glioma cells via exosomes, and it generally downregulates ArfGAP; on the contrary, miRNA's expression within gliomas generally makes them drug resistant [118–120].

11. Conclusions

In summary, exosomal miRNA is a new class of intercellular signaling molecules that can functionally regulate the cellular environment and be positively or negatively correlated with disease progression. Thus, they can play an important role as therapeutics and biomarkers in several diseases, especially cancer, neurobiological disorders, cardiovascular disease, etc. In cancer, exosomal miRNAs demonstrate promising results, as these miRNAs can serve as a signaling molecule between tumor cells and the tumor microenvironment. Over the past few years, exosomal miRNAs' role in the etiology of cancer has attracted a lot of research attention. Exosomal miRNAs appear to regulate cancer growth, angiogenesis, invasion, and metastasis, according to mounting data. To maintain cancer cells' sensitivity to anticancer therapy, it is necessary to understand the mechanisms driving exosomal miRNA-regulated cancer chemoresistance. Exosome synthesis, secretion, and uptake involve complex mechanisms that have yet to be fully understood. A deeper investigation is needed into the regulatory function of exosomal miRNAs.

Similarly, in neurobiological disorders, exosomal miRNA serves as an early biomarker. Research is now being done on the possibility of using chemicals found in peripheral body fluids, such as CSF and serum, to diagnose neurodegenerative disorders. Circulating miRNAs are appealing candidates for disease monitoring because CSF is in direct touch with the extracellular space of the brain and can reflect the biochemical changes that take place there. As a result, it is an ideal source for AD biomarkers. In terms of selecting the best biological fluid for study, serum rather than CSF yields notable results, and circulating miRNAs are intriguing possibilities for disease monitoring. Exosomal miRNAs can rewire immune cells and modify the expression of immune-related genes. However, research on the roles of exosomal miRNAs in regulating host immune responses is still in its early stages.

Circulating exosomal miRNAs have an important role in cardiovascular disease, and stem cell–derived exosomes may also be crucial in the development of heart failure. They might represent a fresh approach to managing or avoiding heart failure in the future. This new tactic is still in its infancy, though there are still several obstacles to overcome before clinical application. The technology for purifying and separating exosomes is not ideal, it is time-consuming, and its efficiency is low. Exosomes can have negative effects or side effects on cardiac tissue because they contain a lot of biologically active substances. Furthermore, based on the characteristics of the donor cells, the heterogenous components in exosomes may indicate a potential risk for tumor formation and unfavorable immune response. The therapeutic effect that exosomes produce during the disease is unknown due to their complex structure. To comprehend the mechanisms behind exosomal miRNA-mediated therapeutics, more research is required.

Acknowledgments

Declarations

Thanks to Dr. Tanushree Ghosh for her valuable insights.

University of Alberta, Edmonton, Canada, and Jubilant Biosys, India, are not directly or indirectly related or contributed to this chapter, and this research article does not have any conflict of interest. This chapter only captures the tremendous potential of exosomal miRNA in disease progression based on existing literature available in public domain. No experiments performed to obtain the data presented in this article.

Consent for publication

Abhishek Bhattacherjee is the main author who contributed to and approved the final manuscript.

References

[1] Lötvall J, Hill AF, Hochberg F, Buzás EI, Di Vizio D, Gardiner C, et al. In: Minimal experimental requirements for definition of extracellular vesicles and their functions: a position statement from the International Society for Extracellular Vesicles. Taylor and Francis; 2014. p. 26913.

[2] Peinado H, Alečković M, Lavotshkin S, Matei I, Costa-Silva B, Moreno-Bueno G, et al. Melanoma exosomes educate bone marrow progenitor cells toward a pro-metastatic phenotype through MET. Nature Medicine 2012;18:883−91.

[3] Boelens MC, Wu TJ, Nabet BY, Xu B, Qiu Y, Yoon T, et al. Exosome transfer from stromal to breast cancer cells regulates therapy resistance pathways. Cell 2014;159:499−513.

[4] Sato-Kuwabara Y, Melo SA, Soares FA, Calin GAJI. The fusion of two worlds: non-coding RNAs and extracellular vesicles-diagnostic and therapeutic implications. International Journal of Oncology 2015;46:17−27.

[5] Hu G, Drescher KM, Chen X-M. Exosomal miRNAs: biological properties and therapeutic potential. Frontiers in Genetics 2012;3:56.

[6] Johnstone RM, Adam M, Hammond J, Orr L, Turbide CJJ. Vesicle formation during reticulocyte maturation. Association of plasma membrane activities with released vesicles (exosomes). Journal of Biological Chemistry 1987;262:9412−20.

[7] Trams EG, Lauter CJ, Salem JN, Heine UJ. Exfoliation of membrane ecto-enzymes in the form of micro-vesicles. Biochima Biophysics Acta 1981;645:63−70.

[8] Simons M, Raposo GJC. Exosomes—vesicular carriers for intercellular communication. Current Opinion in Cell Biology 2009;21:575−81.

[9] Théry C, Ostrowski M, Segura EJNri. Membrane vesicles as conveyors of immune responses. Nature Review of Immunology 2009;9:581−93.

[10] Al-Nedawi K, Meehan B, Micallef J, Lhotak V, May L, Guha A, et al. Intercellular transfer of the oncogenic receptor EGFRvIII by microvesicles derived from tumour cells. Nature of Cell Biology 2008;10:619−24.

[11] Heijnen HF, Schiel AE, Fijnheer R, Geuze HJ, Sixma JJJB. Activated platelets release two types of membrane vesicles: microvesicles by surface shedding and exosomes derived from exocytosis of multivesicular bodies and granules. The Journal of the American Society of Hematology 1999;94:3791−9.

[12] Jaiswal R, Gong J, Sambasivam S, Combes V, Mathys JM, Davey R, et al. Microparticle-associated nucleic acids mediate trait dominance in cancer. FASEB Journal 2012;26:420−9.

[13] Stein JM, Luzio JPJBJ. Ectocytosis caused by sublytic autologous complement attack on human neutrophils. The sorting of endogenous plasma-membrane proteins and lipids into shed vesicles. Biochemical Journal 1991;274:381–6.

[14] Liao J, Liu R, Yin L, Pu YJI. Expression profiling of exosomal miRNAs derived from human esophageal cancer cells by solexa high-throughput sequencing. International Journal of Molecular Science 2014;15:15530–51.

[15] Saunderson SC, Dunn AC, Crocker PR, McLellan ADJB. CD169 mediates the capture of exosomes in spleen and lymph node. Blood 2014;123:208–16.

[16] Wieckowski E, Whiteside TLJI. Human tumor-derived vs dendritic cell-derived exosomes have distinct biologic roles and molecular profiles. Immunology of Research 2006;36:247–54.

[17] Munich S, Sobo-Vujanovic A, Buchser WJ, Beer-Stolz D, Vujanovic NLJO. Dendritic cell exosomes directly kill tumor cells and activate natural killer cells via TNF superfamily ligands. Oncoimmunology 2012;1:1074–83.

[18] Tian T, Zhu YL, Hu FH, Wang YY, Huang NP, Xiao ZDJJocp. Dynamics of exosome internalization and trafficking. Journal of Cell Physiology 2013;228:1487–95.

[19] Mulcahy LA, Pink RC, Carter DRFJ. Routes and mechanisms of extracellular vesicle uptake. Journal of Extracellular Vesicles 2014;3:24641.

[20] Zheng D, Huo M, Li B, Wang W, Piao H, Wang Y, et al. The role of exosomes and exosomal microRNA in cardiovascular disease. Frontiers in Cell and Developmental Biology 2021;8:616161.

[21] Koumangoye RB, Sakwe AM, Goodwin JS, Patel T, Ochieng JJPo. Detachment of breast tumor cells induces rapid secretion of exosomes which subsequently mediate cellular adhesion and spreading. PLoS One 2011;6:e24234.

[22] Escrevente C, Keller S, Altevogt P, Costa JJBc. Interaction and uptake of exosomes by ovarian cancer cells. BMC Cancer 2011;11:1–10.

[23] Goldie BJ, Dun MD, Lin M, Smith ND, Verrills NM, Dayas CV, et al. Activity-associated miRNA are packaged in Map1b-enriched exosomes released from depolarized neurons. Nucleic Acids Research 2014;42:9195–208.

[24] Guduric-Fuchs J, O'Connor A, Camp B, O'Neill CL, Medina RJ, Simpson DAJBg. Selective extracellular vesicle-mediated export of an overlapping set of microRNAs from multiple cell types. BMC Genomics 2012;13:1–14.

[25] Ohshima K, Inoue K, Fujiwara A, Hatakeyama K, Kanto K, Watanabe Y, et al. Let-7 microRNA family is selectively secreted into the extracellular environment via exosomes in a metastatic gastric cancer cell line. PLoS One 2010;5:e13247.

[26] Nilsson J, Skog J, Nordstrand A, Baranov V, Mincheva-Nilsson L, Breakefield X, et al. Prostate cancer-derived urine exosomes: a novel approach to biomarkers for prostate cancer. British Journal of Cancer 2009;100:1603–7.

[27] Silva J, García V, Zaballos A, Provencio M, Lombardía L, Almonacid L, et al. Vesicle-related microRNAs in plasma of nonsmall cell lung cancer patients and correlation with survival. European Respiratory Journal 2011;37:617–23.

[28] Taylor DD, Gercel-Taylor CJGo. MicroRNA signatures of tumor-derived exosomes as diagnostic biomarkers of ovarian cancer. Gynecology and Oncology 2008;110:13–21.

[29] Lee YS, Pressman S, Andress AP, Kim K, White JL, Cassidy JJ, et al. Silencing by small RNAs is linked to endosomal trafficking. Nature Cell Biology 2009;11:1150–6.

[30] Gibbings DJ, Ciaudo C, Erhardt M, Voinnet OJN. Multivesicular bodies associate with components of miRNA effector complexes and modulate miRNA activity. Nature Cell Biology 2009;11:1143–9.

[31] Villarroya-Beltri C, Gutiérrez-Vázquez C, Sánchez-Cabo F, Pérez-Hernández D, Vázquez J, Martin-Cofreces N, et al. Sumoylated hnRNPA2B1 controls the sorting of miRNAs into exosomes through binding to specific motifs. Nature Communications 2013;4:2980.

[32] Zhang J, Li S, Li L, Li M, Guo C, Yao J, et al. proteomics, bioinformatics, exosome and exosomal microRNA: trafficking, sorting, and function. Genomics Proteomics Bioinformatics 2015;13:17–24.

[33] Melo SA, Sugimoto H, O'Connell JT, Kato N, Villanueva A, Vidal A, et al. Cancer exosomes perform cell-independent microRNA biogenesis and promote tumorigenesis. Cancer Cell 2014;26:707–21.

[34] Pefanis E, Wang J, Rothschild G, Lim J, Kazadi D, Sun J, et al. RNA exosome-regulated long non-coding RNA transcription controls super-enhancer activity. Cell 2015;161:774–89.

[35] Thakur BK, Zhang H, Becker A, Matei I, Huang Y, Costa-Silva B, et al. Double-stranded DNA in exosomes: a novel biomarker in cancer detection. Nature 2014;24:766–9.

[36] Bukong TN, Momen-Heravi F, Kodys K, Bala S, Szabo GJPp. Exosomes from hepatitis C infected patients transmit HCV infection and contain replication competent viral RNA in complex with Ago2-miR122-HSP90. PLoS Pathogens 2014;10:e1004424.

[37] Wei W-F, Zhou C-F, Wu X-G, He L-N, Wu L-F, Chen X-J, et al. Disease, retracted article: MicroRNA-221-3p, a TWIST2 target, promotes cervical cancer metastasis by directly targeting THBS2. Cell, Death and Disease 2017;8:3220.

[38] Zhou W, Fong MY, Min Y, Somlo G, Liu L, Palomares MR, et al. Cancer-secreted miR-105 destroys vascular endothelial barriers to promote metastasis. Cancer Cell 2014;25:501–15.

[39] Ma X, Zhou J, Zhong Y, Jiang L, Mu P, Li Y, et al. Expression, regulation and function of microRNAs in multiple sclerosis. International Journal of Medicine Science 2014;11:810.

[40] Fabbri M, Paone A, Calore F, Galli R, Croce CMJ. A new role for microRNAs, as ligands of toll-like receptors. RNA Biology 2013;10:169–74.

[41] Chen X, Liang H, Zhang J, Zen K, Zhang CYJR. microRNAs are ligands of Toll-like receptors. RNA 2013;19:737–9.

[42] Bayraktar R, Bertilaccio MTS, Calin GAJ. The interaction between two worlds: microRNAs and toll-like receptors. Frontiers in Immunology 2019;10:1053.

[43] Bach DH, Hong JY, Park HJ, Lee SKJI. The role of exosomes and miRNAs in drug-resistance of cancer cells. International Journal of Cancer 2017;141:220–30.

[44] Tan A, Rajadas J, Seifalian AMJA. Exosomes as nano-theranostic delivery platforms for gene therapy. Advantages of Drug Delivery Reviews 2013;65:357–67.

[45] Lee S-T, Im W, Ban J-J, Lee M, Jung K-H, Lee SK, et al. Exosome-based delivery of miR-124 in a Huntington's disease model. Journal of Movement Disorder 2017;10:45.

[46] Chen J, Zhao B, Zhao J, Li SJN. Potential roles of exosomal microRNAs as diagnostic biomarkers and therapeutic application in Alzheimer's disease. Neural Plastics 2017:2017.

[47] Bayraktar R, Van Roosbroeck K, Calin GAJ. Cell-to-cell communication: microRNAs as hormones. Molecules of Oncology 2017;11:1673–86.

[48] Ross CA, Tabrizi SJJ. Huntington's disease: from molecular pathogenesis to clinical treatment. Lancet Neurology 2011;10:83–98.

[49] Cao X-Y, Lu J-M, Zhao Z-Q, Li M-C, Lu T, An X-S, et al. MicroRNA biomarkers of Parkinson's disease in serum exosome-like microvesicles. Neuroscience Letters 2017;644:94–9.

[50] Lugli G, Cohen AM, Bennett DA, Shah RC, Fields CJ, Hernandez AG, et al. Plasma exosomal miRNAs in persons with and without Alzheimer disease: altered expression and prospects for biomarkers. PLoS One 2015;10:e0139233.

[51] Huang B, Schiefer J, Sass C, Landwehrmeyer GB, Kosinski CM, Kochanek SJH. High-capacity adenoviral vector-mediated reduction of huntingtin aggregate load in vitro and in vivo. Human Gene Therapy 2007;18:303–11.

[52] Brustovetsky NJMn. Mutant huntingtin and elusive defects in oxidative metabolism and mitochondrial calcium handling. Molecules of Neurobiology 2016;53:2944–53.

[53] Wang M, Yu F, Ding H, Wang Y, Li P, Wang KJMt-Na. Emerging function and clinical values of exosomal microRNAs in cancer. Molecules of Therapeutic Nucleic Acids 2019;16:791–804.

[54] Fabbri M, Paone A, Calore F, Galli R, Gaudio E, Santhanam R, et al. MicroRNAs bind to Toll-like receptors to induce prometastatic inflammatory response. Proceedings of the National Academy of Sciences United States of America 2012;109:E2110–6.

[55] Le MT, Hamar P, Guo C, Basar E, Perdigão-Henriques R, Balaj L, et al. miR-200–containing extracellular vesicles promote breast cancer cell metastasis. Journal of Clinical Investment 2014;124:5109–28.

[56] Donnarumma E, Fiore D, Nappa M, Roscigno G, Adamo A, Iaboni M, et al. Cancer-associated fibroblasts release exosomal microRNAs that dictate an aggressive phenotype in breast cancer. Oncotarget 2017;8:19592–608.

[57] Ono M, Kosaka N, Tominaga N, Yoshioka Y, Takeshita F, Takahashi R-u, et al. Exosomes from bone marrow mesenchymal stem cells contain a microRNA that promotes dormancy in metastatic breast cancer cells. Science Signal 2014;7:ra63.

[58] Challagundla KB, Wise PM, Neviani P, Chava H, Murtadha M, Xu T, et al. Exosome-mediated transfer of microRNAs within the tumor microenvironment and neuroblastoma resistance to chemotherapy. Journal of National Cancer Institute 2015;107.

[59] Buruiană A, Florian ŞI, Florian AI, Timiş T-L, Mihu CM, Miclăuş M, et al. The roles of miRNA in glioblastoma tumor cell communication: diplomatic and aggressive negotiations. International Journal of Molecular Science 2020;21:1950.

[60] Condrat CE, Thompson DC, Barbu MG, Bugnar OL, Boboc A, Cretoiu D, et al. miRNAs as biomarkers in disease: latest findings regarding their role in diagnosis and prognosis. Cell 2020;9:276.

[61] Shen J, Jiang FJEo. Applications of microRNAs in the diagnosis and prognosis of lung cancer. Expert Opinion in Medical Diagnostics 2012;6:197—207.

[62] Jin X, Chen Y, Chen H, Fei S, Chen D, Cai X, et al. Evaluation of tumor-derived exosomal miRNA as potential diagnostic biomarkers for early-stage non—small cell lung cancer using next-generation sequencing exosomal miRNA as early diagnostic biomarkers for NSCLC. Clinical Cancer Research 2017;23:5311—9.

[63] Chirshev E, Oberg KC, Ioffe YJ, Unternaehrer JJJC, medicine t. Let-7 as biomarker, prognostic indicator, and therapy for precision medicine in cancer. Clinical Translative Medicine 2019;8:1—14.

[64] Teng Y, Su X, Zhang X, Zhang Y, Li C, Niu W, et al. miRNA-200a/c as potential biomarker in epithelial ovarian cancer (EOC): evidence based on miRNA meta-signature and clinical investigations. Oncotarget 2016;7:81621.

[65] Hannafon BN, Trigoso YD, Calloway CL, Zhao YD, Lum DH, Welm AL, et al. Plasma exosome microRNAs are indicative of breast cancer. Breast Cancer Research 2016;18:1—14.

[66] Bakr NM, Mahmoud MS, Nabil R, Boushnak H, Swellam MJJoGE. Biotechnology, Impact of circulating miRNA-373 on breast cancer diagnosis through targeting VEGF and cyclin D1 genes. Journal of Genetic Engineering Biotechnology 2021;19:84.

[67] Li C, Zhou T, Chen J, Li R, Chen H, Luo S, et al. The role of Exosomal miRNAs in cancer. Journal of Translative Medicine 2022;20:1—15.

[68] Xi Y, Nakajima G, Gavin E, Morris CG, Kudo K, Hayashi K, et al. Systematic analysis of microRNA expression of RNA extracted from fresh frozen and formalin-fixed paraffin-embedded samples. RNA 2007;13:1668—74.

[69] Bryant R, Pawlowski T, Catto J, Marsden G, Vessella R, Rhees B, et al. Changes in circulating microRNA levels associated with prostate cancer. British Journal of Cancer 2012;106:768—74.

[70] Chen L-j, Li X-y, Zhao Y-q, Liu W-j, Wu H-j, Liu J, et al. Down-regulated microRNA-375 expression as a predictive biomarker in non-small cell lung cancer brain metastasis and its prognostic significance. Pathological Research Practice 2017;213:882—8.

[71] Ogata-Kawata H, Izumiya M, Kurioka D, Honma Y, Yamada Y, Furuta K, et al. Circulating exosomal microRNAs as biomarkers of colon cancer. Plose One 2014;9:e92921.

[72] Yu L, Xiang L, Feng J, Li B, Zhou Z, Li J, et al. miRNA-21 and miRNA-223 expression signature as a predictor for lymph node metastasis, distant metastasis and survival in kidney renal clear cell carcinoma. Journal of Cancer 2018;9:3651.

[73] Mogilyansky E, Rigoutsos IJCD. The miR-17/92 cluster: a comprehensive update on its genomics, genetics, functions and increasingly important and numerous roles in health and disease. Cell, Death and Differentiation 2013;20:1603—14.

[74] Koh S-H, Park H-HJ. Neurogenesis in stroke recovery. Translative Stroke Research 2017;8:3—13.

[75] Liu X, Feng Z, Du L, Huang Y, Ge J, Deng Y, et al. The potential role of MicroRNA-124 in cerebral ischemia injury. International Journal of Molecular Science 2019;21:120.

[76] Smith SD, Eskey CJJRC. Hemorrhagic stroke. National Center for Biotechnology Information 2011;49:27—45.

[77] Jiang M, Wang H, Jin M, Yang X, Ji H, Jiang Y, et al. Exosomes from MiR-30d-5p-ADSCs reverse acute ischemic stroke-induced, autophagy-mediated brain injury by promoting M2 microglial/macrophage polarization. Cell, Physiology of Biochemistry 2018;47:864—78.

[78] Song Y, Li Z, He T, Qu M, Jiang L, Li W, et al. M2 microglia-derived exosomes protect the mouse brain from ischemia-reperfusion injury via exosomal miR-124. Theranostics 2019;9:2910.

[79] Xue R, Tan W, Wu Y, Dong B, Xie Z, Huang P, et al. Role of exosomal miRNAs in heart failure. Frontiers in Cardiovascular Medicine 2020;7:592412.

[80] Xin H, Katakowski M, Wang F, Qian J-Y, Liu XS, Ali MM, et al. MicroRNA-17−92 cluster in exosomes enhance neuroplasticity and functional recovery after stroke in rats. Stroke Vascular Interventional Neurology 2017;48:747−53.

[81] Xin H, Wang F, Li Y, Lu Q-e, Cheung WL, Zhang Y, et al. Secondary release of exosomes from astrocytes contributes to the increase in neural plasticity and improvement of functional recovery after stroke in rats treated with exosomes harvested from microRNA 133b-overexpressing multipotent mesenchymal stromal cells. Cell Transplantation 2017;26:243−57.

[82] Xin H, Li Y, Liu Z, Wang X, Shang X, Cui Y, et al. MiR-133b promotes neural plasticity and functional recovery after treatment of stroke with multipotent mesenchymal stromal cells in rats via transfer of exosome-enriched extracellular particles. Stem Cells 2013;31:2737−46.

[83] Yang Y, Cai Y, Zhang Y, Liu J, Xu ZJJ. Exosomes secreted by adipose-derived stem cells contribute to angiogenesis of brain microvascular endothelial cells following oxygen−glucose deprivation in vitro through microRNA-181b/TRPM7 axis. Journal of Molecular Neuroscience 2018;65:74−83.

[84] Li Q, He Q, Baral S, Mao L, Li Y, Jin H, et al. MicroRNA-493 regulates angiogenesis in a rat model of ischemic stroke by targeting MIF. FEBS Journal 2016;283:1720−33.

[85] Wang W, Li D-B, Li R-Y, Zhou X, Yu D-J, Lan X-Y, et al. Diagnosis of hyperacute and acute ischaemic stroke: the potential utility of exosomal MicroRNA-21-5p and MicroRNA-30a-5p. Cerebrovascular Disease 2018;45:204−12.

[86] Chen F, Du Y, Esposito E, Liu Y, Guo S, Wang X, et al. Effects of focal cerebral ischemia on exosomal versus serum miR126. Translative Stroke Research 2015;6:478−84.

[87] Hergenreider E, Heydt S, Tréguer K, Boettger T, Horrevoets AJ, Zeiher AM, et al. Atheroprotective communication between endothelial cells and smooth muscle cells through miRNAs. Nature Cell Biology 2012;14:249−56.

[88] Lovren F, Pan Y, Quan A, Singh KK, Shukla PC, Gupta N, et al. MicroRNA-145 targeted therapy reduces atherosclerosis. Circulation 2012;126:S81−90.

[89] Vartak T, Kumaresan S, Brennan EJBR. Decoding microRNA drivers in atherosclerosis. Bioscience Reports 2022;42. BSR20212355.

[90] Bi S, Wang C, Jin Y, Lv Z, Xing X, Lu QJIJ, et al. Correlation between serum exosome derived miR-208a and acute coronary syndrome. International Journal of Clinical Experimnet in Medicine 2015;8:4275.

[91] Hou X, Chen X, Shi JJG. Genetic polymorphism of MTHFR C677T and premature coronary artery disease susceptibility: a meta-analysis. Gene 2015;565:39−44.

[92] Wen Z, Mai Z, Zhu X, Wu T, Chen Y, Geng D, et al. Mesenchymal stem cell-derived exosomes ameliorate cardiomyocyte apoptosis in hypoxic conditions through microRNA144 by targeting the PTEN/AKT pathway. Stem Cell Research and Therapy 2020;11:1−17.

[93] Liu C-G, Song J, Zhang Y-Q, Wang P-CJMmr. MicroRNA-193b is a regulator of amyloid precursor protein in the blood and cerebrospinal fluid derived exosomal microRNA-193b is a biomarker of Alzheimer's disease. Molecular Medicine Reports 2014;10:2395−400.

[94] Saraiva C, Paiva J, Santos T, Ferreira L, Bernardino LJJoCR. MicroRNA-124 loaded nanoparticles enhance brain repair in Parkinson's disease. Journal of Control Release 2016;235:291−305.

[95] Xin H, Li Y, Buller B, Katakowski M, Zhang Y, Wang X, et al. Exosome-mediated transfer of miR-133b from multipotent mesenchymal stromal cells to neural cells contributes to neurite outgrowth. Stem Cells 2012;30:1556−64.

[96] Yao Y, Qu M, Li G, Zhang F, Rui HJE. Circulating exosomal miRNAs as diagnostic biomarkers in Parkinson's disease. European Review of Medical Pharmacological Science 2018;22:5278−83.

[97] Deshmukh P, Unni S, Krishnappa G, Padmanabhan BJBr. The Keap1−Nrf2 pathway: promising therapeutic target to counteract ROS-mediated damage in cancers and neurodegenerative diseases. Biophysical Review 2017;9:41−56.

[98] Ba Q, Cui C, Wen L, Feng S, Zhou J, Yang KJB. Schisandrin B shows neuroprotective effect in 6-OHDA-induced Parkinson's disease via inhibiting the negative modulation of miR-34a on Nrf2 pathway. Biomedical Pharmacotherapy 2015;75:165–72.
[99] Finnie JJI. Neuroinflammation: beneficial and detrimental effects after traumatic brain injury. Nature 2013;21:309–20.
[100] Harrison EB, Hochfelder CG, Lamberty BG, Meays BM, Morsey BM, Kelso ML, et al. Traumatic brain injury increases levels of miR-21 in extracellular vesicles: implications for neuroinflammation. FEBS Open Biology 2016;6:835–46.
[101] Long X, Yao X, Jiang Q, Yang Y, He X, Tian W, et al. Astrocyte-derived exosomes enriched with miR-873a-5p inhibit neuroinflammation via microglia phenotype modulation after traumatic brain injury. Journal of Neuroinflammation 2020;17:1–15.
[102] Li D, Huang S, Zhu J, Hu T, Han Z, Zhang S, et al. Exosomes from MiR-21-5p-increased neurons play a role in neuroprotection by suppressing Rab11a-mediated neuronal autophagy in vitro after traumatic brain injury. Medical Science Monitoring 2019;25:1871.
[103] Li D, Huang S, Yin Z, Zhu J, Ge X, Han Z, et al. Increases in miR-124-3p in microglial exosomes confer neuroprotective effects by targeting FIP200-mediated neuronal autophagy following traumatic brain injury. Neurochemical Research 2019;44:1903–23.
[104] Yin Z, Han Z, Hu T, Zhang S, Ge X, Huang S, et al. Neuron-derived exosomes with high miR-21-5p expression promoted polarization of M1 microglia in culture. Brain, Behavior, Immunity 2020;83:270–82.
[105] Yang Y, Ye Y, Kong C, Su X, Zhang X, Bai W, et al. MiR-124 enriched exosomes promoted the M2 polarization of microglia and enhanced hippocampus neurogenesis after traumatic brain injury by inhibiting TLR4 pathway. Neurochemical Research 2019;44:811–28.
[106] Huang S, Ge X, Yu J, Han Z, Yin Z, Li Y, et al. Increased miR-124-3p in microglial exosomes following traumatic brain injury inhibits neuronal inflammation and contributes to neurite outgrowth via their transfer into neurons. FASEB Journal 2018;32:512–28.
[107] Sandrow-Feinberg HR, Houlé JDJB. Exercise after spinal cord injury as an agent for neuroprotection, regeneration and rehabilitation. Brain Research 2015;1619:12–21.
[108] Hombach S, Kretz MJN. Non-coding RNAs: classification, biology and functioning. Advances of Expert in Medical Biology 2016:3–17.
[109] Ji W, Jiang W, Li M, Li J, Li ZJB. miR-21 deficiency contributes to the impaired protective effects of obese rat mesenchymal stem cell-derived exosomes against spinal cord injury. Biochimie 2019;167:171–8.
[110] Lu XC, Zheng JY, Tang LJ, Huang BS, Li K, Tao Y, et al. Biochemistry, MiR-133b Promotes neurite outgrowth by targeting RhoA expression. Cell, Physiology and Biochemistry 2015;35:246–58.
[111] Niu M, Xu R, Wang J, Hou B, Xie AJN. MiR-133b ameliorates axon degeneration induced by MPP+ via targeting RhoA. Neuroscience 2016;325:39–49.
[112] Ostrom QT, Bauchet L, Davis FG, Deltour I, Fisher JL, Langer CE, et al. The epidemiology of glioma in adults: a "state of the science" review. Neuro Onclogy 2014;16:896–913.
[113] Li Q, Wang A-Y, Xu Q-G, Liu D-Y, Xu P-X, Yu DJIJoC, et al. In-vitro inhibitory effect of EGFL7-RNAi on endothelial angiogenesis in glioma. International Journal of Clinical Experts in Pathology 2015;8:12234.
[114] Guo X, Qiu W, Wang J, Liu Q, Qian M, Wang S, et al. Glioma exosomes mediate the expansion and function of myeloid-derived suppressor cells through microRNA-29a/Hbp1 and microRNA-92a/Prkar1a pathways. International Journal of Cancer 2019;144:3111–26.
[115] Guo X, Qiu W, Liu Q, Qian M, Wang S, Zhang Z, et al. Immunosuppressive effects of hypoxia-induced glioma exosomes through myeloid-derived suppressor cells via the miR-10a/Rora and miR-21/Pten pathways. Oncogene 2018;37:4239–59.
[116] Wang Z-F, Liao F, Wu H, Dai JJJoE, Research CC. Glioma stem cells-derived exosomal miR-26a promotes angiogenesis of microvessel endothelial cells in glioma. Journal of Expert in Clinical Cancer Research 2019;38:1–15.
[117] Cai Q, Zhu A, Gong LJB. Exosomes of glioma cells deliver miR-148a to promote proliferation and metastasis of glioblastoma via targeting CADM1. Bulletin Du Cancer 2018;105:643–51.

[118] Yu L, Gui S, Liu Y, Qiu X, Zhang G, Zhang Xa, et al. Exosomes derived from microRNA-199a-overexpressing mesenchymal stem cells inhibit glioma progression by down-regulating AGAP2. Aging 2019;11:5300.

[119] Figueroa J, Phillips LM, Shahar T, Hossain A, Gumin J, Kim H, et al. Exosomes from glioma-associated mesenchymal stem cells increase the tumorigenicity of glioma stem-like cells via transfer of miR-1587GA-hMSCs regulate GSCs via exosomal miRNA. Cancer Research 2017;77:5808–19.

[120] Yue X, Lan F, Xia TJMt. Hypoxic glioma cell-secreted exosomal miR-301a activates Wnt/β-catenin signaling and promotes radiation resistance by targeting TCEAL7. Molecular Therapy 2019;27:1939–49.

CHAPTER NINE

Exosomal RNA in aging-related diseases

Subhadeep Das[1] and Anik Paul[1,2]
[1]Department of Biochemistry, Purdue University, West Lafayette, IN, United States
[2]Department of Biochemistry and Molecular Biology, University of Dhaka, Dhaka, Bangladesh

1. Introduction

Aging is characterized by a gradual and irreversible deterioration in the functioning of physiological organs, culminating in mortality [1]. It is a multifaceted, intrinsic process that involves complex cellular mechanisms. It is characterized by biological degeneration and encompasses various factors such as stem cell depletion, mitochondrial malfunction, cellular senescence, shortening of the telomere, and disrupted metabolic processes [2]. As per a report published in 2010, individuals aged 75–84 years in the United States exhibited a significantly higher likelihood of succumbing to cancer and heart disease, with a 42-fold and 45-fold increase in mortality rates, respectively, as compared with those between the ages of 35–44 years. The mortality risk associated with Alzheimer's disease (AD) exhibits a more than 600-fold increase from ages 50 to 80 [3]. Recent research has indicated that certain noncoding RNAs (ncRNAs) are highly concentrated within exosomes. Exosomal ncRNAs have garnered significant attention for their role in regulating aging and age-related diseases, and diseases associated with age have gained increasing attention [4–6]. When looked at with an electron microscope, exosomes have the shape of a cup and are made up of two layers of phospholipids. Any type of cell can secrete these structures, which have a size range of 30–150 nm [7]. The process of early endosomal formation is initiated through the invagination of cell membranes. After this, endosomal membranes undergo inward budding to make late endosomal structures made up of different intraluminal vesicles (ILVs), which are often called multivesicular bodies (MVBs) [8]. MVBs inside of cells can be broken down by fusing directly with lysosomes or by fusing with cellular membranes and releasing ILVs into the extracellular space. The ILVs that are secreted by cells are commonly referred to as exosomes [9]. The components of exosomes comprise both fundamental and distinct constituents. Exosomes share fundamental constituents such as tetraspanins, namely CD63, CD81, CD9, and CD82 proteins associated with the development of multivesicular bodies [10,11]. Furthermore, there exists an enormous amount of evidence suggesting that exosomes are also enriched with distinct biomolecules, including DNA, mRNA, and different types of

Exosomal RNA
ISBN 978-0-443-14008-2
https://doi.org/10.1016/B978-0-443-14008-2.00002-4

ncRNAs (lncRNAs, circular RNAs [circRNAs], and miRNAs) [12]. Approximately 98.5% of the total human genome is comprised of noncoding sequences. NcRNAs make up a significant portion of eukaryotic transcription, comprising 95% of total RNA. This group comprises lncRNAs, miRNAs, and circRNAs, which are important regulators of gene expression [13–15]. This regulatory function is implicated in numerous pathophysiological processes associated with cellular senescence, aging, and aging-related diseases [16,17]. Existing evidence suggests that lncRNAs, circRNAs, and miRNAs can be transported into host cells through exosomes. These molecules are involved in regulating the process of aging and diseases associated with aging. To date, there has been limited investigation into the correlation between aging and exosomal RNA cargoes. The involvement of exosomal RNA cargoes in the aging process can be inferred based on the identification of various types of lncRNAs with regulatory roles in this biological phenomenon (Fig. 9.1).

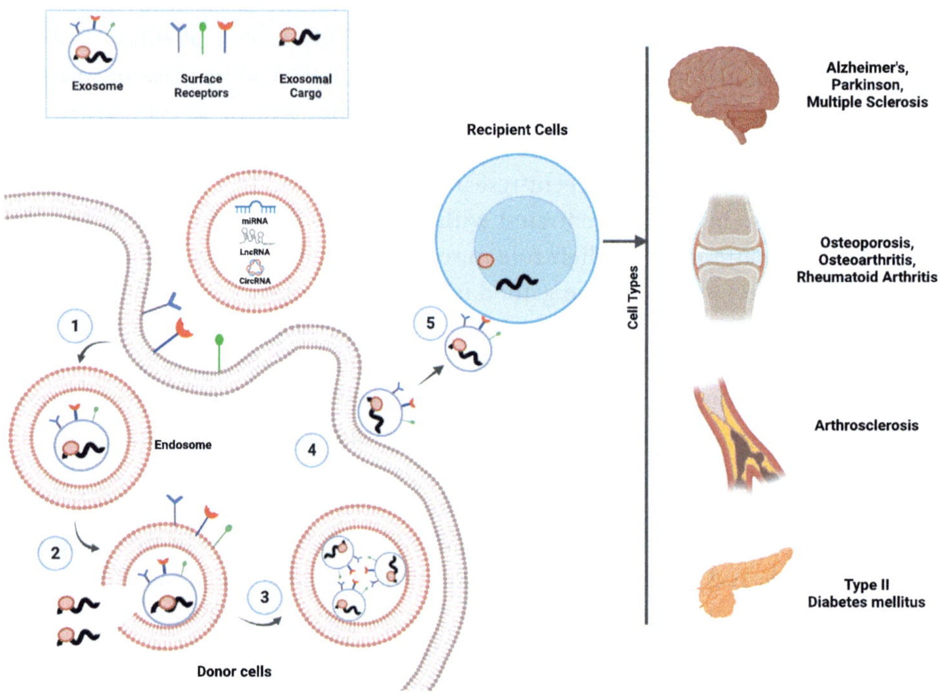

Figure 9.1 ***Steps depicting the biogenesis of exosomes that are released from a cell.*** 1. Endocytosis. 2. RNA and protein being engulfed by invaginating endosomes. 3. Formation of multivesicular bodies (MVBs). 4. Fusion of multivesicular bodies with the membrane. 5. Uptake by the recipient cells. Exosomes deliver different lncRNAs, circRNAs, and miRNAs to the recipient cells, which cause a variety of disorders that are linked to aging. Figure prepared using Biorender.

2. Exosomal LncRNA in aging

Long noncoding RNAs (lncRNAs) are RNA molecules that do not code for proteins and are distinguished by their length exceeding 200 nucleotides. The majority of them are transcribed by RNA polymerase II, leading to the incorporation of a 5′-methylguanosine cap along with a 3′-poly(A) tail [18]. LncRNAs (Table 9.1) participate in regulating diverse cellular processes associated with the aging process [26]. The regulation of miR-181a/Sirt1 by means of antisense ncRNA in the INK4 locus (ANRIL) has the potential to promote cellular viability as well as inhibit senescence in vascular smooth muscle cells (VSMCs). Moreover, suppressing the p53-p21 signaling cascade could potentially relieve cell cycle arrest. The present investigation offers a new perspective on the cellular cross-talk between ANRIL and senescence [27]. The downregulation of SA lncRNA (SALNR) expression has been observed in human fibroblasts undergoing senescence. The protein SALNR engages in interaction with NF90, an RNA-binding protein. This interaction results in the suppression of miRNA biogenesis. Additionally, SALNR regulates miR-22.4 and miR-181, senescence-related miRNAs, thereby preventing premature aging [28,29].

Table 9.1 Representative examples of certain exosomal lncRNAs associated with aging and age-related diseases.

LncRNA	Regulation	Disease	References
BACE 1	Regulates BACE1 mRNA expression, promotes biosynthesis of amyloid-β42 aggregates.	Alzheimer's disease	[19,20]
MEG3	Stimulates the transcription of BMP4 mRNA, reduces bone mass, and decreases osteoblast proliferation and osteoblastogenesis.	Osteoporosis	[21]
XIST	Acts as a ceRNA of miR12775p, enhances the expression of MMP13 and ADAMTS5 which indicates ECM disintegration.	Osteoarthritis	[22]
RNCR3	Upregulates Kruppel-like factor 2 (KLF2) and RNCR3 through sponging miRNA-185-5p, RNCR3, stimulates migration, and proliferation of VSMCs.	Atherosclerosis	[23]
MEG3	Impacts replication of fibroblast-like synoviocytes (FLS), through regulating nucleotide oligomerization domain (NOD)-like receptors 5 (NLRC5).	Rheumatoid arthritis	[24]
RP11-543N12.1	Interacts with miR-324.	Alzheimer's disease	[25]

3. Exosomal miRNA in aging

MicroRNAs (miRNAs) are small, ncRNA molecules that are single-stranded and typically composed of 21–23 nucleotides [29]. miRNAs are present in various organisms, such as plants, animals, and certain viruses. They play a crucial role in RNA silencing and posttranscriptional gene regulation [30,31]. The findings of a particular study indicate that cutaneous miRNAs associated with age showed a rise in the expression of miR-181a/b, miR-138, and miR-130 in keratinocytes during the process of replicative senescence. miRNAs were observed to modulate cell proliferation mechanisms through specific targeting of Sirtuin 1 and mRNAs. Biomarkers linked to aging, such as miR-34a-5p, miR-23a-3p, miR-125a, etc., exhibited elevated levels in dermal fibroblasts. These biomarkers associated with aging that impact the maintenance of telomeres along with various stages of the cellular life cycle [32] (Table 9.2).

4. Exosomal circRNA in aging

Circular RNAs, also known as circRNAs, are a group of single-stranded RNAs that are produced through a process called precursor mRNA back-splicing. These molecules originate from exonic or intronic sequences and are highly conserved. CircRNAs differ from canonical linear RNAs in that they form closed loops that are stable and lack a 5′-end cap and 3′-end poly(A) tail modifications. This unique feature renders them resistant to exonuclease digestion [38]. Upon translocation to the cytoplasmic compartment, circRNAs can interact with microRNAs (miRNAs) or proteins

Table 9.2 Representative examples of certain exosomal microRNAs associated with aging and age-related diseases.

MicroRNA	Regulation	Disease	References
Mir-7, Mir-153	Posttranscriptional regulation of alpha-synuclein levels.	Parkinson's disease	[33]
Mir-31	Downregulates Frizzled-3, inhibition of osteogenic differentiation.	Osteoporosis	[34]
Mir-214	Targets Atf-4, a transcription factor; inhibits osteoblast activity and mineralization.	Osteoporosis	[35]
Mir-92a	Boosts chondrocyte proliferation and matrix formation through the PI3K/AKT/mTOR pathway, mediated by noggin3.	Osteoarthritis	[36]
MiR-20b-5p	Modulates intracellularly insulin-stimulated glucose metabolism through Akt signaling.	Type 2 diabetes mellitus	[37]

Table 9.3 Representative examples of certain exosomal circRNAs associated with aging and age-related diseases.

circRNA	Regulation	Disease	References
CircSLC8A1	Modulate the function and/or activity of miR-128	Parkinson's disease	[41]
CircRtn4-Exos	Sponge miR-146a, attenuates TNF-α induced cytotoxicity and apoptosis	Osteoporosis	[42]
Circ0006393	Sponging miR1455p activity, upregulates FOXO1, stimulates osteogenic genes during bone remodeling	Osteoporosis	[43]
Circ0011269	Sponge miR-122 levels thereby controlling RUNX2 expression	Osteoporosis	[44]
CircRNA-CER	Affects chondrocyte ECM breakdown, influence the production of MMP13 by acting as a competitive endogenous RNA (ceRNA)	Osteoarthritis	[45]

in the cytoplasm, leading to various biological functions [39]. Extracellular vesicle (EV) —mediated release facilitates the extrusion of circRNAs from cells into the extracellular space, potentially enabling cellular circRNA clearance. Moreover, it is important to highlight that EVs possess the capability to be taken up by adjacent cells, suggesting that excreted circRNAs might have a potential role in intercellular communication [40]. According to a study, circRNA, CircCCNB1, has the potential to function as a miR-449a sponge, thereby retarding cellular senescence through targeting CCNE2. Targeting CircCCNB1 may be a promising strategy for interventions aimed at mitigating the effects of aging and age-related illnesses. Multiple reports have shown that circRNAs (Table 9.3) are implicated in the development of age-related diseases and can be used as predictive biomarkers [46]. This chapter will emphasize the significance of exosomal RNAs in the development and advancement of aging-associated illnesses.

5. Exosomes in neurological disorders

5.1 Alzheimer's disease

AD is a prevalent form of dementia characterized by increased inflammation, oxidative damage, and neuronal loss in various regions of the brain. The disease affects over 20% of individuals aged over 80 years. Furthermore, epidemiological data suggest that by the year 2050, the number will increase to 35 million, significantly increasing the social and economic burdens [47,48]. Early identification is crucial for individuals at risk of developing the disorder due to the absence of a cure for advanced stages [49]. It is caused by multiple factors, such as the hyperphosphorylation of tau proteins associated with microtubules in addition to the β-amyloid plaques (Aβ) overexpression. These plaques result from inappropriate splicing of the amyloid precursor protein (APP) [50]. Early indicators of AD include synaptic dysfunction and loss, which are frequently associated with the accumulation of neurotoxic Aβ oligomers in neurons [51–53]. In people with AD, neurons and synapses are gradually lost, and the disease can lead to a fatal drop of 40% in brain mass during the terminal stages [54]. The pathophysiology of AD typically involves a decrease in cholinergic system activity [55]. Extracellular amyloid plaques are prominent pathological features observed in patients with AD. Amyloid-beta (Aβ) molecules aggregate due to the continual division of APP through β-secretase BACE1 and the β-secretase/presenilin complex (PS1, PS2) [56]. Plaque formation in neurons triggers an inflammatory response facilitated by microglia and astrocytes, which in turn induces the immune response and causes macrophages and neutrophils to produce cytokines, interleukins, and TNF-alpha, causing permanent damage to the neurons [57,58]. While this initial reaction provides limited defense to the brain, it has been hypothesized that a persistent state of chronic inflammation is the root cause of neuronal degeneration. Neurofibrillary tangles, resulting from the accumulation of hyperphosphorylated tau protein, play a significant role in the advancement of AD. Detachment from microtubules followed by subsequent phosphorylation events is triggered by the phosphorylation of tau proteins by protein kinases such as GSK-3 and CDK-5. Alternatively, failure to dephosphorylate results in a conformational change that promotes aggregation, neurofibrillary tangle formation, and neuronal death [59]. Depending on their origin, exosomes contain a diverse range of biomolecules that are transported to nearby cells. The recently identified lncRNAs, miRNAs, and circRNAs have the potential to reveal new avenues into the regulatory mechanisms involved in cellular death associated with AD.

5.1.1 Exosomal miRNAs in AD

It is generally known that miRNAs control the activity of proteins involved in AD. Because exosomes may carry miRNAs across biofluids, they remain stable in both CSF and blood [60]. The production and function of tau and APP are regulated by miRNAs that are released from exosomes. To prevent an abnormal accumulation of Aβ, the APP can be modified by miR-16, miR-101, miR-153, and miR-106b. MiRNAs may also have an impact on how APP is translated [61]. The miR-384, miR-193b, and miR-135a combination shows promise for enhancing the efficiency of AD diagnosis. Moreover, mir-384 had the strongest ability to distinguish between vascular dementia, Parkinson's disease (PD), and AD (VaD) [62].

5.1.1.1 Exosomal miRNAs in cerebrospinal fluid

Researchers examined the level of expression of 746 miRNAs in cerebrospinal fluid (CSF) from people with PD and AD. They used the TaqMan miRNA array to evaluate the dysregulation of exosomal miRNAs. The expression levels of several miRNAs were significantly altered in AD patients [63]. Riancho's lab assessed the levels of miRNA in both the total CSF and the CSF exosome-enriched fraction between individuals with AD and healthy individuals. In this study, the researchers analyzed the expression of 760 miRNAs in the CSF of individuals with AD and compared it to that of healthy controls. MiR-598 and miR-9-5p were present in 50% and 75% of the total CSF control specimens, respectively, but were not detected in any CSF samples from patients with AD. However, a significant change in the distribution of these microRNAs was observed in the exosome-enriched fraction. miR-9-5p was detected in 70% of exosome-enriched CSF samples from AD patients, while miR-598 was found in 80% of exosome-enriched CSF samples obtained from AD patients, as compared with the overall CSF specimens collected from individuals with AD [64]. A recent investigation looked at exosomal miRNA expression in the CSF of sporadic individuals suffering from early-onset illness and found that patients suffering from AD had lower amounts of miR-451a, miR-16-5p, and miR-605-5p while having higher amounts of miR-125b-5p. Furthermore, the study found a reduction in miR-605-5p and miR-451a expression along with increased levels of miR-125b-5p. There are also no significant variations in miR-16-5p levels in late-onset AD patients in comparison with healthy individuals. The study found that there were differences in the expression of three CSF-derived exosomal microRNAs (miRNAs) between patients who had an early onset of illness and those who had a late onset [65]. Exosomal miR-16-5p dysregulation was specifically observed in the early-onset group, suggesting its potential as a distinctive biomarker linked to pathological processes in the initial phases of AD.

5.1.1.2 Exosomal miRNAs in blood components

The researchers had to shift their focus toward finding novel biomarkers in the blood of AD patients due to the limitations associated with the CSF. Subsequent studies from the Cheng lab were conducted using NGS to profile miRNA. Their goal was to determine the presence of any difference in the miRNA profiles of exosomes, cell-free plasma or serum, and peripheral blood. Certain miRNAs, such as miR-16-5p, miR-26a-5p, miR-451a, miR-191-5p, miR-223-3p, miR-126-3p, miR-486-5p, miR-484, let-7b-5p, and miR-126-5p, were found to be concentrated in exosomes. As a result, RNAase-resistant miRNAs contained in exosomes appear to be a valuable resource for use as disease biomarkers [61,66]. A separate study indicated that miR-193b potentially contributes to the development of AD. This study measured the levels of Aβ, tau, p-tau, miR-193b, exosomal miR-193b, APOE, and HCY in samples obtained from APP/PS1 double-transgenic mice, individuals who had mild cognitive impairment (MCI), and individuals with dementia of the Alzheimer's type (DAT). Reports pointed out that elevated miR-193b expression can suppress the synthesis of both APP mRNA and APP protein. Exosomal miR-193b levels were found to be significantly lower in the blood of patients with MCI and Alzheimer's disease (DAT) compared with control subjects. However, no significant difference was observed in the levels of total miR-193b. Exosomal miR-193b concentrations in the blood were found to be significantly reduced in patients with DAT compared with those in the MCI group. Additional evidence of decreased expression of exosomal miR-193b was observed in the CSF of patients with DAT. The findings indicate that miR-193b potentially contributes to the progression of AD. Additionally, exosomal miR-193b shows promise as a unique and noninvasive blood-based indicator for individuals with MCI and DAT [67]. A separate study assessed the serum levels of miR-193b, miR-135a, and miR-384 among people with mild cognitive decline, AD, PD, and vascular dementia. The objective was to examine the potential of these miRNAs in AD. In the serum exosomes of individuals with AD, the expression levels of miR-135a and miR-384 were found to be elevated, while no significant change was observed in the expression of miR-193b, when compared with a control group of healthy individuals. The combined activity of miR-384, miR-135a, and miR-193b demonstrated a more robust response toward the early diagnosis and prediction of AD. Exosomal miR-193b exhibited a significant decrease, while miR-384 and miR-135a demonstrated an increase in expression levels within the MCI and AD groups [62,67]. Mir-1306-5p, a microRNA that targets ADAM10, demonstrated the greatest sensitivity and specificity in predicting AD [68].

5.1.2 Exosomal lncRNAs in AD

LncRNA RP11-543N12.1 has been identified as a biological indicator for AD. It exhibits interaction with miR-324 [69]. The lncRNA BACE1 antisense transcript, or BACE1-AS, plays a significant role in the development of AD. It plays a significant role in BACE1 mRNA expression. Lastly, it facilitates the process of biosynthesis of amyloid-β42 aggregates, thereby contributing to the advancement of AD [70,71].

5.1.3 Role of exosomal circRNAs in AD

The aging suppressing activity of CircPVT1 in fibroblast proliferation appears to be responsible for isolating Let-7, which is implicated in AD patients [25]. ciRS-7, also referred to as CDR1as, is a prevalent circRNA present in both the retina and the human brain. It acts as an endocrine inhibitor of miRNA, specifically miRNA-7, by functioning as a "sponge" to restrict its normal activity [19]. An aberrant miR-7-circRNA cascade has been identified in individuals diagnosed with intermittent AD. The ubiquitin E2A (UBE2A) enzyme functions in the elimination of amyloid peptides. Its expression is reduced in the brains of affected patients because of the high regulation of miR-7 resulting from ciRS-7 deficiency. This reduction in UBE2A expression is one of several AD-related mRNA targets [19]. In the rat model, AD was induced by Aβ-142, resulting in a significant increase in 555 circRNAs in the hippocampus [20].

5.2 Parkinson's disease

PD represents the most prevalent neurodegenerative illness, ranking second in prevalence after AD. It has an incidence rate of around 2% among individuals aged 65 and above [72]. Typical clinical signs include characteristic motor symptoms such as resting tremors, slowed movement (bradykinesia), stiffness in the muscles (muscular rigidity), and difficulty maintaining balance (postural instability). Motor symptoms associated with the disease originate from the degeneration of dopaminergic neurons located in the substantia nigra. This leads to an imbalance of the excitatory neurotransmitter acetylcholine and the inhibitory neurotransmitter dopamine (DA) [73].

5.2.1 Exosomal microRNAs in Parkinson's disease

A study found that miR-19b-3p and miR-1 are notably downregulated in exosomes derived from the CSF of individuals with PD. The study also found significant upregulation of miR-153, let-7g-3p, miR-409-3p, and miR-10a-5p in exosomes derived from the CSF of patients with PD [63]. The origin of PD is thought to be influenced by alpha-synuclein buildup. Mir-7 and Mir-153 modulate alpha-synuclein levels through posttranscriptional mechanisms. These microRNAs exert an additional effect on the mRNA and protein levels of alpha-synuclein through binding specifically to the 3′-UTR of the protein. Overexpressing mir-7 and mir-153 significantly reduces the levels of endogenous alpha-synuclein. The translation of a luciferase construct with the alpha-synuclein 3′-UTR is enhanced in primary neurons through the inhibition of mir-7 and mir-153. These observations suggest a novel approach to controlling alpha-synuclein [74]. Individuals diagnosed with PD showed significantly higher levels of circulating exosomal miR-331 expression compared with the control group [75]. The study investigated patients with PD alongside cell models. The study's results indicate that miR-205 has a regulatory function in the expression of LRRK2, a protein linked to PD. Furthermore, miR-205 exhibited inhibitory effects on the impaired neurite outgrowth seen in LRRK2 mutant neurons [76]. In PD models, it was shown that miR-124 and miR-181b were implicated in both the inflammatory response and autophagy processes [77].

5.2.2 Exosomal lncRNAs in Parkinson's disease

NEAT1, a long ncRNA, has been found to be upregulated in PD [78]. In an independent study, it was observed that lncRNAs such as AX747125, AK127687S, MAPT-AS1, and SNCA-AS1 were identified in exosomes obtained from CSF of PD patients, whereas they were absent in the control samples [79]. The expression profiling analysis of lncRNAs reveals that five lncRNAs show statistically significant differential expression in PD. The study demonstrates a stark reduction in lncRNA H19 expression in individuals with PD. In contrast, Malat1, lincRNA-p21, TncRNA, and SNHG1 demonstrate significant upregulation under the same conditions [80]. The study suggests that the lncRNA H19 may have a protective effect against dopaminergic neuronal loss. The observed effect is thought to occur through Wnt/β-catenin activation. This is accomplished by counteracting the suppressive effect of miR-301b-3p on HPRT1 gene expression [81].

5.2.3 Exosomal circular RNAs in Parkinson's disease

According to a study, it was shown that the expression of circDLGAP4 goes down in both the MPTP-induced mouse model of PD and the MPP+-induced cell model of PD. The activity of circDLGAP4 was found to be modulated by miR-134-5p regulation. Additionally, the results suggest that miR-134-5p specifically targets CREB, and it is evident that the circDLGAP4/miR-134-5p axis affects the expression of CREB. The modulation of CREB signaling followed by subsequent expression of CREB-responsive genes such as BDNF, Bcl-2, and PGC-1α in SH-SY5Y and MN9D cells can be influenced by CircDLGAP4/miR-134-5p [82]. This study demonstrates an increase in circSLC8A1 expression in the substantia nigra of brain specimens from individuals with PD. The gene CircSLC8A1 possesses seven binding sites for miR-128 and demonstrates a strong affinity for the microRNA effector, Ago2 protein. The observed increase in RNA targets of miR-128 in individuals with PD suggests that circSLC8A1 may influence the function and/or activity of miR-128 [41]. A recent study has identified an entirely novel circRNA called circzip-2, which originates from the zip-2 gene locus. It was found to affect the aggregation of α-synuclein proteins and lifespan extension [83]. In silico analysis predicted that mmu_circRNA_0001320 and mmu-circRNA-0003292 would interact with miR-124 and miR-132, which are both linked to PD [84].

6. Exosomes in bone-related disorders

6.1 Osteoporosis

Age-related musculoskeletal problems are characterized by a decline in cartilage resilience, a reduction in the strength of the muscles, and increased bone fragility [61]. Bone cells, specifically osteoblasts and osteoclasts, produce exosomes that play a crucial role in coordinating bone remodeling [33]. Osteoporosis is a significant medical and societal issue due to the progressive reduction in density, strength, and microarchitecture of bones, resulting in an increased susceptibility to bone fractures [85]. Deregulation of bone remodeling with age is a crucial pathophysiologic event that drives osteoporosis [61]. Exosomes derived from aged bone marrow exhibit increased expression of the miR-183 cluster and miR-31a-5p. This particular microRNA inhibits the development of mesenchymal stem cells (MSCs), leading to cellular aging and ultimately promoting the formation of osteoclasts, which are responsible for increased bone resorption [86,87]. MicroRNA-21 exerts an antiosteogenic impact, leading to an inhibition of the SMAD7 protein. This leads to reduced expression of osteocalcin (OCN), alkaline phosphatase (ALP), and runt-related transcription factor 2 (RUNX2). The circulating exosomes of individuals with osteoporosis exhibit elevated levels of microRNA-21 [88].

6.1.1 Exosomal microRNAs in osteoporosis

miR-9-5p inhibits bone formation and promotes fat cell development through its interaction with Wnt3a. Additionally, it promotes osteoclastogenesis [89]. In a previous study with 30 patients, it was found that the level of miR-9-5p in the serum of people with osteoporosis was much higher than the level of expression in healthy people [90]. c-Fos, a component involved in osteoclast formation, raises the expression levels of miR-21. This increase in miR-21 further enhances the process of osteoclastogenesis induced by RANKL. Programmed cell death is involved in this mechanism through a positive feedback loop [91]. Jiang and colleagues [92] demonstrated that MiR-21 targets SMAD7, thereby suppressing osteogenesis. Osteoporosis patients exhibited elevated amounts of miR-21 in their blood and bone tissue [93]. In contrast, researchers noted a notable reduction in miR-21-5p expression among women with osteoporosis or osteopenia who experienced vertebral fractures [94]. The role of miR-21 in post-menopausal osteoporosis studies is uncertain, as it has been found to be both upregulated and downregulated [42,93—96]. During various in vitro trials, it was found that miR-21 inhibits osteoclastogenesis in some cases while promoting osteogenesis in others [43,97]. Hence, further experiments need to be conducted to validate miR-21 as a clinical biomarker for osteoporosis. Mice EVs exhibit significantly elevated levels of miR-183-5p [91]. MiR-183-5p plays a role in bone resorption and in the prevention of age-related bone loss. Specifically, miR-183-5p appears to suppress bone marrow—derived stem cells (BMSCs) proliferation along with bone formation. Additionally, it promotes the differentiation of osteoclasts in bone marrow-derived macrophages by interacting with the cooxygenase-1 (Ho1) target [87]. Microvesicles containing endothelial microRNA-31 have been reported to originate from senescent cells. The microvesicles are subsequently internalized by MSCs. This leads to the inhibition of osteogenic differentiation by downregulating Frizzled-3. The potential implication of miR-31 in the plasma of elderly individuals in the pathophysiology of reduced bone formation associated with aging can be deduced. Therefore, it is possible that miR-31 may function as a viable biomarker for the process of aging when present in plasma [44]. MiR-214 suppresses osteoblast function and mineralization through its interaction with the transcription factor Atf-4. Since it also targets Pten, miR-214 is an important factor in the process of osteoclastogenesis. Overexpression of miR-214 in osteoclasts enhances their activity and leads to bone resorption, whereas its inhibition improves bone mass and bone formation [98]. According to another study, miR-214-3p is transferred from osteoclast-derived exosomes to osteoblasts, where it suppresses the formation of bones [99]. Rictor and Hdac9 are the targets of miR-188, which also stimulates the PPARγ transcription factor and promotes BMSC development into adipocytes rather than osteoblasts. MiR-188 thus plays a vital role in regulating the transition between osteogenesis and adipogenesis in bone marrow—derived mesenchymal stem cells (BMSCs) during the aging process. It may also be a potential therapeutic approach for age-related osteoporosis [100]. Researchers found that miRNA-483-5p, by boosting osteoclast formation, may contribute to the development of osteoporosis [101].

6.1.2 Exosomal lncRNAs in osteoporosis

RP11-498C9, XR111050, and ANCR are all lncRNAs that have been linked to processes that change how osteogenic genes are regulated [102]. Among the lncRNAs with different levels of expression, knocking down the myocardial infarction-associated transcript (MIAT) helps human adipose-derived stem cells (ADSCs) become more osteogenic [103]. LncRNA MEG3 was associated with decreased bone mass, osteoblast proliferation, and osteoblastogenesis. Additionally, it is evident that the lncRNA MEG3 stimulated the transcription of BMP4 by causing the transcription factor SOX2 to dissociate [104]. LncRNA XR-111050 upregulates osteopontin (OPN/SPP1), collagen type I alpha 2 chain (COL1A2), bone gamma-carboxyglutamate protein/osteocalcin (BGLAP/OCN), and RUNX2, promoting enhanced osteogenic differentiation of MSCs [105]. By boosting H4 histone acetylation and decreasing HDAC-1.117, the lncRNA AK141205 also stimulates osteogenesis by triggering CXCL13, a proosteogenic chemokine [106]. LncRNA-H19, a known osteogenesis accelerator, is upregulated during osteoblast differentiation. The involvement of lncRNA H19 is mediated by the miR-675/TGF-β1/Smad3/histone deacetylase (HDAC) pathway [107].

6.1.3 Exosomal circRNAs in osteoporosis

It has been suggested that Circ_0001275 can act as a potential biomarker for postmenopausal osteoporosis [108]. Nevertheless, miRNA34a and miRNA146a may interact with circRNA BANP and circRNA ITCH to modulate PDLSC osteogenic development through the MAPK pathway by directly targeting PDGFRA and DUSP1, respectively [34]. Circ 0024097 reduces the progression of osteoporosis by promoting bone formation via the miR-376b-3p/YAP1 axis along with the Wnt/-βcatenin signaling pathway [35]. Exosomes derived from circ-Rtn4-modified bone marrow—derived mesenchymal stromal cells (Rtn4-Exos) exhibit the ability to sponge miR-146a and reduce cytotoxicity and apoptosis induced by TNF-α in murine MC3T3-E1 cells. Thus, Rtn4-Exos may serve as a promising therapeutic approach for osteoporosis treatment [109]. The circRNA_0016624 molecule functions by sponging miR-98 and enhancing its expression, which prevents osteoporosis [110]. The study suggests that circRNA 0006393 plays a role in promoting ossification in glucocorticoid-induced osteoporosis. This is achieved through the upregulation of osteogenic genes during bone remodeling, which is facilitated by the sponging of miR1455p and the upregulation of FOXO1 [111]. Circ 0011269 is responsible for the development of osteoporosis by sponging up miR-122 to control RUNX2 expression [112].

6.2 Osteoarthritis

Osteoarthritis (OA) is characterized by joint pain and dysfunction, ultimately causing joint degeneration. The majority of individuals are impacted by this disorder as compared with any other type of joint disorder [113]. The extracellular matrix (ECM) is crucial for ensuring joint health. OA can arise in aging joints due to an imbalance between ECM production and degradation in addition to an elevation in IL-1 and TNF-α, inflammatory cytokines. Researchers have demonstrated that exosomes enhance the degenerative development of OA by playing a role in the underlying pathological processes of OA [21,114].

6.2.1 Exosomal microRNAs in osteoarthritis

miR-140 is the primary regulator of cartilage homeostasis and is abundantly expressed in cartilage. It was shown that chondrocyte-specific miR-140 expression is downregulated in OA chondrocytes [115]. The upregulation of miR-455-3p has been observed to facilitate histone H3 acetylation and selectively target HDAC2/8. This implies that controlling miR-455-3p could be a potential strategy for regulating chondrogenesis and reducing cartilage degeneration. Exosomal miRNA-92a enhances chondrocyte growth and matrix formation via the PI3K/AKT/mTOR axis, which contributes to the alleviation of OA mediated by noggin3 [116]. MiR-92a-3p affects cartilage growth and maintenance by interacting with WNT5A. Targeting exosomal miR-92a-3p could be a viable approach for the advancement of treatments for OA [117].

6.2.2 Exosomal lncRNAs in osteoarthritis

LncRNAs have been linked to the diagnosis of OA. Moreover, they possess the capacity to function as novel biomarkers for the management of OA. Being the primary mediator of chondrogenesis, in addition to bone and cartilage homeostasis, they can also influence the progression of OA. The expression of HOTAIR, CTD-2574D22.4, PMS2L2, H19, GAS5, and RP11-445H22.4 was found to be higher in osteoarthritic cartilage. This upregulation could affect the pathogenesis of OA by increasing the transcription of ADAMTS5, BMP-2, and MMP-9 mRNAs [118]. This could affect how OA develops. Increased lncRNA H19 is linked to osteogenic differentiation in hBMSCs, which has been linked to the development of musculoskeletal tissues in the past. H19 and the miR-675 it makes have been shown to help human mesenchymal stem cells (hMSCs) turn into bone cells and make new bone tissue [107]. H19 helps bone growth by acting as a competing endogenous RNA that works with miR-22 and miR-141 [107] to help bone growth. There is a notable variation in the levels of exosomal lncRNA PCGEM1 in the synovial fluid of individuals with OA compared with those who are healthy. Therefore, PCGEM1 can serve as a biomarker for differentiating between advanced and early OA [119]. MALAT1 enhances the process of ossification differentiation through positive regulation of Smad4 expression via its interaction with miR-204 [120]. Furthermore, miR-101 together with the distal HOXA transcript at the distal tip (HOTTIP) have been identified as potential bioprotective markers and novel therapeutic candidates for OA [61]. MiR-101 and HOTTIP are two ncRNAs that regulate integrin-1 expression via epigenetic regulation by DNMT-3B and transcriptional regulation by HoxA13 [121]. The lncRNA DANCR exhibits high levels of expression in cartilage affected by OA and is linked to various processes such as chondrogenesis, ossification differentiation, and OA. To regulate the viability of OA chondrocytes, DANCR is used as a competitive endogenous RNA to sequester miR-577 and subsequently target Sphk2 mRNA [122]. The inactive form of the lncRNA XIST promotes the expression of MMP13 and ADAMTS5 in OA by acting as a competing endogenous RNA for miR12775p. The current investigation has introduced a novel objective and mode of operation for the molecular therapy of OA [123].

6.2.3 Exosomal circRNAs in osteoarthritis

Research findings indicate that two circRNAs, namely circRNA-36,866 and mmu-circRNA-30,365, have been identified as emerging molecular indicators for OA [124]. The circRNA Atp9b is responsible for regulating the development of OA through the modulation of ECM degradation and inflammation in chondrocytes [22]. The circRNA-CER played a role in chondrocyte ECM breakdown and affected the production of MMP13 by functioning as a competitive endogenous RNA (ceRNA) [125].

7. Exosomes in heart disease

7.1 Atherosclerosis

Atherosclerosis is a chronic inflammatory condition of major blood vessels that is considered the primary underlying cause of cardiovascular diseases (CVDs) [126]. The process of aging is also a significant contributing factor to the development of atherosclerosis. Atherosclerosis is characterized by plaque formation or fibrofatty lesions in the arterial wall [45,127]. The pathogenesis of atherosclerosis involves a series of events, including endothelial injury, lipid accumulation, infiltration of inflammatory cells, foam cell formation, and the formation of plaque, which collectively contribute to the progressive degeneration of arterial walls [36,128]. Ischemia and stroke eventually occur due to the rupture of susceptible plaques, which induces in situ thrombosis and intramural bleeding [36,45,129]. Myocardial infarction and stroke are only two of the devastating outcomes that can result from thrombus development brought on by unstable atherosclerosis [45,127]. Multiple studies have found that certain exosomal lncRNAs exhibit significantly elevated expression in atherosclerosis, suggesting their potential utility as novel biomarkers for this disease.

7.1.1 Exosomal lncRNAs in atherosclerosis

The expression levels of exosomal long noncoding, HIF1A-AS1 were found to be significantly increased, indicating its use as a biomarker for atherosclerosis [126]. In a separate study, researchers observed a significant upregulation of the lncRNA RNCR3 in cultured endothelial cells (ECs) and vascular smooth muscle cells (VSMCs) treated with oxLDL, as well as in aortic atherosclerotic lesions in both mice and humans. Exosomal RNCR3 promotes VSMC migration and proliferation by upregulating RNCR3 and Kruppel-like factor 2 (KLF2) via the miRNA-185-5p sponge effect [130]. LncRNA AC100865.1 is thought to have originated from monocytes and is predominantly localized in the plasma vesicles of people with coronary heart disease (CHD). This suggests that this lncRNA can be employed as a reliable, sensitive, and accurate diagnostic parameter for CHD [131]. LncRNA GAS5 has also been implicated in atherogenesis. This lncRNA is believed to modulate the apoptosis of ECs and macrophages via exosomes. Studies indicate that targeting the lncRNA GAS5 could be a promising approach for treating atherosclerosis [132]. According to a prior study, GAS5 induced apoptosis in human aortic EC cells and decreased autophagy flux through direct binding to miR-26a [133]. Reports from Zhong's lab pointed out that miR-26a-5p hinders ox-LDL-induced apoptosis in EC cells by interfering with the NF-κB signaling/Toll-like receptor 4 (TLR4) pathway [134]. Hence, it is postulated that exosomal GAS5 induces the activation of the TLR4/NF-kB signaling cascade in ECs by sequestering miR-26a-5p. This would then result in ECs being induced to undergo apoptosis [135].

8. Exosomes in metabolic disorders

8.1 Type 2 diabetes mellitus

β-Cell dysfunction and/or insulin resistance cause glucose homeostasis to be disrupted, leading to the age-related condition known as type 2 diabetes mellitus (T2DM) [135]. This condition is a growing problem that affects people across the world and is intimately connected to obesity worldwide [23]. Diabetes-related insulin resistance along with the production of inflammatory cytokines may be induced by SASP factors secreted by senescent cells [136]. However, the decline in functionality of pancreatic cells due to aging leads to reduced insulin production and an elevated susceptibility to diabetes mellitus [137]. Exosomal RNA is integral to the advancement of T2DM.

8.1.1 Exosomal microRNAs in type 2 diabetes mellitus

The researchers in the Katsuyama lab discovered that individuals with T2DM have higher levels of miR-150-5p and miR-20b-5p in exosome-enriched EVs derived from their serum. However, the total miRNA profile of these miRNAs does not seem to have a substantial difference between normal and glucose-tolerant men. The researchers determined that extracellular miR-20b-5p, produced by exosomes, serves as a circulating biological marker associated with type 2 diabetes (T2D). This biomarker influences intracellular insulin-stimulated glucose metabolism stimulated by insulin through the modulation of AKT signaling [138].

8.1.2 Exosomal lncRNAs in type 2 diabetes mellitus

Expression of lncRNA-p3134, predominantly present in exosomes, exhibited a significant upregulation in the serum of individuals diagnosed with T2D. When β cells are exposed to high levels of glucose, they selectively secrete an lncRNA called p3134. This RNA then affects glucose-stimulated insulin secretion by enhancing crucial regulatory factors in β cells, including MafA, TCF7l2, Pdx-1, and GLUT2. Moreover, it provides additional safeguards to β cells against diminished cellular apoptosis in conjunction with sugar toxicity. The study investigated the PI3K/AKT/mammalian target of rapamycin signaling process and found evidence supporting the protective role of lncRNA-p3134 in glucose-stimulated insulin secretion (GSIS) [139]. These results provide more evidence for the fact that lncRNA-p3134 inhibits the progression of T2D. Diabetic patients with poor glycemic control had higher blood levels of H19-lncRNA. In that same study, it was also observed that serum and serum exosomal MALAT-lncRNA levels were significantly lower in patients with T2D [140].

9. Exosomes in autoimmune disorders

9.1 Rheumatoid arthritis

Rheumatoid arthritis (RA) is an autoimmune condition primarily affecting joints, cartilage, bones, and tissues. The precise etiology of RA remains uncertain [141]. Age is a significant determinant of RA [142]. EVs, which serve as transporters of autoantigens, microRNAs, and matrix-degrading enzymes, contribute to the development of RA [143].

9.1.1 Exosomal microRNAs in RA

The ability of miR-155 to destroy bones is indicative of its impact on the development of RA [37]. Let-7b, an additional miRNA discovered in synovial fluid S exosomes, has been demonstrated to have a notable impact on RA by transforming noninflammatory synovial fluid macrophages into proinflammatory M1 macrophages [144]. This study emphasizes the importance of miR-let-7b in the effector phase of RA. Furthermore, a study found that the serum of patients with RA contained lower levels of miR-548a-3p, enclosed within exosomes [145].

9.1.2 Exosomal lncRNAs in RA

The study employed microarray analysis to examine the levels of lncRNAs in peripheral blood mononuclear cells and serum exosomes of individuals with RA. The findings revealed an increase in the expression of seven lncRNAs that include the taurine up1 gene (TUG1). The levels of LUST, SNHG4, NEAT1, HOTAIR, MEG9, and Anti-NOS2A were also found to be substantially increased. In contrast, the levels of HOXA3, SNHG1, MALAT1, MASCRNA, PS, PR antisensor versions, and HOXA3 are significantly decreased [146]. The NTT/PBOV1 lncRNA axis regulates monocyte inflammation and is crucial for monocyte/macrophage differentiation and the advancement of RA [147]. The potential involvement of lncRNA MEG3 in RA may be due to its influence on the replication of fibroblast-like synoviocytes (FLS), which is activated by nucleotide oligomerization domain (NOD)-like receptor 5 (NLRC5). MEG3 is responsible for the regulation of RA potential by targeting NLRC5 [148]. Another study pointed out that five of the lncRNAs, including RNA143596, HIX0032090, XLOC 002730, RNA143598, and IGHCgamma1, were validated to be highly upregulated in RA clinical samples [149]. The lncRNA HIX003209 regulates TLR4-mediated inflammation in macrophages by targeting miR-6089. By acting as a competing endogenous RNA through sequestering miRNA 6089, HIX003209 leads to the activation of the TLR4/NF-B signaling cascade within macrophages and subsequently enhances the inflammatory response. Therefore, it has the potential to be an innovative therapeutic strategy for treating RA [150].

9.1.3 Exosomal circRNAs in RA

Aberrant expression levels of multiple circRNAs have been detected in individuals diagnosed with RA. CircRNAs may have a significant impact on disease progression. In patients suffering from RA, the expression levels of multiple circRNAs in peripheral blood mononuclear cells (PBMCs) increased significantly [24,151,152]. Data show that circRNA 0140271 has high expression levels in females with RA [153], while circRNA 0000175, circRNA 101328, circRNA 0000396, and circRNA 0130438 had lower levels in the PBMC of RA patients. During the early stages of rheumatoid arthritis, the levels of four circRNAs—circ 0000367, circ 0002715, circ 0035197, and circ 0001947—in the peripheral blood of people with the disease have been shown to rise [154,155]. Conversely, the expression of circ 0044235 is observed to be reduced [156]. Patients with rheumatoid arthritis (RA) as well as fibroblast-like synoviocytes (HFLS-RA) exhibit increased levels of circ 0003353, circ 0088194, circ 0003972, and circ-AFF2, while circ 0130438 and circ 0008360 show decreased expression [157,158]. The plasma samples obtained from individuals who were recently diagnosed with RA exhibit increased levels of circ 0005198, circ 0005008, and circ 0000175 expression. On the other hand, the levels of circ 0044235 were observed to have undergone a significant decrease. The plasma levels of circ 0005198 and circ 0005008 can be used to categorize the degree of severity of RA [159].

9.2 Multiple sclerosis

Multiple sclerosis (MS) is a persistent inflammatory demyelinating ailment of the central nervous system (CNS) that typically manifests in early adulthood. The condition is marked by inflammation leading to the loss of myelin, increased astrocyte proliferation, and the degeneration of axons and neurons [160,161]. The transition to progressive MS is primarily determined by age [162]. Exosomes in circulation have emerged as a potential indicator for patients with MS [161].

9.2.1 Exosomal microRNAs in multiple sclerosis

Compared with healthy people, people with relapsing-remitting MS (RRMS) and progressive MS have different patterns of how their exosomes express miRNA. These patterns include miR-30b-5p, miR-15b-5p, miR-342-3p, and miR-451a in RRMS, and miR-370-3p, miR-127-3p, miR-432-5p, and miR-409-3p in progressive MS. Nine miRNAs have been identified as important in differentiating between progressive MS disease and RRMS. This investigation provides preliminary evidence that microRNAs found in circulating exosomes can serve as valuable biomarkers, not only for detecting MS but also for accurately predicting the specific subtype of the disease [161]. A separate investigation revealed that four exosomal sequences categorized as miRNA exhibited differential expression in individuals with RRMS in comparison with normal individuals. These sequences include hsa-miR-301a-3p, -122-5p, -532-5p, and -196b-5p [163].

9.2.2 Exosomal lncRNAs in multiple sclerosis

The exosomes derived from neurons of individuals affected by MS showed notable upregulation of lncRNAs such as HOTAIR, AIRN, NAMA, TRERNA1, and FAS-AS1 [164]. HOTAIR has been linked to the progression of MS [165]. The study employed a cuprizone-induced demyelination model to examine the impact of sulfasalazine. The findings demonstrated that sulfasalazine effectively blocks the transition of microglia cells into the M1-polarized phenotype, thereby promoting oligodendrocyte differentiation. The study found that cuprizone treatment induces reduced expression of HOTAIR along with elevated expression of miR-136-5p within microglia. This led to the deactivation of the AKT2-NF-κB signaling pathway. The study's findings suggest that sulfasalazine could be beneficial for individuals with MS by potentially modulating epigenetic regulation in microglial cells [166]. It is possible that HOTAIR facilitates the onset and advancement of MS by engaging in exosome-mediated intercellular communication.

10. Concluding remarks and future directions

Presently, there is increasing interest in the area of therapeutics for aging and age-associated illnesses. To delve into this field of research, a comprehensive examination of the biology of aging and age-related diseases, with a particular focus on elucidating the underlying mechanisms, is necessary. Exosomes have garnered significant attention in the etiology of various diseases in the past few years. The inherent capacity of exosomes to transport various RNA, DNA, and protein cargoes to neighboring cells has rendered them a compelling candidate for novel biomarkers. Exosomes have been recognized as potentially valuable biological markers in diverse age-related disorders as well. NcRNA represents a heterogenous class of RNA molecules that perform crucial functions in various cellular processes and diseases. Exosomes and ncRNAs play important roles in complex cellular aging pathways and are considered significant contributors to the aging process. The study of complex exosomal ncRNA networks has provided valuable insights into the cellular progression of age-related diseases in humans. Overall, the investigation of exosomal ncRNA in age-related diseases shows potential for revealing new aspects of disease development and for creating innovative approaches to diagnosis and treatment.

Acknowledgments

The authors would like to acknowledge Department of Biochemistry, Purdue University, United States, and Department of Biochemistry and Molecular Biology, University of Dhaka, Bangladesh. We are thankful to Ms Rajashree Nayak for critically reading the manuscript and giving useful comments and suggestions.

References

[1] Aunan JR, Watson MM, Hagland HR, Søreide K. Molecular and biological hallmarks of ageing. British Journal of Surgery January 15, 2016;103(2):e29–46.
[2] Degirmenci U, Lei S. Role of lncRNAs in cellular aging. Frontiers in Endocrinology December 6, 2016:7.
[3] Murphy SL, Xu J, Kochanek KD. Deaths: final data for 2010. National Vital Statistics Reports May 8, 2013;61(4):1–117.
[4] Machida T, Tomofuji T, Ekuni D, Maruyama T, Yoneda T, Kawabata Y, et al. MicroRNAs in salivary exosome as potential biomarkers of aging. International Journal of Molecular Sciences September 7, 2015;16(9):21294–309.
[5] Fan Q, Yang L, Zhang X, Peng X, Wei S, Su D, et al. The emerging role of exosome-derived non-coding RNAs in cancer biology. Cancer Letters February 2018;414:107–15.
[6] Zamani P, Fereydouni N, Butler AE, Navashenaq JG, Sahebkar A. The therapeutic and diagnostic role of exosomes in cardiovascular diseases. Trends in Cardiovascular Medicine August 2019;29(6):313–23.
[7] Kalra H, Adda CG, Liem M, Ang CS, Mechler A, Simpson RJ, et al. Comparative proteomics evaluation of plasma exosome isolation techniques and assessment of the stability of exosomes in normal human blood plasma. Proteomics November 2013;13(22):3354–64.
[8] Ailawadi S, Wang X, Gu H, Fan GC. Pathologic function and therapeutic potential of exosomes in cardiovascular disease. Biochimica et Biophysica Acta—Molecular Basis of Disease January 2015;1852(1):1–11.
[9] Simpson RJ, Lim JW, Moritz RL, Mathivanan S. Exosomes: proteomic insights and diagnostic potential. Expert Review of Proteomics June 9, 2009;6(3):267–83.
[10] Mathivanan S, Fahner CJ, Reid GE, Simpson RJ. ExoCarta 2012: database of exosomal proteins, RNA and lipids. Nucleic Acids Research January 1, 2012;40(D1):D1241–4.
[11] Simpson RJ, Jensen SS, Lim JWE. Proteomic profiling of exosomes: current perspectives. Proteomics October 2008;8(19):4083–99.
[12] Thakur BK, Zhang H, Becker A, Matei I, Huang Y, Costa-Silva B, et al. Double-stranded DNA in exosomes: a novel biomarker in cancer detection. Cell Research June 8, 2014;24(6):766–9.

[13] Warner JR. The economics of ribosome biosynthesis in yeast. Trends in Biochemical Sciences November 1999;24(11):437–40.
[14] Esteller M. Non-coding RNAs in human disease. Nature Reviews Genetics December 18, 2011;12(12):861–74.
[15] Jeck WR, Sorrentino JA, Wang K, Slevin MK, Burd CE, Liu J, et al. Circular RNAs are abundant, conserved, and associated with ALU repeats. RNA February 2013;19(2):141–57.
[16] Abdelmohsen K, Gorospe M. Noncoding RNA control of cellular senescence. Wiley Interdisciplinary Reviews RNA November 2015;6(6):615–29.
[17] Li Y, Zhu Y, Li G, Xiao J. In: Noncoding RNAs in cardiovascular aging; 2018. p. 37–53.
[18] Spizzo R, Almeida MI, Colombatti A, Calin GA. Long non-coding RNAs and cancer: a new frontier of translational research? Oncogene October 25, 2012;31(43):4577–87.
[19] Lukiw WJ. Circular RNA (circRNA) in Alzheimer's disease (AD). Frontiers in Genetics 2013;4:307.
[20] Wang Z, Xu P, Chen B, Zhang Z, Zhang C, Zhan Q, et al. Identifying circRNA-associated-ceRNA networks in the hippocampus of Aβ1-42-induced Alzheimer's disease-like rats using microarray analysis. Aging April 27, 2018;10(4):775–88.
[21] Chen BY, Sung CWH, Chen C, Cheng CM, Lin DPC, Huang CT, et al. Advances in exosomes technology. Clinica Chimica Acta June 2019;493:14–9.
[22] Liu Q, Zhang X, Hu X, Dai L, Fu X, Zhang J, et al. Circular RNA related to the chondrocyte ECM regulates MMP13 expression by functioning as a MiR-136 'sponge' in human cartilage degradation. Scientific Reports March 2, 2016;6(1):22572.
[23] DeFronzo RA, Ferrannini E, Groop L, Henry RR, Herman WH, Holst JJ, et al. Type 2 diabetes mellitus. Nature Reviews Disease Primers July 23, 2015;1(1):15019.
[24] Ouyang Q, Liu C, Lu X, Liang R, Zhao J, Yang M. Identification of circular RNAs Circ_0005008 and Circ_0005198 in plasma as novel biomarkers for new-onset rheumatoid arthritis. Frontiers in Pharmacology September 1, 2021:12.
[25] Panda AC, Grammatikakis I, Kim KM, De S, Martindale JL, Munk R, et al. Identification of senescence-associated circular RNAs (SAC-RNAs) reveals senescence suppressor CircPVT1. Nucleic Acids Research April 20, 2017;45(7):4021–35.
[26] Gomez-Verjan JC, Vazquez-Martinez ER, Rivero-Segura NA, Medina-Campos RH. The RNA world of human ageing. Human Genetics December 1, 2018;137(11–12):865–79.
[27] Tan P, Guo YH, Zhan JK, Long LM, Xu ML, Ye L, et al. LncRNA-ANRIL inhibits cell senescence of vascular smooth muscle cells by regulating miR-181a/Sirt1. Biochemistry and Cell Biology October 2019;97(5):571–80.
[28] Wu CL, Wang Y, Jin B, Chen H, Xie BS, Mao ZB. Senescence-associated long non-coding RNA (SALNR) delays oncogene-induced senescence through NF90 regulation. Journal of Biological Chemistry December 2015;290(50):30175–92.
[29] Green D, Dalmay T, Chapman T. Microguards and micromessengers of the genome. Heredity February 30, 2016;116(2):125–34.
[30] Bartel DP. Metazoan microRNAs. Cell March 2018;173(1):20–51.
[31] Qureshi A, Thakur N, Monga I, Thakur A, Kumar M. VIRmiRNA: a comprehensive resource for experimentally validated viral miRNAs and their targets. Database (Oxford) 2014;2014:bau103. https://doi.org/10.1093/database/bau103. PMID: 25380780. PMCID: PMC4224276.
[32] Gerasymchuk M, Cherkasova V, Kovalchuk O, Kovalchuk I. The role of microRNAs in organismal and skin aging. International Journal of Molecular Sciences July 25, 2020;21(15):5281.
[33] Liu S, Liu D, Chen C, Hamamura K, Moshaverinia A, Yang R, et al. MSC transplantation improves osteopenia via epigenetic regulation of Notch signaling in Lupus. Cell Metabolism October 2015;22(4):606–18.
[34] Gu X, Li M, Jin Y, Liu D, Wei F. Identification and integrated analysis of differentially expressed lncRNAs and circRNAs reveal the potential ceRNA networks during PDLSC osteogenic differentiation. BMC Genetics December 2, 2017;18(1):100.
[35] Huang Y, Xiao D, Huang S, Zhuang J, Zheng X, Chang Y, et al. Circular RNA YAP1 attenuates osteoporosis through up-regulation of YAP1 and activation of Wnt/β-catenin pathway. Biomedicine and Pharmacotherapy September 2020;129:110365.

[36] Sturtzel C. In: Endothelial cells; 2017. p. 71−91.
[37] Su LC, Huang AF, Jia H, Liu Y, Xu WD. Role of microRNA-155 in rheumatoid arthritis. International Journal of Rheumatic Diseases November 2017;20(11):1631−7.
[38] Tang X, Ren H, Guo M, Qian J, Yang Y, Gu C. Review on circular RNAs and new insights into their roles in cancer. Computational and Structural Biotechnology Journal 2021;19:910−28.
[39] Zang J, Lu D, Xu A. The interaction of circRNAs and RNA binding proteins: an important part of circRNA maintenance and function. Journal of Neuroscience Research January 2020;98(1):87−97.
[40] Lasda E, Parker R. Circular RNAs Co-precipitate with extracellular vesicles: a possible mechanism for circRNA clearance. PLoS One February 5, 2016;11(2):e0148407.
[41] Hanan M, Simchovitz A, Yayon N, Vaknine S, Cohen-Fultheim R, Karmon M, et al. A Parkinson's disease Circ <scp>RNA</scp> s Resource reveals a link between circ <scp>SLC</scp> 8A1 and oxidative stress. EMBO Molecular Medicine September 7, 2020;(9):12.
[42] Weilner S, Skalicky S, Salzer B, Keider V, Wagner M, Hildner F, et al. Differentially circulating miRNAs after recent osteoporotic fractures can influence osteogenic differentiation. Bone October 2015;79:43−51.
[43] Hu CH, Sui BD, Du FY, Shuai Y, Zheng CX, Zhao P, et al. miR-21 deficiency inhibits osteoclast function and prevents bone loss in mice. Scientific Reports February 27, 2017;7:43191.
[44] Weilner S, Schraml E, Wieser M, Messner P, Schneider K, Wassermann K, et al. Secreted microvesicular miR-31 inhibits osteogenic differentiation of mesenchymal stem cells. Aging Cell August 2016;15(4):744−54.
[45] Libby P, Buring JE, Badimon L, Hansson GK, Deanfield J, Bittencourt MS, et al. Atherosclerosis. Nature Reviews Disease Primers August 16, 2019;5(1):56.
[46] Zhang M, Bian Z. The emerging role of circular RNAs in Alzheimer's disease and Parkinson's disease. Frontiers in Aging Neuroscience July 12, 2021:13.
[47] Danborg PB, Simonsen AH, Waldemar G, Heegaard NHH. The potential of microRNAs as biofluid markers of neurodegenerative diseases—a systematic review. Biomarkers June 29, 2014;19(4):259−68.
[48] Van Giau V, An SSA. Emergence of exosomal miRNAs as a diagnostic biomarker for Alzheimer's disease. Journal of Neurological Sciences January 2016;360:141−52.
[49] Fotuhi SN, Khalaj-Kondori M, Hoseinpour Feizi MA, Talebi M. Long non-coding RNA BACE1-AS may serve as an Alzheimer's disease blood-based biomarker. Journal of Molecular Neuroscience November 1, 2019;69(3):351−9.
[50] Šimić G, Babić Leko M, Wray S, Harrington C, Delalle I, Jovanov-Milošević N, et al. Tau protein hyperphosphorylation and aggregation in Alzheimer's disease and other tauopathies, and possible neuroprotective strategies. Biomolecules January 6, 2016;6(1):6.
[51] Clare R, King VG, Wirenfeldt M, Vinters HV. Synapse loss in dementias. Journal of Neuroscience Research April 5, 2010;88(10):2083−90.
[52] Zhang H, Ma Q, Zhang Y, Xu H. Proteolytic processing of Alzheimer's β-amyloid precursor protein. Journal of Neurochemistry January 28, 2012;120(s1):9−21.
[53] Villemagne VL, Burnham S, Bourgeat P, Brown B, Ellis KA, Salvado O, et al. Amyloid β deposition, neurodegeneration, and cognitive decline in sporadic Alzheimer's disease: a prospective cohort study. The Lancet Neurology April 2013;12(4):357−67.
[54] Miller JA, Woltjer RL, Goodenbour JM, Horvath S, Geschwind DH. Genes and pathways underlying regional and cell type changes in Alzheimer's disease. Genome Medicine 2013;5(5):48.
[55] Davies P. Selective loss of central cholinergic neurons in Alzheimer's disease. The Lancet December 1976;308(8000):1403.
[56] Cole SL, Vassar R. The role of amyloid precursor protein processing by BACE1, the β-secretase, in Alzheimer disease pathophysiology. Journal of Biological Chemistry October 2008; 283(44):29621−5.
[57] Silva MVF, Loures C de MG, Alves LCV, de Souza LC, Borges KBG, Carvalho M, et al. Alzheimer's disease: risk factors and potentially protective measures. Journal of Biomedical Sciences December 9, 2019;26(1):33.

[58] Aguzzi A, O'Connor T. Protein aggregation diseases: pathogenicity and therapeutic perspectives. Nature Reviews Drug Discovery March 2010;9(3):237–48.
[59] Brunden KR, Trojanowski JQ, Lee VMY. Advances in tau-focused drug discovery for Alzheimer's disease and related tauopathies. Nature Reviews Drug Discovery October 2009;8(10):783–93.
[60] Basavaraju M, de Lencastre A. Alzheimer's disease: presence and role of microRNAs. Biomolecular Concepts August 1, 2016;7(4):241–52.
[61] Dolati S, Shakouri SK, Dolatkhah N, Yousefi M, Jadidi-Niaragh F, Sanaie S. The role of exosomal non-coding <scp>RNAs</scp> in aging-related diseases. BioFactors May 23, 2021; 47(3):292–310.
[62] Yang TT, Liu CG, Gao SC, Zhang Y, Wang PC. The serum exosome derived MicroRNA-135a, -193b, and -384 were potential Alzheimer's disease biomarkers. Biomedical and Environmental Sciences February 2018;31(2):87–96.
[63] Gui Y, Liu H, Zhang L, Lv W, Hu X. Altered microRNA profiles in cerebrospinal fluid exosome in Parkinson disease and Alzheimer disease. Oncotarget November 10, 2015;6(35):37043–53.
[64] Riancho J, Vázquez-Higuera JL, Pozueta A, Lage C, Kazimierczak M, Bravo M, et al. MicroRNA profile in patients with Alzheimer's disease: analysis of miR-9-5p and miR-598 in raw and exosome enriched cerebrospinal fluid samples. Journal of Alzheimer's Disease March 21, 2017;57(2):483–91.
[65] McKeever PM, Schneider R, Taghdiri F, Weichert A, Multani N, Brown RA, et al. MicroRNA expression levels are altered in the cerebrospinal fluid of patients with young-onset Alzheimer's disease. Molecular Neurobiology December 30, 2018;55(12):8826–41.
[66] Cheng L, Sharples RA, Scicluna BJ, Hill AF. Exosomes provide a protective and enriched source of miRNA for biomarker profiling compared to intracellular and cell-free blood. Journal of Extracellular Vesicles January 26, 2014;3(1):23743.
[67] Liu CG, Song J, Zhang YQ, Wang PC. MicroRNA-193b is a regulator of amyloid precursor protein in the blood and cerebrospinal fluid derived exosomal microRNA-193b is a biomarker of Alzheimer's disease. Molecular Medicine Reports November 2014;10(5):2395–400.
[68] Lei X, Lei L, Zhang Z, Zhang Z, Cheng Y. Downregulated miR-29c correlates with increased BACE1 expression in sporadic Alzheimer's disease. International Journal of Clinical and Experimental Pathology 2015;8(2):1565–74.
[69] Cai M, Wang Y, Xu S, Qiao S, Shu Q, Du J, et al. Regulatory effects of the long non-coding RNA RP11-543N12.1 and microRNA-324-3p axis on the neuronal apoptosis induced by the inflammatory reactions of microglia. International Journal of Molecular Medicine June 20, 2018;42:1741–55.
[70] Liu T, Huang Y, Chen J, Chi H, Yu Z, Wang J, et al. Attenuated ability of BACE1 to cleave the amyloid precursor protein via silencing long noncoding RNA BACE1-AS expression. Molecular Medicine Reports September 2014;10(3):1275–81.
[71] Li F, Wang Y, Yang H, Xu Y, Zhou X, Zhang X, et al. The effect of BACE1-AS on β-amyloid generation by regulating BACE1 mRNA expression. BMC Molecular Biology December 1, 2019;20(1):23.
[72] Lang AE, Lozano AM. Parkinson's disease. New England Journal of Medicine October 8, 1998;339(15):1044–53.
[73] Olanow CW, Stern MB, Sethi K. The scientific and clinical basis for the treatment of Parkinson disease. Neurology 2009;72(Issue 21, Suppl. 4):S1–136. 2009 May 26.
[74] Doxakis E. Post-transcriptional regulation of α-synuclein expression by mir-7 and mir-153. Journal of Biological Chemistry April 2010;285(17):12726–34.
[75] Yao Y-F. European Review for Medical and Pharmacological Sciences 2018;22:5278–83. 5278-5283.
[76] Cho HJ, Liu G, Jin SM, Parisiadou L, Xie C, Yu J, et al. MicroRNA-205 regulates the expression of Parkinson's disease-related leucine-rich repeat kinase 2 protein. Human Molecular Genetics February 1, 2013;22(3):608–20.
[77] Li W, Jiang Y, Wang Y, Yang S, Bi X, Pan X, et al. MiR-181b regulates autophagy in a model of Parkinson's disease by targeting the PTEN/Akt/mTOR signaling pathway. Neuroscience Letters May 2018;675:83–8.

[78] Li K, Wang Z. lncRNA NEAT1: key player in neurodegenerative diseases. Ageing Research Reviews April 2023;86:101878.
[79] Elkouris M, Kouroupi G, Vourvoukelis A, Papagiannakis N, Kaltezioti V, Matsas R, et al. Long non-coding RNAs associated with neurodegeneration-linked genes are reduced in Parkinson's disease patients. Frontiers in Cellular Neuroscience February 22, 2019:13.
[80] Kraus TFJ, Haider M, Spanner J, Steinmaurer M, Dietinger V, Kretzschmar HA. Altered long noncoding RNA expression precedes the course of Parkinson's disease—a preliminary report. Molecular Neurobiology May 28, 2017;54(4):2869—77.
[81] Jiang J, Piao X, Hu S, Gao J, Bao M. LncRNA H19 diminishes dopaminergic neuron loss by mediating microRNA-301b-3p in Parkinson's disease via the HPRT1-mediated Wnt/β-catenin signaling pathway. Aging May 20, 2020;12(10):8820—36.
[82] Feng Z, Zhang L, Wang S, Hong Q. Circular RNA circDLGAP4 exerts neuroprotective effects via modulating miR-134-5p/CREB pathway in Parkinson's disease. Biochemical and Biophysical Research Communications February 2020;522(2):388—94.
[83] Kumar L, Shamsuzzama, Jadiya P, Haque R, Shukla S, Nazir A. Functional characterization of novel circular RNA molecule, circzip-2 and its synthesizing gene zip-2 in *C. elegans* model of Parkinson's disease. Molecular Neurobiology August 23, 2018;55(8):6914—26.
[84] Jia E, Zhou Y, Liu Z, Wang L, Ouyang T, Pan M, et al. Transcriptomic profiling of circular RNA in different brain regions of Parkinson's disease in a mouse model. International Journal of Molecular Sciences April 24, 2020;21(8):3006.
[85] Rachner TD, Khosla S, Hofbauer LC. Osteoporosis: now and the future. The Lancet April 2011;377(9773):1276—87.
[86] Xu R, Shen X, Si Y, Fu Y, Zhu W, Xiao T, et al. MicroRNA-31a-5p from aging BMSCs links bone formation and resorption in the aged bone marrow microenvironment. Aging Cell August 2018;17(4):e12794.
[87] Davis C, Dukes A, Drewry M, Helwa I, Johnson MH, Isales CM, et al. MicroRNA-183-5p increases with age in bone-derived extracellular vesicles, suppresses bone marrow stromal (stem) cell proliferation, and induces stem cell senescence. Tissue Engineering Part A November 2017;23(21—22):1231—40.
[88] Li H, Yang F, Wang Z, Fu Q, Liang A. MicroRNA-21 promotes osteogenic differentiation by targeting small mothers against decapentaplegic 7. Molecular Medicine Reports July 2015;12(1):1561—7.
[89] Wang S, Tang C, Zhang Q, Chen W. Reduced miR-9 and miR-181a expression down-regulates Bim concentration and promote osteoclasts survival. International Journal of Clinical and Experimental Pathology 2014;7(5):2209—18.
[90] Zhang HG, Wang XB, Zhao H, Zhou CN. MicroRNA-9-5p promotes osteoporosis development through inhibiting osteogenesis and promoting adipogenesis via targeting Wnt3a. European Review for Medical and Pharmacological Sciences January 2019;23(2):456—63.
[91] Ko NY, Chen LR, Chen KH. The role of micro RNA and long-non-coding RNA in osteoporosis. International Journal of Molecular Sciences July 10, 2020;21(14):4886.
[92] Jiang LB, Tian L, Zhang CG. Bone marrow stem cells-derived exosomes extracted from osteoporosis patients inhibit osteogenesis via microRNA-21/SMAD7. European Review for Medical and Pharmacological Sciences October 2018;22(19):6221—9.
[93] Seeliger C, Karpinski K, Haug AT, Vester H, Schmitt A, Bauer JS, et al. Five freely circulating miRNAs and bone tissue miRNAs are associated with osteoporotic fractures. Journal of Bone and Mineral Research August 2014;29(8):1718—28.
[94] Yavropoulou MP, Anastasilakis AD, Makras P, Tsalikakis DG, Grammatiki M, Yovos JG. Expression of microRNAs that regulate bone turnover in the serum of postmenopausal women with low bone mass and vertebral fractures. European Journal of Endocrinology February 2017;176(2):169—76.
[95] Kelch S, Balmayor ER, Seeliger C, Vester H, Kirschke JS, van Griensven M. miRNAs in bone tissue correlate to bone mineral density and circulating miRNAs are gender independent in osteoporotic patients. Scientific Reports November 20, 2017;7(1):15861.

[96] Li H, Wang Z, Fu Q, Zhang J. Plasma miRNA levels correlate with sensitivity to bone mineral density in postmenopausal osteoporosis patients. Biomarkers November 1, 2014;19(7):553–6.
[97] Yang N, Wang G, Hu C, Shi Y, Liao L, Shi S, et al. Tumor necrosis factor α suppresses the mesenchymal stem cell osteogenesis promoter miR-21 in estrogen deficiency-induced osteoporosis. Journal of Bone and Mineral Research March 2013;28(3):559–73.
[98] Wang C, Sun W, Ling S, Wang Y, Wang X, Meng H, et al. AAV-Anti-miR-214 prevents collapse of the femoral head in osteonecrosis by regulating osteoblast and osteoclast Activities. Molecular Therapy—Nucleic Acids December 2019;18:841–50.
[99] Li D, Liu J, Guo B, Liang C, Dang L, Lu C, et al. Osteoclast-derived exosomal miR-214-3p inhibits osteoblastic bone formation. Nature Communications March 7, 2016;7(1):10872.
[100] Li CJ, Cheng P, Liang MK, Chen YS, Lu Q, Wang JY, et al. MicroRNA-188 regulates age-related switch between osteoblast and adipocyte differentiation. Journal of Clinical Investigation April 1, 2015;125(4):1509–22.
[101] Li K, Chen S, Cai P, Chen K, Li L, Yang X, et al. MiRNA-483–5p is involved in the pathogenesis of osteoporosis by promoting osteoclast differentiation. Molecular and Cellular Probes February 2020;49:101479.
[102] Teng Z, Zhu Y, Zhang X, Teng Y, Lu S. Osteoporosis is characterized by altered expression of exosomal long non-coding RNAs. Frontiers in Genetics November 12, 2020:11.
[103] Jin C, Zheng Y, Huang Y, Liu Y, Jia L, Zhou Y. Long non-coding RNA *MIAT* knockdown promotes osteogenic differentiation of human adipose-derived stem cells. Cell Biology International January 2017;41(1):33–41.
[104] Sun H, Peng G, Wu H, Liu M, Mao G, Ning X, et al. Long non-coding RNA MEG3 is involved in osteogenic differentiation and bone diseases (Review). Biomedical Reports July 2020;13(1):15–21.
[105] Zhang W, Dong R, Diao S, Du J, Fan Z, Wang F. Differential long noncoding RNA/mRNA expression profiling and functional network analysis during osteogenic differentiation of human bone marrow mesenchymal stem cells. Stem Cell Research and Therapy December 7, 2017;8(1):30.
[106] Li H, Zhang Z, Chen Z, Zhang D. Osteogenic growth peptide promotes osteogenic differentiation of mesenchymal stem cells mediated by LncRNA AK141205-induced upregulation of CXCL13. Biochemical and Biophysical Research Communications October 2015;466(1):82–8.
[107] Wang L, Qi L. The role and mechanism of long non-coding RNA H19 in stem cell osteogenic differentiation. Molecular Medicine December 12, 2021;27(1):86.
[108] Zhao K, Zhao Q, Guo Z, Chen Z, Hu Y, Su J, et al. Hsa_Circ_0001275: a potential novel diagnostic biomarker for postmenopausal osteoporosis. Cellular Physiology and Biochemistry 2018;46(6):2508–16.
[109] Cao G, Meng X, Han X, Li J. Exosomes derived from circRNA Rtn4-modified BMSCs attenuate TNF-α-induced cytotoxicity and apoptosis in murine MC3T3-E1 cells by sponging miR-146a. Bioscience Reports May 29, 2020;40(5).
[110] Yu L, Liu Y. circRNA_0016624 could sponge miR-98 to regulate BMP2 expression in postmenopausal osteoporosis. Biochemical and Biophysical Research Communications August 2019;516(2):546–50.
[111] Wang XB, Li PB, Guo SF, Yang QS, Chen ZX, Wang D, et al. circRNA_0006393 promotes osteogenesis in glucocorticoid-induced osteoporosis by sponging miR-145-5p and upregulating FOXO1. Molecular Medicine Reports July 12, 2019;20(3):2851–8. https://doi.org/10.3892/mmr.2019.10497. PMID: 31322188.
[112] Xu X, Chen Y, Tan B, Wang D, Yuan Z, Wang F. Circular RNA circ_0011269 sponges miR-122 to regulate RUNX2 expression and promotes osteoporosis progression. Journal of Cellular Biochemistry December 3, 2020;121(12):4819–26.
[113] Buckwalter JA, Saltzman C, Brown T. The impact of osteoarthritis. Clinical Orthopaedics and Related Research October 2004;427:S6–15.
[114] Zhou Q fa, Cai Y zhi, jin LX. The dual character of exosomes in osteoarthritis: Antagonists and therapeutic agents. Acta Biomaterialia March 2020;105:15–25.
[115] Asahara H. [miRNAs in cartilage development]. Clinical Calcium May 2012;22(5):653–7.

[116] Mihanfar A, Shakouri SK, Khadem-Ansari MH, Fattahi A, Latifi Z, Nejabati HR, et al. Exosomal miRNAs in osteoarthritis. Molecular Biology Reports June 10, 2020;47(6):4737–48.
[117] Mao G, Zhang Z, Hu S, Zhang Z, Chang Z, Huang Z, et al. Exosomes derived from miR-92a-3p-overexpressing human mesenchymal stem cells enhance chondrogenesis and suppress cartilage degradation via targeting WNT5A. Stem Cell Research and Therapy December 26, 2018;9(1):247.
[118] Jiang S de, Lu J, Deng Z han, sheng LY, hua LG. Long noncoding RNAs in osteoarthritis. Joint Bone Spine October 2017;84(5):553–6.
[119] Kang Y, Song J, Kim D, Ahn C, Park S, Chun CH, et al. PCGEM1 stimulates proliferation of osteoarthritic synoviocytes by acting as a sponge for miR-770. Journal of Orthopaedic Research March 2016;34(3):412–8.
[120] Xiao X, Zhou T, Guo S, Guo C, Zhang Q, Dong N, et al. LncRNA MALAT1 sponges miR-204 to promote osteoblast differentiation of human aortic valve interstitial cells through up-regulating Smad4. International Journal of Cardiology September 2017;243:404–12.
[121] Kim D, Song J, Han J, Kim Y, Chun CH, Jin EJ. Two non-coding RNAs, MicroRNA-101 and HOTTIP contribute cartilage integrity by epigenetic and homeotic regulation of integrin-α1. Cellular Signalling December 2013;25(12):2878–87.
[122] Fan X, Yuan J, Xie J, Pan Z, Yao X, Sun X, et al. Long non-protein coding RNA DANCR functions as a competing endogenous RNA to regulate osteoarthritis progression via miR-577/SphK2 axis. Biochemical and Biophysical Research Communications June 2018;500(3):658–64.
[123] Wang T, Liu Y, Wang Y, Huang X, Zhao W, Zhao Z. Long non-coding RNA XIST promotes extracellular matrix degradation by functioning as a competing endogenous RNA of miR-1277-5p in osteoarthritis. International Journal of Molecular Medicine June 12, 2019;44(2):630–42. https://doi.org/10.3892/ijmm.2019.4240. PMID: 31198977. PMCID: PMC6605283.
[124] Zhou Z, Du D, Chen A, Zhu L. Circular RNA expression profile of articular chondrocytes in an IL-1β-induced mouse model of osteoarthritis. Gene February 2018;644:20–6.
[125] Zhou Z bin, Du D, xiang HG, Chen A, Zhu L. Circular RNA Atp9b, a competing endogenous RNA, regulates the progression of osteoarthritis by targeting miR-138-5p. Gene March 2018;646:203–9.
[126] Wang Y, Liang J, Xu J, Wang X, Zhang X, Wang W, et al. Circulating exosomes and exosomal lncRNA HIF1A-AS1 in atherosclerosis. International Journal of Clinical and Experimental Pathology 2017;10(8):8383–8.
[127] Basatemur GL, Jørgensen HF, Clarke MCH, Bennett MR, Mallat Z. Vascular smooth muscle cells in atherosclerosis. Nature Reviews Cardiology December 26, 2019;16(12):727–44.
[128] Libby P, Ridker PM, Hansson GK. Progress and challenges in translating the biology of atherosclerosis. Nature May 18, 2011;473(7347):317–25.
[129] Gomez D, Owens GK. Smooth muscle cell phenotypic switching in atherosclerosis. Cardiovascular Research July 15, 2012;95(2):156–64.
[130] Shan K, Jiang Q, Wang XQ, Wang YNZ, Yang H, Yao MD, et al. Role of long non-coding RNA-RNCR3 in atherosclerosis-related vascular dysfunction. Cell Death and Disease June 2, 2016;7(6):e2248. e2248.
[131] Yang Y, Cai Y, Wu G, Chen X, Liu Y, Wang X, et al. Plasma long non-coding RNA, CoroMarker, a novel biomarker for diagnosis of coronary artery disease. Clinical Science October 1, 2015;129(8):675–85.
[132] Chen L, Yang W, Guo Y, Chen W, Zheng P, Zeng J, et al. Exosomal lncRNA GAS5 regulates the apoptosis of macrophages and vascular endothelial cells in atherosclerosis. PLoS One September 25, 2017;12(9):e0185406.
[133] Liang W, Fan T, Liu L, Zhang L. Knockdown of growth-arrest specific transcript 5 restores oxidized low-density lipoprotein-induced impaired autophagy flux via upregulating miR-26a in human endothelial cells. European Journal of Pharmacology January 2019;843:154–61.
[134] Che J, Liang B, Zhang Y, Wang Y, Tang J, Shi G. Kaempferol alleviates ox-LDL-induced apoptosis by up-regulation of autophagy via inhibiting PI3K/Akt/mTOR pathway in human endothelial cells. Cardiovascular Pathology 2017;31:57–62.

[135] Cao Q, Guo Z, Yan Y, Wu J, Song C. Exosomal long noncoding RNAs in aging and age-related diseases. IUBMB Life December 6, 2019;71(12):1846–56.

[136] Palmer AK, Tchkonia T, LeBrasseur NK, Chini EN, Xu M, Kirkland JL. Cellular senescence in type 2 diabetes: a therapeutic opportunity. Diabetes July 1, 2015;64(7):2289–98.

[137] Sone H, Kagawa Y. Pancreatic beta cell senescence contributes to the pathogenesis of type 2 diabetes in high-fat diet-induced diabetic mice. Diabetologia January 29, 2005;48(1):58–67.

[138] Katayama M, Wiklander OPB, Fritz T, Caidahl K, El-Andaloussi S, Zierath JR, et al. Circulating exosomal miR-20b-5p is elevated in type 2 diabetes and could impair insulin action in human skeletal muscle. Diabetes March 2019;68(3):515–26.

[139] Ruan Y, Lin N, Ma Q, Chen R, Zhang Z, Wen W, et al. Circulating LncRNAs analysis in patients with type 2 diabetes reveals novel genes influencing glucose metabolism and islet β-cell function. Cellular Physiology and Biochemistry 2018;46(1):335–50.

[140] Tello-Flores VA, Valladares-Salgado A, Ramírez-Vargas MA, Cruz M, del-Moral-Hernández O, Cahua-Pablo JÁ, et al. Altered levels of MALAT1 and H19 derived from serum or serum exosomes associated with type-2 diabetes. Noncoding RNA Research June 2020;5(2):71–6.

[141] Ceribelli A, Nahid MA, Satoh M, Chan EKL. MicroRNAs in rheumatoid arthritis. FEBS Letters December 1, 2011;585(23):3667–74.

[142] Guidelines for the management of rheumatoid arthritis: 2002 Update. Arthritis and Rheumatism February 2002;46(2):328–46.

[143] Withrow J, Murphy C, Liu Y, Hunter M, Fulzele S, Hamrick MW. Extracellular vesicles in the pathogenesis of rheumatoid arthritis and osteoarthritis. Arthritis Research and Therapy December 1, 2016;18(1):286.

[144] Kim SJ, Chen Z, Essani AB, Elshabrawy HA, Volin MV, Volkov S, et al. Identification of a novel TLR7 endogenous ligand in RA synovial fluid that can provoke arthritic joint inflammation. Arthritis and Rheumatology May 2016;68(5):1099–110. https://doi.org/10.1002/art.39544. PMID: 26662519. PMCID: PMC5221504.

[145] Wang Y, Zheng F, Gao G, Yan S, Zhang L, Wang L, et al. MiR-548a-3p regulates inflammatory response via TLR4/NF-κB signaling pathway in rheumatoid arthritis. Journal of Cellular Biochemistry February 2019;120(2):1133–40.

[146] Qin W, Wang TH, Xie BH, Sun QQ, Huang H, Zhao BJ, et al. Plasma long non-coding RNA expression profiles in patients with rheumatoid arthritis. Clinical Laboratory 2019;65(08/2019).

[147] Yang CA, Li JP, Yen JC, Lai IL, Ho YC, Chen YC, et al. lncRNA NTT/PBOV1 Axis promotes monocyte differentiation and is elevated in rheumatoid arthritis. International Journal of Molecular Sciences September 18, 2018;19(9):2806.

[148] Liu Y, Yang L, Xu Q, Lu X, Ma T, Huang C, et al. Long noncoding RNA MEG3 regulates rheumatoid arthritis by targeting NLRC5. Journal of Cellular Physiology August 15, 2019;234(8):14270–84.

[149] Xu D, Jiang Y, Yang L, Hou X, Wang J, Gu W, et al. Long noncoding RNAs expression profile and functional networks in rheumatoid arthritis. Oncotarget November 10, 2017;8(56):95280–92.

[150] Yan S, Wang P, Wang J, Yang J, Lu H, Jin C, et al. Long non-coding RNA HIX003209 promotes inflammation by sponging miR-6089 via TLR4/NF-κB signaling pathway in rheumatoid arthritis. Frontiers in Immunology September 18, 2019:10.

[151] Luo Q, Zeng L, Zeng L, Rao J, Zhang L, Guo Y, et al. Expression and clinical significance of circular RNAs hsa_circ_0000175 and hsa_circ_0008410 in peripheral blood mononuclear cells from patients with rheumatoid arthritis. International Journal of Molecular Medicine February 13, 2020;45(4):1203–12. https://doi.org/10.3892/ijmm.2020.4498. PMID: 32124964.

[152] Yang J, Qi M, Fei X, Wang X, Wang K. Hsa_circRNA_0088036 acts as a ceRNA to promote bladder cancer progression by sponging miR-140-3p. Cell Death and Disease April 8, 2022;13(4):322.

[153] Chen Y, Xu X, Li X, Zhong J, Wu B, Shang J, et al. Identification of circular RNAs hsa_-circ_0140271 in peripheral blood mononuclear cells as a novel diagnostic biomarker for female rheumatoid arthritis. Journal of Orthopaedic Surgery and Research December 30, 2021;16(1):647.

[154] Luo Q, Liu J, Fu B, Zhang L, Guo Y, Huang Z, et al. Circular RNAs Hsa_circ_0002715 and Hsa_circ_0035197 in peripheral blood are novel potential biomarkers for new-onset rheumatoid arthritis. Disease Markers 2019:1–12. 2019 Oct 29.

[155] Yang Y, Lin S, Yang Z, Huang Y, Zhan F. Circ_0001947 promotes cell proliferation, invasion, migration and inflammation and inhibits apoptosis in human rheumatoid arthritis fibroblast-like synoviocytes through miR-671-5p/STAT3 axis. Journal of Orthopaedic Surgery and Research December 29, 2022;17(1):54.

[156] Chen S, Luo Z, Chen X. Hsa_circ_0044235 regulates the pyroptosis of rheumatoid arthritis via MiR-135b-5p-SIRT1 axis. Cell Cycle June 18, 2021;20(12):1107–21.

[157] Feng L, Jing W, Jin S, Wang B. Circ_0088194 regulates proliferation, migration, apoptosis, and inflammation by miR-30a-3p/ADAM10 Axis in rheumatoid arthritis fibroblastic synovial cells. Inflammation February 2, 2023;46(1):161–74.

[158] Wen J ting, Liu J, Wan L, Xin L, Guo J chen, Sun Y qiu, et al. Triptolide inhibits cell growth and inflammatory response of fibroblast-like synoviocytes by modulating hsa-circ-0003353/microRNA-31-5p/CDK1 axis in rheumatoid arthritis. International Immunopharmacology May 2022;106:108616.

[159] Li Z, Wang J, Lin Y, Fang J, Xie K, Guan Z, et al. Newly discovered circRNAs in rheumatoid arthritis, with special emphasis on functional roles in inflammatory immunity. Frontiers in Pharmacology October 7, 2022:13.

[160] Compston A, Coles A. Multiple sclerosis. The Lancet October 2008;372(9648):1502–17.

[161] Ebrahimkhani S, Vafaee F, Young PE, Hur SSJ, Hawke S, Devenney E, et al. Exosomal microRNA signatures in multiple sclerosis reflect disease status. Scientific Reports October 30, 2017;7(1):14293.

[162] Scalfari A, Neuhaus A, Daumer M, Ebers GC, Muraro PA. Age and disability accumulation in multiple sclerosis. Neurology September 27, 2011;77(13):1246–52.

[163] Selmaj I, Cichalewska M, Namiecinska M, Galazka G, Horzelski W, Selmaj KW, et al. Global exosome transcriptome profiling reveals biomarkers for multiple sclerosis. Annals of Neurology May 2017;81(5):703–17.

[164] Oldoni E. The role of cellular and exosomal neural-derived long non-coding RNA (lncRNA) in multiple sclerosis: potential biomarkers of disease susceptibility and progression. 2016.

[165] Pahlevan Kakhki M, Nikravesh A, Shirvani Farsani Z, Sahraian MA, Behmanesh M. *HOTAIR* but not *ANRIL* long non-coding RNA contributes to the pathogenesis of multiple sclerosis. Immunology April 2018;153(4):479–87.

[166] Duan C, Liu Y, Li Y, Chen H, Liu X, Chen X, et al. Sulfasalazine alters microglia phenotype by competing endogenous RNA effect of miR-136-5p and long non-coding RNA HOTAIR in cuprizone-induced demyelination. Biochemical Pharmacology September 2018;155:110–23.

CHAPTER TEN

Exosomal RNA in cardiovascular diseases

Paramjeet Singh, Suparna Ghosh, Sampriti Sarkar, Avik Acharya Chowdhury
Department of Biosciences, JIS University, Kolkata, West Bengal, India

1. Role of exosomal RNAs in cardiovascular diseases

Despite advances in therapy, cardiovascular disorders and diseases remain one of the primary causes of fatalities and morbidity worldwide. Some of the prevalent of these ailments are atherosclerosis, hypertension, heart failure (HF), acute myocardial infarction (AMI), and atrial fibrillation (AF) [1,2]. In recent studies, it has been found that exosomal noncoding RNAs (ncRNAs) build up with the cardiovascular disease (CVD) pathogenesis, and this positive correlation provides novel insights into the mechanisms and the target for diagnosis, therapy, and treatment of these ailments. Furthermore, exosomal RNAs have evolved as regulators of dyslipidemia, which leads to an increased risk of atherosclerotic CVDs. An example is the miR-26a that shows a downregulation in the exosomes in circulation in obese mice and humans [3–8]. On the other hand, a lower level of plasma cholesterol, hepatic triglycerides, lipid droplets as well as low-density lipoprotein (LDL) and high-density lipoprotein (HDL) is reported in the adipose tissues when there is an upregulation of miR-26a [9–12].

1.1 Exosomal RNA in atherosclerosis

Atherosclerosis is a disorder associated with age with concurrent chronic inflammation caused by the immune system with a characteristic accumulation of lipid-rich plaques in the walls of the artery. Atherosclerosis is a predominant cause of the occurrence of CVDs that may give rise to serious clinical conditions such as strokes and myocardial infarctions. Several studies have gathered enough evidence to show that exosomal RNAs play a major role in the regulation of the pathophysiological processes of atherosclerosis [13,14]. Exosomal miRNAs have the ability to control inflammation within the artery wall. Examples include the ability of some miRNAs transported by exosomes to target and inhibit the expression of proinflammatory genes, hence reducing inflammation that fuels atherosclerosis. Cellular dysfunction: Within blood artery walls, endothelial cells (ECs), smooth muscle cells, and macrophages may behave differently as a result of exosomal miRNAs. The development of atherosclerotic plaques depends heavily on

Exosomal RNA
ISBN 978-0-443-14008-2
https://doi.org/10.1016/B978-0-443-14008-2.00007-3

these cells. Besides, they can affect the proliferation and migration of smooth muscle cells, which are implicated in the creation of the fibrous cap that covers atherosclerotic plaques. This fibrous cap's resilience is critical for preventing plaque rupture and has the potential to influence angiogenesis, or the development of new blood vessels, which can be important in the progression of atherosclerosis and the delivery of nutrients to developing plaques [15].

Several investigations have reported that exosomal miRNAs secreted by different cell types are detectable in the blood and play an important factor in the regulation of pathogenic atherosclerosis. The levels of thrombin-activated platelet-derived exosomal miR-223, miR-339, and miR-21 were shown to be highly upregulated. They were able to be transported into vascular smooth muscle cells (VSMCs) and reduced the proliferation of VSMCs triggered by platelet-derived growth factor. Another study revealed that miR-223 reduced inflammation by the TNF-stimulated EC by lowering ICAM-1 expression. These data indicate that exosomal miR-223 may protect against atherosclerosis by reducing the vascular inflammatory response [16—18].

The interaction of exosomal miRNAs and ECs is critical in the progression of atherosclerosis. Exosomal miR-92a, which was increased by a combinatorial effect of oxidized LDL and low shear stress in an atherosclerotic mouse model, encouraged endothelial inflammation and the development of atherosclerotic plaques [19]. Furthermore, it was discovered that exosomal miR-342-5p produced by adipose-derived MSCs had an antiatherosclerotic impact by boosting H_2O_2-induced EC apoptosis and protecting against EC damage. On the contrary, miR-155 produced from VSMCs damaged the tight junction and integrity of ECs, resulting in EC injury and perhaps promoting atherosclerosis progression [20—22].

The malfunctioning of VSMCs caused by macrophage-derived exosomal miRNAs is also significant in the progression of atherosclerosis. According to one study, nicotine-stimulated macrophages expressed more exosomal miR-21-3p, which might be ingested by nearby VSMCs. This action increased the ability of VSMCs to proliferate and migrate, hastening the development of atherosclerosis. Exosomal miR-106a-3p generated from oxidized LDL-incubated macrophages was also shown to increase VSMC proliferation and block apoptosis, potentially protecting against atherosclerosis [23—25].

Macrophages primarily are responsible for the inflammatory response in atherosclerosis. Exosomal miRNAs influence macrophage function, which has received a lot of interest. The miR-let7 family was discovered to be substantially enriched with exosomes generated by mesenchymal stem cells (MSCs) and to suppress macrophage invasion via the IGF2BP1 pathway. Furthermore, it stimulates the polarization of M2 macrophages through HMGA2/NF-B pathway, which slows the development of atherosclerosis [26,27].

Several studies have documented the involvement of exosomal ncRNAs in the pathogenesis of atherosclerosis. Additionally, the expression of long ncRNA GAS5 has been observed to be significantly elevated in exosomes obtained from animals or patients with atherosclerosis. Later on, in vitro experiments have identified that GAS5 promotes apoptotic death of macrophages and ECs induced by oxidized LDL, thereby exacerbating the development of atherosclerosis. Conversely, the expression MALAT1, a lncRNA known for its role in tumorigenesis, is downregulated in oxidized LDL-induced ECs. Increased MALAT1 expression in dendritic cells due to overexpression of exosomal MALAT1 derived from ECs and sensitization of the NRF2 signaling pathway, resulting in the inhibition of dendritic cell maturation and reduction of reactive oxygen species production. This shows that downregulation of exosomal MALAT1 from oxidized LDL-induced ECs may hasten atherosclerosis progression by increasing dendritic cell maturation. Interestingly, two conflicting results were found in two different studies. One of the studies reported that increased expression of EC-derived exosomal MALAT1 stimulates the development of neutrophil extracellular traps and accelerates atherosclerosis progression; another study found that the same exosomal MALAT1 enhances M2 macrophage polarization and may protect people against atherosclerosis [26,28,29].

Several studies have investigated the involvement of exosomal circular RNAs (circRNAs) in the pathogenesis of atherosclerosis. One of these studies revealed that in people with unstable plaque atherosclerosis, the expression of exosomal circRNA-0006896 in the blood is positively correlated with levels of triglycerides, LDL, and C-reactive protein. Moreover, the study found that increased levels of circulating exosomal circRNA-0006896 lead to increased expression of the same circRNA in ECs, resulting in enhanced capacities for cell proliferation and migration. This effect is associated with the upregulation of the DNMT1 gene and phosphorylation of the STAT3 protein, which may prevent vulnerable plaque formation. These findings suggest a potential role for exosomal circRNAs in the development and advancement of atherosclerosis [28,30].

1.2 Exosomal RNAs in hypertension

Hypertension is a critical risk factor for CVD and total mortality. Exosomal ncRNAs have been implicated in hypertension development by modulating various cellular and molecular events, such as exosomal miRNAs have the ability to affect the function of ECs, which line blood arteries. Endothelial dysfunction is a major factor to high blood pressure in hypertension. Exosomal miRNAs may influence endothelium health by influencing mechanisms that affect blood vessel tone and blood pressure, such as inflammation, oxidative stress, and nitric oxide synthesis. Also, these miRNAs have the ability to influence components of the renin—angiotensin system (RAS), a hormonal system that is important in blood pressure regulation. Hypertension is exacerbated by RAS dysfunction. Exosomal miRNAs may influence the balance of vasoconstriction and vasodilation by targeting genes implicated in RAS components. Besides, exosomal miRNAs may also influence the sympathetic nervous system, which regulates blood pressure through its effects on heart rate and blood vessel tone [31].

Exosomal miRNAs from various sources have been revealed to have an impact on the onset and progression of hypertension, according to recent studies. Exosomal miR-17-5p and miR-425-5p were found to be expressed at higher levels in the plasma of spontaneously hypertensive rats (SHRs) and normotensive Wistar—Kyoto rats (WKYs), respectively, and were associated with inflammation and blood pressure homeostasis [9—11].

In contrast, low expression of exosomal miR-155-5p, derived from aortic adventitial fibroblast, enhanced vascular angiotensin-converting enzyme and angiotensin II expression, initiated vascular smooth muscle cell proliferation and remodeling, and increased blood pressure. Macrophages-derived exosomal miRNAs were also found to play crucial roles in hypertension, with exosomal miR-27a from THP-1 cells impairing vasodilation and increasing blood pressure, and exosomal miR-106b-5p derived from macrophages promotes inflammation-induced hypertension by blocking transcription factors E2f1 and Pde3b [32,33].

1.3 Exosomal RNAs in acute myocardial infarction

AMI is a severe disorder caused by the sudden blockage of the coronary artery, leading to myocardial necrosis and cardiomyocyte loss. This can have serious clinical implications. Emerging research suggests that exosomal ncRNAs play a vital part in the pathophysiology of AMI and may be considered as valuable biomarkers for diagnosing and treating this condition [34—36]. Exosomal miRNAs generated by injured heart tissue can be detected in the bloodstream, indicating tissue damage and inflammation. These miRNAs can indicate the severity of myocardial injury and inflammation produced by a heart attack. Certain miRNAs delivered by exosomes are linked to inflammatory and cell death pathways, adding to the overall harm. Following an AMI, exosomal miRNAs can influence heart tissue repair and remodeling. Exosomes emitted by healthy cells, stem cells, or progenitor cells may include miRNAs that enhance angiogenesis (the production of new blood vessels), minimize fibrosis (the formation of scar tissue), and encourage cardiomyocyte regeneration [37].

AMI is associated with increased expression of various exosomal miRNAs, including miR-125b, miR-499, miR-133, miR-22, miR-21, and miR-301. It is becoming evident that exosomal miRNAs play a crucial role in the development of AMI by modulating various pathological pathways. Therefore, these exosomal miRNAs may serve as potential biomarkers for AMI diagnosis and treatment [38,39].

Numerous exosomal miRNAs (miR-125b, miR-25-3p, miR-144, miR-126, and miR-146a) play a decisive part in AMI through the suppression of atherosclerotic effect via the inhibition of myocardial apoptosis and promoting ischemic cardiac repair. However, the levels of cardiomyocytes-derived exosomal miR-499 and miR-133a are upregulated in AMI patients, indicating that these miRNAs may act as determining factor in the progression of AMI [38,40,41]. Other cell types including MSCs also release exosomal miRNAs that hold immense importance on AMI. For example, exosomal miR-22 from MSCs protects cardiomyocytes against apoptosis, reduces cardiac fibrosis, and promotes cardiomyocyte survival. Exosomal miR-21 from MSCs also functions to protect cardiac cells by suppressing apoptosis and increasing EC angiogenesis [42–44].

Exosomal miRNAs have demonstrated potential as indicators for the early diagnosis and prognosis of AMI. Changes in certain miRNA profiles carried by exosomes can be detected in the blood quickly after a heart attack, assisting in the rapid diagnosis and assessment of the event's severity. Exosomal miRNAs may also be involved in ischemia–reperfusion damage, which happens when blood flow is restored to previously ischemic (oxygen-deprived) tissue. Exosomal miRNAs can influence cell survival, inflammation, and oxidative stress in response to reperfusion [37].

Recent studies have also gathered significant evidence on exosomal long noncoding RNAs (lncRNAs) that justify their critical role in AMI. Exosomal MALAT1 from hyperbaric oxygen-induced cardiomyocytes increases after AMI and enhances neovascularization by suppressing miR-92a expression. In addition, it was discovered that lncRNA H19 generated from exosomes produced by MSCs following atorvastatin treatment enhanced cardiac function recovery, decreased infarct size, and decreased cardiomyocyte apoptosis. In contrast, exosomal lncRNA AK139128 produced by cardiomyocytes in a hypoxic environment encourages apoptotic cell death while preventing the invasion, migration, and proliferation of cardiac fibroblasts. The study of exosomal ncRNAs offers a promising avenue for the diagnosis as well as the treatment of AMI [45–47].

1.4 Exosomal RNAs in heart failure

HF is caused by the dysfunction of either myocardial systole or diastole, leading to a reduction in cardiac blood output and inadequate blood flow in either the pulmonary or systemic circulation. This condition is characterized by several mechanisms, such as myocardial inflammation, remodeling, autophagy, and apoptosis. Several research studies have elucidated that the pathogenesis of HF is regulated by exosomal ncRNAs [48,49]. Exosomal miRNAs have been identified as possible indicators for HF. Exosome miRNA profiles from blood or other body fluids can indicate the existence and severity of HF. Changes in these miRNA patterns may aid in illness detection and monitoring also. Exosomal miRNAs play a role in the regulation of cardiac remodeling, which happens when the heart undergoes structural changes as a result of stress or damage. Certain miRNAs carried by exosomes can influence fibrosis, hypertrophy (heart muscle cell enlargement), and apoptosis (cell death), all of which are hallmarks of cardiac remodeling in HF [50].

Serum levels of different exosomal miRNAs have been found to be changed throughout the etiology of HF. In particular, the exosomal miR-425 and miR-744 levels were found to be decreased in HF patients.

Exosomal miRNAs can affect inflammation and immunological responses, which contribute to the progression of HF. These miRNAs have the potential to influence cytokine production, macrophage activation, and other immune-related processes associated with HF and have the ability to regulate angiogenesis. Impaired blood supply adds to tissue damage in HF. Exosomal miRNAs have the capacity to control angiogenic pathways, encouraging blood vessel formation and improving tissue perfusion.

These miRNAs can enhance communication and signaling between different cell types in the heart. Exosomes released by cardiac cells, for example, might carry miRNAs that influence neighboring cells, such as fibroblasts or ECs, and affect their function [50]. Angiotensin II-treated cardiac fibroblasts have also been found to downregulate exosomal miR-744 and miR-425, and this downregulation has been positively correlated with the production of collagen 1 and -SMA, leading to cardiac fibroblast fibrosis. Additional research has demonstrated that exosomal miR-425 alongside miR-744 suppresses collagen and fibrogenesis induced by angiotensin-II by targeting TGFβ1. This prevents myocardial remodeling. Overall, these observations clearly indicate that exosomal miRNAs, particularly miR-425 and miR-744, may serve as potential biomarkers as well as therapeutic targets in the treatment of HF [41,51,52].

1.5 Exosomal RNAs in atrial fibrillation

AF is defined as a type of cardiac arrhythmia associated with structural and electrical remodeling. Recent research suggests that ncRNAs, particularly miRNAs, play a major part in the pathological processes involved in AF. Exosomal miRNAs have been involved in the molecular mechanisms underpinning AF pathogenesis and substrate formation. Exosome-borne miRNAs can target genes involved in ion channel modulation, calcium handling, fibrosis, and inflammation. Dysregulation of these mechanisms contributes to substrate formation in the atria, resulting in an environment favorable to the development of AF. Exosomal miRNAs can influence electrical remodeling in atrial cells, which is a defining feature of AF. These miRNAs may influence ion channel expression and function, hence influencing the atrial cell's capacity to generate and conduct electrical signals correctly [53].

One study discovered that AF cardiomyocytes exhibit low expression levels of miR-320d, which is linked to increased apoptosis and reduced level of cell viability reportedly in cardiomyocytes. However, the transfection of adipose tissue—derived MSC exosomes with miR-320d significantly reversed the effects of AF on cardiomyocytes by inhibiting STAT3. In another study, it was shown that AF patients' exosomal miR-107 expression was considerably higher than that of patients without AF [54,55]. Exosomes produced from AF or miR-107 mimics decreased cell survival and migration when incubated with ECs. Cell apoptosis was promoted by the same via the regulation of the miR-107/USP14 pathway. Moreover, studies have demonstrated that miR-21-3p loaded in myofibroblast-derived exosomes increases the susceptibility of AF by enhancing the L-type calcium channel Cav1.2 expression [56].

Exosomal ncRNAs have become important regulators in a variety of cardiovascular conditions. Among these ncRNAs, exosomal microRNAs (miRNAs) have been extensively studied in the context of CVDs. These miRNAs exhibit a wide range of functions, which includes the suppression of inflammatory response in the vascular system and alteration cellular pathophysiology pathways, and more. Additionally, exosomal lncRNAs, namely GAS5, MALAT1, H19, and AK139128, have been implicated in AS and AMI. In contrast, the role of exosomal circular RNAs (circRNAs) in the pathophysiology of AS is not as well understood, with only a limited number of studies examining their involvement.

2. Conclusion

Based on extensive literature survey, we have come across the fact that exosomes hold a huge potential as important biomarker of a large number of diseased conditions in human biology. Exosomes aid in the complicated cellular communication that can contribute to CVD, alongside the disease of immunological system, metabolism, central nervous system, and renal system by transporting a variety of information via RNA, proteins, lipids, etc. Exosomal RNA content including micro (mi)RNA, lncRNA, circular RNA, and mRNA has been implicated in different diseeases (Fig. 10.1). We have focused on the role of the Exosomal RNA in different CVDs. Still a great deal of information is lacking to understand the exact role of exosomal RNA in CVDs, which will contribute in the future therapeutic interventions.

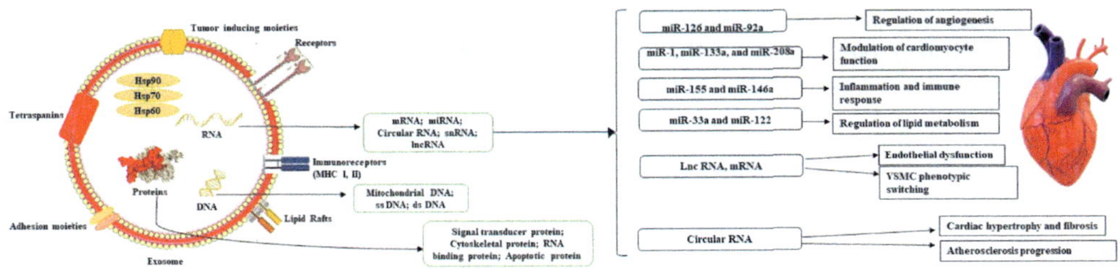

Figure 10.1 Exosomal RNA's interplay in cardiovascular disorders.

References

[1] Zheng D-H, Han Z-Q, Wang X-X, Ma D, Zhang J. Erythropoietin attenuates high glucose-induced oxidative stress and inhibition of osteogenic differentiation in periodontal ligament stem cell (PDLSCs). Chemico-Biological Interactions 2019;305:40–7. https://doi.org/10.1016/j.cbi.2019.03.007.

[2] Libby P, Ridker PM, Hansson GK. Inflammation in Atherosclerosis. Journal of the American College of Cardiology 2009;54:2129–38. https://doi.org/10.1016/j.jacc.2009.09.009.

[3] Barile L, Lionetti V, Cervio E, Matteucci M, Gherghiceanu M, Popescu LM, et al. Extracellular vesicles from human cardiac progenitor cells inhibit cardiomyocyte apoptosis and improve cardiac function after myocardial infarction. Cardiovascular Research 2014;103:530–41. https://doi.org/10.1093/cvr/cvu167.

[4] Davidson SM, Andreadou I, Barile L, Birnbaum Y, Cabrera-Fuentes HA, Cohen MV, et al. Circulating blood cells and extracellular vesicles in acute cardioprotection. Cardiovascular Research 2019;115:1156–66. https://doi.org/10.1093/cvr/cvy314.

[5] Libby P, Ridker PM, Hansson GK. Progress and challenges in translating the biology of atherosclerosis. Nature 2011;473:317–25. https://doi.org/10.1038/nature10146.

[6] Pfeifer P, Werner N, Jansen F. Role and function of MicroRNAs in extracellular vesicles in cardiovascular biology. BioMed Research International 2015:1–11. https://doi.org/10.1155/2015/161393. 2015.

[7] Qiao S, Zhang W, Yin Y, Wei Z, Chen F, Zhao J, et al. Extracellular vesicles derived from Krüppel-Like Factor 2-overexpressing endothelial cells attenuate myocardial ischemia-reperfusion injury by preventing Ly6C high monocyte recruitment. Theranostics 2020;10:11562–79. https://doi.org/10.7150/thno.45459.

[8] Rezaie J, Rahbarghazi R, Pezeshki M, Mazhar M, Yekani F, Khaksar M, et al. Cardioprotective role of extracellular vesicles: A highlight on exosome beneficial effects in cardiovascular diseases. Journal of Cellular Physiology 2019;234:21732–45. https://doi.org/10.1002/jcp.28894.

[9] Bress AP, Colantonio LD, Cooper RS, Kramer H, Booth JN, Odden MC, et al. Potential cardiovascular disease events prevented with adoption of the 2017 American College of Cardiology/American Heart Association Blood Pressure Guideline. Circulation 2019;139:24–36. https://doi.org/10.1161/CIRCULATIONAHA.118.035640.

[10] Brunström M, Carlberg B. Association of blood pressure lowering with mortality and cardiovascular disease across blood pressure levels: a systematic review and meta-analysis. JAMA International Medicine 2018;178:28–36. https://doi.org/10.1001/jamainternmed.2017.6015.

[11] Fuchs FD, Whelton PK. high blood pressure and cardiovascular disease. Hypertension 2020;75:285–92. https://doi.org/10.1161/HYPERTENSIONAHA.119.14240.

[12] Kopin L, Lowenstein CJ. Dyslipidemia. Annals of International Medicine 2017;167:ITC81. https://doi.org/10.7326/AITC201712050.

[13] O'Malley PG, Arnold MJ, Kelley C, Spacek L, Buelt A, Natarajan S, et al. Management of dyslipidemia for cardiovascular disease risk reduction: Synopsis of the 2020 updated U.S. Department of Veterans Affairs and U.S. Department of Defense Clinical Practice Guideline. Annals of International Medicine 2020;173:822–9. https://doi.org/10.7326/M20-4648.

[14] Plakht Y, Shiyovich A, Gilutz H. Predictors of long-term (10-year) mortality postmyocardial infarction: Age-related differences. Soroka Acute Myocardial Infarction (SAMI) Project. Journal of Cardiology 2015;65:216–23. https://doi.org/10.1016/j.jjcc.2014.06.001.

[15] Yang K, Xiao Q, Niu M, Pan X, Zhu X. Exosomes in atherosclerosis: Convergence on macrophages. International Journal of Biological Science 2022;18(8):3266–81. https://doi.org/10.7150/ijbs.71862.

[16] Benjamin EJ, Blaha MJ, Chiuve SE, Cushman M, Das SR, Deo R, et al. Heart disease and stroke statistics—2017 update: A report from the American Heart Association. Circulation 2017;135. https://doi.org/10.1161/CIR.0000000000000485.

[17] Zheng B, Yin WN, Suzuki T, Zhang XH, Zhang Y, Song L-L, et al. Exosome-mediated miR-155 transfer from smooth muscle cells to endothelial cells induces endothelial injury and promotes atherosclerosis. Molecular Therapautics 2017;25:1279–94. https://doi.org/10.1016/j.ymthe.2017.03.031.

[18] Zhang Z, Yi D, Zhou J, Zheng Y, Gao Z, Hu X, et al. Exosomal LINC01005 derived from oxidized low-density lipoprotein-treated endothelial cells regulates vascular smooth muscle cell phenotypic switch. BioFactors 2020;46:743–53. https://doi.org/10.1002/biof.1665.

[19] Zhu J, Liu B, Wang Z, Wang D, Ni H, Zhang L, et al. Exosomes from nicotine-stimulated macrophages accelerate atherosclerosis through miR-21-3p/PTEN-mediated VSMC migration and proliferation. Theranostics 2019;9:6901–19. https://doi.org/10.7150/thno.37357.

[20] Iaconetti C, Sorrentino S, De Rosa S, Indolfi C. Exosomal miRNAs in heart disease. Physiology 2016;31:16–24. https://doi.org/10.1152/physiol.00029.2015.

[21] Tan M, Yan H-B, Li J-N, Li W-K, Fu Y-Y, Chen W, et al. thrombin stimulated platelet-derived exosomes inhibit platelet-derived growth factor receptor-beta expression in vascular smooth muscle cells. Cell Physiology and Biochemistry 2016;38:2348–65. https://doi.org/10.1159/000445588.

[22] Li J, Tan M, Xiang Q, Zhou Z, Yan H. Thrombin-activated platelet-derived exosomes regulate endothelial cell expression of ICAM-1 via microRNA-223 during the thrombosis-inflammation response. Thrombosis Research 2017;154:96–105. https://doi.org/10.1016/j.thromres.2017.04.016.

[23] Liu J, Wu J, Li L, Li T, Wang J. The Role of Exosomal non-coding RNAs in coronary artery disease. Frontiers in Pharmacology 2020;11:603104. https://doi.org/10.3389/fphar.2020.603104.

[24] Loyer X, Potteaux S, Vion A-C, Guérin CL, Boulkroun S, Rautou P-E, et al. Inhibition of MicroRNA-92a prevents endothelial dysfunction and atherosclerosis in mice. Circular Research 2014;114:434–43. https://doi.org/10.1161/CIRCRESAHA.114.302213.

[25] Xing X, Li Z, Yang X, Li M, Liu C, Pang Y, et al. Adipose-derived mesenchymal stem cells-derived exosome-mediated microRNA-342-5p protects endothelial cells against atherosclerosis. Aging 2020;12:3880–98. https://doi.org/10.18632/aging.102857.

[26] Chen L, Yang W, Guo Y, Chen W, Zheng P, Zeng J, et al. Exosomal lncRNA GAS5 regulates the apoptosis of macrophages and vascular endothelial cells in atherosclerosis. PLoS ONE 2017;12:e0185406. https://doi.org/10.1371/journal.pone.0185406.

[27] Liu Y, Zhang W-L, Gu J-J, Sun Y-Q, Cui H-Z, Bu J-Q, et al. Exosome-mediated miR-106a-3p derived from ox-LDL exposed macrophages accelerated cell proliferation and repressed cell apoptosis of human vascular smooth muscle cells. European Review for Medical and Pharmacological Sciences 2020;24:7039–50. https://doi.org/10.26355/eurrev_202006_21697.

[28] Gao H, Wang X, Lin C, An Z, Yu J, Cao H, et al. Exosomal MALAT1 derived from ox-LDL-treated endothelial cells induce neutrophil extracellular traps to aggravate atherosclerosis. Biological Chemistry 2020;401:367–76. https://doi.org/10.1515/hsz-2019-0219.

[29] Huang C, Han J, Wu Y, Li S, Wang Q, Lin W, et al. Exosomal MALAT1 derived from oxidized low-density lipoprotein-treated endothelial cells promotes M2 macrophage polarization. Molecular Medical Report 2018. https://doi.org/10.3892/mmr.2018.8982.

[30] Wen Y, Chun Y, Lian Z, Yong Z, Lan Y, Huan L, et al. circRNA-0006896-miR1264-DNMT1 axis plays an important role in carotid plaque destabilization by regulating the behavior of endothelial cells in atherosclerosis. Molecular Medical Report 2021;23:311. https://doi.org/10.3892/mmr.2021.11950.

[31] Nemecz M, Alexandru N, Tanko G, Georgescu A. Role of MicroRNA in endothelial dysfunction and hypertension. Current Hypertension Reports 2016;18(12):87. https://doi.org/10.1007/s11906-016-0696-8.

[32] Jayaseelan VP, Arumugam P. Exosomal microRNAs as a promising theragnostic tool for essential hypertension. Hypertension Research 2020;43:74–5. https://doi.org/10.1038/s41440-019-0343-2.

[33] Tan PPS, Hall D, Chilian WM, Chia YC, Mohd Zain S, Lim HM, et al. Exosomal microRNAs in the development of essential hypertension and its potential as biomarkers. American Journal of Physiology-Heart and Circulatory Physiology 2021;320:H1486–97. https://doi.org/10.1152/ajpheart.00888.2020.

[34] Huang P, Wang L, Li Q, Tian X, Xu Jun, Xu Junyan, et al. Atorvastatin enhances the therapeutic efficacy of mesenchymal stem cells-derived exosomes in acute myocardial infarction via up-regulating long non-coding RNA H19. Cardiovascular Research 2020;116:353–67. https://doi.org/10.1093/cvr/cvz139.
[35] Ong S-G, Lee WH, Huang M, Dey D, Kodo K, Sanchez-Freire V, et al. Cross talk of combined gene and cell therapy in ischemic heart disease: role of exosomal microRNA transfer. Circulation 2014;130:S60–69. https://doi.org/10.1161/CIRCULATIONAHA.113.007917.
[36] Shyu K-G, Wang B-W, Fang W-J, Pan C-M, Lin C-M. Hyperbaric oxygen-induced long non-coding RNA MALAT1 exosomes suppress MicroRNA-92a expression in a rat model of acute myocardial infarction. Journal of Cell Molecular Medicine 2020;24:12945–54. https://doi.org/10.1111/jcmm.15889.
[37] He GD, Huang YQ, Liu L, Huang JY, Lo K, Yu YL, et al. Association of circulating, inflammatory-response exosomal mRNAs with acute myocardial infarction. Frontiers in cardiovascular medicine 2021;8:712061. https://doi.org/10.3389/fcvm.2021.712061.
[38] De Rosa S, Fichtlscherer S, Lehmann R, Assmus B, Dimmeler S, Zeiher AM. Transcoronary concentration gradients of circulating MicroRNAs. Circulation 2011;124:1936–44. https://doi.org/10.1161/CIRCULATIONAHA.111.037572.
[39] Zhu L-P, Tian T, Wang J-Y, He J-N, Chen T, Pan M, et al. Hypoxia-elicited mesenchymal stem cell-derived exosomes facilitates cardiac repair through miR-125b-mediated prevention of cell death in myocardial infarction. Theranostics 2018;8:6163–77. https://doi.org/10.7150/thno.28021.
[40] Feng Y, Huang W, Wani M, Yu X, Ashraf M. Ischemic preconditioning potentiates the protective effect of stem cells through secretion of exosomes by targeting Mecp2 via miR-22. PLoS One 2014;9:e88685. https://doi.org/10.1371/journal.pone.0088685.
[41] Wang J, Yang K, Yuan W, Gao Z. Determination of serum exosomal H19 as a noninvasive biomarker for bladder cancer diagnosis and prognosis. Medical Science Monitoring 2018;24:9307–16. https://doi.org/10.12659/MSM.912018.
[42] Khan M, Nickoloff E, Abramova T, Johnson J, Verma SK, Krishnamurthy P, et al. Embryonic stem cell–derived exosomes promote endogenous repair mechanisms and enhance cardiac function following myocardial infarction. Circular Research 2015;117:52–64. https://doi.org/10.1161/CIRCRESAHA.117.305990.
[43] Li Yong, Yang R, Guo B, Zhang Hongbo, Zhang Hui, Liu S, et al. Exosomal miR-301 derived from mesenchymal stem cells protects myocardial infarction by inhibiting myocardial autophagy. Biochemical and Biophysical Research Communications 2019;514:323–8. https://doi.org/10.1016/j.bbrc.2019.04.138.
[44] Sahoo S, Losordo DW. Exosomes and cardiac repair after myocardial infarction. Circular Research 2014;114:333–44. https://doi.org/10.1161/CIRCRESAHA.114.300639.
[45] Peng Y, Zhao J-L, Peng Z-Y, Xu W-F, Yu G-L. Exosomal miR-25-3p from mesenchymal stem cells alleviates myocardial infarction by targeting pro-apoptotic proteins and EZH2. Cell Death Disease 2020;11:317. https://doi.org/10.1038/s41419-020-2545-6.
[46] Saparov A, Ogay V, Nurgozhin T, Chen WCW, Mansurov N, Issabekova A, et al. Role of the immune system in cardiac tissue damage and repair following myocardial infarction. Inflammation Research 2017;66:739–51. https://doi.org/10.1007/s00011-017-1060-4.
[47] Wen Z, Mai Z, Zhu X, Wu T, Chen Y, Geng D, et al. Mesenchymal stem cell-derived exosomes ameliorate cardiomyocyte apoptosis in hypoxic conditions through microRNA144 by targeting the PTEN/AKT pathway. Stem Cell Research Ther 2020;11:36. https://doi.org/10.1186/s13287-020-1563-8.
[48] Chen Z, Yan Y, Wu J, Qi C, Liu J, Wang J. Expression level and diagnostic value of exosomal NEAT1/miR-204/MMP-9 in acute ST-segment elevation myocardial infarction. IUBMB Life 2020;72:2499–507. https://doi.org/10.1002/iub.2376.
[49] Wang L, Zhang J. Exosomal lncRNA AK139128 derived from hypoxic cardiomyocytes promotes apoptosis and inhibits cell proliferation in cardiac fibroblasts. International Journal of Nanomedicine 2020;15:3363–76. https://doi.org/10.2147/IJN.S240660.

[50] Das S, Halushka MK. Extracellular vesicle microRNA transfer in cardiovascular disease. Cardiovascular Pathology: the Official Journal of the Society for Cardiovascular Pathology 2015;24(4):199–206. https://doi.org/10.1016/j.carpath.2015.04.007.
[51] Liew C-C, Dzau VJ. Molecular genetics and genomics of heart failure. Nature Review of Genetics 2004;5:811–25. https://doi.org/10.1038/nrg1470.
[52] Melman YF, Shah R, Das S. MicroRNAs in heart failure: is the picture becoming less miRky? Circular Heart Failure 2014;7:203–14. https://doi.org/10.1161/CIRCHEARTFAILURE.113.000266.
[53] McManus D, Freedman J. MicroRNAs in platelet function and cardiovascular disease. Nature Review of Cardiology 2015;12:711–7. https://doi.org/10.1038/nrcardio.2015.101.
[54] Li S, Gao Y, Liu Y, Li J, Yang Xiyan, Hu R, et al. Myofibroblast-derived exosomes contribute to development of a susceptible substrate for atrial fibrillation. Cardiology 2020;145:324–32. https://doi.org/10.1159/000505641.
[55] Wang S, Li L, Hu X, Liu T, Jiang W, Wu R, et al. Effects of atrial fibrillation-derived exosome delivery of miR-107 to human umbilical vein endothelial cells. DNA and Cell Biology 2021;40:568–79. https://doi.org/10.1089/dna.2020.6356.
[56] Li C, Ni Y-Q, Xu H, Xiang Q-Y, Zhao Y, Zhan J-K, et al. Roles and mechanisms of exosomal non-coding RNAs in human health and diseases. Significant of Transduction Target Therapy 2021;6:383. https://doi.org/10.1038/s41392-021-00779-x.

CHAPTER ELEVEN

Exosomal RNA in renal diseases

Ana K. González Palomo[1], Eneida Turiján Espinoza[2], Jorge A. Jiménez Avalos[3] and Juan D. Cortés García[4]

[1]Coordinación para la Innovación y Aplicación de la Ciencia y Tecnología (CIACYT), Universidad Autónoma de San Luis Potosí (UASLP), San Luis Potosí, San Luis Potosí, México
[2]Instituto de Investigación en Biociencias, Centro Universitario de Los Altos, Universidad de Guadalajara, Guadalajara, Jalisco, México
[3]Centro de Investigación y Desarrollo Oncológico, S.A. de C.V. (CIDO), San Luis Potosí, San Luis Potosí, México
[4]Facultad de Medicina, Universidad Autónoma de San Luis Potosí (UASLP), San Luis Potosí, San Luis Potosí, México

Abbreviations

AAV associated vasculitis
ANCA antineutrophil cytoplasmic antibody
AP-1 activator protein–1
AQP1 Aquaporin 1
AQP2 Aquaporin 2
AXL AXL receptor tyrosine kinase
BUN blood urea nitrogen
c-MET C-mesenchymal–epithelial transition factor
CAFs cancer-associated fibroblasts
CCL2 chemokine (C–C motif) ligand 2
ccRCC clear cell carcinoma
CD2AP CD2-associated protein
circRNAs circular RNAs
CKD chronic kidney disease
CSCs cancer stem cells
DNA deoxyribonucleic acid
EMT epithelial–mesenchymal transition
ESRD end-stage renal disease
FSGS focal segmental glomerulosclerosis
GFR glomerular filtration rate
GN glomerulonephritis
HbA1c glycated hemoglobin
HG high glucose
HIF1α/2α hypoxia-inducible factors 1α/2α
HK2 human proximal tubular epithelial cells
hUSCs human urine-derived stem cells
IC immune complex
IgA Immunoglobulin A
IRI ischemia/reperfusion injury
ISN/RPS International Society of Nephrology/Renal Pathology Society
LN lupus nephritis
lncARSR activated in RCC sunitinib resistance

Exosomal RNA
ISBN 978-0-443-14008-2
https://doi.org/10.1016/B978-0-443-14008-2.00005-X

lncRNA long noncoding RNA
MAPK mitogen-activated protein kinase
miRs microRNAs
MPC5 immortalized mouse podocyte cell line MPC5
mRNAs messenger RNAs
MVB multivesicular bodies
ncRNAs noncoding RNAs
NETs neutrophil extracellular traps
NGAL neutrophil gelatinase-associated lipocalin
PAHO Pan-American Health Organization
PDCD4 programmed cell death 4
PERMANOVA permutational multivariate analysis of variance
PPAR-α peroxisome proliferator—activated receptor alpha
PTEN phosphatase and tensin homolog protein
PTEN phosphatase and tensin homolog
qPCR quantitative polymerase chain reaction
RCCs renal cell carcinomas
RNA ribonucleic acid
ROC receiver operating characteristic
ROS reactive oxygen species
Scr serum creatinine
Shh sonic hedgehog
SIMPER similarity percentage
SLE systemic lupus erythematosus
Smad3/TGF mothers against decapentaplegic homolog 3/transforming growth factor
SP1 specificity protein 1
STAT3 signal transducer and activator of transcription 3
TD1 type 1 diabetes
TD2 type 2 diabetes
TGF-β1 transforming growth factor β1
TKIs tyrosine kinase inhibitors
TMA tumor-associated macrophages
TME tumor microenvironment
TSG101 tumor susceptibility gene 101
TSP-1 thrombospondin target 1
US United States
VEGF vascular endothelial growth factor
VHL Von Hippel—Lindau
WHO World Health Organization

1. Introduction

Exosomes are nanoscale vesicles (40—70 nm) constituted by a lipid bilayer of endosomal origin. Their biogenesis occurs through mechanisms that include invagination of the cell membrane and the formation of multivesicular bodies (MVBs). Eventually, MVBs are excreted into the extracellular space by exocytosis after the fusion of their outer membrane with the cell plasma membrane [1,2]. Exosome phenotyping includes the expression of different markers such as CD9, CD63, CD81, flotillin, and tumor susceptibility gene 101 (TSG101). As a result, exosomes participate in cell–cell communication by carrying specific cargoes of bioactive molecules such as lipids, proteins, deoxyribonucleic acid (DNA), and ribonucleic acid (RNA) that can regulate the fate of target cells [3].

Recent studies have evidenced that exosomes can be secreted by all renal cells, including podocytes, glomerular endothelial cells, tubular cells, and glomerular mesangial cells, under normal and pathological conditions. In addition, exosomes secreted by damaged renal cells can be transferred to normal renal cells and change their phenotype due to their cargo, especially for the wide variety of RNA types, ranging from messenger RNAs (mRNAs) to noncoding RNAs (ncRNAs) [4]. Exosomal RNAs are promising diagnostic tools because they are biomarkers of early events and disease development. Also, studying exosomal RNAs offers different advantages due to their stability and accessibility obtained through noninvasive processes. Besides, exosomes are present in various biological fluids, such as serum, plasma, synovial fluid, breast milk, saliva, and urine. Recently, proteomic studies demonstrated that the proteins in urinary exosomes express typical markers from different nephron segments, which are highly relevant to the renal physiology and pathophysiology of kidney diseases [5,6]. However, the study of exosomal RNAs in other biofluids has also been examined.

Currently, the therapeutic potential of mesenchymal stem cell–derived exosomes is widely used in kidney diseases. They have been designed as deliverers to specifically bioactive molecules, such as some microRNAs (miRs), showing promising efficacy in alleviating kidney injury in different experimental models of kidney diseases [7,8]. Therefore, in this chapter, we focus on the current knowledge of the role played by exosomal RNAs in diagnosing and treating different renal diseases, particularly renal cancer, diabetic kidney disease, lupus nephritis, and chronic kidney disease (Fig. 11.1).

1.1 Exosomal RNA in renal cancer

Worldwide, kidney cancer represents a significant public health problem, representing the 14th most frequently diagnosed cancer in men and women, according to 2020 estimates from the World Health Organization (WHO) [9]. Only in the United States (US), near to 80,000 new cases of kidney cancer and almost 14,000 deaths in both sexes were estimated in 2022 [10]. More than 90% of these kidney neoplasms are renal cell carcinomas (RCCs) originating in renal tubular epithelial cells [11]. The most prevalent RCC histological subtype is clear cell carcinoma (ccRCC) with 70%–80% of cases, followed by papillary (10%–15%) and chromophobe (5%–10%) [11,12].

The ccRCC subtype is characterized by well-delimited tumor cells with apparently empty cytoplasm containing abundant lipids and glycogen that are washed out during

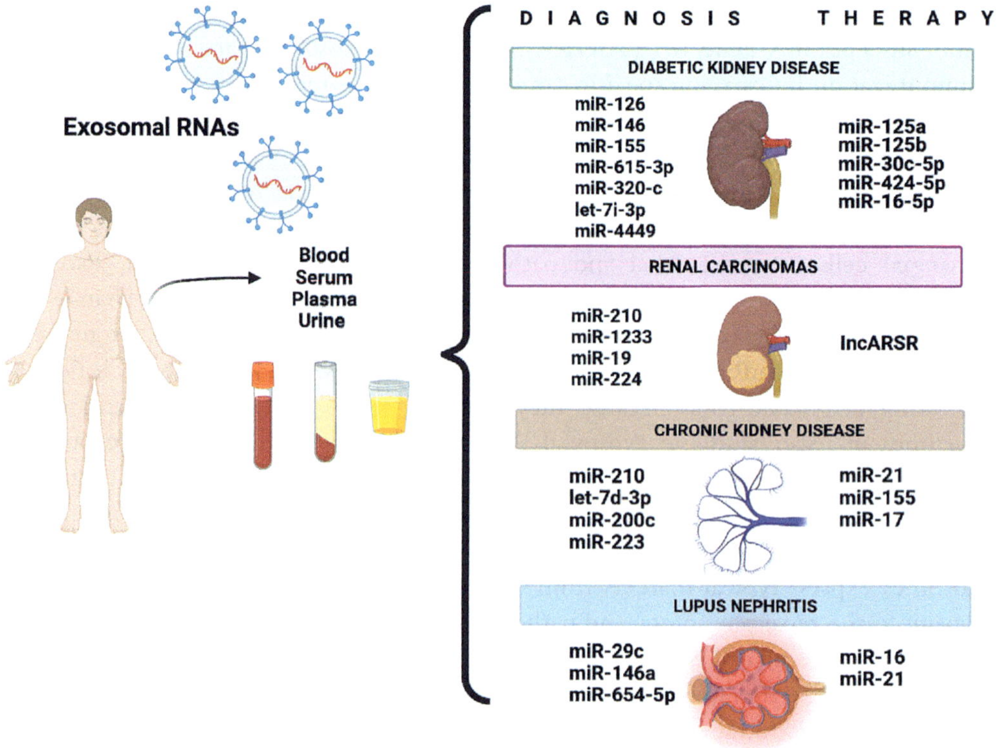

Figure 11.1 ***Exosomal RNAs in renal diseases*** Exosomes as biomarkers of diagnosis and therapy of renal diseases (diabetic kidney disease, renal carcinoma, chronic kidney disease, and lupus nephritis). Created by BioRender. miR, microRNA; lncARSR, activated in RCC sunitinib resistance.

histological processing, originating the characteristic clear cytoplasm, where its name comes from Ref. [13]. Also, determinant genetic and epigenetic features of ccRCC constitute the earliest truncal oncogenic events. ccRCC comprises a very high frequency of biallelic inactivation or loss of the Von Hippel–Lindau (*VHL*) tumor suppressor caused by the short-arm chromosome 3p loss (~90%) [14,15], gene mutation (~50%) [16,17], or promoter hypermethylation (5%–10%) [18]. The VHL protein is an E3 ubiquitin ligase, best known for its role as a negative regulator of the hypoxia-inducible factors 1α/2α (HIF1α/2α) via ubiquitylation and further proteasomal degradation [19]. The HIFs drive the transcription of several genes associated with angiogenesis, pH regulation, glycolysis, aberrant fatty acid metabolism, and chromatin remodeling [20,21]. Thus, loss of VHL results in constitutive HIFs activation in ccRCC, which leads to highly vascularized tumors with characteristic glycogen and lipid deposits, despite an adequate oxygenated tissue microenvironment [20].

Although most RCC discoveries are in the early stages, mainly due to incidental detection on routine imaging for other diseases or nonspecific complaints, many patients may present with an advanced stage of the disease. Currently, partial or radical kidney resection is the standard treatment for localized RCC. However, ~30% of patients eventually develop metastases even after nephrectomy [22]. In addition, most RCC subtypes are chemotherapy- and radiotherapy-resistant tumors [23]. Drug resistance has become an important concern, even for targeted therapies such as mTOR inhibitors, tyrosine kinase inhibitors (TKIs), and immunotherapies against vascular endothelial growth factor (VEGF) and immune checkpoint inhibition [11].

Chemo-, radio-, and targeted therapy resistance involves not fully understood molecular mechanisms associated with genetic and epigenetic alterations that lead to shifts in the regulation of cell survival, invasion, proliferation, and resistance to apoptotic mechanisms and expression of drug efflux pumps, as well as epithelial—mesenchymal transition (EMT) and dedifferentiation to cancer stem cells (CSCs) [23]. Alterations in epigenetic mechanisms such as histone posttranslational modifications, DNA methylation, and regulatory noncoding RNAs are involved in the pathogenesis of RCC [24—29]. In addition, the dysregulation of noncoding RNAs has been associated with the diagnosis, prognosis, and treatment response among the distinct subtypes of RCC [29,30].

RNAs can be secreted from tumor cells encapsulated in exosomes. Recent evidence supports the idea that tumor-derived exosomes act as signaling molecules between cancer cells and surrounding and distant tissues. According to their cargo, they may suppress or promote cancer progression and serve as a prognostic marker or therapeutic target. Thus, ncRNAs participate actively in the cell-to-cell communication between cancer and normal cells, regulating distinct cellular processes such as proliferation, apoptosis, metastasis, and chemoresistance [31]. For instance, miRs are RNA molecules of 20—25 posttranscriptional regulatory nucleotides related to physiological and pathological processes [32].

One of the most commonly used miRs as an RCC biomarker is miR-210. In ccRCC, the serum miR-210 has been used as a biomarker to discriminate ccRCC patients from healthy individuals. Zhao et al. found a statistical increase in serum levels of miR-210 in cRCC patients compared with controls, with a significant decrease postoperatively [33]. Similarly, other research centered on miR-210 evaluation in urine, plasma, and serum samples, finding higher expression of miR-210 in patients with ccRCC that declines after nephrectomy [34—37]. However, these studies focused on measuring total miR-210 (free and within exosomes). Interestingly, in a study by Wang et al. [38], from the six miRs measured in serum, tumor tissue, and serum exosomes, the authors reported an upregulation of miR-155 and miR-210 in serum samples from ccRCC patients compared with healthy individuals. Moreover, in exosomes, only the miR-210 was significantly overexpressed at advanced ccRCC stages and metastatic status with consecutive diminishing after surgery [38]. Similarly, Zhang et al. [39] found that exosomal miR-210 and miR-1233 are significantly higher in ccRCC patients than in controls and are lowered postoperatively [39]. These results suggest that the only exosomal-derived miR-210 is a valuable biomarker of ccRCC disease.

Besides, exosomal-derived ncRNAs have been reported to play a critical role in RCC progression. A study performed by Wang et al. demonstrated that CSC exosomes derived from ccRCC patients transporting significantly higher levels of miR-19-3p induce EMT transition and increased migration and invasion in recipient cells and rodents. The initiation of EMT is promoted by miR-19b targeting the phosphatase and tensin homolog protein (PTEN), causing the downregulation of E-cadherin and upregulation of vimentin and N-, promoting metastasis [40]. Besides the cancer cells, in the tumor microenvironment (TME), a heterogenous population of stromal cells, nonmalignant cells, such as immune cells, endothelial cells, and fibroblasts, closely interacts with the cancer cells. In ccRCC patients, it has been reported that cancer-associated fibroblasts (CAFs) can secrete and transfer exosomes to ccRCC cells. The CFA-derived exosomes significantly express miR-224-5p that can be transferred to ccRCC, increasing the malignancy of ccRCC cells by promoting their proliferation, invasion, and migration and inhibiting apoptosis. Thus, miR-224-5p from CAF-derived exosomes is involved in ccRCC progression and could serve as a diagnostic or prognostic marker [41].

Other stromal cells that modulate nonspecific and specific immune defense mechanisms in response to the internal conditions of the TME are tumor-associated macrophages (TMAs). It has been reported that in the TME, the dominant phenotype is M2. Unlike the proinflammatory and antitumor M1-like macrophages, the M2-like macrophages provide an immunosuppressive microenvironment through the release of antiinflammatory cytokines such as interleukin 10 (IL-10) and transforming growth factor β (TGF-β) that eventually contribute to angiogenesis, hypoxia, tumor cell proliferation, and metastasis [42]. Recent evidence suggests that ccRCC-derived exosomes can promote the M2 polarization of macrophages in vitro and in vivo, secreting IL-10 and TGF-β1 and inducing angiogenesis. This M2 polarization is originated by the long noncoding RNA (lncRNA) activated in RCC sunitinib resistance (lncARSR) carried in ccRCC-derived exosomes via the activation of the signal transducer and activator of transcription 3 (STAT3) pathway. The knockdown of the lncARSR significantly affected M2-macrophage polarization and tumor progression. Thus, the exosome-derived lncARSR can potentially be a therapeutic target for the treatment of ccRCC [43].

Also, lncRNAs are reported to be involved in chemoresistance development in RCC. Resistance to RTKs, such as the multitarget agent sunitinib, a key molecule in the treatment of RCC, is exhibited by the lncARSR. The lncARSR binds competitively to miR-449 and miR-34, inducing the upregulation of C-mesenchymal—epithelial transition factor (c-MET) and AXL receptor tyrosine kinase (AXL). Also, the authors showed that lncARSR was upregulated in sunitinib-resistant RCC cells and can be transferred into exosomes from sunitinib-resistant cells to sunitinib-sensitive cells, which could contribute to acquiring a drug-resistance phenotype [44].

1.2 Exosomal RNA in diabetic kidney disease

Diabetic kidney disease (DKD) or diabetic nephropathy is one of the main and most serious complications of type 1 (TD1) and type 2 (TD2) diabetes. Approximately 20%−50% of patients with diabetes develop DKD. The WHO has estimated that 10% of the world's population may be affected by DKD, and the prevalence is expected to increase with the increase in diabetes [45]. In the United States, the number of patients receiving treatment for end-stage renal disease increased significantly in 2014. Also, DKD patients have risen to 24.3 million in China. In Latin America, an average of 613 patients per million inhabitants had access to renal replacement therapy in 2011 [46,47].

DKD is a microvascular complication associated with poor glycemic control, characterized by loss of renal function with persistent albuminuria, decreased glomerular filtration rate (GFR), and glomerular hypertrophy. Structural and morphological changes include mesangial matrix expansion, progressive thickening of the glomerular basement membrane, tubulointerstitial fibrosis, loss of podocytes, and glomerular hyperfiltration [48]. Different mechanisms and pathways have been related to the development of DKD, such as oxidative stress, the renin−angiotensin−aldosterone system, and inflammatory processes [47]. In the case of TD1, an association between DKD progression and albuminuria levels has been described [4]. Thus, DKD can occur in five stages; stages 1 and 2 are characterized by normal albuminuria levels (<30 mg/day); from the third stage onward, an initial subclinical phase of hyperfiltration is described, followed by a loss of albuminuria (30−300 mg/day). As DKD progresses to stages 4 and 5, albumin losses increase to develop macroalbuminuria (>300 mg/day), considering stage 5 as end-stage renal damage [49,50]. However, this progression model cannot be applied to patients with TD2, where a considerable percentage of patients have DKD without changes in their albuminuria and GFR levels, even though these events do not always occur [51]. In this regard, albuminuria levels can be modified by other factors such as blood pressure, dyslipidemias, and the use of some drugs. Therefore, albuminuria levels cannot predict DKD because they lack sensitivity and specificity, which makes it necessary to identify new biomarkers and develop early diagnosis and treatment strategies to avoid the development of the disease.

Although the gold standard for diagnosing DKD is mainly based on renal biopsy, an invasive procedure where various adverse effects may occur. At present, disease research has involved the study of exosomes, which may play an essential role as noninvasive biomarkers for predicting DKD [52]. In this context, it has been shown that high glucose

(HG) conditions promote the release of exosomes containing miRs, circular RNAs (circRNAs), or mRNA, leading to activation, proliferation, and overproduction of extracellular matrix proteins, and dysfunction of glomerular endothelial cells and podocytes [3]. Specifically, urinary exosomes can transport and provide information about structural changes in the kidney since urinary exosomes originate from nephrons and collecting ducts, whereas blood-derived exosomes cannot cross the glomerular membrane. Therefore, urinary exosomal RNAs are used as biomarkers of DKD [6].

In the following, we will give some examples of these biomarkers. The expression profile of urinary exosomal miRNAs in TD2 patients, with and without albuminuria, a significant increase in miR-126, miR-146, and miR-155 in both groups was shown. These miRs have been related to the inflammatory process and early glomerular endothelial damage. Interestingly, in this study, a permutational multivariate analysis of variance (PERMANOVA) was used to analyze an association between clinical and molecular variables, followed by a similarity percentage (SIMPER) analysis to establish the main variables contributing to the difference between groups, with the following variables being the main contributors: glycated hemoglobin (HbA1c), albumin/creatinine ratio, glucose, urinary albumin, creatinine, body mass index, miR-126, miR-155, and miR-146. Furthermore, this model developed by the authors may provide a platform for assessing the risk of a person with TD2 developing kidney damage [53].

Similarly, the expression of miR-615-3p in urinary exosomes of DKD patients was significantly higher compared with the TD2 and control groups. Also, the level of miR-615-3p correlated positively with cystatin C, an indicator of microinflammation status in DKD patients that reflects renal damage. This miR also positively correlated with a TGF-β1 marker of the renal tissue fibrosis process [54]. Continuing work on urinary exosomes, Delić et al. [55] showed by microarray technique the downregulation of 16 miRs related to renal damage progression in DKD patients with microalbuminuria as opposed to DKD patients with normoalbuminuria. Authors confirmed by qPCR the overexpression of miR-320-c, a regulator of thrombospondin target 1 (TSP-1) involved in the TGF-β signaling pathway [55]. In addition, the expression of other urinary exosomal miRs with noninvasive biomarker potential in this disease have been described: miR-21-5p, miR-30b-5p, and miR-451-5p [56,57].

Also, exosomal RNAs isolated from other fluids have been studied in DKD. For example, exosomal circ_DLGAP4 was increased in DKD patients with high eGFR and macroalbuminuria and in cell culture supernatants of mesangial cells treated with high glucose concentrations (25 mM glucose). Of interest, this study showed that circ_DLGAP4 directly targeted miR-143 and repressed the proliferation and fibrosis of mesangial cells, which were induced by circ_DLGAP4 [58]. Likewise, in serum, the profile of circulating exosomal miRs has been identified through RNA sequencing, showing a difference of 95 miRs between patients with DKD and healthy subjects. From these, eight miRs were found to be upregulated (miR-4449, miR-642a-3p, miR-1246, miR-

150-3p, miR-5010-5p, miR-1255b-5p, let-7c-5p, and let-7i-3p), probably related to angiogenesis and regulation of mitogen-activated protein kinase (MAPK) signaling and the transcription factor activator protein-1 (AP-1) [59].

Moreover, in vitro studies have been employed to elucidate the role of exosomes in DKD. For example, in exosomes derived from human proximal tubular epithelial cells (HK2) treated with and without HG, changes in the expression of 152 miRNAs, 169 lncRNAs, 885 mRNAs, and 3 circRNAs were found by RNA sequencing analysis. Such exosomal RNAs can be explored in future studies to determine their implications as therapeutic targets or biomarkers of DKD [60]. In this regard, miR-486-5p, which was found to be downregulated in cells treated with high glucose, has been proposed as a key paracrine factor that protects against ischemic cell death and loss of renal function [61]. The functional role of miR-4449 derived from serum exosomes of DKD patients on the regulation of inflammatory cytokines and reactive oxygen species (ROS) in cultured HK2-cells was also investigated. It was found that HK-2 cells could uptake serum exosomes obtained from patients and that miR-4449 contained in these exosomes increased the generation of ROS and the cytokines IL-1β and IL-18 [62].

Stem cell—derived exosomes have been used as therapeutic tools. They offer several advantages because they target a specific tissue and can be internalized through a ligand/receptor recognition pathway in the target cell [8]. Exosomes derived from different types of stem cells can have beneficial effects on DKD by transporting renoprotective RNAs. Duan et al. transferred miR-16-5p through human urine—derived stem cells to human podocyte cultures under HG conditions and in diabetic rats, intending to regulate the expression of vascular endothelial growth factor A (VEFGA). miR-16-5p transference resulted in the suppression of VEGFA and apoptosis, probably ameliorating HG-induced podocyte injury [63]. Also, it has been investigated that miR-486 transported by exosomes from adipose-derived stem cells can be transferred to podocytes participating in decreasing the damage of immortalized mouse podocyte cell line (MPC5) also and reducing levels of urine protein, serum creatinine (Scr), blood urea nitrogen (BUN), and podocyte apoptosis in mice [64]. A listing of miRs contained within stem cell—derived exosomes applied to ameliorate diabetic kidney disease is shown in Table 11.1.

1.3 Evidence of exosomal RNAs in chronic kidney disease

Chronic kidney disease (CKD), also known as chronic kidney failure, is the irreversible and progressive loss of kidney function, altering the filtering capacity of waste and fluids from the blood to excrete in the urine. Although CKD is highly prevalent in older of black and South Asian origin, reports from the Pan American Health Organization (PAHO) indicate that CKD was the leading cause of mortality and disease burden in the Region of the Americas in 2019 with 254,028 total deaths [69]. CKD could result from

Table 11.1 Mesenchymal stem cell–derived exosomal microRNAs (miRs) applied to ameliorate diabetic kidney disease.

Exosomal miR	Type of stem cell	Cell target in culture	In vivo or in vitro effect	References
miR-424-5p	Mesenchymal stem cell	HK2	Decreased apoptosis and inhibited EMT in db/db mice and YAP1 in HK2 cells.	[65]
miR-30e-5p	Bone marrow mesenchymal stem cell	HK2	Inhibited caspase-1-mediated pyroptosis by targeting ELAVL1 in HG condition.	[66]
miR-125b	Bone marrow mesenchymal stem cell	Human embryonic kidney epithelial cell	Inhibition of apoptosis by TRAF-6/Akt axis	[67]
miR-16-5p	Human urine–derived stem cells	Human podocytes	Suppression of VEGFA and podocyte apoptosis, decreasing HG-induced podocyte injury	[63]
miR-125a	Adipose-derived mesenchymal stem cells	Rat glomerular mesangial cell	miR-125a directly regulates HDAC1 and contributed to suppressed mesangial hyperplasia and kidney fibrosis.	[68]
miR-486	Adipose-derived mesenchymal stem cells	MCP5	Decreased cell damage and apoptosis of podocytes. Reducing levels of urine protein in mice.	[64]

Akt, serine/threonine-protein kinase; ELAV1, ELAV-like RNA-binding protein; EMT, epithelial-to-mesenchymal transition; HDCA1, histone deacetylase 1; HG, high glucose; HK2, proximal tubular cells; MCP5, immortalized mouse podocyte cell line; TRAF-6, tumor necrosis factor receptor–associated factor 6; VEGFA, vascular endothelial growth factor A; YAP1, Yes-associated protein 1.

different conditions that alter kidney function such as high blood pressure, diabetes, high cholesterol levels, kidney infections, glomerulonephritis, kidney inflammation, polycystic kidney disease, or blockages in the flow of urine. Also, these conditions increase the risk of death by heart attack or stroke, although only 1 in 50 patients develop kidney failure [70].

The diagnosis of CKD is performed through blood and urine tests to measure substances and indexes of renal function, such as albuminuria and glomerular filtration rate (GFR). These parameters indicate the stage of kidney disease; for example, a GFR of less than 60 mL/min and albuminuria are markers of kidney damage, while a GFR of less than 15 mL/min indicates the need for dialysis or replacement therapy [71]. Symptoms at the advanced stages include tiredness, feeling sick, or the presence of blood in the urine; however, in the early stages, no symptoms of CKD are shown. Thus, it is necessary to identify new techniques or markers that define the early stage of CKD with a high sensibility to slow its progression, prevent complications, and diminish the years of life lived with disability in the patients [72].

In this sense, exosomes have played a significant role as potential applications of markers of development, progression, and treatment of renal diseases [73]. In CKD, it has been reported that exosomes are involved in the damage of tubular epithelial cells, polycystic kidney, fibrosis, inflammation, and calcification, through inducing the increase of sonic hedgehog (Shh) protein in mice; the augment of miR-200b, miR-200c, and miR-429 in 3D cell cultures; and the upregulation of miR-21 and IL-8 in activated monocyte cultures [74]. In contrast, exosomes derived from mesenchymal stem cells containing miR-30, miR-15, and miR-16 participate in kidney injury repair through their antiapoptotic, antiinflammatory, and antioxidant effects [73]. In addition, exosome transference in vitro and in vivo improves polycystic kidney by repairing the structure and function of the kidney [75], while the increase of miR-29c promotes eGFR that slow the tubulointerstitial fibrosis [76]. Also, exosome RNAs molecules could be potential biomarkers of CKD, such as the decrease of CD2-associated protein (CD2AP) mRNA in glomerular disease patients or the upregulation of miR-29c and miR-200c levels in focal segmental glomerulosclerosis (FSGS) and immunoglobulin A (IgA) nephropathy patients [75]. Thus, exosomes' role and their cargo are vast and should not be generalized to all CKD types.

Regarding IgA nephropathy (IgAN) and primary kidney diseases characterized by IgA deposits in glomeruli that progress to glomerulonephritis, urinary exosome levels increase with the pathogenesis [77]. The chemokine (C—C motif) ligand 2 (CCL2) could be considered the central molecule carried by these exosomes. A CCL2 augment was related to GFR decrease and tubulointerstitial inflammation; hence, it may be viewed as a biomarker [78]. Along with chemokines, miRs from urinary exosomes show differential expression. A downregulation of miR-29c, miR-205, and miR-204 was related to the pathogenesis, progression, and prognostic risk of IgAN disease. In contrast, upregulation of miR-215-5p, miR-378i, miR-365b-3p, and miR-135b-5p was shown in IgAN patients. Also, miR-4639 and miR-210 participate in the histological activity in progressive IgAN patients and could be noninvasive molecular biomarkers such as miR-451a and let-7d-3p used by the Oxford classification [77]. In pediatric patients with acute and chronic glomerular diseases, exosome excretion was reduced during kidney damage, but it was restored after recovery; thus, exosomes may differentiate between health and glomerular disease [79].

In kidney transplants, urinary exosomes may predict the clinical outcomes after transplantation because low exosome levels translate into slow graft function and vascular damage. Likewise, exosome-containing neutrophil gelatinase-associated lipocalin (NGAL) and IL-18 levels were associated with creatinine reduction and delayed graft function [80]. Also, in a rat model of ischemia/reperfusion injury (IRI), postoperative days showed a reduction of urinary aquaporin 1 (AQP1) and aquaporin 2 (AQP2). Low AQP1 amounts were found in kidney transplant recipients, while acute diuresis was caused by low AQP2 levels [81]. Immunosuppressive drugs are required to avoid allograft injury. In this context, exosomes could be an attractive therapy monitoring tool; for instance, an increase of the exosomal miR-155-5p and miR-223-3p and a diminished expression of miR-223-3p and miR-1228-3p would allow clinical monitoring of tacrolimus dose, serum creatinine, and blood leukocytes, respectively [82]. Also, copy numbers of DNA from exosomes could differentiate normal histology, rejection injury, and nonrejection injury in transplanted patients [83].

Vascular inflammation affects renal function in antineutrophil cytoplasmic antibody (ANCA)—associated vasculitis (AAV), a category of autoimmune vascular disorders with significant mortality and mainly unidentified etiology. It has been described that the exosome-upregulated expression of miR-30a-5p, miR-31-3p, miR-99a-5p, miR-106b-5p, and miR-182-5p may be involved in cell surface processes, signaling events, disease activity, and progression [84]. Therefore, future investigations with exosomal RNAs could help increase awareness of AAV pathogenesis.

Although it is evident that communication cell-to-cell mediated by exosomes promotes the diverse renal damage that progresses to CKD, exosomes are involved in diagnosing the early stage of CKD. The current investigation describes its role in the treatment of CKD. In this sense, different reports indicate that exosomes derived from healthy renal sections transfer information to damaged renal sections to try to improve them [85]. Also, CKD models have demonstrated that exosomes released from healthy or stem cells can be administered to promote tissue regeneration and improve kidney damage [86]. In contrast, exosomes released from injured kidney cells induce disease progression. The aforementioned depends on the kidney's functional state and its metabolic and physiological conditions [87]. In the same context, miRs contained in exosomes could modulate the function of their target cells, inducing the development and progression of CKD; hence, blocking specific miRs could revert or slow the progression of CKD [88]. It has been reported that anti-miR-17 slows cystic growth by decreasing the peroxisome proliferator—activated receptor alpha (PPAR-α), attenuating the disease development; silencing miR-21 improves inflammation, glomerulosclerosis, and interstitial fibrosis by increasing mitochondrial function; and the in vitro depletion of miR-155 expression prevents renal damage mediated by oxidative stress and autophagy [89]. Therefore, exosomal miRs could be a future tool to stop the progression of CKD or improve the disease, so it is necessary to carry out new protocols that verify this hypothesis, along with the discovery of new biomarkers to identify early stages of CKD, allowing opportune treatments focused on the improvement of CKD and the patient's quality of life.

1.4 Roles of exosomal RNAs in lupus nephritis

Lupus nephritis (LN) is a form of glomerulonephritis caused by an autoimmune disease, systemic lupus erythematosus (SLE), characterized by loss of immunological tolerance and which can lead to end-stage renal disease (ESRD) in approximately 50% of patients. The etiologic factors of LN involve neutrophil extracellular traps (NETs), immune complex (IC) deposition, autoantibodies against endogenous nucleic acids, and innate and adaptive immune activation [90].

Around 30%—60% of adults and up to 70% of children with SLE present LN within the first 5 years of diagnosis. Frequently, the LN is the clinical characteristic that allows the diagnosis of SLE [91]. The diagnosis requires histological evaluation of a kidney biopsy to determine the severity of LN and the need for specific therapy. According to the International Society of Nephrology/Renal Pathology Society (ISN/RPS) classification, more than 50% of renal biopsies are diagnosed as active LN-class III or IV A, the most serious in terms of renal prognosis and mortality [92].

In recent years, urinary RNAs have been reported as stable molecules secreted within exosomes by nephron cells. Due to their stability, urinary RNAs have been widely proposed as a promising source of noninvasive biomarkers for diagnostics, monitoring of disease activity, and therapeutic efficacy of various kidney diseases, including LN. Evidence suggests podocyte injury is involved in the pathogenesis of autoimmune glomerulonephritis (GN) in LN [93]. Podocytes are epithelial cells in the capillaries of the glomerulus that regulate blood filtration [94,95]. Data indicate that miR-26a in urinary exosomes regulates podocyte differentiation and cytoskeletal integrity. In patients with LN, urinary exosomes containing miR-26a were significantly higher compared with the healthy control; therefore, they could serve as a valuable biomarker of injured podocytes in LN [94].

As mentioned before, up to 30% of patients with LN often progress to ESRD, a poor long-term prognosis that increases patient mortality by up to four times [96]. Recently, the expression of miR-29c in urinary exosomes was inversely correlated with the renal-biopsy chronicity index of patients with LN. Therefore, miR-29c is proposed as an important noninvasive biomarker to assess the progression and development of renal fibrosis in LN [97]. Pérez-Hernández et al. showed that among the exosomal miRs, urinary miR-146a was increased and had a significant diagnostic role for active LN compared with SLE in the absence of LN with high sensitivity and specificity (100% and 90%, respectively). In addition, between active and inactive LN, authors reported an optimal 80% sensitivity and 89% specificity, which might discriminate the presence of active LN from SLE against healthy controls [98].

According to the ISN/RPS classification, type IV LN with cellular crescent (LN IV-CC) is characterized by diffuse proliferative lesions. The cellular crescents are formed by two or more layers of abnormally proliferating cells, increasing the risk of poor prognosis in LN patients [99]. The authors observed that the high expression of miR-654-5p urinary exosomal is a characteristic of LNIV-CC compared with LNIV active, LNIV inactive, and the control group and has a significant diagnostic role in LNIV-CC compared with LNIV inactive with a specificity of 96%. Thereby, miR-654-5p is suggested to promote irregular cell proliferation and may be used as a biomarker of LNIV-CC. On the other hand, for LNIV active, the expression of miR-3135b exosomal is higher than in other groups (LNIV inactive, LNIV active, and control group) and has a specificity of prediction of 83.33% [99].

Another study conducted in 13 patients with active LN and 18 with inactive LN showed significant downregulation of 2 miRs in urinary exosomes from patients with active LN compared with inactive disease: let-7a and miR-21. It suggests that different miRs are selectively packed in urinary exosomes at different disease stages, and both miRs could be involved in functional defects of renal cells to control inflammation [100]. Given the diversity of miRs that may be implicated in different stages of LN development, researchers are currently focused on finding a group of exosomal miRs with high specificity and sensitivity that can be used as biomarkers called fibromiRs. In this sense, Cristina Solé et al. identified for the first time that the exosomal expression of miR-21, miR-150, and miR-29c allows discrimination between low and moderate-high renal chronicity in patients with LN. However, as a panel, miR-29c/miR-150/miR-21 allow prediction of progression to ESRD thanks to their role in renal fibrogenesis, which increases the synthesis of profibrotic molecules through different pathways. Collagen production in tubular epithelial cells in the early stages of LN is regulated by miR-29c and specificity protein 1 (SP1). At the same time, via the Smad3/TGF pathway (mothers against decapentaplegic homolog 3/transforming growth factor), independent of SP1, miR-150, and miR-21 contribute to the fibrotic process in late stages [101].

Macrophages play an important role in developing SLE, its clinical consequences as LN and its relationship with these immune cells are lately being addressed. In a recent in vivo study with pristane-induced murine LN, ~40 nm exosomes isolated from bone marrow mesenchymal stem cells and internalized into macrophages were shown for the first time to play an important role in alleviating SLE-related nephritis. These results indicated that exosomal delivery of miR-16 and miR-21 promotes antiinflammatory polarization of kidney-infiltrated M2 macrophages by targeting PDCD4 and PTEN, respectively. At the same time, increased antiinflammatory cytokine production and decreased ROS, proinflammatory cytokines, and renal T cell infiltration alleviated SLE-related nephritis in lupus mice [102]. Therefore, this study proposes exosomes as a new therapeutic strategy for autoimmune diseases.

Transfer RNA-derived small RNAs (tsRNAs) are other regulatory noncoding RNAs present in a higher proportion (~70%) than miRNAs [103]. tsRNAs participate actively in gene expression regulation, antiapoptotic, translation inhibition, and virus infection and have been involved mainly in the occurrence and development of cancer [104,105]. It has been found that the deregulation of some tsRNAs in serum can function as diagnostic and prognostic biomarkers in LN [106]; however, there are limited studies of tsRNA-containing exosomes. Only a recent study shows increased levels of tRF3-Ile-AAT-1 and tiRNA5-Lys-CTT-1 in urinary exosomes as well as high discriminatory power (by ROC curve analysis) among LN patients with SLE compared with SLE without LN and healthy controls [107].

Circulating circRNA expression profiles have been reported in renal diseases such as LN. In fact, circRNA has been identified in exosomes secreted by the kidneys; nevertheless, the role of exosomal circRNA in renal LN remains unknown [108].

On the other hand, lncRNAs have been reported to play a critical role in the pathogenesis of autoimmune diseases. Over the past few years, several studies have reported that some circulating lncRNAs in serum and plasma have different expressions in SLE patients with and without LN [109]. However, so far, no published papers have directly studied the presence of IncRNA in exosomes in LN. Based on these findings, circRNA and incRNA are promising targets in LE that need further investigation.

2. Concluding remarks

Studying exosomal RNAs is a novel and constantly expanding field in cell biology. Since exosomes are released by broad cell types as intercellular messengers, their specific set of RNAs can be transferred to recipient cells, playing a key role in regulating kidney physiological and pathological processes. Due to the stability and widespread distribution of human body fluid—derived exosomal RNAs, they are considered a great source of noninvasive biomarkers for early and differential diagnosis, response to therapy, and the progression and severity of renal diseases. Interestingly, circulating exosomal RNAs provide a promising approach to therapeutic interventions by directly targeting critical molecules and their potential as engineered platforms to deliver specific information to alter the processes that lead to physiological abnormalities. However, ongoing knowledge in this area, along with the development of high-throughput techniques focusing on determining and validating the potential of integral components of exosomal RNAs derived from patient fluids, will contribute to a better understanding of the molecules involved in the pathogenesis of renal diseases, impacting in new and more effective therapeutic approaches.

References

[1] Dai J, Escara-Wilke J, Keller JM, Jung Y, Taichman RS, Pienta KJ, et al. Primary prostate cancer educates bone stroma through exosomal pyruvate kinase M2 to promote bone metastasis. Journal of Experimental Medicine 2019;216(12):2883–99. https://doi.org/10.1084/JEM.20190158.

[2] Jiménez-Avalos, Armando J, Fernández-Macías JC, González-Palomo AK. Circulating exosomal microRNAs: new non-invasive biomarkers of non-communicable disease. Molecular Biology Reports 2020. December. https://doi.org/10.1007/s11033-020-06050-w.

[3] Chen J, Zhang Q, Liu D, Liu Z. Exosomes: advances, development and potential therapeutic strategies in diabetic nephropathy. Metabolism 2021;122:154834. https://doi.org/10.1016/J.METABOL.2021.154834. 154834.

[4] Lu Y, Liu D, Qi F, Liu Z. Diabetic nephropathy: perspective on extracellular vesicles. Frontiers in Immunology 2020;11(943):1–7. https://doi.org/10.3389/fimmu.2020.00943.

[5] Grange C, Bussolati B. Extracellular vesicles in kidney disease. Nature Reviews Nephrology 2022;18(8):499–513. https://doi.org/10.1038/s41581-022-00586-9.

[6] Thongboonkerd V. Roles for exosome in various kidney diseases and disorders. Frontiers in Pharmacology 2020;10(January):1655. https://doi.org/10.3389/FPHAR.2019.01655/BIBTEX.

[7] Cao Q, Huang C, Chen XM, Carol A, Pollock. Mesenchymal stem cell-derived exosomes: toward cell-free therapeutic strategies in chronic kidney disease. Frontiers of Medicine 2022;9(March):715. https://doi.org/10.3389/FMED.2022.816656/BIBTEX.

[8] Peng L, Chen Y, Shi S, Wen H. Stem cell-derived and circulating exosomal MicroRNAs as new potential tools for diabetic nephropathy management. Stem Cell Research and Therapy 2022;13(1):1–19. https://doi.org/10.1186/S13287-021-02696-W. 25.

[9] WHO. Globocan 2020—kidney cancer. International Agency for Research on Cancer; 2020. p. 2020.

[10] Siegel RL, Miller KD, Fuchs HE, Jemal A. Cancer statistics, 2022. CA: A Cancer Journal for Clinicians 2022;72(1). https://doi.org/10.3322/caac.21708.

[11] Hsieh JJ, Purdue MP, Signoretti S, Swanton C, Albiges L, Schmidinger M, et al. Renal cell carcinoma. Nature Reviews Disease Primers 2017;3(1):17009. https://doi.org/10.1038/nrdp.2017.9.

[12] Sanchez DJ, Celeste Simon M. Genetic and metabolic hallmarks of clear cell renal cell carcinoma. Biochimica et Biophysica Acta, Reviews on Cancer 2018. https://doi.org/10.1016/j.bbcan.2018.06.003.

[13] Grignon DJ, Che M. Clear cell renal cell carcinoma. Clinics in Laboratory Medicine 2005. https://doi.org/10.1016/j.cll.2005.01.012.

[14] Gnarra JR, Tory K, Weng Y, Schmidt L, Wei MH, Li H, et al. Mutations of the VHL tumour suppressor gene in renal carcinoma. Nature Genetics 1994;7(1). https://doi.org/10.1038/ng0594-85.

[15] Hsieh JJ, Valerie H Le, Oyama T, Ricketts CJ, Ho TH, Emily H, et al. Chromosome 3p loss—orchestrated VHL, HIF, and epigenetic deregulation in clear cell renal cell carcinoma. Journal of Clinical Oncology 2018;36(36). https://doi.org/10.1200/JCO.2018.79.2549.

[16] Gallou C, Joly D, Méjean A, Staroz F, Martin N, Tarlet G, et al. Mutations of the VHL gene in sporadic renal cell carcinoma: definition of a risk factor for VHL patients to develop an RCC. Human Mutation 1999;13(6). https://doi.org/10.1002/(SICI)1098-1004(1999)13:6<464::AID-HUMU6>3.0.CO;2-A.

[17] Schraml P, Struckmann K, Hatz F, Sonnet S, Kully C, Gasser T, et al. VHL mutations and their correlation with tumour cell proliferation, microvessel density, and patient prognosis in clear cell renal cell carcinoma. The Journal of Pathology 2002;196(2). https://doi.org/10.1002/path.1034.

[18] Herman JG, Latif F, Weng Y, Lerman MI, Zbar B, Liu S, et al. Silencing of the VHL tumor-suppressor gene by DNA methylation in renal carcinoma. Proceedings of the National Academy of Sciences of the United States of America 1994;91(21). https://doi.org/10.1073/pnas.91.21.9700.
[19] Krieg M, Haas R, Brauch H, Acker T, Flamme I, Karl H, et al. Up-regulation of hypoxia-inducible factors HIF-1α and HIF-2α under normoxic conditions in renal carcinoma cells by von Hippel-Lindau tumor suppressor gene loss of function. Oncogene 2000;19(48). https://doi.org/10.1038/sj.onc.1203938.
[20] Hakimi AA, Reznik E, Lee CH, Creighton CJ, Brannon AR, Luna A, et al. An integrated metabolic atlas of clear cell renal cell carcinoma. Cancer Cell 2016;29(1):104–16. https://doi.org/10.1016/j.ccell.2015.12.004.
[21] Maxwell PH, Wlesener MS, Chang GW, Clifford SC, Vaux EC, Cockman ME, et al. The tumour suppressor protein VHL targets hypoxia-inducible factors for oxygen-dependent proteolysis. Nature 1999;399(6733). https://doi.org/10.1038/20459.
[22] Kroeger N, Choueiri TK, Lee JL, Bjarnason GA, Knox JJ, Mackenzie MJ, et al. Survival outcome and treatment response of patients with late relapse from renal cell carcinoma in the era of targeted therapy. European Urology 2014;65(6). https://doi.org/10.1016/j.eururo.2013.07.031.
[23] Acharya N, Singh KP. Recent advances in the molecular basis of chemotherapy resistance and potential application of epigenetic therapeutics in chemorefractory renal cell carcinoma. WIREs Mechanisms of Disease 2022;14(6):e1575. https://doi.org/10.1002/wsbm.1575.
[24] Chen W, Jia Z, Wang PP, Jiang J, Lin C, Zeng P, et al. DNA methylation-based classification and identification of renal cell carcinoma prognosis-subgroups. Cancer Cell International 2019;19(1). https://doi.org/10.1186/s12935-019-0900-4.
[25] Guo R, Zou B, Liang Y, Bian J, Xu J, Zhou Q, et al. LncRNA RCAT1 promotes tumor progression and metastasis via MiR-214-5p/E2F2 axis in renal cell carcinoma. Cell Death and Disease 2021;12(7). https://doi.org/10.1038/s41419-021-03955-7.
[26] Kumar A, Kumari N, Sharma U, Ram S, Singh SK, Kakkar N, et al. Reduction in H3K4me patterns due to aberrant expression of methyltransferases and demethylases in renal cell carcinoma: prognostic and therapeutic implications. Scientific Reports 2019;9(1). https://doi.org/10.1038/s41598-019-44733-y.
[27] Liu XP, Ju L, Chen C, Liu T, Li S, Wang X. DNA methylation-based panel predicts survival of patients with clear cell renal cell carcinoma and its correlations with genomic metrics and tumor immune cell infiltration. Frontiers in Cell and Developmental Biology 2020;8. https://doi.org/10.3389/fcell.2020.572628.
[28] Wang S, Xiang T, Yu L, Wen J, Liu F, Dong Y, et al. Novel molecular subtypes and related score based on histone acetylation modification in renal clear cell carcinoma. Frontiers in Cell and Developmental Biology 2021;9. https://doi.org/10.3389/fcell.2021.668810.
[29] Wang Y, Zhang Y, Wang P, Fu X, Lin W. Circular RNAs in renal cell carcinoma: implications for tumorigenesis, diagnosis, and therapy. Molecular Cancer; 2020. https://doi.org/10.1186/s12943-020-01266-7.
[30] Zhong W, Chen B, Zhong H, Huang C, Lin J, Zhu M, et al. Identification of 12 immune-related LncRNAs and molecular subtypes for the clear cell renal cell carcinoma based on RNA sequencing data. Scientific Reports 2020;10(1). https://doi.org/10.1038/s41598-020-71150-3.
[31] Zhong Y, Li H, Li P, Chen Y, Zhang M, Yuan Z, et al. Exosomes: a new pathway for cancer drug resistance. Frontiers in Oncology 2021. https://doi.org/10.3389/fonc.2021.743556.
[32] MacFarlane L-A, Murphy PR. MicroRNA: biogenesis, function and role in cancer. Current Genomics 2010;11(7):537–61. https://doi.org/10.2174/138920210793175895.
[33] Zhao A, Li G, Péoc'h M, Genin C, Gigante M. Serum MiR-210 as a novel biomarker for molecular diagnosis of clear cell renal cell carcinoma. Experimental and Molecular Pathology 2013;94(1). https://doi.org/10.1016/j.yexmp.2012.10.005.
[34] Dias F, , Ana Luísa Teixeira, Ferreira M, Adem B, Bastos N, Vieira J, Fernandes M, et al. Plasmatic MiR-210, MiR-221 and MiR-1233 profile: potential liquid biopsies candidates for renal cell carcinoma. Oncotarget 2017;8(61). https://doi.org/10.18632/oncotarget.21733.

[35] Fedorko M, Stanik M, Iliev R, Redova-Lojova M, Machackova T, Svoboda M, et al. Combination of MiR-378 and MiR-210 serum levels enables sensitive detection of renal cell carcinoma. International Journal of Molecular Sciences 2015;16(10). https://doi.org/10.3390/ijms161023382.

[36] Iwamoto H, Kanda Y, Sejima T, Osaki M, Okada F, Takenaka A. Serum MiR-210 as a potential biomarker of early clear cell renal cell carcinoma. International Journal of Oncology 2014;44(1). https://doi.org/10.3892/ijo.2013.2169.

[37] Petrozza V, Costantini M, Tito C, Maria Giammusso L, Sorrentino V, Cacciotti J, et al. Emerging role of secreted MiR-210-3p as potential biomarker for clear cell renal cell carcinoma metastasis. Cancer Biomarkers 2020;27(2). https://doi.org/10.3233/CBM-190242.

[38] Wang X, Wang T, Chen C, Wu Z, Bai P, Li S, et al. Serum exosomal miR-210 as a potential biomarker for clear cell renal cell carcinoma. Journal of Cellular Biochemistry 2019;120(2):1492–502. https://doi.org/10.1002/jcb.27347.

[39] Zhang W, Ni M, Su Y, Wang H, Zhu S, Zhao A, et al. MicroRNAs in serum exosomes as potential biomarkers in clear-cell renal cell carcinoma. European Urology Focus 2018;4(3). https://doi.org/10.1016/j.euf.2016.09.007.

[40] Wang L, Yang G, Zhao D, Wang J, Yang B, Peng Q, et al. CD103-positive CSC exosome promotes EMT of clear cell renal cell carcinoma: role of remote MiR-19b-3p. Molecular Cancer 2019;18(1). https://doi.org/10.1186/s12943-019-0997-z.

[41] Liu Y, Fu W, Cao X, Li S, Xiong T, Zhang X, et al. Delivery of MiR-224-5p by exosomes from cancer-associated fibroblasts potentiates progression of clear cell renal cell carcinoma. Computational and Mathematical Methods in Medicine 2021;2021:5517747. https://doi.org/10.1155/2021/5517747.

[42] Mantovani A, Allavena P, Federica Marchesi, Garlanda C. Macrophages as tools and targets in cancer therapy. Nature Reviews Drug Discovery. Nature Research 2022. https://doi.org/10.1038/s41573-022-00520-5.

[43] Zhang W, Zheng X, Yu Y, Zheng L, Lan J, Wu Y, et al. Renal cell carcinoma-derived exosomes deliver LncARSR to induce macrophage polarization and promote tumor progression via STAT3 pathway. International Journal of Biological Sciences 2022;18(8):3209–22. https://doi.org/10.7150/ijbs.70289.

[44] Qu Le, Ding J, Chen C, Wu ZJ, Liu B, Gao Y, et al. Exosome-transmitted LncARSR promotes sunitinib resistance in renal cancer by acting as a competing endogenous RNA. Cancer Cell 2016;29(5). https://doi.org/10.1016/j.ccell.2016.03.004.

[45] Hoogeveen EK. The epidemiology of diabetic kidney disease. Kidney and Dialysis 2022;2(3):433–42. https://doi.org/10.3390/KIDNEYDIAL2030038.

[46] OPS/OMS. OPS/OMS | La OPS/OMS y la sociedad latinoamericana de Nefrología llaman a prevenir la enfermedad renal y a mejorar el acceso al Tratamiento. 2020. p. 2020. https://www3.paho.org/hq/index.php?option=com_content&view=article&id=10542:2015-opsoms-sociedad-latinoamericana-nefrologia-enfermedad-renal-mejorar-tratamiento&Itemid=0&lang=es#gsc.tab=0.

[47] Samsu N. Diabetic nephropathy: challenges in pathogenesis, diagnosis, and treatment. BioMed Research International 2021;2021. https://doi.org/10.1155/2021/1497449.

[48] Campion CG, Sanchez-Ferras O, Batchu SN. Potential role of serum and urinary biomarkers in diagnosis and prognosis of diabetic nephropathy. Canadian Journal of Kidney Health and Disease 2017;4(January). https://doi.org/10.1177/2054358117705371.

[49] Arana, Ander Vergara, Castelao AM, Luis J, Górriz Teruel, De F, et al. Enfermedad diabética renal: albuminuria y progresión. 2020.

[50] Kidney Disease: Improving Global Outcomes (KDIGO) Diabetes Work Group. KDIGO 2022 clinical practice guideline for diabetes management in chronic kidney disease. Kidney International 2022;102(5):S1–127. https://doi.org/10.1016/J.KINT.2022.06.008/ATTACHMENT/C74021A0-2B56-4CE4-9E31-AAB1A43C225C/MMC1.PDF.

[51] Porrini E, Ruggenenti P, Mogensen CE, Barlovic DP, Praga M, Josep MC, et al. Non-proteinuric pathways in loss of renal function in patients with type 2 diabetes. Lancet Diabetes and Endocrinology 2015;3(5):382–91. https://doi.org/10.1016/S2213-8587(15)00094-7.

[52] Wei J, Wang B, Shen Fjie, Zhang Tting, Duan Z, Dong mei Z. Diagnostic value of triglyceride and cystatin C ratio in diabetic kidney disease: a retrospective and prospective cohort study based on renal biopsy. BMC Nephrology 2022;23(1):1–8. https://doi.org/10.1186/S12882-022-02888-3/TABLES/5.

[53] González-Palomo AK, Francisco J, Pérez-Vázquez, Karen B, Méndez-Rodríguez, Ilizaliturri-Hernández CA, et al. Profile of urinary exosomal MicroRNAs and their contribution to diabetic kidney disease through a predictive classification model. Nephrology 2022;27(6):484–93. https://doi.org/10.1111/NEP.14039.

[54] Wang J, Tao Y, Fan Z, Tong L, Shen X, Zhou L. Expression of urinary exosomal miRNA-615-3p and miRNA-3147 in diabetic kidney disease and their association with inflammation and fibrosis. 2023. https://doi.org/10.1080/0886022X.2022.2121929. 45 (1).

[55] Delić D, Eisele C, Schmid R, Baum P, Wiech F, Gerl M, et al. Urinary exosomal MiRNA signature in type II diabetic nephropathy patients. PLoS One 2016;11(3):1–16. https://doi.org/10.1371/journal.pone.0150154.

[56] Mohan A, Singh RS, Kumari M, Garg D, Upadhyay A, Ecelbarger CM, et al. Urinary exosomal MicroRNA-451-5p is a potential early biomarker of diabetic nephropathy in rats. PLoS One 2016;11(4):e0154055. https://doi.org/10.1371/JOURNAL.PONE.0154055.

[57] Zang J, Maxwell AP, Simpson DA, Mckay GJ. Differential expression of urinary exosomal MicroRNAs MiR-21-5p and MiR-30b-5p in individuals with diabetic kidney disease. Scientific Reports 2019;(July):1–10. https://doi.org/10.1038/s41598-019-47504-x.

[58] Bai S, Xiong X, Tang B, Ji T, Li X, Qu X, et al. Exosomal circ_DLGAP4 promotes diabetic kidney disease progression by sponging MiR-143 and targeting ERBB3/NF-KB/MMP-2 Axis. Cell Death and Disease 2020;11(11):1–13. https://doi.org/10.1038/s41419-020-03169-3.

[59] Kim H, Bae Y-U, Jeon JS, Noh H, Park HK, Byun DW, et al. The circulating exosomal MicroRNAs related to albuminuria in patients with diabetic nephropathy. Journal of Translational Medicine 2019;17(1):1–11. https://doi.org/10.1186/S12967-019-1983-3/TABLES/4. 236.

[60] Zhou S, Fang J, Hu M, Pan S, Liu D, Xing G, et al. Determining the influence of high glucose on exosomal LncRNAs, MRNAs, CircRNAs and MiRNAs derived from human renal tubular epithelial cells. Aging 2021;13(6):8467–80. https://doi.org/10.18632/AGING.202656.

[61] Livingston MJ, Wei Q. MicroRNAs in extracellular vesicles protect kidney from ischemic injury: from endothelial to tubular epithelial. Kidney International 2016;90(6):1150–2. https://doi.org/10.1016/J.KINT.2016.08.032.

[62] Gao C, Wang B, Chen Q, Wang M, Xiao F, Zhao N. Serum exosomes from diabetic kidney disease patients promote pyroptosis and oxidative stress through the MiR-4449/HIC1 pathway. Nutrition and Diabetes 2021;11(1):1–9. https://doi.org/10.1038/s41387-021-00175-y.

[63] Duan YR, Chen BP, Chen F, Yang SX, Yang Zhu C, Ma YL, et al. Exosomal MicroRNA-16-5p from human urine-derived stem cells ameliorates diabetic nephropathy through protection of podocyte. Journal of Cellular and Molecular Medicine 2021;25(23):10798–813. https://doi.org/10.1111/JCMM.14558.

[64] Jin J, Shi Y, Gong J, Zhao L, Li Y, He Q, et al. Exosome secreted from adipose-derived stem cells attenuates diabetic nephropathy by promoting autophagy flux and inhibiting apoptosis in podocyte. Stem Cell Research and Therapy 2019;10(1):1–15. https://doi.org/10.1186/S13287-019-1177-1/FIGURES/6.

[65] Cui C, Zang N, Jia S, Guo X, Qin H, Hu H, et al. Exosomes derived from mesenchymal stem cells attenuate diabetic kidney disease by inhibiting cell apoptosis and epithelial-to-mesenchymal transition via MiR-424-5p. The FASEB Journal 2022;36(10):e22517. https://doi.org/10.1096/FJ.2022 00488R.

[66] Lv J, Ya Ning H, Wang XP, Lu WH, Xie LY, Niu D. Bone marrow mesenchymal stem cell-derived exosomal MiR-30e-5p ameliorates high-glucose induced renal proximal tubular cell pyroptosis by inhibiting ELAVL1. Renal Failure 2023;45(1):2177082. https://doi.org/10.1080/0886022X.2023.2177082.

[67] Cai X, Lai X-Y. Exosomes from mesenchymal stem cells expressing microribonucleic acid-125b inhibit the progression of diabetic nephropathy via the tumour necrosis factor receptor-associated factor 6/Akt axis. 2021.

[68] Hao Y, Miao J, Liu W, Cai K, Huang X, Peng L. Mesenchymal stem cell-derived exosomes carry microRNA-125a to protect against diabetic nephropathy by targeting histone deacetylase 1 and downregulating endothelin-1. Diabetes, Metabolic Syndrome and Obesity: Targets and Therapy 2021;14:1405. https://doi.org/10.2147/DMSO.S286191.

[69] Pan American Health Organization—World Health Organization. n.d. "Burden of Kidney Diseases." Accessed March 29, 2023. https://www.paho.org/en/enlace/burden-kidney-diseases.

[70] Centers for Disease Control and Prevention. n.d. "Chronic kidney disease basics." Accessed March 29, 2023. https://www.cdc.gov/kidneydisease/basics.html#:~:text=CKD%20is%20a%20condition%20in,as%20heart%20disease%20and%20stroke.

[71] Charles C, Ferris AH. Chronic kidney disease. Primary Care—Clinics in Office Practice 2020. https://doi.org/10.1016/j.pop.2020.08.001.

[72] Ammirati AL. Chronic kidney disease. Revista Da Associacao Medica Brasileira. Associacao Medica Brasileira 2020. https://doi.org/10.1590/1806-9282.66.S1.3.

[73] Jin C, Wu P, Li L, Xu W, Qian H. Exosomes: emerging therapy delivery tools and biomarkers for kidney diseases. Stem Cells International 2021;2021. https://doi.org/10.1155/2021/7844455.

[74] Xiang H, Zhang C, Xiong J. Emerging role of extracellular vesicles in kidney diseases. Frontiers in Pharmacology. Frontiers Media S.A 2022. https://doi.org/10.3389/fphar.2022.985030.

[75] Merchant ML, Rood IM, Deegens JKJ, Klein JB. Isolation and characterization of urinary extracellular vesicles: implications for biomarker discovery. Nature Reviews Nephrology 2017. https://doi.org/10.1038/nrneph.2017.148.

[76] Lv LL, Cao YH, Ni HF, Xu M, Liu D, Liu H, et al. MicroRNA-29c in urinary exosome/microvesicle as a biomarker of renal fibrosis. American Journal of Physiology—Renal Physiology 2013;305(8). https://doi.org/10.1152/ajprenal.00148.2013.

[77] Delrue C, De Bruyne S, Speeckaert R, Marijn M, Speeckaert. Urinary extracellular vesicles in chronic kidney disease: from bench to bedside? Diagnostics 2023. https://doi.org/10.3390/diagnostics13030443.

[78] Feng Y, Lv LL, Wu WJ, Li ZL, Chen J, Ni HF, et al. Urinary exosomes and exosomal CCL2 mRNA as biomarkers of active histologic injury in IgA nephropathy. The American Journal Of Pathology 2018;188(11):2542–52. https://doi.org/10.1016/j.ajpath.2018.07.017.

[79] Dimuccio V, Peruzzi L, Felice Brizzi M, Cocchi E, Fop F, Alberto B, et al. Acute and chronic glomerular damage is associated with reduced CD133 expression in urinary extracellular vesicles. American Journal of Physiology—Renal Physiology 2020;318:486–95. https://doi.org/10.1152/ajprenal.00404.2019.-Extracellular.

[80] Alvarez S, Suazo C, Boltansky A, Ursu M, Carvajal D, Innocenti G, et al. Urinary exosomes as a source of kidney dysfunction biomarker in renal transplantation. Transplantation Proceedings 2013;45(10):3719–23. https://doi.org/10.1016/j.transproceed.2013.08.079.

[81] Asvapromtada S, Sonoda H, Kinouchi M, Oshikawa S, Takahashi S, Hoshino Y, et al. Characterization of urinary exosomal release of aquaporin-1 and-2 after renal ischemia-reperfusion in rats. American Journal of Physiology—Renal Physiology 2018;314:584–601. https://doi.org/10.1152/ajprenal.00184.2017.-Acute.

[82] Costa De Freitas RC, Raul HB, Fabiana Dalla VG, Bonezi V, Thiago Dominguez CH, Rosso Felipe C, et al. Differentially expressed urinary exo-MiRs and clinical outcomes in kidney recipients on short-term tacrolimus therapy: a pilot study. Epigenomics 2020;12(22):2019–34. https://doi.org/10.2217/epi-2020-0160.

[83] Sedej I, Štalekar M, Žnidarič MT, Goričar K, Kojc N, Kogovšek P, et al. Extracellular vesicle-bound DNA in urine is indicative of kidney allograft injury. Journal of Extracellular Vesicles 2022;11(9). https://doi.org/10.1002/jev2.12268.

[84] Frydlova J, Zednikova I, Satrapova V, Pazourkova E, Santorova S, Hruskova Z, et al. Analysis of MicroRNAs in small urinary extracellular vesicles and their potential roles in pathogenesis of renal ANCA-associated vasculitis. International Journal of Molecular Sciences 2022;23(8). https://doi.org/10.3390/ijms23084344.

[85] Prunotto M, Farina A, Lane L, Pernin A, Schifferli J, Hochstrasser DF, et al. Proteomic analysis of podocyte exosome-enriched fraction from normal human urine. Journal of Proteomics 2013;82(April):193–229. https://doi.org/10.1016/j.jprot.2013.01.012.
[86] Karpman D, Ståhl AL, Arvidsson I. Extracellular vesicles in renal disease. Nature Reviews Nephrology 2017. https://doi.org/10.1038/nrneph.2017.98. Nature Publishing Group.
[87] Zhou X, Zhang W, Yao Q, Zhang H, Dong G, Zhang M, et al. Exosome production and its regulation of EGFR during wound healing in renal tubular cells. American Journal of Physiology—Renal Physiology 2017;312:963–70. https://doi.org/10.1152/ajprenal.00078.2017.-Kidney.
[88] Zhao H, Ma SX, Shang YQ, Zhang HQ, Su W. MicroRNAs in chronic kidney disease. Clinica Chimica Acta 2019. https://doi.org/10.1016/j.cca.2019.01.008. Elsevier B.V.
[89] Aparicio-Trejo, Omar E, Ana Karina Aranda-Rivera, Osorio-Alonso H, Martínez-Klimova E, Gabriela Sánchez-Lozada L, et al. Extracellular vesicles in redox signaling and metabolic regulation in chronic kidney disease. Antioxidants 2022. https://doi.org/10.3390/antiox11020356.
[90] Wang W, Yue C, Gao S, Li S, Zhou J, Chen J, et al. Promising roles of exosomal microRNAs in systemic lupus erythematosus. Frontiers in Immunology 2021;12. https://doi.org/10.3389/fimmu.2021.757096.
[91] Yu C, Li P, Dang X, Zhang X, Mao Y, Chen X. Lupus nephritis: new progress in diagnosis and treatment. Journal of Autoimmunity 2022;132(October):102871. https://doi.org/10.1016/J.JAUT.2022.102871.
[92] Raimbourg Q, Daugas É. Atteintes rénales du lupus. Néphrologie and Thérapeutique 2019;15(3):174–89. https://doi.org/10.1016/J.NEPHRO.2018.11.003.
[93] Nagata, Michio. 1-S2.0-S0085253816002933-Main-3. Kidney International 2016;89(6):1221–30. https://doi.org/10.1016/j.kint.2016.01.012.
[94] Ichii O, Otsuka-Kanazawa S, Horino T, Kimura J, Nakamura T, Matsumoto M, et al. Decreased MiR-26a expression correlates with the progression of podocyte injury in autoimmune glomerulonephritis. PLoS One 2014;9(10):e110383. https://doi.org/10.1371/journal.pone.0110383.
[95] Sakhi H, Moktefi A, Bouachi K, Vincent A, Hénique C, Remy P, et al. Podocyte injury in lupus nephritis. Journal of Clinical Medicine 2019;8(9):1–14. https://doi.org/10.3390/jcm8091340.1340.
[96] Cheng C, Guo F, Yang H, Ma J, Li H, Yin L, et al. Identification and analysis of the predictive urinary exosomal MiR-195-5p in lupus nephritis based on renal MiRNA-MRNA co-expression network. Lupus 2022;31(14):1786–99. https://doi.org/10.1177/09612033221133684.
[97] Solé C, Cortés-Hernández J, Felip ML, Vidal M, Ordi-Ros J. MiR-29c in urinary exosomes as predictor of early renal fibrosis in lupus nephritis. Nephrology Dialysis Transplantation 2015;30(9):1488–96. https://doi.org/10.1093/ndt/gfv128.
[98] Perez-Hernandez J, Forner MJ, Carolina P, Chaves FJ, Cortes R, Redon J. Increased urinary exosomal MicroRNAs in patients with systemic lupus erythematosus. PLoS One 2015;10(9):e0138618. https://doi.org/10.1371/journal.pone.0138618.
[99] Li Y, Xu X, Tang X, Bian X, Shen B, Zhao H, et al. MicroRNA expression profile of urinary exosomes in type IV lupus nephritis complicated by cellular crescent. Journal of Biological Research-Thessaloniki 2018;25(1). https://doi.org/10.1186/s40709-018-0088-0.
[100] Tangtanatakul P, Klinchanhom S, Sodsai P, Sutichet T, Promjeen C, Avihingsanon Y, et al. Down-regulation of let-7a and MiR-21 in urine exosomes from lupus nephritis patients during disease flare. Asian Pacific Journal of Allergy and Immunology 2019;37(4):189–97. https://doi.org/10.12932/AP-130318-0280.
[101] Solé C, Moliné T, Vidal M, Ordi-ros J, Cortés-hernández J. An exosomal urinary MiRNA signature for early diagnosis of renal fibrosis in lupus nephritis. Cells 2019;8(8):1–17. https://doi.org/10.3390/cells8080773.
[102] Zhang M, Takerra K, Johnson-Stephenson, Wang W, Wang Y, Jing L, et al. Mesenchymal stem cell-derived exosome-educated macrophages alleviate systemic lupus erythematosus by promoting efferocytosis and recruitment of IL-17+ regulatory T cell. Stem Cell Research and Therapy 2022;13(1). https://doi.org/10.1186/s13287-022-03174-7.

[103] Zhang Y, Zhang Y, Shi J, Zhang H, Cao Z, Gao X, et al. Identification and characterization of an ancient class of small RNAs enriched in serum associating with active infection. Journal of Molecular Cell Biology 2014;6(2):172–4. https://doi.org/10.1093/jmcb/mjt052.

[104] Wen J-tao, Huang Z-hao, Li Q-hui, Chen X, Qin H-lei, Zhao Y. Research progress on the TsRNA classification, function, and application in gynecological malignant tumors. Cell Death Discovery 2021;7(1). https://doi.org/10.1038/s41420-021-00789-2.

[105] Xie Y, Yao L, Yu X, Yao R, Li Z, Guo J. Action mechanisms and research methods of tRNA-derived small RNAs. Signal Transduction and Targeted Therapy 2020;5(1):2–9. https://doi.org/10.1038/s41392-020-00217-4. 109.

[106] Yang P, Zhang X, Chen S, Yue T, Ning M, Zhu Y, et al. A novel serum TsRNA for diagnosis and prediction of nephritis in SLE. Frontiers in Immunology 2021;12. https://doi.org/10.3389/fimmu.2021.735105.

[107] Chen S, Zhang X, Meng K, Sun Y, Shu R, Han Y, et al. Urinary exosome tsRNAs as novel markers for diagnosis and prediction of lupus nephritis. Frontiers in Immunology 2023;14(1):1–11. https://doi.org/10.3389/fimmu.2023.1077645. 1077645.

[108] Jin J, Sun H, Shi C, Yang H, Wu Y, Li W, et al. Circular RNA in renal diseases. Journal of Cellular and Molecular Medicine 2020;24(12):6523–33. https://doi.org/10.1111/jcmm.15295.

[109] Mihaylova G, Vasilev V, Kosturkova MB, George SS, Radanova M. Long non-coding RNAs as new biomarkers in lupus nephritis: a connection between present and future. Cureus; 2020. https://doi.org/10.7759/cureus.9003.

CHAPTER TWELVE

Exosomal RNA in bacterial infections

Rajarshi Das[1] and Soumyadev Sarkar[2]
[1]Amity Institute of Biotechnology, Amity University, Noida, UP, India
[2]Center for Fundamental and Applied Microbiomics, Biodesign Institute, Arizona State University, Tempe, AZ, United States

1. Introduction

1.1 Characteristics of exosomes

Exosomes are extracellular vesicles (EVs) generated by the invagination of endosome membranes. Exosomes form multivesicular bodies (MVBs) containing intraluminal vesicles (ILVs) [1]. Transmission electron microscopy (TEM) examination of exosomes reveals a cup-shaped morphology [1]. Typically, exosomes range from 50 to 100 nm, with densities from 1.13 to 1.19 g/mL in a sucrose density gradient [1]. EVs are well studied as having important cellular properties, from cellular communications with the neighboring cellular environment to their explicit transport properties in delivering cargo. The exosomes carry different cellular transmembrane and cytosolic proteins and nucleic acids [2].

1.2 Biogenesis of exosomes

The biogenesis of exosomes involves several steps, which include (1) initiation, (2) endocytosis, (3) the formation of MVBs, and (4) secretion [3]. The secreted membrane vesicles of variable size and density produce from the plasma membrane by budding or shedding.

During plasma membrane biogenesis, the cell membrane undergoes membranous infoldings where the early endosomes bud off. The nascent endosomes, over time, develop into late endosomes [4]. During exosome biogenesis, proteins, nucleic acids, and other membrane-associated molecules are internalized by endocytic vesicles caused by plasma membrane invagination [1]. Endocytic vesicles fuse with early endosomes to deliver the materials, which ultimately mature into late endosomes. The late endosomes are often called MVBs [5]. The MVBs can then fuse with lysosomes, where the materials degrade, or merge with the plasma membrane, releasing materials into the extracellular domain and termed exosomes [6] (Fig. 12.1).

Exosomes originate from the fusion of several MVBs, typically having a lipoidal plasma membrane composition. These can be distinctly different from exosomes derived from various types of cells. These EVs express transmembrane proteins on the lipid membrane, which are positioned in the extracellular environment [7].

Exosomal RNA
ISBN 978-0-443-14008-2
https://doi.org/10.1016/B978-0-443-14008-2.00011-5

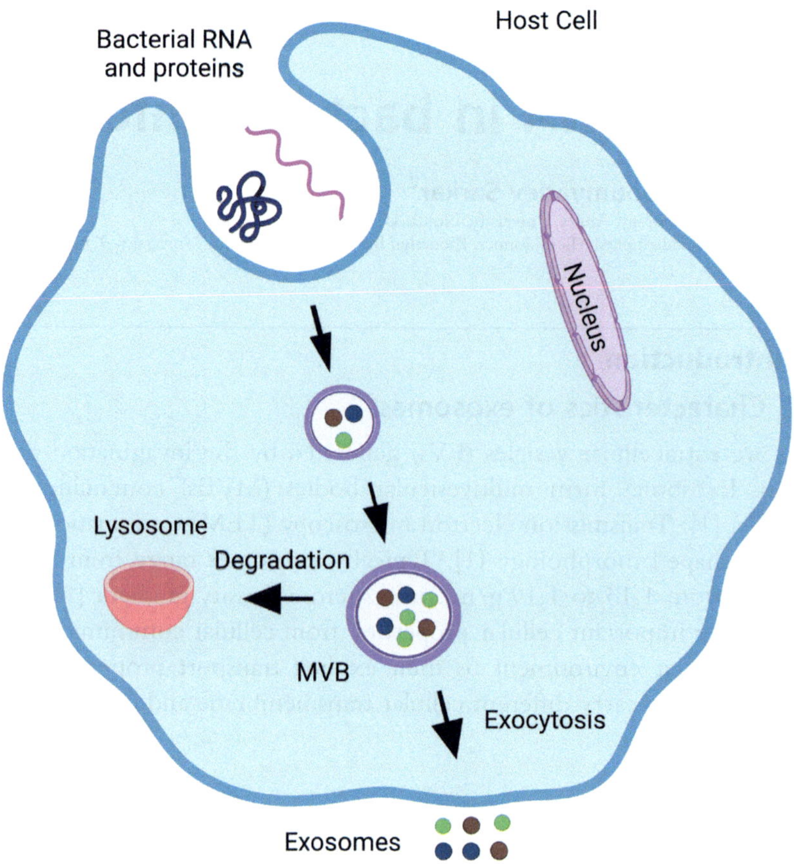

Figure 12.1 Mechanism of exosomes production.

1.3 Role of exosomes in innate immune response

The system of innate immune has a significance in the defense of the host against bacterial infection. The exosomes efficiently transfer functional components to recipient cells due to the ability that protects the materials from getting degraded, which eludes detection and finally elimination by immune system. The exosomes formed from bacteria enable communication between invaders and the immune cells that are innate in nature and are crucial in spreading the pathogens and cell-derived chemicals from donors, influencing host's responses of innate immune in nature [8].

Recent investigations about the exosomes, the immunity, and also their interactions in the human illnesses imply much attention in this arena. The immune system is a network comprising cells that mediates the first encounters between host and pathogen, including dendritic cells, macrophages/monocytes, neutrophils, and the natural killer cells [9].

1.4 Exosomal RNA

Exosomal RNA can be coding or noncoding RNA in nature or a combination of both [10]. Exosomal RNA mostly contains RNAs that are short in nature, including mRNA and microRNA (miRNA), and traces of ribosomal RNA [11]. The RNAs that are mostly enriched are not always the most numerous in the cells, suggesting that particular mRNA and the miRNA are sorted in the lumen of the exosome [11—17]. RNAs can be loaded into intracellular vesicles with the help of RNA-binding proteins [10]. The process underlying this sorting is unknown; however, it demonstrated that RNA-induced silencing complex (RISC) is required for the miRNA and also small interfering RNA (siRNA) to inhibit mRNAs linked to endosomal route [18,19]. It has been proposed that the MVB is where the RISC complex and miRNA are assembled [18]. MVB supports entry of miRNA to the RISC [13]. These results cast insight into the potential mechanisms in sorting RNA into exosomes, as exosomes are formed in MVBs.

2. Alterations in exosomes upon bacterial infection

Exosome content, quantity, and membrane structure can alter upon pathogen infection [1]. Also, the number of exosomes produced by the host cells differs when infected by pathogens compared with uninfected cells. This is primarily due to alterations in the cellular activity of infected cells, and the pathogenic organisms utilize the endocytic pathways of the host cells [1]. *Mycobacterium bovis* infection in mice induced secretion of exosomes, which was consistent with the bacterial load [20]. There are alterations in the membrane structure, such as in the number of structural lipids and proteins [21]. Upon infection by *Mycobacterium tuberculosis*, there has been more abundance of exosomes in macrophages [22]. Also, there have been reports of exosome content changes such as protein kinase G, miRNA-155, and miRNA-132 upon infection by *M. tuberculosis* [23,24]. Similarly, there was a change in lipopolysaccharide in infections by gram-negative bacteria [25].

3. Role of exosomes when pathogens infect

Exosomes have the potency to play essential roles in infection and promoting events like cell death [25], sharing of antibiotic resistance genes [26], and protection of formation of biofilms [27], among others.

Exosomes can function in both pathogen infection and antiinfection in the host. Pathogens-derived exosomes having RNA can infect in the following ways: (1) causing intense infection by transmitting molecules that are pathogen-related, (2) helping the pathogen to implement mechanisms related to immune escape, and (3) counteract responses of the immune system by initiating apoptosis of immune cells [1]. Exosomes can

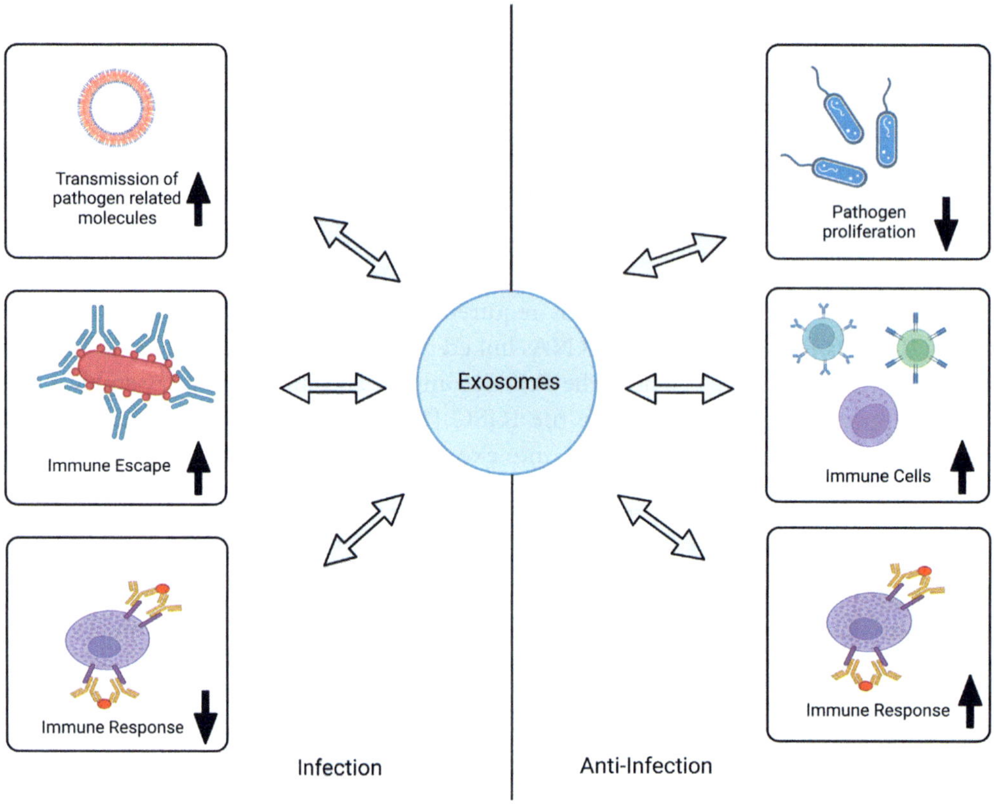

Figure 12.2 Role of exosomes in infection and antiinfection.

also trigger antiinfective roles by (1) directly inhibiting the proliferation of pathogens and subsequent infections and (2) activating the immune system by accelerating the function of natural killer cells, B cells, and T cells. It is expected to have a balance between the processes of infection and antiinfection, with exosomes playing a pivotal role in maintaining the balance [1] (Fig. 12.2).

4. The role and delivery of bacterial exosomal RNA during pathogen infections

RNAs in exosomes were first reported in 1989 after nucleic acids were extracted from *Neisseria gonorrhoeae* [28]. Very recently, the development of the latest RNA sequencing technologies and protocols has paved the way to study different bacterial RNAs in exosomes [29].

It is the potential of exosomal RNAs and how they can interact with the host targets, and also the different mechanisms for uptake by the host cells that will determine the efficiency of the delivery of the RNA [29]. The size of the exosomes can often impact the uptake [30]. For *Pseudomonas aeruginosa*, the exosomes must cross the cell membrane before reaching the host [29]. Bacterium such as *Listeria* can survive in the host, whereas *Salmonella* forms specialized vacuoles for infection [29].

Human and animal health is constantly threatened by infectious diseases caused by pathogenic bacteria [29]. If assessed at the molecular level, mammalian host infection needs a well-drilled program at the genetic level that would enable it to avoid the innate immune response [31].

Pathogenic bacteria depend on rapid modifications of gene expressions that enable them to invade the mammalian host cells. Often, it is observed that the bacteria release effector RNAs that suppress the innate immune responses [31]. Small RNA (sRNA) such as PinT has been a master regulator for inducing *Salmonella typhimurium* virulence during the infection [32]. Similar observations are reported in Enterobacteriaceae where there is evidence of protein-mediated noncoding RNA and mRNA interactions [33,34]. Also, RNAIII in *Staphylococcus aureus* can induce a switch from the mode of expected growth to infection in the bacterial pathogen [35].

The innate immune response is the first barrier mammalian cells provide to fight off bacterial infection [29]. There are pieces of evidence of the involvement of transcriptional modulation in such a response [29]. Mammalian cells infected by bacteria show alterations in microRNA levels [36]. Tristetraprolin, an RNA-binding protein, could arrest the translational capabilities of mRNAs. Also, in infected macrophages, T cell–restricted intracellular antigen proteins can stall the preinitiation of ribosomal complexes [37,38].

So, RNA has been identified as the new regulator that can modulate the balance between the virulence capacities of the bacterial pathogens on one end, and on the other end, it is also critical for the infected host cells to generate a substantial immune response [29].

5. Case studies

Several bacterial pathogens are discussed in the following section, where the involvement of the miRNA and secreted exosomal RNA are listed along with their mechanisms of action in the host (Table 12.1).

5.1 *Helicobacter pylori*

The spiral-shaped gram-negative bacterium *Helicobacter pylori* is present in the stomach epithelium of more than 50% of persons worldwide [39]. Its infestation has been related to several extra gastrointestinal (GI) disorders and a broad spectrum of GI conditions, such as chronic gastritis and gastric cancer (GC). Small noncoding RNAs (sncRNAs) of

Table 12.1 microRNAs and secreted exosomal RNAs identified during bacterial infections and mechanisms of action.

Pathogenic bacteria	miRNA	Mechanism
	miR-155	Decreases MyD88 synthesis and proinflammatory cytokine IL-8 levels. Differentiates Th1 or Th17 cells during microbial infections.
Helicobacter pylori	miR-25	Increases the expression of interleukin 6 (IL6), inflammatory factors, vascular cell adhesion molecule-1 (VCAM-1), and the monocyte chemoattractant protein-1 (MCP-1).
	sR-2509025, and sR-989262	Reduces IL-8 secretion.
	miR-206, miR-147, and miR-148a	Regulates host's production of a variety of inflammatory cytokines.
	miR-20a-3p, miR-99b, and miR-1178	Reduction in the host's immune system response.
Mycobacterium tuberculosis	miR-155	Suppresses T cells bacterial clearance and cytokine production.
	miR-27	Prevents autophagosome formation.
	miR-1958	Prevents autophagic flow.
	miR-146a	Assists the pathogen to infiltrate macrophages intracellularly.
Listeria monocytogenes	miR-21	Regulates macrophage polarization.
	rli32	Induces IFN-β expressions.
Pseudomonas aeruginosa	sRNA52320	Induces immune responses.

H. pylori typically range in size from 15 to 55 nucleotides [40]. By increasing the expression of miR-155 in exosomes of the infected cells, *H. pylori* may control inflammation [36,41]. As a result, MyD88 synthesis and cytokine IL-8 levels are decreased [42].

Conversely, a persistent *H. pylori* infection might result in other issues, such as coronary artery disease [43]. MicroRNAs have a role in controlling this as well. In particular, miR-25 is stimulated and produced by exosomes, which heightens vascular endothelial cells inflammatory factors expression [44]. Moreover, miR-223, miR-218, and miR-21 are upregulated in cases of gastric cancer following a *H. pylori* infection and are probably carcinogenic [45]. As a result, these miRNAs are being employed as markers for stomach inflammation and cancer caused by *H. pylori*. These findings provide insights on the relationship between infections associated with bacteria and illnesses that are related to miRNA changes.

Recently, it was found that miR-155 is crucial for regulating a normal immune response or inflammatory response [46,47]. When it comes to adaptive and innate immune responses, miRNA-155 is essential. Activated macrophages, T cells, and dendritic cells can express miR-155. Recent research has shown that miR-155 plays specific functions in differentiating Th1 or Th17 cells during microbial infections. This miR-155 expression is elevated in T cells in response to the pathogenic microbe challenge [48].

In a study, the levels of miR-25 were assessed using qRT-PCR [49]. Protein concentrations were determined by Western blotting, exosomes by electron microscopy. The miR-25 target gene was determined by luciferase report technique. According to the findings, infection by *H. pylori* increased miR-25 expression in gastric epithelial cells and linked with miR-25 in the human blood. Kruppel-like factor 2 (KLF2) was discovered to be a target of vascular endothelial cells miR-25. Additionally, the monocyte chemoattractant protein-1 (MCP-1), interleukin 6 (IL6), and vascular cell adhesion molecule-1 (VCAM-1) levels were increased because of the miR-25/KLF2 axis controlling the intercellular adhesion molecule-1 (ICAM-1) and NF-B signaling cascade (VCAM-1). This study concluded that miR-25/KLF2 pathway might qualify as a therapeutic target for coronary heart disease linked to *H. pylori*. In addition, increased levels of miR-25, which is transported by exosomes, in blood may be a factor, which is risky for coronary heart disease [49]. Moreover, miR-155 is a component of the negative feedback loop that causes other inflammatory mediators to be downregulated [50].

The transcript of miR-155 can be increased in stomach epithelial cells and mucosal organs by *H. pylori* infection. As miR-155 has upregulated in *H. pylori*—infected macrophages, this suggests that miR-155 may function as a novel negative regulator in the inflammation caused by *H. pylori* [48]. The study also found that *H. pylori*, repressed in vitro on a growth medium, might be harmed by macrophages transfected with miR-155 packaged in exosomes. Nonetheless, the cell culture supernatant of macrophages transfected with miR-155 packaged in exosomes may likewise limit *Helicobacter pylori* proliferation, indicating that macrophage cytokines are involved in the suppression [48].

Two secreted exosomal RNAs, sR-2509025 and sR-989262, had higher expressions in outer membrane vesicles and could reduce IL-8 secretion in cultured human gastric adenocarcinoma cells. These data indicate the critical role of *H. pylori*'s outer membrane vesicles and secreted RNAs in impacting the hosts' immune responses [40].

5.2 *Mycobacterium tuberculosis*

Tuberculosis (TB) infections are a global issue. To help slow the spread of TB, we need to have better diagnostic techniques and medications. Research on the roles of exosomes related to tuberculosis is still in its early stages. Typically, *M. tuberculosis* secretes short RNAs in the hosts' macrophages that range in size from 15 to 43 nucleotides [51]. In recent studies, exosomes from *M. tuberculosis* (Mtb) have been isolated and characterized both in vivo and in vitro. Mtb-derived exosomes (Mtbexo), which deliver mycobacterial components to target cells, may be crucial in the development of TB. The TB bacteria utilize many proteins and miRNAs for pathogenesis, according to Mtbexo proteomic and transcriptome investigations. Exosome study requires strict standardization of exosome extraction from biological fluids, essential for tuberculosis research [52].

M. tuberculosis is the most often used model to study miRNAs and bacteria during infection [53]. Ruptures in phagosomes are common in infections associated with mycobacteria, and miRNAs have a vital contribution in this process. For instance, the *Mycobacterium* TB virulence factors EsxA and EsxB work as miRNA regulators and phagosome development inhibitors. As a result of esxBA genes deletion, expression of exosomal RNAs such as miR-148a, miR-147, and mir-206, and increases, all of which are crucial for the host's production of a range of inflammatory cytokines [54—56].

Also, the immune system's response was reduced by miR-99b, miR-1178, and miR-20a-3p, overexpression in cells infected by TB, which allowed pathogen to become established in the host. miR-20a-3p regulates host defense by preventing proinflammatory mediators productions through IKK/NF-κB pathway [57]. Via MyD88 signaling, an increase in miR-99b simultaneously reduces the production of proinflammatory cytokines [58]. Finally, miR-1178 increases the survival rate of *M. tuberculosis* by reducing the immune response in cells that are infected with the disease [59]. However, several miRNAs' roles in TB are unknown, and their expression is cell-type dependent [60]. For instance, the miR-155 promotes bacterial survival and cell growth in macrophages at one end. In contrast, on the other end, using various metabolic pathways, miR-155 suppresses clearances of bacteria and production of cytokines in T cells [61].

During the intracellular infection with *M. tuberculosis*, autophagy is crucial, and miRNAs regulate this molecular process. *M. tuberculosis* is less likely to colonize cells when miR-155 and miR-17-5p are expressed. This is accomplished by controlling several metabolic pathways that cause macrophage autophagy to be activated [61,62]. The increase of miR-155 results in the stimulation of autophagy and the eradication of mycobacteria. miR-155 specifically binds the Rheb genes' 3′UTR region, which promotes the phagosomes maturation, binding of the lysosomes, and ultimately eradicating cells [63]. To reduce autophagy, *M. tuberculosis* can control the synthesis of several miRNAs. As an illustration, miR-27 is elevated and suppresses calcium-associated autophagy during *M. tuberculosis* infection [64].

The miR-27 target is Ca^{2+}-transporting CAC-NA2D3, which is located at the endoplasmic reticulum (ER), and whose underexpression prevents formation of autophagosomes [64]. The miR-1958 expression, which binds to Atg5 3′UTR region and the silencing prevents autophagic flow, is brought on by *M. tuberculosis* infection [65].

Ataxia–telangiectasia-mutated (ATM) gene suppression increases *M. tuberculosis* infection by lowering levels of LC3-II in cells, which are infected and inhibiting xenophagy [66]. miR-33 is upregulated during infection of *M. tuberculosis*, which targets several genes in the host cells involved in cholesterol transfer and oxidation of fatty acids, including HADHB, ABCA1, CPT1, PRKAA1, and CROT. *M. tuberculosis*'s extremely lipid-dependent metabolism causes infected host cells to undergo lipid catabolism, which aids in bacterial colonization [67]. Hence, anti-miR-33 antibody might promote maturation of phagosomes and removal of bacteria by preventing contaminated host cell's lipid metabolism.

Secreted RNA from *Mycobacterium* has also been identified to induce apoptosis [68]. The induced apoptosis occurs in human monocytes through caspase-1, TNF-α-independent pathway, and caspase-8-dependent pathway [69]. The host, on the contrary, can stimulate the killing of bacterial pathogens by inducing a robust immune response. The mechanism involves the trafficking of the *Mycobacterium* into Lamp-1 and LC3 vesicles. This action often follows the RNA sensing pathway from the host that involves players such as RIG-1/MAVS/TBK1/IRF3 [70].

5.3 *Listeria monocytogenes*

Listeria monocytogenes disrupt the host cell's structure by using virulence factors. Previous research has shown that *L. monocytogenes* utilize microRNAs to exploit cells [71]. To construct an immune-tolerant milieu and host pathways to their benefit, *Listeria* has developed sophisticated techniques for manipulating host miRNAs. This has increased the viability, multiplication, and longevity of bacteria. These miRNAs are essential for proper host immune response, infection prevention, and eradication [72].

A wide variety of miRNAs can be up- or downregulated by *L. monocytogenes*, and each one may affect different cellular pathways or genes. While *L. monocytogenes* may influence miRNAs differently if these effector proteins are absent, internalins and listeriolysin are principally responsible for miRNA expression in Caco-2 epithelial cells [72].

Exosomal RNAs such as miR-146a expression are more in listeriosis, and its silencing decreases the ability of the pathogen to infiltrate macrophages [73]. miR-21 is induced by infection and regulates macrophage polarization, ultimately reducing *L. monocytogenes* intracellular survival [74,75]. By regulating the ephrin receptor tyrosine kinase 2 (EphA2), miR-26a controls *L. monocytogenes* infection by preventing the pathogen's internalization or phagosomal escape [76].

Noncoding RNAs in *Listeria* can trigger type I interferons (IFN) and IFN-β expressions. Specifically, noncoding RNA rli32 from the pathogen can induce IFN-β expressions, allowing the bacteria to propagate inside the host cells [77]. rli32 RIG-I (retinoic acid–inducible gene I)-dependent induction of IFN-β expression may involve Zea. This RNA-binding protein is responsible for RNA accumulation in the extracellular medium for *L. monocytogenes*.

5.4 *Salmonella* sp.

Salmonella is a gram-negative bacterium enteric in character, which can induce infections foodborne in nature in people, known as salmonellosis. *Salmonella enterica* serovar Typhimurium showed secretion of RNAs in outer membrane vesicles when cultured in a medium simulating interactions in host pathogens [78]. The research also revealed a distinct degradation or processing process, as transcripts kept the similar read coverage profiles in outer-membrane vesicles while others displayed a different pattern. *Salmonella* RNAs are also safeguarded by outer membrane vesicles, as transcripts that are full length (CsrC, SsrS, rnpB, and 10Sa) contained in outer membrane vesicles stay intact after digestion [78].

Salmonella enteritidis can release RNAs when in HT-29 cells of gut epithelium [79]. Furthermore, those RNAs can simply take their mature form after being processed by the host cell machinery to create microRNA-like fragment structures, which is the situation with Sal-1, the product of the *Salmonella* 5′-leader ribosomal RNA transcript being processed. The host cell Argonaute 2 (AGO2) is required to mature Sal-1 intermediates. The authors depleted Sal-1 and demonstrated that *Salmonella* resistance to host defense decreased, revealing a new *Salmonella* strategy for evading human immune clearance [79].

5.5 *Escherichia coli*

As a gram-negative bacterium model, *Escherichia coli* was one of the first bacteria to be used to explore the existence of secreted RNAs. A study investigated the RNA composition of outer-membrane vesicles and their free form in the extracellular region using the model bacteria *E. coli* K-12 substrain MG1655 [80]. RNA molecules that are extracellular in nature in *E. coli* vary in size from 15 to 40 nucleotides. When intracellular medium, outer-membrane vesicles-free, and outer-membrane vesicles were examined, different expression patterns in RNA were detected. RNAs that are extracellular are enriched in products that are cleaved of structurally essential noncoding RNAs such as tmRNA (transfer-messenger RNA), 6S RNA, 4.5S RNA, and tRNAs. However, the reason the bacteria secrete these RNAs is yet to be explained [80].

A study evaluated the ability of 1% of RNA cargo from *Escherichia coli* (UPEC) strain 536 to reach the epithelial cells' intracellular space [81]. Some studies also confirmed the uptake of *E. coli* noncoding RNA CsrC into the epithelial cells and involved in carbon storage [81,82].

5.6 *Pseudomonas aeruginosa*

P. aeruginosa is a gram-negative opportunistic organism that often displays antibiotic resistance mechanisms, forming multicellular biofilms that are detrimental to patients with chronic obstructive pulmonary disease (COPD) and cystic fibrosis [83]. The small RNAs from *P. aeruginosa* are 15–45 nucleotides long [84].

sRNA52320 has been identified to induce immune responses in primary human bronchial epithelial (HBE) cells [84]. Lipopolysaccharides and outer-membrane vesicles—induced interleukin 8 (IL-8) release by cultivated HBE cells can be reduced by sRNA52320. The secreted RNA also inhibits outer-membrane vesicle—induced KC cytokine release and subsequent infiltration of neutrophils in rodent lungs [84].

Another study identified RNA that includes 16S and 23S in membrane vesicles produced from *P. aeruginosa* PAO1 catastrophic cell lysis as a part of SOS response [85]. The researchers also identified small RNAs in membrane vesicles, including PA3305.1 (PhrS) [85]. PhrS is known to stimulate the production of *Pseudomonas* quinolone signal (PQS) [86].

5.7 *Streptococcus sanguinis, Porphyromonas gingivalis,* and *Aggregatibacter actinomycetemcomitans*

Streptococcus sanguinis, Porphyromonas gingivalis, and Aggregatibacter actinomycetemcomitans have all contributed to the development of periodontitis. There was a reduction in antiinflammatory cytokines such as IL-5, IL-13, and IL-15 when Jurkat T cells were transferred with miRNA-size RNAs from the periodontal pathogens [87]. *Porphyromonas* vesicles, secreted RNAs from *Aggregatibacter*, and mi-RNA-size RNA were able to modulate the host cell immune response [88—90]. Secreted RNA from *Aggregatibacter* would increase TNF-α production through the NF-κB and TLR-8 pathways.

5.8 *Haemophilus influenzae*

Haemophilus influenzae is a gram-negative bacterium, well known to colonize human nasopharynx [91]. RNA secreted from *H. influenzae* has been observed to form biofilms in vitro [91]. Researchers have questioned the significance of NTHi (nontypeable *Haemophilus influenzae*) biofilms in diseases in the past years, even though NTHi develop form of aggregates compatible with formation of a biofilm, which impacts virulence [92—94]. Extracellular DNA (eDNA) is a significant component of NTHi biofilms, as suggested by studies in vivo [95]. Although DNase I treatment makes the NTHi present more susceptible to specific antibiotics [96], it is not known if the eDNA in question (or any EPS present) is of bacterial, host, or both origins [97]. In addition to proteins and eDNA, two NTHi lipooligosaccharide (LOS) components are implicated in the formation of biofilm: phosphorylcholine and sialic acid (Neu5Ac) [98]. As NTHi has been a Neu5Ac auxotroph, the molecule should be absorbed from host, and mutants defective in Neu5Ac incorporation into LOS are less capable of forming biofilms in vitro [99]. In an experiment conducted by Domenech et al. [91], it was observed that proteolytic enzymes caused the NTHi biofilms to disperse, suggesting that their matrix contains proteins essential to their preservation. DNase I treatment of the biofilms verified the presence of eDNA in the matrix and its importance in its maintenance. The growing biofilms when nuclease treated firmly indicated the involvement of extracellular RNA and not DNA in the development of biofilms [91]. Extracellular RNA is needed for the biofilm initial attachment, and not for its maintenance. Certainly, the presence of RNase reduced the formation of biofilms significantly but had no impact on already-formed biofilms. The findings are not typical of all clinical isolates corresponding to NTHi; necessity of extracellular RNA for biofilm formation has never been documented. Although the release of RNA into the medium is usually ascribed to processes autolytic in nature, *E. coli* and *M. tuberculosis* are able to synthesize RNA fragments in the culture medium even when no autolysis is present. In comparison, the role of eDNA in forming and preserving bacterial biofilms is well understood [91].

5.9 *Vibrio cholerae*

Sjöström et al. [100] separated outer-membrane vesicles from *Vibrio cholerae* strain A1552 and examined them ultrastructurally using cryo-electron tomography and TEM [100]. TEM has revealed outer-membrane vesicles with sizes varying from 25 to 200 nm. Cryo-electron tomography was used to examine the purified vesicles' 3D volume, which showed that bigger vesicles had double membranes [100]. These outer membrane vesicles may resemble bacterial minicells, particles achromosomal in nature produced from mutated bacterial cells that inactivate genes that regulate normal growth of bacterial cells. Outer membrane vesicles, on the other hand, are produced during normal bacterial development, and *V. cholerae* that has been used with no abnormalities in cell division. Also, outer-membrane vesicles are noticeably reduced in the area when compared with minicells. To explore if outer membrane vesicles may have RNA [100], submitted *V. cholerae* strain A1552 vesicle samples for extraction and purification of hot phenol RNA [100]. Because no bands were identified in the samples treated with RNase, the results firmly indicated that bands from untreated sample were mostly made up of RNA. RNA was also linked with outer-membrane vesicles isolated from bacteria during their exponential development period. It indicated that RNA release from cells associated to outer-membrane vesicles was not contingent on the growth phase. Also, nucleic acids from bacterial cell disintegration did not likely contaminate outer-membrane vesicle—associated RNA [100].

5.10 *Borrelia burgdorferi*

Lyme disease is caused by *Borrelia burgdorferi*, which belongs to spirochete family. *B. burgdorferi* has a linear genome and several linear and circular plasmids. *B. burgdorferi* secretes cylindrical outer-membrane vesicles, which transport secretory products including metabolites, proteins, and nucleic acids [101]. *B. burgdorferi* has been demonstrated to produce outer-membrane vesicles (20—250 nm) containing virulence factors that can attach to human endothelium cells in vitro and cause responses in B cell in mice models [101]. The outer-membrane vesicles include proteins, endotoxic lipopolysaccharides, RNA, and DNA. Outer-membrane vesicles are proven in studies to influence host immunological responses and in development of disease by transporting poisons and several virulence factors. Malge et al. [101] unraveled the transcriptomes of *B. burgdorferi* and outer-membrane vesicles to (1) elucidate the differences in the transcript that may exist between the cells and the secreted outer membrane vesicles and (2) investigate the role of outer membrane vesicles in the Lyme disease manifestation by considering functions associated with the outer-membrane vesicle—enriched transcripts. It was observed that the genetic source of transcripts caused a biasness in transcripts distribution from outer-membrane vesicles and bacteria. Transcripts that are plasmid-encoded are prone to be concentrated in outer membrane vesicles than in bacterial cells (concentration differences ~1.5 fold), but transcripts that are chromosome-encoded are more numerous in bacteria. Additional to transcripts, *B. burgdorferi* short noncoding RNAs distribution in the bacterial cell and outer-membrane vesicles was investigated. These RNAs were shown to be more likely to be enriched in outer-membrane vesicles, and there was a bias in the RNAs distribution from bacteria and outer-membrane vesicles, with bulk of the RNAs that are plasmid derived being higher in outer-membrane vesicles rather than bacteria. The study identified a distribution similar in nature for short RNAs, which is intriguing since short noncoding RNAs have been linked to the regulation of the enzootic cycle in *B. burgdorferi*, albeit precise role of several RNAs is unclear [102]. Malge et al. [101] observed that *B. burgdorferi* vesicles contain a diverse range of RNAs, many preferentially derived from plasmids and abundant in outer membrane vesicles. According to the findings of functional enrichment analysis, outer-membrane vesicles may have key involvements in *Borrelia* infection. Furthermore, outer-membrane vesicles may be important during Lyme disease. Further studies would examine whether *B. burgdorferi* outer-membrane vesicles can be identified, whether isolated from patients' bodily fluids, and whether outer-membrane vesicles can influence disease progression. This technique will open up new possibilities for investigating the functional role of outer-membrane vesicles in disease progression and treatment response [101].

5.11 *Chlamydia trachomatis*

Chlamydia trachomatis is an intracellular bacterium, which causes both serious sexually transmitted infections and a potentially fatal eye condition known as *Chlamydia* conjunctivitis or trachoma [103]. In an experiment conducted by Furuse et al. [51], *C. trachomatis* was infected into HeLa cells at an MOI of 20, from which total RNA was extracted 42 h later. In-depth sequencing of the short RNA transcriptome in the range 15—43 nucleotide size range generated ~11 million reads that could be aligned to the bacterial or human genome, with the bacterial genome accounting for 24.9% of reads [51].

To enrich short RNAs involved with the cellular RISC, Furuse et al. [51] used RNA immunoprecipitation (RIP) with the specific antibody for the crucial Argonaute (Ago) of RISC [51]. In-depth sequencing of these RISC-associated short RNAs (RIP-seq) revealed the two most prevalent bacterial small RNAs (CT-A and CT-B) obtained from bacterial tRNAs and was too long to be miRNAs, measuring 36 nucleotides and 32 nucleotides, respectively [51]. However, no reported short RNAs from bacteria account >0.09% of the overall miRNA pool, making them not to have a phenotypic influence that could be significant when loaded into RISC [51]. The findings also suggested against *C. trachomatis* expressing miRNAs in human cells that are infected [51].

5.12 *Legionella pneumophila*

Legionella pneumophila is a gram-negative bacterium that lives and replicates inside free-living ameba in soil and freshwater sources [51]. Even though *L. pneumophila* is predominantly a protozoan pathogen, bacteria inhaled and aerosolized are able to infect alveolar macrophages and produce infections opportunistic in nature in animal models and also in people [104]. *L. pneumophila* can effectively infect a wide range of species, which is due to *Legionella's* ability to inject over 300 unique bacterial proteins into the host cytoplasm [105]. Furuse et al. [51] tested whether *L. pneumophila* can create and transmit pre-miRNAs from bacteria into the host cells by infecting macrophage-like cell line RAW264.7 from murine with *L. pneumophila* at an MOI of 0.1 and isolating RNA 48 h later [51]. Sequencing of the whole population of short RNA (15—43 nucleotide) produced 14 million reads, which could be mapped to the bacterial or mouse genome. 9.5% were of *L. pneumophila* origin [51].

The immunoprecipitation small RNAs that are RISC-associated with a pan-Ago antibody provided ~20 million reads and boosted the fraction of reads corresponding to known murine miRNAs from 37.6% to 77.4% of the overall short RNA population. Furthermore, the reads of murine acquired by RIP-seq were nearly entirely inside the size window of anticipated $22 + 2$ nucleotides [51]. Small RNA readings from the *L. pneumophila* genome, on the other hand, fell from 9.5% of total small RNA reads to 1.4% of RISC-associated short RNA reads and were not clustered at the estimated 2262 nucleotide size expected for legitimate miRNAs. The RISC-associated small RNA library significantly boosted the 10 most common small RNA readings of *L. pneumophila*, dubbed LP-A to LP-J. However, 7 of the 10 were generated from bacterial tRNAs. LP-D, together with neighboring bacterial genomic sequences, is projected to fold into stem—loop structure of the three remaining reads. However, the length of this small RNA, 15 nucleotides, is short for a legitimate miRNA, and all 10 of the dominating *L. pneumophila* small RNAs not only sit outside the predicted size range for a functioning miRNA but are also present at low levels so that those are not functionally important [106]. According to the findings of Furuse et al. [51], infected murine cells by *L. pneumophila* do not express miRNAs of bacterial origin [51].

6. Bacterial exosomal RNA in therapy development

In the previous sections, we have summarized how bacterial exosomal RNA can trigger the host's miRNA. The host's miRNA further regulates the innate immune pathways, which ultimately decide the inflammatory response from the host [107]. Host miRNAs have the potential to be used as biomarkers and develop new strategies for therapies against diverse pathogenic microorganisms, including bacteria [36,60,108–110] (Fig. 12.3).

For *H. pylori*, miR-21, miR-218, and miR-223 have been used as biomarkers of gastric cancer and inflammation [45]. In *M. tuberculosis*, the upregulation of miR-18a and miR20a-3p is linked to the downregulation of ATM and IKKβ. The downregulation of ATM and IKKβ can be linked with facilitating intracellular survivability [57,66]. miRNAs during *M. tuberculosis* infection play a crucial role in regulating autophagy [107]. The miR-21, miR-26a, and miR-29 have critical roles in *L. monocytogenes* infection. Their targets include MARCKS, RhoB, EPHA2, and IFN-γ

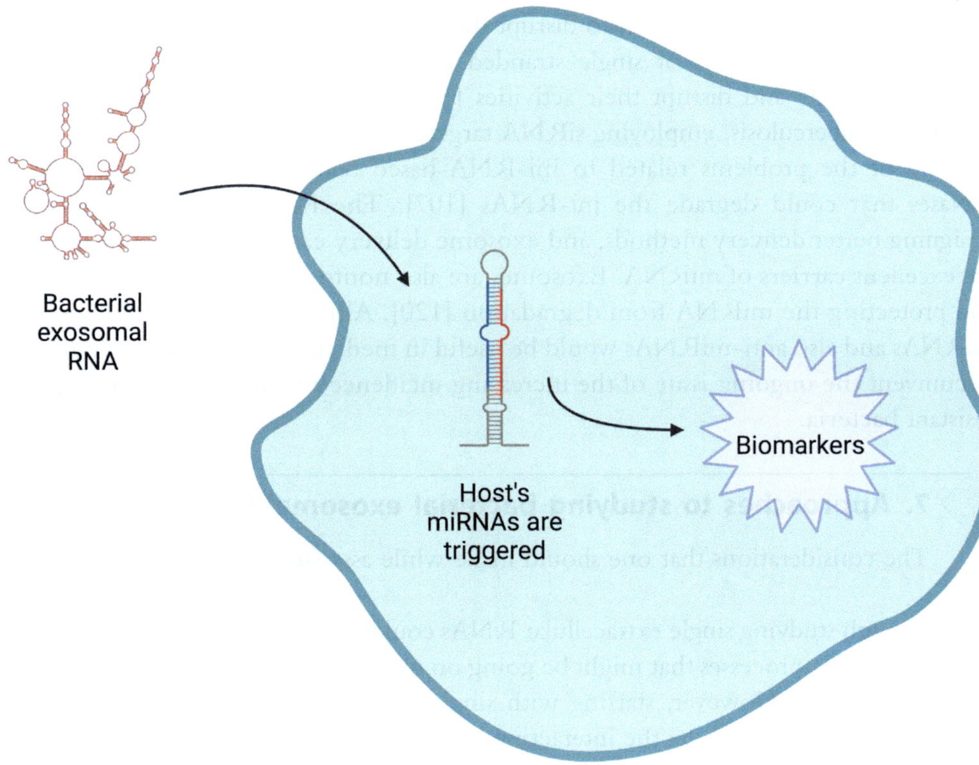

Figure 12.3 The potential role of the host's miRNAs as biomarkers when triggered by bacterial exosomal RNA.

[107]. MARCKS and RhoB prevent the internalization of pathogens [75], EPHA2 hinders intracellular survival [76], and IFN-γ lowers the immune response [111]. For *Salmonella* sp., RGS2 protein is targeted by miR-let-7i-3p of the host, controls the trafficking of the vacuoles during *Salmonella*, and prevents the pathogenesis [112]. In adherent-invasive *Escherichia coli*, exosomes that contain miR-30c and miR-130a are released to silence ATG5 and ATG16L1 in noninfected cells. That inhibits autophagic flux, allowing easy intracellular replication of *E. coli* [113]. In *C. trachomatis*, miR-19, miR-451, and miR-9, controlling the pathway corresponding to NF-κB, are overexpressed during infection [114]. Similarly, the overexpression of miR-155 and down-expression of miR-184 are related to patients' inflammation with trichoma [115]. So, the expression patterns of specific miRNA can be excellent markers of genital infections [116]. Upregulation of genes such as miR-125b, miR-579, and miR-221 in *L. pneumophila* results in the clearance of bacteria [117]. On the contrary, the over-expression of miR-221, miR-125b, and miR-579 promotes the intracellular propagation of the microorganisms [117]. These miRNAs are essential in antibacterial and host cell immune responses [107].

Strategies involve designing anti-miRNAs that target miRNAs that aid infection of the bacteria. The ultimate goal is to disrupt the pathogen's ability to colonize the host. Anti-miRNAs are RNAs of single stranded in nature that selectively complement specific miRNAs and disrupt their activities [118]. The strategy is successful against pulmonary tuberculosis, employing siRNA targeting tfgb1 [119].

One of the problems related to mi-RNA-based technology is the presence of RNases that could degrade the mi-RNAs [107]. The issue could be avoided by designing better delivery methods, and exosome delivery can be an option. Exosomes are excellent carriers of miRNA. Exosomes are also nontoxic, having low antigenicity and protecting the miRNA from degradation [120]. Although very new technology, miRNAs and also anti-miRNAs would be useful in medicinal sectors. That would also circumvent the ongoing issue of the increasing incidence of antimicrobial multidrug-resistant bacteria.

7. Approaches to studying bacterial exosomal RNAs

The considerations that one should make while assessing the extracellular RNAs are:

(1) Although studying single extracellular RNAs could be misleading and oversimplifies the complex processes that might be going on in vesicles that consist of thousands of RNA species. However, starting with single extracellular RNAs is an intelligent approach to determining the interacting host cell proteins before considering other complex physiological parameters.

(2) Another potential approach could be identifying the bacterial RNA's cellular localization in the host cells. Identifying whether bacterial RNAs are in the nucleus, cytosol, or membrane-bound should indicate their interacting cellular partners.

(3) It is essential to highlight that the exosomal RNAs are in two forms: (1) one form when it is packaged and (2) another form when the RNA is delivered into the host cells. Some RNA functions may alter during the transition from the packaging to the delivery stage. The RNA regulation should be different in exosomes in eukaryotic hosts than when packaged and should be considered when studying bacterial RNAs.

8. Conclusions

The studies on the role of exosomal RNA in bacterial infections are still in their infancy. There are outstanding questions in this field. Formal studies on the role of bacterial exosomal RNAs will yield significant biological implications. We see that exosomal RNAs aid the bacteria in infecting the host cells. On the contrary, the host cells generate an immune response that could modulate the infective properties of the bacteria. The host—bacterial infective—antiinfective interactions can be critical in understanding the dynamics. The outstanding questions include the following: (1) What are the mechanisms behind the sorting and secretion of exosomal RNAs? (2) How do bacteria use the exosomal RNA to infect the hosts, and what are the mechanisms? (3) How does the host respond to the infection by bacterial exosomal RNA? (4) What are the roles of bacterial exosomal RNAs in the severity of diseases in the host? (5) Is there any way to target these bacterial exosomal RNAs to develop therapy? (6) Could this field be exploited to develop methods to diagnose human diseases? (7) There is a vast pool of unexplored bacterial populations that could contain exosomal RNAs that shed new light on the knowledge that we possess currently. The outstanding questions invite more studies and research to ensure a proper understanding of the role of exosomal RNA in bacterial infections.

Acknowledgments

Soumyadev Sarkar would like to acknowledge the support of Prof. Ferran Garcia-Pichel of the Center for Fundamental and Applied Microbiomics, Arizona State University, and all the members of Garcia-Pichel Lab. The figures are created with BioRender.com.

References

[1] Zhang W, Jiang X, Bao J, Wang Y, Liu H, Tang L. Exosomes in pathogen infections: a bridge to deliver molecules and link functions. Frontiers in Immunology 2018;9:90. https://doi.org/10.3389/fimmu.2018.00090.

[2] Tkach M, Théry C. Communication by extracellular vesicles: where we are and where we need to go. Cell 2016;164(6):1226—32. https://doi.org/10.1016/j.cell.2016.01.043.

[3] Théry C, Zitvogel L, Amigorena S. Exosomes: composition, biogenesis and function. Nature Reviews Immunology 2002;2(8):569—79. https://doi.org/10.1038/nri855.

[4] Wang J, Yao Y, Chen X, Wu J, Gu T, Tang X. Host derived exosomes-pathogens interactions: potential functions of exosomes in pathogen infection. Biomedicine and Pharmacotherapy 2018;108:1451−9. https://doi.org/10.1016/j.biopha.2018.09.174.

[5] H Rashed M, Bayraktar E, K Helal G, Abd-Ellah MF, Amero P, Chavez-Reyes A, et al. Exosomes: from garbage bins to promising therapeutic targets. International Journal of Molecular Sciences 2017;18(3). https://doi.org/10.3390/ijms18030538.

[6] Guay C, Regazzi R. Exosomes as new players in metabolic organ cross-talk. Diabetes, Obesity and Metabolism 2017;19:137−46. https://doi.org/10.1111/dom.13027.

[7] Singh PP, Li L, Schorey JS. Exosomal RNA from Mycobacterium tuberculosis-infected cells is functional in recipient macrophages. Traffic 2015;16(6):555−71. https://onlinelibrary.wiley.com/doi/abs/10.1111/tra.12278.

[8] Zhou X, Xie F, Wang L, Zhang L, Zhang S, Fang M, et al. The function and clinical application of extracellular vesicles in innate immune regulation. Cellular and Molecular Immunology 2020;17(4):323−34. https://doi.org/10.1038/s41423-020-0391-1.

[9] Shaw AC, Joshi S, Greenwood H, Panda A, Lord JM. Aging of the innate immune system. Current Opinion in Immunology 2010;22(4):507−13. https://doi.org/10.1016/j.coi.2010.05.003.

[10] Narang P, Shah M, Beljanski V. Exosomal RNAs in diagnosis and therapies. Non-Coding RNA Research 2022;7(1):7−15. https://doi.org/10.1016/j.ncrna.2022.01.001.

[11] Valadi H, Ekström K, Bossios A, Sjöstrand M, Lee JJ, Lötvall JO. Exosome-mediated transfer of mRNAs and microRNAs is a novel mechanism of genetic exchange between cells. Nature Cell Biology 2007;9(6):654−9. https://doi.org/10.1038/ncb1596.

[12] Chaput N, Taïeb J, André F, Zitvogel L. The potential of exosomes in immunotherapy. Expert Opinion on Biological Therapy 2005;5(6):737−47. https://doi.org/10.1517/14712598.5.6.737.

[13] Escola JM, Kleijmeer MJ, Stoorvogel W, Griffith JM, Yoshie O, Geuze HJ. Selective enrichment of tetraspan proteins on the internal vesicles of multivesicular endosomes and on exosomes secreted by human B-lymphocytes. Journal of Biological Chemistry 1998;273(32):20121−7. https://doi.org/10.1074/jbc.273.32.20121.

[14] Krek A, Grün D, Poy MN, Wolf R, Rosenberg L, Epstein EJ, et al. Combinatorial microRNA target predictions. Nature Genetics 2005;37(5):495−500. https://doi.org/10.1038/ng1536.

[15] Lim LP, Lau NC, Garrett-Engele P, Grimson A, Schelter JM, Castle J, et al. Microarray analysis shows that some microRNAs downregulate large numbers of target mRNAs. Nature 2005;433(7027):769−73. https://doi.org/10.1038/nature03315.

[16] Skog J, Würdinger T, van Rijn S, Meijer DH, Gainche L, Sena-Esteves M, et al. Glioblastoma microvesicles transport RNA and proteins that promote tumour growth and provide diagnostic biomarkers. Nature Cell Biology 2008;10(12):1470−6. https://doi.org/10.1038/ncb1800.

[17] Subra C, Laulagnier K, Perret B, Record M. Exosome lipidomics unravels lipid sorting at the level of multivesicular bodies. Biochimie 2007;89(2):205−12. https://doi.org/10.1016/j.biochi.2006.10.014.

[18] Gibbings DJ, Ciaudo C, Erhardt M, Voinnet O. Multivesicular bodies associate with components of miRNA effector complexes and modulate miRNA activity. Nature Cell Biology 2009;11(9):1143−9. https://doi.org/10.1038/ncb1929.

[19] Lee YS, Pressman S, Andress AP, Kim K, White JL, Cassidy JJ, et al. Silencing by small RNAs is linked to endosomal trafficking. Nature Cell Biology 2009;11(9):1150−6. https://doi.org/10.1038/ncb1930.

[20] Yang C, Chalasani G, Ng Y-H, Robbins PD. Exosomes released from Mycoplasma infected tumor cells activate inhibitory B cells. PLoS One 2012;7(4):e36138. https://doi.org/10.1371/journal.pone.0036138.

[21] Zhang J, Li S, Li L, Li M, Guo C, Yao J, et al. Exosome and exosomal microRNA: trafficking, sorting, and function. Genomics, Proteomics and Bioinformatics 2015;13(1):17−24. https://doi.org/10.1016/j.gpb.2015.02.001.

[22] Diaz G, Wolfe LM, Kruh-Garcia NA, Dobos KM. Changes in the membrane-associated proteins of exosomes released from human macrophages after *Mycobacterium tuberculosis* infection. Scientific Reports 2016;6:37975. https://doi.org/10.1038/srep37975.

[23] Khan MZ, Bhaskar A, Upadhyay S, Kumari P, Rajmani RS, Jain P, et al. Protein kinase G confers survival advantage to *Mycobacterium tuberculosis* during latency-like conditions. Journal of Biological Chemistry 2017;292(39):16093–108. https://doi.org/10.1074/jbc.M117.797563.
[24] Zheng M-L, Zhou N-K, Luo C-H. MiRNA-155 and miRNA-132 as potential diagnostic biomarkers for pulmonary tuberculosis: a preliminary study. Microbial Pathogenesis 2016;100:78–83. https://doi.org/10.1016/j.micpath.2016.09.005.
[25] Vanaja SK, Russo AJ, Behl B, Banerjee I, Yankova M, Deshmukh SD, et al. Bacterial outer membrane vesicles mediate cytosolic localization of LPS and caspase-11 activation. Cell 2016;165(5):1106–19. https://doi.org/10.1016/j.cell.2016.04.015.
[26] Yaron S, Kolling GL, Simon L, Matthews KR. Vesicle-mediated transfer of virulence genes from *Escherichia coli* O157:H7 to other enteric bacteria. Applied and Environmental Microbiology 2000;66(10):4414–20. https://doi.org/10.1128/AEM.66.10.4414-4420.2000.
[27] Schooling SR, Beveridge TJ. Membrane vesicles: an overlooked component of the matrices of biofilms. Journal of Bacteriology 2006;188(16):5945–57. https://doi.org/10.1128/JB.00257-06.
[28] Dorward DW, Garon CF. DNA-binding proteins in cells and membrane blebs of Neisseria gonorrhoeae. Journal of Bacteriology 1989;171(8):4196–201. https://doi.org/10.1128/jb.171.8.4196-4201.1989.
[29] Lécrivain A-L, Beckmann BM. Bacterial RNA in extracellular vesicles: a new regulator of host-pathogen interactions? Biochimica et Biophysica Acta, Gene Regulatory Mechanisms 2020;1863(7):194519. https://doi.org/10.1016/j.bbagrm.2020.194519.
[30] Turner L, Bitto NJ, Steer DL, Lo C, D'Costa K, Ramm G, et al. *Helicobacter pylori* outer membrane vesicle size determines their mechanisms of host cell entry and protein content. Frontiers in Immunology 2018;9:1466. https://doi.org/10.3389/fimmu.2018.01466.
[31] Baxt LA, Garza-Mayers AC, Goldberg MB. Bacterial subversion of host innate immune pathways. Science 2013;340(6133):697–701. https://doi.org/10.1126/science.1235771.
[32] Westermann AJ, Förstner KU, Amman F, Barquist L, Chao Y, Schulte LN, et al. Dual RNA-seq unveils noncoding RNA functions in host–pathogen interactions. Nature 2016;529(7587):496–501. https://doi.org/10.1038/nature16547.
[33] Chao Y, Vogel J. The role of Hfq in bacterial pathogens. Current Opinion in Microbiology 2010;13(1):24–33. https://doi.org/10.1016/j.mib.2010.01.001.
[34] Chen J, Morita T, Gottesman S. Regulation of transcription termination of small RNAs and by small RNAs: molecular mechanisms and biological functions. Frontiers in Cellular and Infection Microbiology 2019;9:201. https://doi.org/10.3389/fcimb.2019.00201.
[35] Boisset S, Geissmann T, Huntzinger E, Fechter P, Bendridi N, Possedko M, et al. *Staphylococcus aureus* RNAIII coordinately represses the synthesis of virulence factors and the transcription regulator Rot by an antisense mechanism. Genes and Development 2007;21(11):1353–66. https://doi.org/10.1101/gad.423507.
[36] Eulalio A, Schulte L, Vogel J. The mammalian microRNA response to bacterial infections. RNA Biology 2012;9(6):742–50. https://doi.org/10.4161/rna.20018.
[37] Carpenter S, Ricci EP, Mercier BC, Moore MJ, Fitzgerald KA. Post-transcriptional regulation of gene expression in innate immunity. Nature Reviews Immunology 2014;14(6):361–76. https://doi.org/10.1038/nri3682.
[38] Kafasla P, Skliris A, Kontoyiannis DL. Post-transcriptional coordination of immunological responses by RNA-binding proteins. Nature Immunology 2014;15(6):492–502. https://doi.org/10.1038/ni.2884.
[39] Jones KR, Whitmire JM, Merrell DS. A tale of two toxins: *Helicobacter pylori* CagA and VacA modulate host pathways that impact disease. Frontiers in Microbiology 2010;1:115. https://doi.org/10.3389/fmicb.2010.00115.
[40] Zhang H, Zhang Y, Song Z, Li R, Ruan H, Liu Q, et al. sncRNAs packaged by *Helicobacter pylori* outer membrane vesicles attenuate IL-8 secretion in human cells. International Journal of Medical Microbiology 2020;310(1):151356. https://doi.org/10.1016/j.ijmm.2019.151356.

[41] Săsăran MO, Meliţ LE, Dobru ED. MicroRNA modulation of host immune response and inflammation triggered by *Helicobacter pylori*. International Journal of Molecular Sciences 2021;22(3). https://doi.org/10.3390/ijms22031406.
[42] Tang B, Xiao B, Liu Z, Li N, Zhu E-D, Li B-S, et al. Identification of MyD88 as a novel target of miR-155, involved in negative regulation of *Helicobacter pylori*-induced inflammation. FEBS Letters 2010;584(8):1481–6. https://doi.org/10.1016/j.febslet.2010.02.063.
[43] Lee M, Baek H, Park JS, Kim S, Kyung C, Baik SJ, et al. Current *Helicobacter pylori* infection is significantly associated with subclinical coronary atherosclerosis in healthy subjects: a cross-sectional study. PLoS One 2018;13(3):e0193646. https://doi.org/10.1371/journal.pone.0193646.
[44] Ceppi M, Pereira PM, Dunand-Sauthier I, Barras E, Reith W, Santos MA, et al. MicroRNA-155 modulates the interleukin-1 signaling pathway in activated human monocyte-derived dendritic cells. Proceedings of the National Academy of Sciences of the United States of America 2009;106(8):2735–40. https://doi.org/10.1073/pnas.0811073106.
[45] Li B-S, Zhao Y-L, Guo G, Li W, Zhu E-D, Luo X, et al. Plasma microRNAs, miR-223, miR-21 and miR-218, as novel potential biomarkers for gastric cancer detection. PLoS One 2012;7(7):e41629. https://doi.org/10.1371/journal.pone.0041629.
[46] Matta R, Wang X, Ge H, Ray W, Nelin LD, Liu Y. Triptolide induces anti-inflammatory cellular responses. American Journal of Tourism Research 2009;1(3):267–82. https://www.ncbi.nlm.nih.gov/pubmed/19956437.
[47] Worm J, Stenvang J, Petri A, Frederiksen KS, Obad S, Elmén J, et al. Silencing of microRNA-155 in mice during acute inflammatory response leads to derepression of c/ebp Beta and down-regulation of G-CSF. Nucleic Acids Research 2009;37(17):5784–92. https://doi.org/10.1093/nar/gkp577.
[48] Wang J, Deng Z, Wang Z, Wu J, Gu T, Jiang Y, et al. MicroRNA-155 in exosomes secreted from *Helicobacter pylori* infection macrophages immunomodulates inflammatory response. American Journal of Tourism Research 2016;8(9):3700–9. https://www.ncbi.nlm.nih.gov/pubmed/27725852.
[49] Li N, Liu S-F, Dong K, Zhang G-C, Huang J, Wang Z-H, et al. Exosome-transmitted miR-25 induced by *H. pylori* promotes vascular endothelial cell injury by targeting KLF2. Frontiers in Cellular and Infection Microbiology 2019;9:366. https://doi.org/10.3389/fcimb.2019.00366.
[50] Risch HA, Lu L, Kidd MS, Wang J, Zhang W, Ni Q, et al. *Helicobacter pylori* seropositivities and risk of pancreatic carcinoma. Cancer Epidemiology, Biomarkers and Prevention: A Publication of the American Association for Cancer Research, Cosponsored by the American Society of Preventive Oncology 2014;23(1):172–8. https://doi.org/10.1158/1055-9965.EPI-13-0447.
[51] Furuse Y, Finethy R, Saka HA, Xet-Mull AM, Sisk DM, Smith KLJ, et al. Search for microRNAs expressed by intracellular bacterial pathogens in infected mammalian cells. PLoS One 2014;9(9):e106434. https://doi.org/10.1371/journal.pone.0106434.
[52] Biadglegne F, König B, Rodloff AC, Dorhoi A, Sack U. Composition and clinical significance of exosomes in tuberculosis: a systematic literature review. Journal of Clinical Medicine Research 2021;10(1). https://doi.org/10.3390/jcm10010145.
[53] Cullen BR. Viruses and microRNAs: RISCy interactions with serious consequences. Genes and Development 2011;25(18):1881–94. https://doi.org/10.1101/gad.17352611.
[54] Wright K, de Silva K, Plain KM, Purdie AC, Blair TA, Duggin IG, et al. Mycobacterial infection-induced miR-206 inhibits protective neutrophil recruitment via the CXCL12/CXCR4 signalling axis. PLoS Pathogens 2021;17(4):e1009186. https://doi.org/10.1371/journal.ppat.1009186.
[55] Wu H, Bao Y, Wang L, Li X, Sun J. Mycobacterium marinum down-regulates miR-148a in macrophages in an EsxA-dependent manner. International Immunopharmacology 2019;73:41–8. https://doi.org/10.1016/j.intimp.2019.04.056.
[56] Zuo X, Wang L, Bao Y, Sun J. The ESX-1 virulence factors downregulate miR-147-3p in Mycobacterium marinum-infected macrophages. Infection and Immunity 2020;88(6). https://doi.org/10.1128/IAI.00088-20.
[57] Cui J, Li Z, Cui K, Gao Y, Zhang B, Niu J, et al. MicroRNA-20a-3p regulates the host immune response to facilitate the mycobacterium tuberculosis infection by targeting IKKβ/NF-κB pathway. International Immunopharmacology 2021;91:107286. https://doi.org/10.1016/j.intimp.2020.107286.

[58] Singh Y, Kaul V, Mehra A, Chatterjee S, Tousif S, Dwivedi VP, et al. *Mycobacterium tuberculosis* controls microRNA-99b (miR-99b) expression in infected murine dendritic cells to modulate host immunity. Journal of Biological Chemistry 2013;288(7):5056–61. https://doi.org/10.1074/jbc.C112.439778.
[59] Shi G, Mao G, Xie K, Wu D, Wang W. MiR-1178 regulates mycobacterial survival and inflammatory responses in Mycobacterium tuberculosis-infected macrophages partly via TLR4. Journal of Cellular Biochemistry 2018;119(9):7449–57. https://doi.org/10.1002/jcb.27054.
[60] Silwal P, Kim YS, Basu J, Jo E-K. The roles of microRNAs in regulation of autophagy during bacterial infection. Seminars in Cell and Developmental Biology 2020;101:51–8. https://doi.org/10.1016/j.semcdb.2019.07.011.
[61] Rothchild AC, Sissons JR, Shafiani S, Plaisier C, Min D, Mai D, et al. MiR-155-regulated molecular network orchestrates cell fate in the innate and adaptive immune response to *Mycobacterium tuberculosis*. Proceedings of the National Academy of Sciences of the United States of America 2016;113(41):E6172–81. https://doi.org/10.1073/pnas.1608255113.
[62] Kumar R, Sahu SK, Kumar M, Jana K, Gupta P, Gupta UD, et al. MicroRNA 17-5p regulates autophagy in Mycobacterium tuberculosis-infected macrophages by targeting Mcl-1 and STAT3. Cellular Microbiology 2016;18(5):679–91. https://doi.org/10.1111/cmi.12540.
[63] Wang J, Yang K, Zhou L, Minhaowu, Wu Y, Zhu M, et al. MicroRNA-155 promotes autophagy to eliminate intracellular mycobacteria by targeting Rheb. PLoS Pathogens 2013;9(10):e1003697. https://doi.org/10.1371/journal.ppat.1003697.
[64] Liu F, Chen J, Wang P, Li H, Zhou Y, Liu H, et al. MicroRNA-27a controls the intracellular survival of *Mycobacterium tuberculosis* by regulating calcium-associated autophagy. Nature Communications 2018;9(1):4295. https://doi.org/10.1038/s41467-018-06836-4.
[65] Ding S, Qu Y, Yang S, Zhao Y, Xu G. Novel miR-1958 promotes *Mycobacterium tuberculosis* survival in RAW264.7 cells by inhibiting autophagy via Atg5. Journal of Microbiology and Biotechnology 2019;29(6):989–98. https://doi.org/10.4014/jmb.1811.11062.
[66] Yuan Q, Chen H, Yang Y, Fu Y, Yi Z. miR-18a promotes Mycobacterial survival in macrophages via inhibiting autophagy by down-regulation of ATM. Journal of Cellular and Molecular Medicine 2020;24(2):2004–12. https://doi.org/10.1111/jcmm.14899.
[67] Ouimet M, Koster S, Sakowski E, Ramkhelawon B, van Solingen C, Oldebeken S, et al. *Mycobacterium tuberculosis* induces the miR-33 locus to reprogram autophagy and host lipid metabolism. Nature Immunology 2016;17(6):677–86. https://doi.org/10.1038/ni.3434.
[68] Pita T, Feliciano JR, Leitão JH. Extracellular RNAs in bacterial infections: from emerging key players on host-pathogen interactions to exploitable biomarkers and therapeutic targets. International Journal of Molecular Sciences 2020;21(24). https://doi.org/10.3390/ijms21249634.
[69] Obregón-Henao A, Duque-Correa MA, Rojas M, García LF, Brennan PJ, Ortiz BL, et al. Stable extracellular RNA fragments of *Mycobacterium tuberculosis* induce early apoptosis in human monocytes via a caspase-8 dependent mechanism. PLoS One 2012;7(1):e29970. https://doi.org/10.1371/journal.pone.0029970.
[70] Cheng Y, Schorey JS. Extracellular vesicles deliver Mycobacterium RNA to promote host immunity and bacterial killing. EMBO Reports 2019;20(3). https://doi.org/10.15252/embr.201846613.
[71] Lu L-F, Boldin MP, Chaudhry A, Lin L-L, Taganov KD, Hanada T, et al. Function of miR-146a in controlling Treg cell-mediated regulation of Th1 responses. Cell 2010;142(6):914–29. https://doi.org/10.1016/j.cell.2010.08.012.
[72] Allahverdy J, Rashidi N. MicroRNAs induced by Listeria monocytogenes and their role in cells. Microbial Pathogenesis 2023;175:105997. https://doi.org/10.1016/j.micpath.2023.105997.
[73] Du C-T, Gao W, Ma K, Yu S-X, Li N, Yan S-Q, et al. MicroRNA-146a deficiency protects against Listeria monocytogenes infection by modulating the gut microbiota. International Journal of Molecular Sciences 2018;19(4). https://doi.org/10.3390/ijms19040993.
[74] Wang Z, Brandt S, Medeiros A, Wang S, Wu H, Dent A, et al. MicroRNA 21 is a homeostatic regulator of macrophage polarization and prevents prostaglandin E2-mediated M2 generation. PLoS One 2015;10(2):e0115855. https://doi.org/10.1371/journal.pone.0115855.

[75] Johnston DGW, Kearney J, Zasłona Z, Williams MA, O'Neill LAJ, Corr SC. MicroRNA-21 limits uptake of Listeria monocytogenes by macrophages to reduce the intracellular niche and control infection. Frontiers in Cellular and Infection Microbiology 2017;7:201. https://doi.org/10.3389/fcimb.2017.00201.

[76] Zhang J, Yuan J, Wang L, Zheng Z, Ran H, Liu F, et al. MiR-26a targets EphA2 to resist intracellular Listeria monocytogenes in macrophages. Molecular Immunology 2020;128:69–78. https://doi.org/10.1016/j.molimm.2020.09.016.

[77] Frantz R, Teubner L, Schultze T, La Pietra L, Müller C, Gwozdzinski K, et al. The secRNome of Listeria monocytogenes harbors small noncoding RNAs that are potent inducers of beta interferon. mBio 2019;10(5). https://doi.org/10.1128/mbio.01223-19.

[78] Malabirade A, Habier J, Heintz-Buschart A, May P, Godet J, Halder R, et al. The RNA complement of outer membrane vesicles from *Salmonella enterica* serovar typhimurium under distinct culture conditions. Frontiers in Microbiology 2018;9:2015. https://doi.org/10.3389/fmicb.2018.02015.

[79] Gu H, Zhao C, Zhang T, Liang H, Wang X-M, Pan Y, et al. Salmonella produce microRNA-like RNA fragment Sal-1 in the infected cells to facilitate intracellular survival. Scientific Reports 2017;7(1):2392. https://doi.org/10.1038/s41598-017-02669-1.

[80] Ghosal A, Upadhyaya BB, Fritz JV, Heintz-Buschart A, Desai MS, Yusuf D, et al. The extracellular RNA complement of *Escherichia coli*. MicrobiologyOpen 2015;4(2):252–66. https://doi.org/10.1002/mbo3.235.

[81] Blenkiron C, Simonov D, Muthukaruppan A, Tsai P, Dauros P, Green S, et al. Uropathogenic *Escherichia coli* releases extracellular vesicles that are associated with RNA. PLoS One 2016;11(8):e0160440. https://doi.org/10.1371/journal.pone.0160440.

[82] Heroven AK, Sest M, Pisano F, Scheb-Wetzel M, Steinmann R, Böhme K, et al. Crp induces switching of the CsrB and CsrC RNAs in Yersinia pseudotuberculosis and links nutritional status to virulence. Frontiers in Cellular and Infection Microbiology 2012;2:158. https://doi.org/10.3389/fcimb.2012.00158.

[83] Diggle SP, Whiteley M. Microbe Profile: *Pseudomonas aeruginosa*: opportunistic pathogen and lab rat. Microbiology 2020;166(1):30–3. https://doi.org/10.1099/mic.0.000860.

[84] Koeppen K, Hampton TH, Jarek M, Scharfe M, Gerber SA, Mielcarz DW, et al. A novel mechanism of host-pathogen interaction through sRNA in bacterial outer membrane vesicles. PLoS Pathogens 2016;12(6):e1005672. https://doi.org/10.1371/journal.ppat.1005672.

[85] Turnbull L, Toyofuku M, Hynen AL, Kurosawa M, Pessi G, Petty NK, et al. Explosive cell lysis as a mechanism for the biogenesis of bacterial membrane vesicles and biofilms. Nature Communications 2016;7:11220. https://doi.org/10.1038/ncomms11220.

[86] Sonnleitner E, Gonzalez N, Sorger-Domenigg T, Heeb S, Richter AS, Backofen R, et al. The small RNA PhrS stimulates synthesis of the *Pseudomonas aeruginosa* quinolone signal. Molecular Microbiology 2011;80(4):868–85. https://doi.org/10.1111/j.1365-2958.2011.07620.x.

[87] Choi J-W, Kim S-C, Hong S-H, Lee H-J. Secretable small RNAs via outer membrane vesicles in periodontal pathogens. Journal of Dental Research 2017;96(4):458–66. https://doi.org/10.1177/0022034516685071.

[88] Choi J-W, Kwon T-Y, Hong S-H, Lee H-J. Correction to: isolation and characterization of a microRNA-size secretable small RNA in Streptococcus sanguinis. Cell Biochemistry and Biophysics 2018;76(3):441. https://doi.org/10.1007/s12013-018-0841-x.

[89] Han E-C, Choi S-Y, Lee Y, Park J-W, Hong S-H, Lee H-J. Extracellular RNAs in periodontopathogenic outer membrane vesicles promote TNF-α production in human macrophages and cross the blood-brain barrier in mice. Federation of American Societies for Experimental Biology Journal: Official Publication of the Federation of American Societies for Experimental Biology 2019;33(12):13412–22. https://doi.org/10.1096/fj.201901575R.

[90] Ho M-H, Chen C-H, Goodwin JS, Wang B-Y, Xie H. Functional advantages of Porphyromonas gingivalis vesicles. PLoS One 2015;10(4):e0123448. https://doi.org/10.1371/journal.pone.0123448.

[91] Domenech M, Pedrero-Vega E, Prieto A, García E. Evidence of the presence of nucleic acids and β-glucan in the matrix of non-typeable Haemophilus influenzae in vitro biofilms. Scientific Reports 2016;6:36424. https://doi.org/10.1038/srep36424.

[92] Dohar JE. Evidence that otitis media is not a biofilm disease. Ear, Nose, and Throat Journal 2007;86(11_Suppl):8–12. https://doi.org/10.1177/014556130708611s04.

[93] Erwin AL, Smith AL. Nontypeable Haemophilus influenzae: understanding virulence and commensal behavior. Trends in Microbiology 2007;15(8):355–62. https://doi.org/10.1016/j.tim.2007.06.004.

[94] Swords WE. Nontypeable Haemophilus influenzae biofilms: role in chronic airway infections. Frontiers in Cellular and Infection Microbiology 2012;2:97. https://doi.org/10.3389/fcimb.2012.00097.

[95] Jurcisek JA, Bakaletz LO. Biofilms formed by nontypeable Haemophilus influenzae in vivo contain both double-stranded DNA and type IV pilin protein. Journal of Bacteriology 2007;189(10):3868–75. https://doi.org/10.1128/JB.01935-06.

[96] Cavaliere R, Ball JL, Turnbull L, Whitchurch CB. The biofilm matrix destabilizers, EDTA and DNaseI, enhance the susceptibility of nontypeable *Hemophilus influenzae* biofilms to treatment with ampicillin and ciprofloxacin. MicrobiologyOpen 2014;3(4):557–67. https://doi.org/10.1002/mbo3.187.

[97] Moxon ER, Sweetman WA, Deadman ME, Ferguson DJP, Hood DW. *Haemophilus influenzae* biofilms: hypothesis or fact? Trends in Microbiology 2008;16(3):95–100. https://doi.org/10.1016/j.tim.2007.12.005.

[98] Langereis JD, Hermans PWM. Novel concepts in nontypeable Haemophilus influenzae biofilm formation. FEMS Microbiology Letters 2013;346(2):81–9. https://doi.org/10.1111/1574-6968.12203.

[99] Swords WE, Moore ML, Godzicki L, Bukofzer G, Mitten MJ, VonCannon J. Sialylation of lipooligosaccharides promotes biofilm formation by nontypeable Haemophilus influenzae. Infection and Immunity 2004;72(1):106–13. https://doi.org/10.1128/IAI.72.1.106-113.2004.

[100] Sjöström AE, Sandblad L, Uhlin BE, Wai SN. Membrane vesicle-mediated release of bacterial RNA. Scientific Reports 2015;5:15329. https://doi.org/10.1038/srep15329.

[101] Malge A, Ghai V, Reddy PJ, Baxter D, Kim T-K, Moritz RL, et al. mRNA transcript distribution bias between Borrelia burgdorferi bacteria and their outer membrane vesicles. FEMS Microbiology Letters 2018;365(13). https://doi.org/10.1093/femsle/fny135.

[102] Lybecker MC, Samuels DS. Small RNAs of Borrelia burgdorferi: characterizing functional regulators in a sea of sRNAs. Yale Journal of Biology & Medicine 2017;90(2):317–23. https://www.ncbi.nlm.nih.gov/pubmed/28656017.

[103] Byrne GI, Ojcius DM. Chlamydia and apoptosis: life and death decisions of an intracellular pathogen. Nature Reviews Microbiology 2004;2(10):802–8. https://doi.org/10.1038/nrmicro1007.

[104] Friedman H, Yamamoto Y, Klein TW. *Legionella pneumophila* pathogenesis and immunity. Seminars in Pediatric Infectious Diseases 2002;13(4):273–9. https://doi.org/10.1053/spid.2002.127206.

[105] Isaac DT, Isberg R. Master manipulators: an update on *Legionella pneumophila* Icm/Dot translocated substrates and their host targets. Future Microbiology 2014;9(3):343–59. https://doi.org/10.2217/fmb.13.162.

[106] Mullokandov G, Baccarini A, Ruzo A, Jayaprakash AD, Tung N, Israelow B, et al. High-throughput assessment of microRNA activity and function using microRNA sensor and decoy libraries. Nature Methods 2012;9(8):840–6. https://doi.org/10.1038/nmeth.2078.

[107] Mourenza Á, Lorente-Torres B, Durante E, Llano-Verdeja J, Aparicio JF, Fernández-López A, et al. Understanding microRNAs in the context of infection to find new treatments against human bacterial pathogens. Antibiotics 2022;11(3):356. https://doi.org/10.3390/antibiotics11030356.

[108] Das K, Garnica O, Dhandayuthapani S. Modulation of host miRNAs by intracellular bacterial pathogens. Frontiers in Cellular and Infection Microbiology 2016;6:79. https://doi.org/10.3389/fcimb.2016.00079.

[109] Fang Y, Chen H, Hu Y, Li Q, Hu Z, Ma T, et al. Burkholderia pseudomallei-derived miR-3473 enhances NF-κB via targeting TRAF3 and is associated with different inflammatory responses compared to Burkholderia thailandensis in murine macrophages. BMC Microbiology 2016;16(1). https://doi.org/10.1186/s12866-016-0901-6.
[110] Zhou X, Li X, Wu M. miRNAs reshape immunity and inflammatory responses in bacterial infection. Signal Transduction and Targeted Therapy 2018;3:14. https://doi.org/10.1038/s41392-018-0006-9.
[111] Ma F, Xu S, Liu X, Zhang Q, Xu X, Liu M, et al. The microRNA miR-29 controls innate and adaptive immune responses to intracellular bacterial infection by targeting interferon-γ. Nature Immunology 2011;12(9):861–9. https://doi.org/10.1038/ni.2073.
[112] Aguilar C, Cruz AR, Rodrigues Lopes I, Maudet C, Sunkavalli U, Silva RJ, et al. Functional screenings reveal different requirements for host microRNAs in Salmonella and Shigella infection. Nature Microbiology 2020;5(1):192–205. https://doi.org/10.1038/s41564-019-0614-3.
[113] Larabi A, Dalmasso G, Delmas J, Barnich N, Nguyen HTT. Exosomes transfer miRNAs from cell-to-cell to inhibit autophagy during infection with Crohn's disease-associated adherent-invasive *E. coli*. Gut Microbes 2020;11(6):1677–94. https://doi.org/10.1080/19490976.2020.1771985.
[114] Igietseme JU, Omosun Y, Stuchlik O, Reed MS, Partin J, He Q, et al. Role of epithelial-mesenchyme transition in Chlamydia pathogenesis. PLoS One 2015;10(12):e0145198. https://doi.org/10.1371/journal.pone.0145198.
[115] Derrick T, Last AR, Burr SE, Roberts CH, Nabicassa M, Cassama E, et al. Inverse relationship between microRNA-155 and -184 expression with increasing conjunctival inflammation during ocular *Chlamydia trachomatis* infection. BMC Infectious Diseases 2016;16:60. https://doi.org/10.1186/s12879-016-1367-8.
[116] Yeruva L, Myers GSA, Spencer N, Creasy HH, Adams NE, Maurelli AT, et al. Erratum for Yeruva et al., Early MicroRNA Expression Profile as a Prognostic Biomarker for the Development of Pelvic Inflammatory Disease in a Mouse Model of Chlamydial Genital Infection. mBio 2014;5(4). https://doi.org/10.1128/mbio.01575-14.
[117] Herkt, C. E., Caffrey, B. E., Surmann, K., & Blankenburg, S. (n.d.). A microRNA network controls *Legionella pneumophila* replication in human macrophages via LGALS8 and MX1. mBio 11: 031555-19. Retrieved March 28, 2023, from https://www.academia.edu/download/82994026/e03155-19.full.pdf.
[118] Iannaccone M, Dorhoi A, Kaufmann SHE. Host-directed therapy of tuberculosis: what is in it for microRNA? Expert Opinion on Therapeutic Targets 2014;18(5):491–4. https://doi.org/10.1517/14728222.2014.897696.
[119] Rosas-Taraco AG, Higgins DM, Sánchez-Campillo J, Lee EJ, Orme IM, González-Juarrero M. Local pulmonary immunotherapy with siRNA targeting TGFβ1 enhances antimicrobial capacity in *Mycobacterium tuberculosis* infected mice. Tuberculosis 2011;91(1):98–106. https://doi.org/10.1016/j.tube.2010.11.004.
[120] Van Niel G, d'Angelo G, Raposo G. Shedding light on the cell biology of extracellular vesicles. Nature Reviews Molecular Cell Biology 2018;19(4):213–28. https://www.nature.com/articles/nrm.2017.125.

CHAPTER THIRTEEN

Role of exosomal RNA in wound healing and tissue repair

Avirup Chakraborty[1,*] Ravindra V. Badhe[2,*] Muhammad Abbas[1,*] Anjali Chauhan[1], Abhishek Jaiswal[3], Rameesha Fareed[4], Vinay Kumar[5], Yongtao Duan[6] and Nalok Dutta[7]

[1]Department of Neurosurgery, University of Florida, Gainesville, FL, United States
[2]Pharmaceutical Chemistry Department, Marathwada Mitra Mandal's College of Pharmacy, Pune, Maharashtra, India
[3]Viral Research and Diagnostic Laboratory (VRDL), Department of Microbiology, Midnapore Medical College and Hospital, Midnapore, West Bengal, India
[4]Experimental Therapeutics Group, University of Florida Health Cancer Center, University of Florida, Gainesville, FL, United States
[5]Department of Physiology and Cell Biology, The Ohio State University Wexner Medical Centre, Columbus, OH, United States
[6]Henan Provincial Key Laboratory of Pediatric Hematology, Children's Hospital Affiliated to Zhengzhou University, Zhengzhou University, Zhengzhou, China
[7]Biochemical Engineering Department, University College London, London, United Kingdom

1. Introduction

Inflammation, cell proliferation and migration, angiogenesis, and vascularization, as well as tissue remodeling, are some of the fundamental cellular and molecular processes that take place in injured or tissues during wound healing [1]. So, it is evident that wound healing is a biologically critical process that may result in either chronic non-healing or delayed healing. In this context, studies reported that exosomes have played significant role via intercellular communication. It should be noted that exosomes are endosomal-derived extracellular small vesicles (EVs) that measures in from 40 to 160 nm in diameter. These vesicles contain DNA, RNA, lipids, metabolites, and proteins found in the cytosol and on the surface of cells [2]. The composite materials of the EVs are all signature characteristics of the cells from which they are derived. Exosomal RNA has been implicated in and is being studied for a variety of functions, including the removal of stale molecules, enhancing immunity, angiogenesis, antigen presentation, apoptosis, inflammation, coagulation, etc. Additionally, exosomes perform a prime role in the transfer of viral pathogens and oncogenes from tumor cells and normal cells to other cells [3].

The intermediary biomolecule messenger RNA (mRNA), convey genetic inheritance from DNA to ribosomes for protein weaving, has been considered a potential tool in cutting-edge therapeutic strategies for the treatment of many illnesses, wound healing

*Equal contribution.

Exosomal RNA
ISBN 978-0-443-14008-2
https://doi.org/10.1016/B978-0-443-14008-2.00001-2

or tissue repair and malignancies [4]. Exosomes provide highly effective cell-cell signaling and connection by carrying a variety of RNAs to nearby cells. The matter of concern for all RNA types is the early RNA breakdown in the body's internal environment. Because of its poor stability, bare exogenous mRNA has half-lives in the body that range from 2 to 25 min [5]. This vulnerability is attributed to the occurrence of 5' capping, or the size of the 3′ poly-A tail (secondary structure) [6]. Vehicles for RNA delivery that are effective must fulfill a number of criteria. They should be fairly sturdy, biodegradable, and non-toxic. They should bind RNA with high affinity and shield it from nuclease cleavage. So, exosome mediated delivery solved all the aspects of concern. As a result, researchers are now paying close attention to exosomes because they are highly intriguing nanocarriers for the transport of both short and long RNAs. A critical discovery in immune cells revealed that miRNA and mRNA may be carried in exosomes. As a result of this finding, exosomal RNA is thought to execute a key role in wound healing and tissue repair [7].

The cells implicated in wound recovery, such as macrophages, fibroblasts, and keratinocytes, can undergonegative changes in the context of various disease environments. These changes lead to chronic inflammation, poor angiogenesis, ischaemia, hypoxia, and decreased collagen synthesis, all of which are

extremely detrimental to wound healing. Here, exosomal RNA can be of significant help in changing the course of wound healing.

Exosomes from various cells have the potential to revamp the activity of cells that are engaged in wound repair, increase neovascular proliferation [8], decrease inflammation [9], and induce collagen deposition. The difficulties of healing wounds can also be somewhat reduced by inhibiting the overexpressed microRNAs (miRNAs) in exosomes, as proven in diabetic patients [10].

Since exosomal RNA contains both coding and noncoding RNA and is reliant on the microenvironment of the secreting cells, the particular material can serve as a marker for the cell or cellular system that produced it. While the exact processes by which RNA is packed into exosomes are still obscure, it has been demonstrated that a few RNA-binding proteins have the ability to specifically bind RNA molecules using certain patterns and carry them into intracellular vesicles. The utilization of exosomal RNAs for either wound healing or tissue mending is incredibly little understood. This chapter's first goal is to provide a detailed biogenesis of exosomes and associated RNA, and the its second goal is to investigate the roles of RNA molecules supplied by exosomes and possible clinical uses for such RNA molecules in tissue repair or treatments.

2. Background of regenerative medicine and tissue engineering

Regenerative medicine is an interdisciplinary field that encompasses the development for biological substitutes of tissues and organs in-vitro for restoring, maintaining, or improving the body function and also focuses on research in self-healing (regenerate). Tissue engineering is a part of regenerative medicine and is based on four main pillars: cells, scaffolds, signaling molecules, and bioreactors. Bioreactors provide a sealed and sterile environment along with a continuous supply of fresh nutrient media, gaseous exchange, and mechanical movement if needed [11,12]. Scaffolds are 3D architectures developed from various biomaterials using traditional or modular approaches. These scaffolds are generally sacrificial and allow the seeded/laden cells to proliferate in 3D architecture and ultimately develop into a tissue. Cells used in the majorities of tissue engineering experiments nowadays are stem cells that are expected to differentiate into different cell type in the presence of different signaling molecules to develop multilineage cells containing tissue. Thus, the role of signaling molecules becomes very important in tissue engineering experiments and therapies [13,14].

Signaling molecules are the biological agents (proteins, growth factors, RNA, vitamins, enzymes, hormones, minerals, etc.) that are responsible for the divergence of stem cell lineage into specific type of cells or tissue and are significantly responsible for supporting the proliferation of differentiated cells in the culture/scaffold. For tissue engineering applications, the signaling molecules can be directly incorporated/immobilized in scaffolds or 3D architecture, or first they are loaded in nanostructures and incorporated/immobilized in the scaffolds or 3D architecture; in some experiments, the signaling molecules were supplied through nutrient media and every method has its own advantages based on the cell types used in the experiment. In regenerative therapies, the molecules were supplied through injectable hydrogels, cell suspensions, emulsions, nanoparticulate solutions, etc. at the diseased site [15,16].

One of the methods that are used to protect, carry, and deliver the signaling molecules in the scaffold or at the diseased site is the use of lipid nanocages or carriers (Liposomes, Solid lipid nanoparticles, Microemulsion). Liposomes are small artificial spherical vesicles synthesized using cholesterol and natural, non-toxic phospholipids. They can carry hydrophilic or hydrophobic loads with them. Solid lipid nanoparticles are surfactant stabilized lipids and can solubilize lipophilic loads. Micro-emulsions are clear, thermodynamically steady isotropic fusions of surfactant, water, oil, and co-surfactant [17,18].

Exosomes are also considered in the same category; however the main difference between exosomes and lipid nanocages or carriers is that exosomes are generated by live cells as compared to the synthetic origin of the latter.

2.1 Exosomes, biogenesis and isolation procedure

Exosomes develop in the cytoplasm, inside the endosomal compartment architecture, as opposed to other extracellular vesicles like apoptotic micro-vesicles and apoptotic bodies, which sprout straight from the plasma membrane [19]. Exosomes are generated from the internal membranes of multivesicular bodies, which are formed by early endosomes generated with the aid of the plasma membrane through endocytosis. The early endosomes sprout into the cytosol and mature before being packed with multivesicular bodies or multiple vesicular endosomes in a certain protein-cargo associated phenomenon. The concave sprout of the endosomal associated membrane in multivesicular bodies or multivesicular endosomes leads to the accumulation of intraluminal associated vesicles, which are the predecessor of exosomes. When multivesicular bodies or multivesicular endosomes merge into the plasma membrane, the accumulated intraluminal vesicles are delivered into the extracellular environment as exosomes [20]. It is reported that exosomes carry DNA [21], RNA [22], mitochondria, proteins, cytokines, lipids, and metabolites as well as certain receptors. Naturally they act as intercellular communication mediators and modulate near- and long-distance massages between cells [23]. They are generated by the cells due to the 'Endosomal Sorting Complex Required for Transport (ESCRT)' protein complex's dependent and independent pathways [24].

Exosome biogenesis is currently divided into two distinct categories: Endosomal Sorting Complex Required for Transport (ESCRT)-independent and ESCRT-dependent pathways, although it is difficult to separate them into clearly defined pathways. It should be mentioned here that ESCRT-dependent pathway-dependent exosomal secretion is also influenced by proteins such as the ESCRT-0, −I, −II, and −III, as well as mechanistic target of rapamycin complex 1 (mTORC1) and HSP90 [25]. Exosomes are around 40−100 nm in size and secreted by all types of cells in the body. They are present in plasma, milk, urine, malignant ascites, saliva, and amniotic fluid [26,27]. Additionally, exosomes can be easily isolated using ultracentrifugation (Fig. 13.1) [28], size exclusion (filtration or chromatography), immune affinity capture,

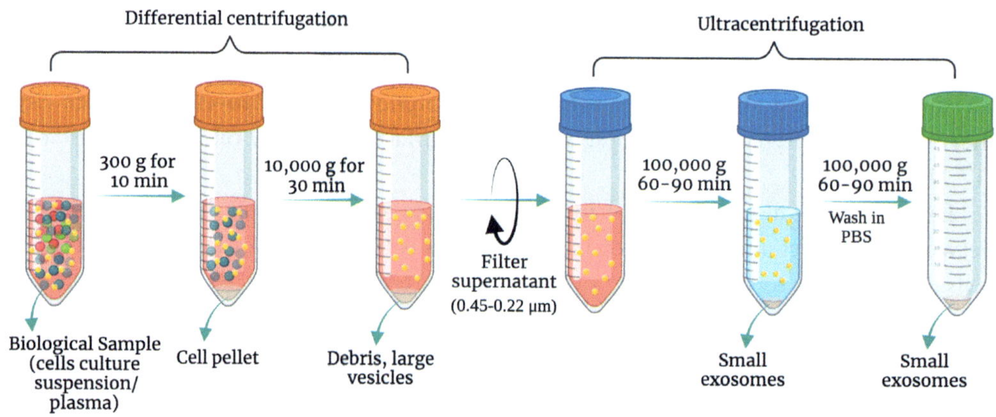

Figure 13.1 Schematic representation showing the isolation of exosomes using ultracentrifugation method. Adopted from Li P, Kaslan M, Lee SH, Yao J, Gao Z. Progress in exosome isolation techniques. Theranostics 2017; 7(3):789−804. https://doi.org/10.7150/thno.18133.

precipitation methods, and microfluidic technologies. The exosomes can be identified and quantified by nanoparticle tracing investigation, dynamic light dissipation, and surface plasmon resonance technologies [29].

Being cell generated vesicles, exosomes can have proteins like GTPases, annexin, flotillin (proteins required for membrane transport and fusion) [30]; receptors like CD9, CD63, CD81, and CD82 that are tetrameric proteins required for cell targeting and adhesion; Hsp70 and Hsp90 that are heat shock proteins; Alix and TSG101 that are multivesicular synthesis proteins (Fig. 13.2) [31,32]. Exosome tetrastatins like CD81,

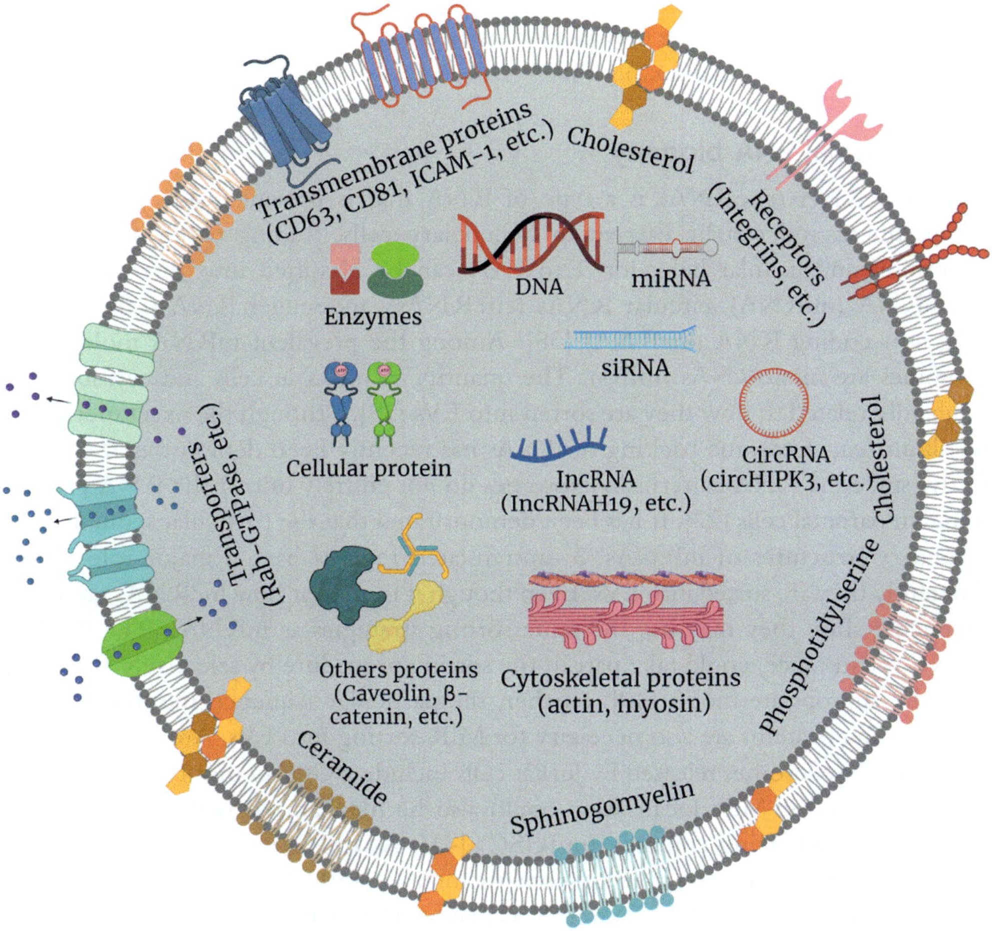

Figure 13.2 Schematic diagram showing the general structure of the exosomes. Adopted from Jan AT, Rahman S, Badierah R, Lee EJ, Mattar EH, Redwan EM, Choi I. Expedition into exosome biology: a perspective of progress from discovery to therapeutic development. Cancers 2021; 13: 1157. https://doi.org/10.3390/cancers13051157.

CD63, and CD9 can be employed as biomarkers for specific identification, isolation, and analysis [33]. Some studies reported the presence of metabolic enzymes like peroxidase, pyruvate kinase, glyceraldehyde-3-phosphate dehydrogenase, β-enolase and fatty acid synthase in exosomes [2].

Exosomes also carry lipids like cholesterol, ceramide, and sphingolipids as lipid rafts [34]. They also carry lipid configured signaling biomolecules like prostaglandins (PGs), arachidonic acid (AA) and phospholipase A2 (PLA2) are also reported to carry lipid metabolizing enzymes [35]. The lipid-based receptors like Phosphatidylserine (PS) and lysophosphatidylcholine (LPC) on the exosomes get detected by immunoglobulin or mucin [T-cell immunoglobulin and mucin domain (TIM-4)] and G protein-coupled receptor (G2A) present on the target cell respectively (Fig. 13.2) [32,36]. The PS–TIM receptor forms a bridge between two structures, thus deciding the immunological fate of the exosomes (Fig. 13.2) [37].

2.2 Exosomal RNA biogeny

Exosomal RNA (exoRNA) is a type of RNA that is present inside the exosomal compartment and can be carried to beneficiary cells, leading to changes in gene expression and cellular behavior. ExoRNA can be classified into three categories: microRNA (miRNA), circular RNAs (circRNAs), messenger RNA (mRNA), and long non-coding RNA (lncRNA) [38]. Among the prevalent mRNA molecules in exosomes are microRNAs (miRs). The quantity of miRs in cells and cytosol is undoubtedly related to how they are sorted into EVs [39], although the exact mechanism is still unclear. Exosome packing of RNAs has recently been demonstrated to be selective since the RNA patterns in exosomes do not entirely mirror the RNA profiling found in parental cells [22]. It has been demonstrated that the particular sequences and secondary structures of mRNAs' 3′-untranslated regions have a major role in how differently they are sorted into EVs. Even though it is unclear how lncRNAs target EV-producing sites, they may use the same sorting strategies as mRNAs. The lncRNA YBX1, for instance, could take part in the sorting procedure by selectively associating with the appropriate motifs [40]. Further, the particular sequence/motif associations with binding proteins are also necessary for MiR sorting into EVs. Around 70% of the exo-miRs in exosomes released by Jurkat cells include a GGAG motif in the 3′ part of the miR. Besides, MiR localization could also be mechanized with the help of the miR-induced silencing complex (miRISC) [41]. As miRNAs are common in exosomes and circRNAs have been shown to interact with them in cells, it has been hypothesized that miRNAs and circRNAs may work together in producer cells to sort circRNAs into exosomes. The findings revealed that variations in the levels of associated miRNAs in producer cells were at least partially responsible for controlling the trafficking of circRNAs to exosomes [42].

3. Role of different exosomal RNA in wound healing and tissue repair

Exosomal RNA has been demonstrated to be essential for several phases of tissue repair and wound healing. Exosomal RNA generated by immune cells like macrophages and others during the initial phases of wound healing can promote angiogenesis, which encourages the growth of fresh blood vessels.

ExoRNA is involved in encouraging cellular migration and proliferation during wound healing. ExoRNA molecules produced by mesenchymal stem cells, fibroblasts, and other types of cells encourage the migration and proliferation of adjacent cells [43]. Exosomal RNA not only supports cellular functions, but also controls immune reactivity during wound healing. Immune cells can produce exosomes, which contain RNA molecules that control the behavior of other immune cells, which in turn modulates the inflammatory process [44].

3.1 Exosomal mRNAs in wound healing

Exosomal mRNAs' role in wound healing has been explored in very little research. Among them, a study found that UCMSC-derived microvesicles accelerated renal rejuvenation and postponed fibrosis by delivering several hepatocyte growth inducing factor (HGF) mRNAs, which facilitated the dedifferentiation and multiplication of injured tubular associated cells [45]. In another study, it was discovered that mRNAs linked to the endothelial NOS and PI3K/AKT signaling pathways were transferred by endothelial progenitor cell-derived micro-vesicles, significantly improving angiogenesis [46].

3.2 Exosomal miRNAs in wound healing

Some of the miRNAs have been found to stimulate angiogenesis by targeting different vascular development signaling pathways. For instance, it has been demonstrated that exosomal miR-21 produced by fibroblasts promotes fibroblast migration and proliferation, speeding up the healing of wounds. Controlling the activity of T cells and B cells with the aid of exosomal miR-155 produced by macrophages can alter the adaptive immune response. In the case of miR-221-3p, increased angiogenesis and cell proliferation can be seen at the skin wound site [8].

MSC-derived exosomal miRs (exo-miRs) have been considered to be the key bioactive components underlying those promising outcomes in tissue repair. Exo-miR-150-3p enhances osteoblast development and divergence in osteoporosis, establishing promising milestones for osteoporosis therapy [47]. According to one study, the phosphatase and tension (PTEN) homologs, ERK and PI3K/AKT, signaling pathways are stimulated by miR-146a-3p, miR-21-3p, miR-21-5p, and miR-126-3p. So, these miRNAs augment the growth, metastasis, and diaspora of endothelial and fibroblast cells, promoting human endothelial angiogenesis [48–51]. In another, study it was proven, by down regulating factor Keap1 against simulated nuclear factor erythroid2-related factor 2, MiR-200a-enhanced embryonic stem cell exosomal vesicles were capable of reversing vascular origin endothelial cell aging and restoring defective growth, tube formation, and migration [52]. A study in human ADSC-derived exosomes overexpressing miR-29a reduced human hypertrophic scar fibroblast growth, emigration, fibrosis, and scar hyperplasia following scald trauma in mice by inhibiting the TGF-2/Smad3 signaling pathway [53].

A study that used high-throughput sequencing to analyze the exosomal miRNAs in UCMSCs discovered that a particular miRNA cluster (miR-145, miR-23a, miR-21, and miR-125b) was crucial for the development of myofibroblasts. Clinical and laboratory-based experiments have confirmed that certain miRNAs suppress the production of smooth muscle actin and the accumulation of collagen by regulating the TGF/Smad2 axis (Fig. 13.3) [54].

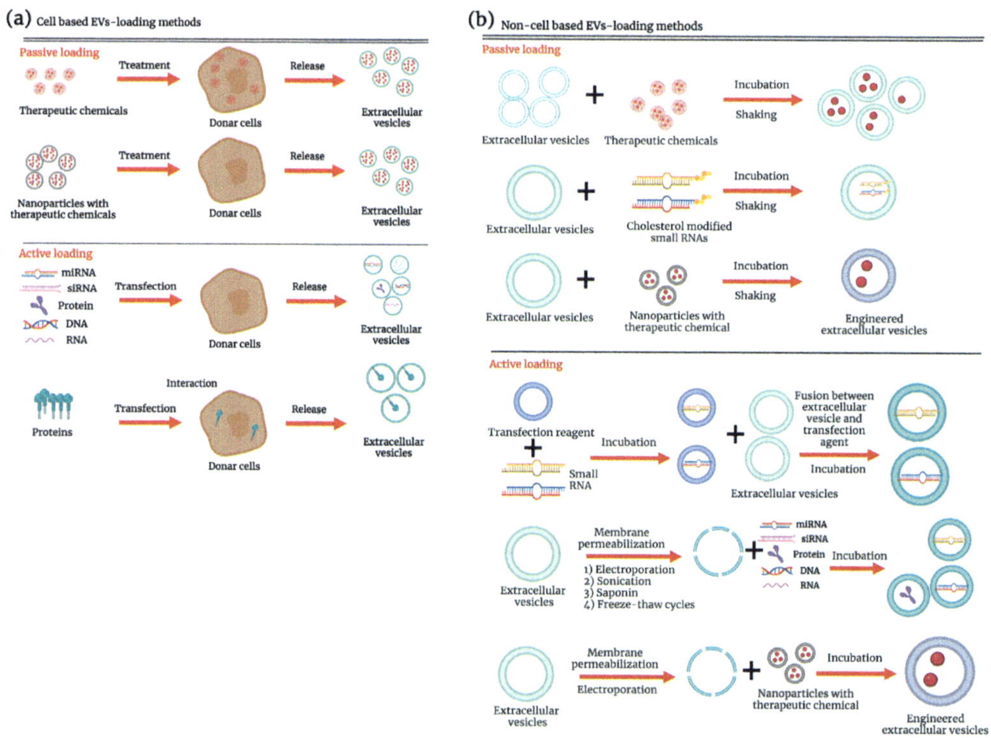

Figure 13.3 Schematic diagram showing the cargo loading methods via (A) cell-based (B) non-cell based. Adopted from Han Y, Jones TW, Dutta S, Zhu Y, Wang X, Narayanan SP, Fagan SC, Zhang D. Overview and update on methods for cargo loading into extracellular vesicles. Processes 2021; 9: 356. https://doi.org/10.3390/pr9020356.

3.3 Exosomal lncRNA in wound healing

Exosomal lncRNAs improve wound healing by encouraging cell growth, multiplication, anti-fibrosis, anti-inflammatory, and angiogenesis responses. lncRNA are crucial exosome constituents as well. Exosomal long noncoding RNA KLF3-AS1 (lncRNA) produced from mesenchymal stem cells of bone marrow increased blood vessel formation, decreased inflammation, decreased miR-383 expression, and increased VEGFA to enhance cutaneous wound healing in diabetic mice, according to research [55]. One of the first lncRNA detected was lncRNA H19 (H19), which was reported to be abundant in mesenchymal stem cell (MSC)-exosomes. Human skin fibroblast growth, migration, and infiltration were hastened by ADSC exosomes' inhibition of miR-19b expression via H19 lncRNA, which also upregulated SOX9 and activated the Wnt signaling catenin pathway [56].

3.4 Exosomal circRNA in wound healing

Research has proved that circRNAs are important in the biochemical mechanism of wound reversal, which has increased interest in the functions that circRNAs play in healing wounds. In a rodent wound model, Circulating Gcap14 from preconditioned hypoxic ADSCs boosted production of angiogenic growth components and hasten the closure of diabetic wounds by impeding downstream miR-18a-5p generation and boosting the secretion of HIF-1 [57]. Exosomes abundant in mmu_circ_0,000,250 promoted skin angiogenesis in diabetic mouse injuries and prevented apoptosis by activating autophagy through downregulation of miR-128-3p and SIRT1 upregulation [58].

4. Applications of exosomes

4.1 Targeted drug delivery vehicles

Exosomes are being extensively explored for drug delivery applications due to their indigenous nature and ease of drug cargo loading (Fig. 13.3) [59,60]. The main techniques that can be used to produce drug-cargoloaded exosomes are: a. incubation, where exosome-secreting cells or exosomes were incubated with drug cargo; b transfection where nucleic acid producing the desired cargo is transfected either in exosome-secreting cells or exosomes; c. Physical treatments like sonication, surfactant treatment, and electroporation to directly load cargo in exosomes and d. In-situ assembly and synthesis where cargo is loaded in the exosome when the vesicle membrane is reassembled, and the techniques are freeze—thaw, extrusion, and dialysis [61].

The main advantages of exosomes are higher bioavailability, bio-distribution, good drug encapsulation capacity, lessened toxicity, increased cell-specific tailored delivery, lengthy circulation half-life, antigen presentation, antitumor programming, the capacity to elicit an immune response toward infections, capability of crossing the BBB, and controlled release [62–65]. There are many exosome base formulations that are currently under clinical assessment, and few of them are even approved for clinical use [66]. Some studies reported successful loading and delivery of dexamethasone, curcumin, neurotoxin-I and paclitaxel-like molecules in exosomes [67–70] and the disease models on which exosome-based treatments were tried are wound healing, bone and cartilage regeneration, diabetes, cancers, Parkinson's, prions, and Alzheimer's [71,72]. Exosomes are also reported to act as vehicles for CRISPR-Cas9 expression vectors for in vivo gene editing and gene knockdown experiments [73]. Exosomes are also reported for immunomodulation activity by carrying and presenting MHC I and II to CD4+ and CD8+ cell receptors via cross- and direct-presentation [74]. It is reported that dendritic cells derived from exosomes show angiogenic, cancer tumor destruction, and metastasis inhibition properties [75]. There are some reports that stem cell-derived exosomes inhibit inflammation and support wound healing [76]. It was reported that if exosomes are used for chemotherapeutic applications, the dose of the chemotherapeutic agents is very low and targeted delivery can be achieved, thus minimizing the toxicity [77]. It is reported that siRNA delivered to brain cells through exosomes successfully knocks down BACE1 gene to modulate Alzheimer's disease [78]. Exosomes were also reported for vaccine development and delivery. In one experiment, TNF-ϒ is transfected to the cells using exosomes to induce antiviral activity [79]. The dendritic cell (DC) -derived exosomes are reported to carry MHC I/II, CD40/80/86 receptors that lead to activation of cytotoxic T- and natural killer (NK) cells [80].

4.2 Non-invasive diagnostic biomarkers

According to the NIH, a biomarker is a quantifiable amount of a biological molecule found in blood, other body fluids, or tissues for any normal or abnormal biochemical process or conditions or diseases happening in the body. They can be used for progress evaluation of a therapy/treatment [81]. Exosomes are MHC-expressing bodies; they also carry cell-specific surface markers, proteins and miR/mRNA cargoes, they are highly stable during storage (at −80 °C for 1 week to 1–2 years, depending on the source) [82]. Thus, exosomes are being studied as diagnostic biomarkers for different conditions like metabolic disorders, cancer, neurodegeneration, liver diseases, infections, and kidney disorders. Generally, for diagnostic applications biomarkers are extracted from tissues, blood, cerebrospinal fluid, ascites, amniotic fluid, urine, and saliva. The techniques used for the exosome-based biomarker analysis are flow cytometry (FACS), principal component analysis (PCA), biochemical investigation, including immunohistochemistry, surface-enhanced Raman spectroscopy (SERS), Western blotting, and microarray research [83].

The invasive techniques for isolating samples of exosomes involve withdrawal of blood, cerebrospinal fluid, ascites, amniotic fluid or a biopsy of the tissues. The most common are the isolation of blood and separation of the plasma. In plasma, isolated exosomes having CD63+ protein on the surface were reported for melanoma [84]; presence of CD81 was reported for hepatitis C [85]; presence of CD171, CD151, CD171, and tetraspanin eight were proven for lung cancer [86]; the presence of Glypican-1 (GPC1) was reported for pancreatic cancer [87]; the presence of CD-97 was reported for gastric cancer; the presence of microRNAs miR-320, miR-574−3p were reported for glioblastoma (GBM); the presence of microRNAs miR-21 and miR-1246 were reported for human breast cancer [88].

The non-invasive techniques are painless and require the collection of urine, saliva, milk, tears, sweat, or semen. In urine, isolated exosomes having CD10, CD26, CD81, and S1C3A1 were reported for hepatic damage in rats. Similarly, the presence of CD9, CD24, HSP70, ALG2-interacting protein X (ALIX), lysosome-associated membrane protein 1 (LAMP1), MME neprilysin, gp330 precursor, flotillin-1, tumor susceptibility gene 101 (TSG101), beta-galactosidase precursor, pro-epidermal, uromodulin, and growth factor precursor were reported for different diseases like bladder and Kidney cancer [89,90]. Exosome-based non-invasive biomarkers for diagnostic applications have a promising future and extensive work is being reported in this area.

4.3 Tissue engineering and regenerative medicine applications

The most important advantages of exosomes in tissue regeneration are that they can be isolated from patients own body fluids, they can be modified with biological molecules and drugs and delivered back to the patients to provide regenerative or therapeutic advantages [23,91,92]. It is reported that stem cell-based tissue regeneration is driven by proteins, RNAs, and cytokines released by exosomes and consequent gene expression. The microenvironment developed by exosomes secreted by stem cells under the influence of differentiating factors promotes cell survival and proliferation. An important role in this process is played by miR, as it modulates the tissue regeneration process by governing multiple downstream biological pathways and host immune responses. The exosomal miRs were studied and reported for skin, cardiovascular, cartilage/bone, and neuron tissue regeneration and repair studies [93]. Generally, it is possible to separate mesenchymal stem cells (MSCs) from bone marrow, adipose tissue, cord blood, dental pulp, and other sources. It was reported that specific stem cell derived exosomes produce a range of biochemical entities that are responsible for multiple functions in tissue engineering or regenerative medicine [94,95]. In addition to isolated exosomes, for specific applications, exosomes can be loaded with DNAs, mRNAs, miRs, rRNAs, lnRNAs, and circRNAs [96,97]. These exosomes loaded with specific nucleic acid cargoes can be used for many therapeutic applications like regenerative medicine, gene editing, gene silencing, immune system modulation, etc.

In recent years, applications of exosomes loaded with specialized miRs have gained a lot of interest in regenerative medicine applications. The miRs are a group of nucleotides consisting of small single -stranded non-coding RNA that is about 18–22 nucleotides in length. It has a capacity to modulate (enhancing or inhibiting) gene expression as it can bind to the 3′-untranslated region (3′UTR) of the mRNA [98,99].

4.4 Skin regeneration

Skin is the outermost protective layer of the body that protects us from infections, temperatures, radiations (UV) and mechanical damage [100]. Naturally, the skin can repair minor damages through skin regeneration process. When the damage is severe (like burns, ulcers, or larger wounds), currently only wound closures are available, and the majority of times develop scar tissue and defective wound healing. Skin regenerative therapies are therefore the solution that is being explored nowadays. The wound healing process is a well-orchestrated mechanism that includes following steps: a. angiogenesis, b. cell proliferation and movement, c. ECM synthesis and deposition,d. re-epithelialization, and e. wound remodeling [101].

Exosomes can play a vital role in faster wound healing by facilitating angiogenesis as well as activating endothelial cells, keratinocytes and fibroblasts migration, differentiation, and proliferation (Fig. 13.4) [102]. They can be used for modulating the immune system, and it

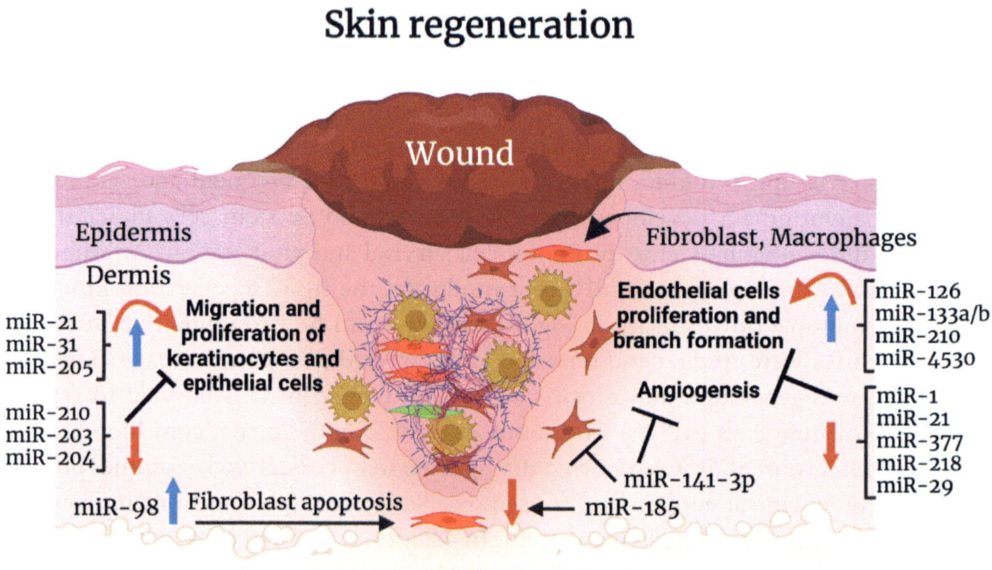

Figure 13.4 Schematic diagram showing the expression of various miRNAs in wound healing. Adopted from Singhvi G, Manchanda P, Rapalli VK, Dubey SK, Gupta G, Dua K. MicroRNAs as biological regulators in skin disorders. Biomedicine and Pharmacotherapy 2018; 108: 996–1004. https://doi.org/10.1016/j.biopha.2018.09.090.

was reported that important mechanisms involved in achieving all these functions are Wnt/β-catenin signaling pathways and AKT/extracellular regulated protein kinases (ERK)/STAT-3. These systems also impede the apoptosis of skin cells during wound healing [103—105]. The exosome isolated from different stem cells (like iPSCs, MSCs from synovial and adipose tissue) promotes the different growth factor releases, including hepatocyte growth factor (HGF), stromal cell-derived factor 1 (SDF-1), insulin-like growth factors (IGF) and nerve growth factor (NGF), that overall promote remodeling of the skin tissue [50,106]. It was reported that exosomes can promote expression of IL-10 which is an inflammatory inhibitor, and at the same time it can inhibit TNF-α and IL-1β to prevent inflammatory responses, specifically in burn and diabetic wound patients. The MSC derived exosomes can enhance Wnt/β-catenin signal and also interrupt yes-associated protein (YAP) signals to overall enhance the wound healing process. In burn patients, the same exosomes promote the synthesis of collagen (types I and III) in the early stages and inhibits collagen synthesis at the late stages of wound healing to promote scar free wound healing [107]. The roles of micro-RNA (miR)- loaded exosomes were also studied in the wound healing process. It was reported that miR-21-3p acts as a mediator for exosome induced RTK signaling antagonist 1 (SPRY1) and phosphatase and tensin homolog (PTEN) inhibitor and activity to control cutaneous regeneration in burn wounds [49,108]. It was reported that exosomes containing miR-126, miR-130a, and miR-132 can promote angiogenesis. Exosome rich in miR124a and miR125b show anti-analgesic activity and a few of them are rich in miR-21, which regulates collagen formation. Thus, exosomes loaded with different miRs can collectively contribute to wound healing [24,109]. It was reported that miR-142-5p and miR-142-3p enhanced the migration of neutrophils, which enhanced infection resistance. At the same time, knocking down the miR-139-5p in animals improved neutrophil activation and enhanced infection resistance [110,111]. At the same time, knocking down miR-139-5p in animals improved neutrophil activation and enhanced infection resistance. It is known that macrophages migrate to wounds and release inflammatory factors as body's defense mechanism. The abundance of miR-181c and miR-132 in exosomes can significantly reduce the macrophage-stimulated inflammatory factors. At the same time, it also inhibits the inflammation signaling pathways TLR-4 and NF-κB/p65 to suppress the inflammation caused by burns and inhibit NFκB [112,113]. It was also reported that miR-223 can induce macrophage polarization in M2 at the wound site [114] and miR-155 promotes the differentiation of Th17 and Th9 cells [104]. The presence of excess mount of miR-31 was reported to promote the proliferation and migration of keratinocytes [38]. Recently, miR-149, miR-497, and miR-23b were reported to increase the anti-inflammatory factor IL-10 [115—117]. Yu et al. reported that miR-26a and miR-26b can inhibit COX-2 expression and suppress allergic inflammation [118]. miR-27b, miR-375 and miR-5591-5p are expressed in large amount in exosomes and can reduce mitochondrial ROS production, leading to faster wound healing in diabetic mice (Fig. 13.4) [119].

4.5 Cardiovascular protection

Cardiac ailments like heart attacks, angina, high blood pressure, heart failure, and arrhythmias, are mainly caused by blocked Coronary arteries that supply blood to the heart muscle. It was reported that 'Sonic Hedgehog (Shh)' rich exosomes can regulate angiogenesis and neovascularization in nutrient and oxygen deprived heart muscles [120]. The exosomes were reported to suppress myocardial injuries and induce cardiomyocyte functions in Heart attacks (myocardial infarction) (Fig. 13.5) [121,122].

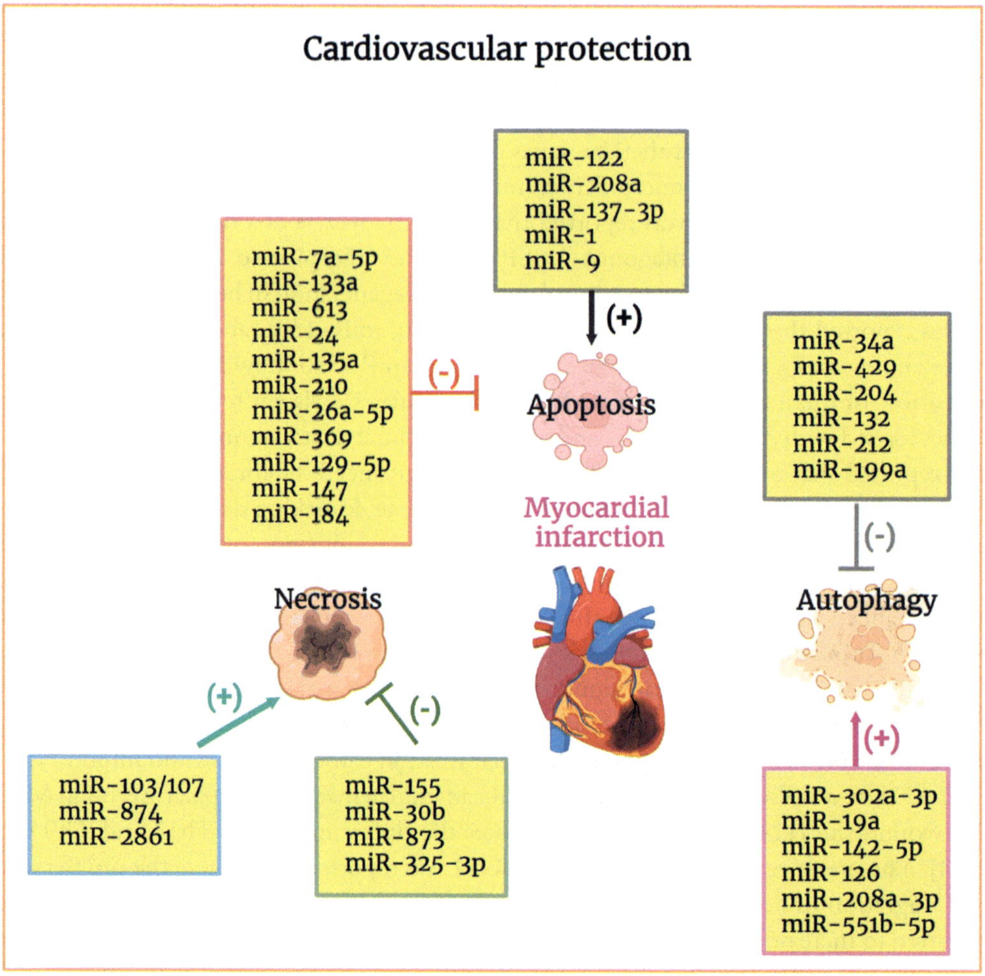

Figure 13.5 Schematic diagram showing the functional role of various miRNAs in regulatory cardiomyocyte death. Adopted from Kansakar U, Varzideh F, Mone P, Jankauskas SS, Santulli G. Functional role of microRNAs in regulating cardiomyocyte death. Cells 2022 March 12; 11(6):983. https://doi.org/10.3390/cells11060983.

With growing evidence of the usefulness of exosomes in cardiac treatments, miR also attracted the attention of the researchers [123]. In some recent studies, miRs were isolated from MSC. The miR-19a, miR-125b-5p, miR-21, miR-21a-5p, miR-21-5p, miR-22, miR-24, miR-26a, miR-29, miR-210, miR-126, miR-201, and miR-294 were reported to promote the survival of cardiomyocytes and reduce cardiac fibrosis, thus exhibiting cardio-protective effects [122]. miR-210, miR-126, miR-130a-3p, miR-21, and miR-23a-3p were reported to show myocardial infarction protection by angiogenesis induction in the cardiac tissue [124]. Wang et al., explored the role of miR-103/107, which stimulate cardiomyoblast necrosis by attaching to Fas-associated protein with death domain (FADD). Thus, blocking the miR-103/107 binding sites using long non-coding RNA H19 (lncRNA-H19) can inhibit cardiomyoblast necrosis [125]. The miR-325-3p, miR-30b, miR-873, and miR-155 were reported to target cyclophilin D (CypD), silencing the translation of RIPK1/RIPK3, and inhibiting RIP1, respectively. In addition, miR-613 targets programmed cell death 10 (PDCD10), miR-24 increases Bcl-2, miR-135a modulates Toll-like receptor 4 (TLR4), miR-210 and miR-369 decrease caspase-3 activity, miR-26a-5p activates the PI3K/AKT signaling pathway, miR-129-5p targets HMGB1, miR-223 silences PARP-1, miR-142-3p inhibits HMGB1 and Rac1, and miR-29b-3p reduces the TGF-β1/SMAD pathway, so upregulation of these miRs can inhibit Apoptosis [126–128].

The roles of miR-1 downregulation to increase Bcl-2, expression of miR-133a to repress transgelin 2 (TAGLN2), and suppression of miR-122 via targeting GATA-4 were reported to inhibit apoptosis. At the same time, downregulation of miR-137-3p, miR-208a, miR-7a-5p, miR-9, miR-147, and miR-184 will help suppress apoptosis [122].

Cardiac autophagy is a natural mechanism in the body necessary to maintain homeostasis and is regulated by the mTOR (mammalian or mechanistic target of rapamycin) and Bcl2 pathways [129]. There are many miRs reported to protect against myocardial injury through autophagy. Some examples include miR-34a, which decreases secretion of Lc3-II, p62, and TNF-α, miR-542-5p upregulates Autophagy-related 7 (ATG7), miR-208a-3p upregulates Autophagy-related 5 (ATG5); miR-145 targets fibroblast growth factor receptor substrate 2 (FRS2), miR-142-5p and miR-126 increase beclin-1 levels; miR-204 targets beclin-1, miR-19a increases repression of Bim; miR-30e-3p expresses expression of Egr-1 (Early growth response-1) and miR-26a target ubiquitin specific peptidase 15 (Usp15). All these miRs will promote cardiac autophagy. At the same time, inhibition/downregulation of miRs like miR-429, miR-199a, miR-212, miR-132, miR-302a-3p, miR-17-5p, miR-551b-5p and miR-188-3p will promote cardiac autophagy [122].

4.6 Bone and cartilage regeneration

Bone and cartilage tissue engineering is one of the most researched areas which involves cells, scaffolds, different biomaterials for scaffolds, methods of 3D architecture development, stimulators, and biological factors [130]. Despite rigorous work in the field, researchers were unable to provide promising solutions. The major problems associated with in vitro tissue engineering are insufficient vascularization and immune rejection of the engineered body parts [131]. Many recent research works suggest the use of exosomes for regenerating damaged bone and cartilage invivo (Fig. 13.6) [132–135]. The most common disease for which exosome applications are reported is Osteoarthritis. Osteoarthritis is the disruption of osteochondral cartilage tissues due to cellular inflammation. Initially, whole mesenchymal stem cell therapy was tested, but it has limitations like large = scale production of MSCs, post injection reduction in regeneration capacity, and dedifferentiation of MSCs. Researchers found a way out of these limitations through the use of MSC derived exosomes. It was reported that the

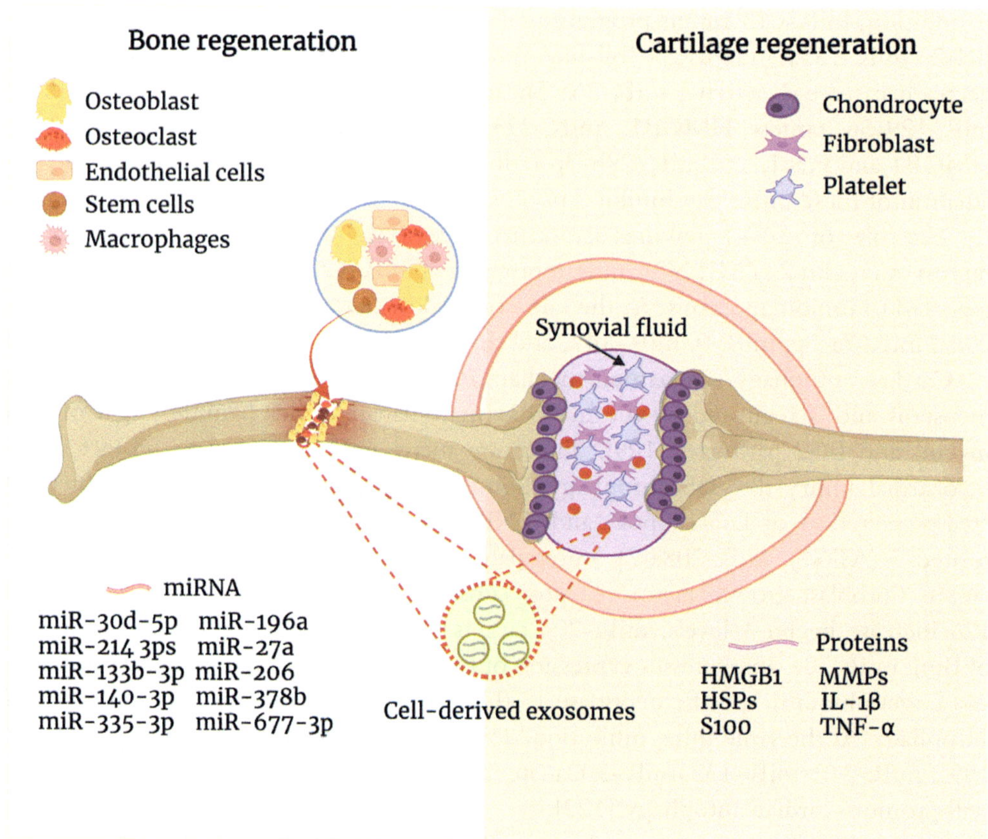

Figure 13.6 Schematic diagram showing the novel miRNAs as a therapeutic agent for cartilage and bone tissue regeneration. Adopted from Liu Y, Ma Y, Zhang J, Yuan Y, Wang J. Exosomes: a novel therapeutic agent for cartilage and bone tissue regeneration. Dose-Response 2019; 17(4). https://doi.org/10.1177/1559325819892702.

production of exosomes and their delivery as therapeutic agents for osteoarthritis treatment have advantages like reduction of inflammatory cytokines, increased cartilage ECM expression components, and overall cartilage tissue regeneration [136]. Similarly, cell free invivo bone injury repair was also attempted using exosomes. The use of exosomes was claimed to overcome immune rejection inspite of their cell like functions, and they can modulate osteoblast, macrophage, and angioblast polarization to ultimately give invivo bone regeneration (Fig. 13.6) [137].

The expression of miRs like miR-23b inhibit PKA signaling, miR-92a targets Noggin3, Targets Noggin3 suppresses IL1, miR-320 downregulates MMP-13, miR-145 downregulates Sox9, miR-221 downregulates MDM2, miR-22 inhibits BMP-7 and PPARA, miR-92a-3p targets Wnt5a, miR-135b down regulates Sp1a, miR-100-5p inhibits mTOR, miR-140-5p activates YAP, miR-26a-5p downregulates PTGS2, miR-136-5p downregulatesELF3, miR-127-3p inhibits CDH11, miR-9-5p inhibits SDC-1, miR-29b-3p inhibits TGF-β1, miR-27a inhibits LPS-induced chondrocyte apoptosis and reported to provide chondrocytes and cartilage regeneration [118,138,139], expression of miRs like miR-200c Targets Sox2, miR-31a-5p suppresses special AT-rich sequence-binding protein 2 (SATB2), miR-29a increases Dickkopf-1 (DKK1), miR-26a inhibits mTOR pathway, miR-31 inhibits RhoA, miR-199a-3p targets Kdm3a, miR-34a regulates NAD + -Sirt1 pathway supports bone regeneration. At the same time, miRs like miR-148a, miR-103a, miR-22-5p, miR-214-3p, miR-146a-5p, miR-415a, and miR-138-5p need to be suppressed to avoid bone resorption [140].

4.7 Neuronal tissues regeneration

Exosomes were also reported to have superior blood-brain barrier crossing capacity [141]. Thus, they can deliver important biochemical molecules to the brain to fight neurodegenerative diseases, and the exosomes entering the blood from the brain can provide important diagnostic clues [142,143]. The exosomes isolated from MSC are very promising candidates for many cell-free therapies as they inherently contain the biochemicals that promote angiogenesis, anti-inflammation, and anti-apoptosis functions. The exosomes isolated from MSC were reported to be used for the treatment of Parkinson's disease, stroke, Perinatal brain injury (PBI), hypoxic-ischemic brain damage, prevent changes in gray and white matter, axonal regeneration, neurovascular dysfunction, axon demyelination and diabetic peripheral neuropathy in animal experimental models [23]. Exosomes are communication mediators between cells and have the capacity to modulate multiple biochemical reactions responsible for disease conditions. It was also reported that miRs present in exosomes are responsible for modulations. Thus, loading specialized miR cargoes in exosomes and delivering it at diseased site can positively modulate the disease state. This technique was used to get a protective and regenerative effect in many neuronal disease conditions [144,145]. The upregulation of miRs like miR-204, miR-323, miR-365, miR-672 miR-376b-5p, miR-760-5p, miR-369-5p, miR-133a, miR-463, miR-466d, miR-505, miR-665, miR-132, in ischemia-reperfusion spinal cord injury and miR-15b, miR-1, miR-20b-5p, miR-20a, miR-21, miR-30a, miR-31, miR-92a, miR-92b, miR-93, miR-98, miR-106b, miR-145, miR-146b, miR-152, miR-199a-3p, miR-203, miR-206, miR-214, miR-221, miR-290, miR-223, miR-333, miR-378, miR-374, miR-674-5p, miR-672, miR-17, miR-872, miR-146a, miR-486, miR-124, in mechanical spinal cord injury were reported to regain/improve central nervous functions (Fig. 13.7) [112].

At the same time, downregulation of miR-146a, miR-210, miR-199a-3p in ischemia-reperfusion spinal cord injury and miR-30b-5p, miR-30d, miR-30c, miR-129, miR-34a, miR-138, miR-137, miR-219-2-3p, miR-219-5p, miR-325-3p, miR-338, miR-323, miR-384-5p,miR-379, miR-543, miR-708, miR-495, miR-125-3p, miR-let-7b, miR-126, miR-129-1, miR-129-2, miR-342, miR-129-3p, in mechanical spinal cord injury were reported to regain/improve central nervous functions [112]. For central nervous system, it was reported that upregulation of miRs like miR-138, miR-137, miR-124, miR-21, miR-100, miR-296-5p, miR-148a, miR-218, miR-378a-3p, and down regulation of miR-221, miR-222, miR-10b effectively suppresses glioblastoma (GBM). Similarly, upregulation of miR-21, miR-134, miR-124,

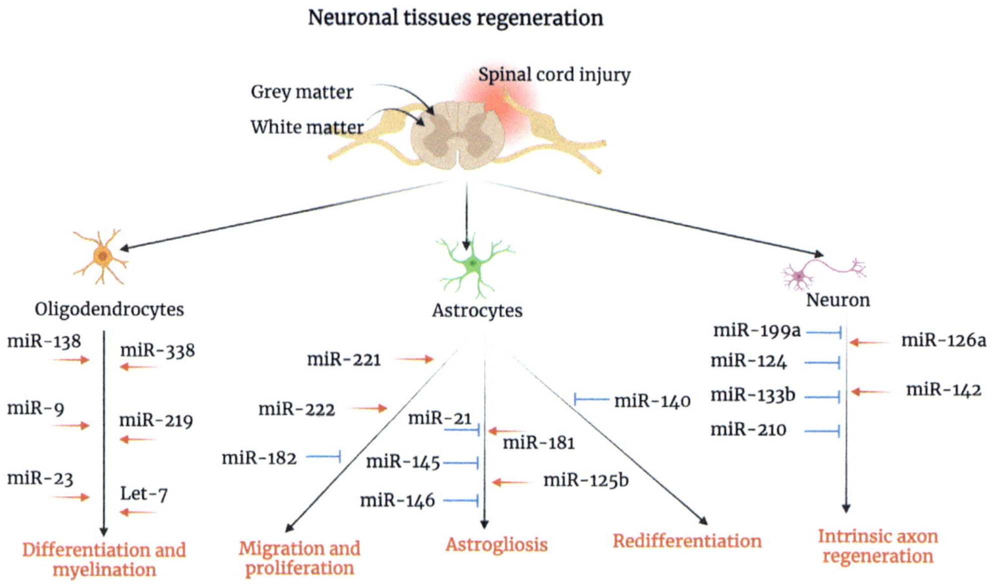

Figure 13.7 Schematic diagram showing the role of various miRNAs in the extrinsic and intrinsic regulation of axon regeneration following the spinal cord injury. Adopted from Li P., Teng Z-Q, Liu C-M. Extrinsic and intrinsic regulation of axon regeneration by micrornas after spinal cord injury. 2016; 2016: 1279051. https://doi.org/10.1155/2016/1279051.

miR-145-5p, miR-155, miR-7116-5p, and miR-let-7a, and downregulating miR-181a, miR-301b, miR-181c, miR-150, miR-125b, effectively suppress Parkinson's disease [146,147]. The ischemic brain injury condition can be improved by upregulating miR-29, miR-2115, miR-424, miR-101, miR-27b and downregulating miR-15a/16-1, miR-326 [148,149]. In spinal cord injury upregulation of miR-186-5p, miR-21, miR-711, miR-132, miR-431, miR-222, miR-16, miR-124, miR-15, miR-96, miR-200, miR-129, miR146a and downregulation of miR-221, miR-29b, miR-223, miR-20a, miR-107, miR-486, miR-1 can improve the regeneration and restore the function [150,151]. The task of miRs in bordering nervous system regeneration involves upregulation of miR-21, miR-222, miR-142-3p, miR-431, miR-511-3p, miR-192-5p, miR-431, miR-210, miR-192-5p, miR-26a, miR-210, miR-26a, miR-500, miR-138, miR-7239, miR-6538, and downregulation of miR-9, miR-152-3p, miR-24-3p, and miR-331-3p, can improve Sciatic nerve injury specifically DRG neurons. Similarly, upregulation of miR-129, miR-9, miR-327, miR-7, miR-221, miR-sc14, miR-222, miR-3099, miR-21, miR-34a, miR-30c, and downregulation of miR-let-7, miR-1b, miR-29a-3p, miR-485-5p, miR-221-3p, and miR-9, can improve Sciatic nerve injury, specifically Schwann cells [152].

Additionally, it was also reported that the use of miRs like miR-124a, miR-124 as nanoparticle delivery; miR-25, miR-92a, miR-19b, miR-125a-5p, miR-351, miR-219, miR-338, and miR-339-5p as aligned nanofibers, and miR-20, miR-7, as 3D scaffolds substantially improved the chances of functional nerve tissue development in vitro [147].

In summary, exosomes having demonstrated encouraging regeneration potentials in preclinical studies of several diseases, such as, regulate inflammatory responses, autophagy regulation; myocardial infraction, axon regeneration and damaged blood vessel etc. Consequently, the use of exosomes in regenerative medicine is seen as a potential option.

5. Future prospects

The right diagnosis of disease is essential in clinical practice. The detection of soluble biomarkers in biofluids has become a critical step in the early prognosis of disease. In this regard, exosomes have been under close study owing to their extensive presence in body fluids. As the multivesicular bodies merge with the plasma membrane, the exosomes, which are nanoscale biovesicles, are discharged from the cell into the neighboring fluid. They carry lipids, proteins, and genetic material that is very specific to the cell, and is selectively endocytosed by nearby or distant cells. Upon entry, exosomes reprogram the beneficiary cells via their bioactive complexes and hence reflect the physiological and pathological conditions present in the body. Exosomes have shown potential therapeutic outcomes in soft tissue mending and regeneration. The therapeutic effects of exosomal RNA nanoparticles mimic those of cell therapies and lack many of their sideeffects. Preclinical studies have reported the safety and effectiveness of exosomes. At this stage, exosomal medicine is still in its infancy and has not yet seen widespread clinical use. The freshly created PEP from the Mayo Clinic has impressively cleared the clinical studies to examine its use in tendon restoration [153–155], peripheral nerve regeneration [156], myocardial infarction recovery, and vaginal tissue regeneration [157]. There is a need to elucidate the signaling pathways of exosome functions. The underlying principles of exosomes' stimulating effects on the regeneration of soft tissues are also unknown. Exosome isolation, purification, administration, and exosome transporters lack a standardized method or procedure. For the effective clinical translation of exosome usage, it is essential to identify an acceptable route for exosomal injection as well as methodologies for their separation, qualification, and purification. Exosomal therapies' safety and negative consequences have been examined in a few studies; for instance, plasma-derived exosomes can transfer components of hazardous medications to sick tissues [91]. Exosomes from tumors have the ability to respectively, in vitro and in vivo produce chemotherapeutics for the beneficiary cells [158]. Cossetti et al., have demonstrated that exosomes can transfer materials from tumor cells to beneficiary cells [159]. Additional studies should be done to detect safer sources of exosomes, and a step-by-step approach should be followed to address the safety concerns of exosomes. Finding a benign and active source of exosomes is the initial step, followed by the determination of the mechanism through which exosomes act therapeutically. The final step can consist of modifying the exosomes to render them safe. This may be accomplished by acquiring them, filling them with a known payload, and then reinfusing them into the respective patient. Oblique isolation and manipulation techniques are used to modify them in order to produce exosomes that carry therapeutic molecules, such as those from mesenchymal stem cells (MSCs) or juvenile dendritic cells (iDCs). The patient can be engaged with these modified exosomes. Overall, exosomes have a high therapeutic potential in regenerative medicine, and the next important step is to take their application to clinical trials. There are multiple mechanisms and signaling pathways through which exosomes exert their repair and regeneration effects in soft tissue [91].

PEP has been known to promote peripheral nerve [156] and tendon [153] rejuvenation but has not yet been reviewed for skeletal muscles. Future research on its role in muscle repair can be fruitful. Their most readily available sources are urine-derived exosomes [160] and bovine milk [161], but they have only been analyzed on skeletal muscles. Further research on the effects of exosomes produced from cow urine and milk on tendon and peripheral nerve tissue recovery is important. However, currently, the majority of the research is still in the preclinical stage, and the related drugs are still under review to be approved for use. Future prospects include discovering the scope of applications for exosomes and large-scale production routes for clinical uses. Moving forward with exosome research may pave the way to address diabetes-related wounds in the future [43,162].

6. Conclusion

Exosomal RNA produced from several cell varieties has been revealed in several studies to be useful for wound restorative in recent years. Exosomal RNA targets on destination cells are not clearly understood, so further study of their molecular processes is required. There are currently very few studies looking at mRNAs, circRNAs, and IncRNAs, with the majority of ies focusing on how miRNAs from stem cell generated exosomal transcriptomes influence fibroblast multiplication and movement in wound remediation. Exosomal RNA generated from progenitor stem cells transmits signals and participates in the reorganization of the extracellular matrix through a paracrine mechanism. As exo RNA is less immunogenic, more durable, and easier to maintain than donor stem cells, it can be used as a substitute for stem cells in the management of wound healing. Clinical studies using exosome RNA made from stem cells to heal diabetic wounds have been filed on the ClinicalTrials.gov website [NCT05243368]. Moreover, patents [CN114376989A and CN114377194A] on the use of exosomes produced from stem cells as wound dressings have been published. Besides, Exosomal RNAs have demonstrated high safety and cytocompatibility in contrast to the significant dangers offered by viral vectors, including mutagenesis, cytotoxicity, and a restricted potential for genetic cargo.

Exosomes have remarkable therapeutic promise for the treatment of wound healing and tissue regeneration. Exosome clinical translation, however, still confronts several obstacles. A problem still exists in determining the circumstances under which naturally occurring exosomes containing indicated active components can be produced, as well as in scaling up exosome synthesis for therapeutic therapy. Exosomes, for instance, may have dramatically different therapeutic potential relying on the health of the cells from which they are produced (e.g., age, functional status, and disease state). RNAs, whether coding or non-coding, include a lot of exosomes. It is important to investigate any possible interactions between RNAs. Researchers should concentrate their efforts in the future on the biomaterial that causes the therapy effect. The majority of current investigations are still in the preclinical phase, and the relevant medications have not yet received FDA approval. Further study will be needed to ascertain the range of exosomal RNA applications and the production processes for a wide range of therapeutic purposes. Further research on several types of exosomal RNA might lead to advanced wound healing and tissue repair technologies in the near future.

References

[1] Rodriguez-Menocal L, Salgado M, Ford D, Van Badiavas E. Stimulation of skin and wound fibroblast migration by mesenchymal stem cells derived from normal donors and chronic wound patients. Stem Cells Translational Medicine 2012;1:221–9.

[2] Kalluri R, LeBleu VS. The biology, function, and biomedical applications of exosomes'. Science 2020;367.

[3] György B, Szabó TG, Pásztói M, Pál Z, Misják P, Aradi B, et al. Membrane vesicles, current state-of-the-art: emerging role of extracellular vesicles. Cellular and Molecular Life Sciences 2011;68:2667–88.
[4] Liang X, Li D, Leng S, Zhu X. RNA-based pharmacotherapy for tumors: from bench to clinic and back. Biomedicine and Pharmacotherapy 2020;125:109997.
[5] Wada T, Becskei A. Impact of methods on the measurement of mRNA turnover'. International Journal of Molecular Sciences 2017;18.
[6] Mauger DM, Cabral BJ, Presnyak V, Su SV, Reid DW, Goodman B, et al. mRNA structure regulates protein expression through changes in functional half-life'. Proceedings of the National Academy of Sciences of the United States of America 2019;116:24075–83.
[7] Gangoda L, Boukouris S, Liem M, Kalra H, Mathivanan S. Extracellular vesicles including exosomes are mediators of signal transduction: are they protective or pathogenic? Proteomics 2015;15:260–71.
[8] Xu J, Bai S, Cao Y, Liu L, Fang Y, Du J, et al. miRNA-221-3p in endothelial progenitor cell-derived exosomes accelerates skin wound healing in diabetic mice. Diabetes, Metabolic Syndrome and Obesity 2020;13:1259–70.
[9] Ti D, Hao H, Tong C, Liu J, Dong L, Zheng J, et al. LPS-preconditioned mesenchymal stromal cells modify macrophage polarization for resolution of chronic inflammation via exosome-shuttled let-7b. Journal of Translational Medicine 2015;13:308.
[10] Xiong Y, Chen L, Yan C, Wu Z, Endo Y, Liu J, et al. Circulating exosomal miR-20b-5p inhibition restores Wnt9b signaling and reverses diabetes-associated Impaired wound healing. Small 2019;16:1904044.
[11] Badhe RV, Bijukumar D, Mesquita P, Cheng KY, Ramachandran RA, Lin Y, et al. Dynamic microfluidic bioreactor-Hip simulator (DMBH) system for implant toxicity monitoring. Biotechnology and Bioengineering 2021;118:4829–39.
[12] Badhe RV, Bijukumar D, Chejara DR, Mabrouk M, Choonara YE, Kumar P, et al. A composite chitosan-gelatin bi-layered, biomimetic macroporous scaffold for blood vessel tissue engineering. Carbohydrate Polymers 2017;157:1215–25.
[13] Gupta AA, Kheur S, Badhe RV, Raj AT, Bhonde R, Jaisinghani A, et al. Assessing the potential use of chitosan scaffolds for the sustained localized delivery of vitamin D. Saudi Journal of Biological Sciences 2021;28:2210–5.
[14] Badhe RV, Chatterjee A, Bijukumar D, Mathew MT. Current advancements in bio-ink technology for cartilage and bone tissue engineering'. Bone 2023:116746.
[15] Zardad A, Mabrouk M, Marimuthu T, du Toit LC, Kumar P, Choonara YE, et al. Synthesis and biocompatibility of dual-responsive thermosonic injectable organogels based on crosslinked N-(isopropyl acrylamide) for tumour microenvironment targeting. Materials Science and Engineering C, Materials for Biological Applications 2018;90:148–58.
[16] Badhe RV, Nipate SS. Low-intensity current (LIC) stimulation of subcutaneous adipose derived stem cells (ADSCs)—a missing link in the course of LIC based wound healing. Medical Hypotheses 2019;125:79–83.
[17] Mabrouk M, Mulla JA, Kumar P, Chejara DR, Badhe RV, Choonara YE, et al. Intestinal targeting of ganciclovir release employing a novel HEC-PAA blended lyomatrix. AAPS PharmSciTech 2016;17:1120–30.
[18] Natsaridis E, Mouzoura P, Gkartziou F, Marazioti A, Antimisiaris SG. Development of growth factor-incorporating liposomes for integration into scaffolds as a method to improve tissue regeneration. International Journal of Developmental Biology 2022;66:137–54.
[19] Clancy JW, Sedgwick A, Rosse C, Muralidharan-Chari V, Raposo G, Method M, et al. Regulated delivery of molecular cargo to invasive tumour-derived microvesicles. Nature Communications 2015;6:6919.
[20] Gurung S, Perocheau D, Touramanidou L, Baruteau J. The exosome journey: from biogenesis to uptake and intracellular signalling. Cell Communication and Signaling 2021;19:47.
[21] Guescini M, Genedani S, Stocchi V, Agnati LF. Astrocytes and Glioblastoma cells release exosomes carrying mtDNA. Journal of Neural Transmission 2010;117:1–4.

[22] Valadi H, Ekström K, Bossios A, Sjöstrand M, Lee JJ, Lötvall JO. Exosome-mediated transfer of mRNAs and microRNAs is a novel mechanism of genetic exchange between cells. Nature Cell Biology 2007;9:654–9.

[23] Xie M, Wu Di, Li G, Yang J, Zhang YS. Exosomes targeted towards applications in regenerative medicine. Nano Select 2021;2:880–908.

[24] Zeng EZ, Chen I, Chen X, Yuan X. Exosomal microRNAs as novel cell-free therapeutics in tissue engineering and regenerative medicine. Biomedicines 2022;10.

[25] Lauwers E, Wang YC, Gallardo R, Van der Kant R, Michiels E, Swerts J, et al. Hsp90 mediates membrane deformation and exosome release. Molecular Cell 2018;71:689–702.e9.

[26] Keller S, Ridinger J, Rupp A-K, Johannes W, Janssen G, Peter A. Body fluid derived exosomes as a novel template for clinical diagnostics. Journal of Translational Medicine 2011;9:86.

[27] Warren MR, Zhang C, Vedadghavami A, Bokvist K, Dhal PK, Bajpayee AG. Milk exosomes with enhanced mucus penetrability for oral delivery of siRNA. Biomaterials Science 2021;9:4260–77.

[28] Li P, Kaslan M, Han Lee S, Yao J, Gao Z. Progress in exosome isolation techniques. Theranostics 2017;7:789–804.

[29] Zhang Y, Liu Y, Liu H, Tang WH. Exosomes: biogenesis, biologic function and clinical potential. Cell and Bioscience 2019;9:19.

[30] Kowal J, Arras G, Colombo M, Jouve M, Morath JP, Primdal-Bengtson B, et al. Proteomic comparison defines novel markers to characterize heterogeneous populations of extracellular vesicle subtypes. Proceedings of the National Academy of Sciences of the United States of America 2016;113:E968–77.

[31] Hu Q, Su H, Li J, Lyon C, Tang W, Wan M, et al. Clinical applications of exosome membrane proteins. Precision Clinical Medicine 2020;3:54–66.

[32] Jan AT, Rahman S, Badierah R, Lee EJ, Mattar EH, Redwan EM, et al. Expedition into exosome biology: a perspective of progress from discovery to therapeutic development. Cancers 2021;13:1157.

[33] Jing H, He X, Zheng J. Exosomes and regenerative medicine: state of the art and perspectives. Translational Research 2018;196:1–16.

[34] Plebanek MP, Kannan Mutharasan R, Volpert O, Matov A, Gatlin JC, Shad Thaxton C. Nanoparticle targeting and cholesterol flux through scavenger receptor type B-1 inhibits cellular exosome uptake. Scientific Reports 2015;5:15724.

[35] Skotland T, Sandvig K, Llorente A. Lipids in exosomes: current knowledge and the way forward. Progress in Lipid Research 2017;66:30–41.

[36] Shi Y, Du L, Lv D, Li Y, Zhang Z, Huang X, et al. Emerging role and therapeutic application of exosome in hepatitis virus infection and associated diseases. Journal of Gastroenterology 2021;56:336–49.

[37] Liu S, Xu X, Liang S, Chen Z, Zhang Y, Qian A, et al. The application of MSCs-derived extracellular vesicles in bone disorders: novel cell-free therapeutic strategy. Frontiers in Cell and Developmental Biology 2020;8:619.

[38] Li D, Li XI, Wang A, Meisgen F, Pivarcsi A, Sonkoly E, et al. MicroRNA-31 promotes skin wound healing by enhancing keratinocyte proliferation and migration. Journal of Investigative Dermatology 2015;135:1676–85.

[39] Squadrito ML, Baer C, Burdet F, Maderna C, Gilfillan GD, Lyle R, et al. Endogenous RNAs modulate microRNA sorting to exosomes and transfer to acceptor cells. Cell Reports 2014;8:1432–46.

[40] Yanshina DD, Kossinova OA, Gopanenko AV, Krasheninina OA, Malygin AA, Venyaminova AG, et al. Structural features of the interaction of the 3'-untranslated region of mRNA containing exosomal RNA-specific motifs with YB-1, a potential mediator of mRNA sorting. Biochimie 2018;144:134–43.

[41] Villarroya-Beltri C, Gutiérrez-Vázquez C, Sánchez-Cabo F, Pérez-Hernández D, Vázquez J, Martin-Cofreces N, et al. Sumoylated hnRNPA2B1 controls the sorting of miRNAs into exosomes through binding to specific motifs. Nature Communications 2013;4:2980.

[42] Bao C, Lyu D, Huang S. Circular RNA expands its territory. Molecular and Cellular Oncology 2016;3:e1084443.

[43] Wan R, Hussain A, Behfar A, Moran SL, Zhao C. The therapeutic potential of exosomes in soft tissue repair and regeneration. International Journal of Molecular Sciences 2022;23.

[44] Heo JS. Selenium-Stimulated exosomes enhance wound healing by modulating inflammation and angiogenesis. International Journal of Molecular Sciences 2022;23.

[45] Ju GQ, Cheng J, Zhong L, Wu S, Zou XY, Zhang GY, et al. Microvesicles derived from human umbilical cord mesenchymal stem cells facilitate tubular epithelial cell dedifferentiation and growth via hepatocyte growth factor induction. PLoS One 2015;10:e0121534.

[46] Deregibus MC, Cantaluppi V, Calogero R, Lo Iacono M, Tetta C, Biancone L, et al. Endothelial progenitor cell derived microvesicles activate an angiogenic program in endothelial cells by a horizontal transfer of mRNA. Blood 2007;110:2440–8.

[47] Qiu M, Zhai S, Fu Q, Liu D. Bone marrow mesenchymal stem cells-derived exosomal MicroRNA-150-3p promotes osteoblast proliferation and differentiation in osteoporosis. Human Gene Therapy 2021;32:717–29.

[48] Chen Md G, Wu Md Y, Zou Md L, Zeng Md Y. Effect of MicroRNA-146a modified adipose-derived stem cell exosomes on rat back wound healing. The International Journal of Lower Extremity Wounds 2021. 15347346211038092.

[49] Hu Y, Rao SS, Wang ZX, Cao J, Tan YJ, Luo J, et al. Exosomes from human umbilical cord blood accelerate cutaneous wound healing through miR-21-3p-mediated promotion of angiogenesis and fibroblast function. Theranostics 2018;8:169–84.

[50] Tao SC, Guo SC, Li M, Ke QF, Guo YP, Zhang CQ. Chitosan wound dressings incorporating exosomes derived from MicroRNA-126-overexpressing synovium mesenchymal stem cells provide sustained release of exosomes and heal full-thickness skin defects in a diabetic rat model. Stem Cells Translational Medicine 2017;6:736–47.

[51] Wu D, Kang L, Tian J, Wu Y, Liu J, Li Z, et al. Exosomes derived from bone mesenchymal stem cells with the stimulation of Fe(3)O(4) nanoparticles and static magnetic field enhance wound healing through upregulated miR-21-5p. International Journal of Nanomedicine 2020;15:7979–93.

[52] Chen B, Sun Y, Zhang J, Zhu Q, Yang Y, Niu X, et al. Human embryonic stem cell-derived exosomes promote pressure ulcer healing in aged mice by rejuvenating senescent endothelial cells. Stem Cell Research and Therapy 2019;10:142.

[53] Yuan R, Dai X, Li Y, Li C, Liu L. Exosomes from miR-29a-modified adipose-derived mesenchymal stem cells reduce excessive scar formation by inhibiting TGF-β2/Smad3 signaling. Molecular Medicine Reports 2021;24.

[54] Fang S, Xu C, Zhang Y, Xue C, Yang C, Bi H, et al. Umbilical cord-derived mesenchymal stem cell-derived exosomal MicroRNAs suppress myofibroblast differentiation by inhibiting the transforming growth factor-β/SMAD2 pathway during wound healing. Stem Cells Translational Medicine 2016;5:1425–39.

[55] Han ZF, Cao JH, Liu ZY, Yang Z, Qi RX, Xu HL. Exosomal lncRNA KLF3-AS1 derived from bone marrow mesenchymal stem cells stimulates angiogenesis to promote diabetic cutaneous wound healing. Diabetes Research and Clinical Practice 2022;183:109126.

[56] Qian L, Pi L, Fang BR, Meng XX. Adipose mesenchymal stem cell-derived exosomes accelerate skin wound healing via the lncRNA H19/miR-19b/SOX9 axis. Laboratory Investigation 2021;101:1254–66.

[57] Wang Z, Feng C, Liu H, Meng T, Huang W, Long X, et al. Hypoxic pretreatment of adipose-derived stem cells accelerates diabetic wound healing via circ-Gcap14 and HIF-1α/VEGF mediated angiopoiesis. International Journal of Stem Cells 2021;14:447–54.

[58] Shi R, Jin Y, Hu W, Lian W, Cao C, Han S, et al. Exosomes derived from mmu_circ_0000250-modified adipose-derived mesenchymal stem cells promote wound healing in diabetic mice by inducing miR-128-3p/SIRT1-mediated autophagy. American Journal of Physiology—Cell Physiology 2020;318:c56–C848.

[59] Aryani A, Denecke B. Exosomes as a nanodelivery system: a key to the future of neuromedicine? Molecular Neurobiology 2016;53:818–34.

[60] Han Y, Jones TW, Dutta S, Zhu Y, Wang X, Priya Narayanan S, et al. Overview and update on methods for cargo loading into extracellular vesicles. Processes 2021;9:356.
[61] Fu S, Wang Y, Xia X, Zheng J. Exosome engineering: current progress in cargo loading and targeted delivery. NanoImpact 2020;20:100261.
[62] Kumari A, Yadav SK, Yadav SC. Biodegradable polymeric nanoparticles based drug delivery systems. Colloids and Surfaces B: Biointerfaces 2010;75:1−18.
[63] Agrawal AK, Aqil F, Jeyabalan J, Spencer WA, Beck J, Gachuki BW, et al. Milk-derived exosomes for oral delivery of paclitaxel. Nanomedicine 2017;13:1627−36.
[64] Turturici G, Tinnirello R, Sconzo G, Geraci F. Extracellular membrane vesicles as a mechanism of cell-to-cell communication: advantages and disadvantages. American Journal of Physiology—Cell Physiology 2014;306:C621−33.
[65] van den Boorn, Jasper G, Martin S, Coch C, Hartmann G. SiRNA delivery with exosome nanoparticles. Nature Biotechnology 2011;29:325−6.
[66] Rezaie J, Feghhi M, Etemadi T. A review on exosomes application in clinical trials: perspective, questions, and challenges. Cell Communication and Signaling 2022;20:145.
[67] Chen C, Johnston TD, Jeon H, Gedaly R, McHugh PP, Burke TG, et al. An in vitro study of liposomal curcumin: stability, toxicity and biological activity in human lymphocytes and Epstein-Barr virus-transformed human B-cells. International Journal of Pharmaceutics 2009;366:133−9.
[68] Feng SS, Mu L, Win KY, Huang G. Nanoparticles of biodegradable polymers for clinical administration of paclitaxel. Current Medicinal Chemistry 2004;11:413−24.
[69] Cheng Q, Feng J, Chen J, Zhu X, Li F. Brain transport of neurotoxin-I with PLA nanoparticles through intranasal administration in rats: a microdialysis study. Biopharmaceutics and Drug Disposition 2008;29:431−9.
[70] Gómez-Gaete C, Tsapis N, Besnard M, Bochot A, Fattal E. Encapsulation of dexamethasone into biodegradable polymeric nanoparticles. International Journal of Pharmaceutics 2007;331:153−9.
[71] Vázquez-Ríos, Abi J, Molina-Crespo Á, Bouzo BL, López-López R, Moreno-Bueno G, et al. Exosome-mimetic nanoplatforms for targeted cancer drug delivery. Journal of Nanobiotechnology 2019;17:85.
[72] Goetzl EJ, Boxer A, Schwartz JB, Abner EL, Petersen RC, Miller BL, et al. Altered lysosomal proteins in neural-derived plasma exosomes in preclinical Alzheimer disease. Neurology 2015;85:40−7.
[73] Lin Y, Wu J, Gu W, Huang Y, Tong Z, Huang L, et al. Exosome−liposome hybrid nanoparticles deliver CRISPR/Cas9 system in MSCs. Advanced Science 2018;5:1700611.
[74] Chen W, Huang Y, Han J, Yu L, Li Y, Lu Z, et al. Immunomodulatory effects of mesenchymal stromal cells-derived exosome. Immunologic Research 2016;64:831−40.
[75] Kaur S, Singh SP, Elkahloun AG, Wu W, Abu-Asab MS, Roberts DD. CD47-dependent immunomodulatory and angiogenic activities of extracellular vesicles produced by T cells. Matrix Biology 2014;37:49−59.
[76] Vu NB, Nguyen HT, Palumbo R, Pellicano R, Fagoonee S, Pham PV. Stem cell-derived exosomes for wound healing: current status and promising directions. Minerva Medica 2021;112:384−400.
[77] Yan L, Shen J, Wang J, Yang X, Dong S, Lu S. Nanoparticle-Based drug delivery system: a patient-friendly chemotherapy for oncology. Dose-Response 2020;18. 1559325820936161.
[78] Alvarez-Erviti L, Seow Y, Yin HF, Betts C, Lakhal S, Matthew J, et al. Delivery of siRNA to the mouse brain by systemic injection of targeted exosomes. Nature Biotechnology 2011;29:341−5.
[79] Li J, Liu K, Liu Y, Xu Y, Zhang F, Yang H, et al. Exosomes mediate the cell-to-cell transmission of IFN-α-induced antiviral activity. Nature Immunology 2013;14:793−803.
[80] Théry C, Regnault A, Garin J, Wolfers J, Zitvogel L, Ricciardi-Castagnoli P, et al. Molecular characterization of dendritic cell-derived exosomes. Selective accumulation of the heat shock protein hsc73. The Journal of Cell Biology 1999;147:599−610.
[81] Califf RM. Biomarker definitions and their applications. Experimental Biology and Medicine (Maywood) 2018;243:213−21.
[82] Jeyaram A, Jay S. Preservation and storage stability of extracellular vesicles for therapeutic applications. The AAPS Journal 2018;20.

[83] Makler A, Asghar W. Exosomal biomarkers for cancer diagnosis and patient monitoring. Expert Review of Molecular Diagnostics 2020;20:387–400.
[84] Lu J, Li J, Liu S, Wang T, Ianni A, Bober E, et al. Exosomal tetraspanins mediate cancer metastasis by altering host microenvironment. Oncotarget 2017;8:62803–15.
[85] Ashraf Malik M, Ishtiyaq Ali Mirza J, Umar M, Manzoor S. CD81(+) exosomes play a pivotal role in the establishment of hepatitis C persistent infection and contribute toward the progression of hepatocellular carcinoma. Viral Immunology 2019;32:453–62.
[86] Sandfeld-Paulsen B, Jakobsen KR, Bæk R, Folkersen BH, Rasmussen TR, Meldgaard P, et al. Exosomal proteins as diagnostic biomarkers in lung cancer. Journal of Thoracic Oncology 2016;11:1701–10.
[87] Melo SA, Luecke LB, Kahlert C, Fernandez AF, Gammon ST, Kaye J, et al. Glypican-1 identifies cancer exosomes and detects early pancreatic cancer. Nature 2015;523:177–82.
[88] Rodríguez-Martínez A, de Miguel-Pérez D, Ortega FG, García-Puche JL, Robles-Fernández I, Exposito J, et al. Exosomal miRNA profile as complementary tool in the diagnostic and prediction of treatment response in localized breast cancer under neoadjuvant chemotherapy. Breast Cancer Research 2019;21:21.
[89] Lee N, Canagasingham A, Bajaj M, Shanmugasundaram R, Hutton A, Bucci J, et al. Urine exosomes as biomarkers in bladder cancer diagnosis and prognosis: from functional roles to clinical significance. Frontiers in Oncology 2022;12:1019391.
[90] Huda MN, Nafiujjaman M, Deaguero IG, Okonkwo J, Hill ML, Kim T, et al. Potential use of exosomes as diagnostic biomarkers and in targeted drug delivery: progress in clinical and preclinical applications. ACS Biomaterials Science and Engineering 2021;7:2106–49.
[91] Campanella C, Caruso Bavisotto C, Logozzi M, Marino Gammazza A, Mizzoni D, Cappello F, et al. On the Choice of the extracellular vesicles for therapeutic Purposes. International Journal of Molecular Sciences 2019;20.
[92] Shafiei M, Ansari MNM, Razak SIA, Khan MUA. A comprehensive review on the applications of exosomes and liposomes in regenerative medicine and tissue engineering'. Polymers 2021;13.
[93] Bjørge IM, Kim SY, Mano JF, Kalionis B, Chrzanowski W. Extracellular vesicles, exosomes and shedding vesicles in regenerative medicine—a new paradigm for tissue repair. Biomaterials Science 2017;6:60–78.
[94] Bhujel B, Shin HE, Choi DJ, Han I. Mesenchymal stem cell-derived exosomes and Intervertebral Disc regeneration: review. International Journal of Molecular Sciences 2022;23.
[95] Zhang B, Tian X, Hao J, Xu G, Zhang W. Mesenchymal stem cell-derived extracellular vesicles in tissue regeneration. Cell Transplantation 2020;29. 963689720908500.
[96] Han S, Qi Y, Luo Y, Chen X, Liang H. Exosomal long non-coding RNA: interaction between cancer cells and non-cancer cells. Frontiers in Oncology 2020;10:617837.
[97] Geng X, Lin X, Zhang Y, Li Q, Guo Y, Fang C, et al. Exosomal circular RNA sorting mechanisms and their function in promoting or inhibiting cancer. Oncology Letters 2020;19:3369–80.
[98] Peng Y, Croce CM. The role of MicroRNAs in human cancer. Signal Transduction and Targeted Therapy 2016;1:15004.
[99] Yoshida T, Asano Y, Ui-Tei K. Modulation of MicroRNA processing by Dicer via its associated dsRNA binding proteins. Noncoding RNA 2021;7.
[100] Gravitz L. Skin. Nature 2018;563:S83.
[101] Rodrigues M, Kosaric N, Bonham CA, Gurtner GC. Wound healing: a cellular perspective. Physiological Reviews 2019;99:665–706.
[102] Singhvi G, Manchanda P, Krishna Rapalli V, Kumar Dubey S, Gupta G, Dua K. MicroRNAs as biological regulators in skin disorders. Biomedicine and Pharmacotherapy 2018;108:996–1004.
[103] Muthu S, Bapat A, Jain R, Jeyaraman N, Jeyaraman M. Exosomal therapy—a new frontier in regenerative medicine. Stem Cell Investigation 2021;8:7.
[104] Wang CR, Zhu HF, Zhu Y. Knockout of MicroRNA-155 ameliorates the Th17/Th9 immune response and promotes wound healing. Current Medical Science 2019;39:954–64.

[105] Wang X, Jiao Y, Pan Y, Zhang L, Gong H, Qi Y, et al. Fetal dermal mesenchymal stem cell-derived exosomes accelerate cutaneous wound healing by activating Notch signaling. Stem Cells International 2019;2019:2402916.
[106] Lv H, Liu H, Sun T, Wang H, Zhang X, Xu W. Exosome derived from stem cell: a promising therapeutics for wound healing. Frontiers in Pharmacology 2022;13:957771.
[107] Nallakumarasamy A, Jeyaraman M, Maffulli N, Jeyaraman N, Suresh V, Ravichandran S, et al. Mesenchymal stromal cell-derived extracellular vesicles in wound healing'. Life (Basel) 2022:12.
[108] Xie J, Wu W, Zheng L, Lin X, Tai Y, Wang Y, et al. Roles of MicroRNA-21 in skin wound healing: a Comprehensive review. Frontiers in Pharmacology 2022;13:828627.
[109] Ti D, Hao H, Fu X, Han W. Mesenchymal stem cells-derived exosomal microRNAs contribute to wound inflammation. Science China Life Sciences 2016;59:1305–12.
[110] Tanaka K, Kim SE, Yano H, Matsumoto G, Ohuchida R, Ishikura Y, et al. MiR-142 is required for *Staphylococcus aureus* clearance at skin wound sites via small GTPase-mediated regulation of the neutrophil actin cytoskeleton. Journal of Investigative Dermatology 2017;137:931–40.
[111] Zhang W, Xu Q, Zhang Z, Wang L, Qi Q, Zhou P, et al. Inhibition of miR-139-5p by topical JTXK gel promotes healing of Staphylococcus aureus-infected skin wounds. Cells and Development 2021;166:203658.
[112] Li P, Teng ZQ, Liu CM. Extrinsic and intrinsic regulation of axon regeneration by microRNAs after spinal cord injury'. Neural Plasticity 2016;2016:1279051.
[113] Li X, Liu L, Yang J, Yu Y, Chai J, Wang L, et al. Exosome derived from human umbilical cord mesenchymal stem cell mediates MiR-181c attenuating burn-induced excessive inflammation. EBioMedicine 2016;8:72–82.
[114] Saleh B, Dhaliwal H, Portillo Lara R, Shirzaei E, Abdi R, Amiji M, et al. Local immunomodulation using an adhesive hydrogel loaded with miRNA-laden nanoparticles promotes wound healing. Small 2019;15.
[115] Li H, Han X, Zuo K, Li L, Liu J, Yuan X, et al. miR-23b promotes cutaneous wound healing through inhibition of the inflammatory responses by targeting ASK1. Acta Biochimica et Biophysica Sinica 2018;50:1104–13.
[116] Ban E, Jeong S, Park M, Kwon H, Park J, Song EJ, et al. Accelerated wound healing in diabetic mice by miRNA-497 and its anti-inflammatory activity. Biomedicine and Pharmacotherapy 2020;121:109613.
[117] Lang H, Zhao F, Zhang T, Liu X, Wang Z, Wang R, et al. MicroRNA-149 contributes to scarless wound healing by attenuating inflammatory response. Molecular Medicine Reports 2017;16:2156–62.
[118] Yu L, Zhang X, Liu X, Li G, Chen M, Liu Z, et al. CircTMOD3 promotes lipopolysaccharide-induced chondrocyte apoptosis in osteoarthritis by sponging miR-27a. Journal of Bone and Mineral Metabolism 2022;40:415–21.
[119] Jiang Y, Xu X, Xiao L, Wang L, Qiang S. The role of microRNA in the inflammatory response of wound healing. Frontiers in Immunology 2022;13:852419.
[120] Song M, Han L, Chen FF, Wang D, Wang F, Zhang L, et al. Adipocyte-derived exosomes carrying sonic hedgehog mediate M1 macrophage polarization-induced insulin resistance via Ptch and PI3K pathways. Cellular Physiology and Biochemistry 2018;48:1416–32.
[121] Wu X, Iroegbu CD, Peng J, Guo J, Yang J, Fan C. Cell death and exosomes regulation after myocardial infarction and ischemia-reperfusion. Frontiers in Cell and Developmental Biology 2021;9:673677.
[122] Kansakar U, Varzideh F, Mone P, Jankauskas SS, Santulli G. Functional role of microRNAs in regulating cardiomyocyte death. Cells 2022:11.
[123] Kir D, Schnettler E, Modi S, Ramakrishnan S. Regulation of angiogenesis by microRNAs in cardiovascular diseases. Angiogenesis 2018;21:699–710.
[124] Moghaddam AS, Afshari JT, Esmaeili SA, Saburi E, Joneidi Z, Momtazi-Borojeni AA. Cardioprotective microRNAs: lessons from stem cell-derived exosomal microRNAs to treat cardiovascular disease. Atherosclerosis 2019;285:1–9.

[125] Wang JX, Zhang XJ, Li Q, Wang K, Wang Y, Jiao JQ, et al. MicroRNA-103/107 regulate programmed necrosis and myocardial ischemia/reperfusion injury through targeting FADD. Circulation Research 2015;117:352–63.
[126] Zhang DY, Wang BJ, Ma M, Yu K, Zhang Q, Zhang XW. MicroRNA-325-3p protects the heart after myocardial infarction by inhibiting RIPK3 and programmed necrosis in mice. BMC Molecular Biology 2019;20:17.
[127] Wang K, Liu F, Liu CY, An T, Zhang J, Zhou LY, et al. The long noncoding RNA NRF regulates programmed necrosis and myocardial injury during ischemia and reperfusion by targeting miR-873. Cell Death Differentiation 2016;23:1394–405.
[128] Wang K, An T, Zhou LY, Liu CY, Zhang XJ, Feng C, et al. E2F1-regulated miR-30b suppresses cyclophilin D and protects heart from ischemia/reperfusion injury and necrotic cell death. Cell Death Differentiation 2015;22:743–54.
[129] Tungsukruthai S, Reamtong O, Roytrakul S, Sukrong S, Vinayanwattikun C, Chanvorachote P. Targeting AKT/mTOR and Bcl-2 for autophagic and apoptosis cell death in lung cancer: novel activity of a polyphenol compound'. Antioxidants 2021:10.
[130] Zhang Q, Zhou J, Zhi P, Liu L, Liu C, Fang A, et al. 3D printing method for bone tissue engineering scaffold. Medicine in Novel Technology and Devices 2023;17 [None].
[131] Devillard CD, Marquette CA. Vascular tissue engineering: challenges and requirements for an ideal large scale blood vessel. Frontiers in Bioengineering and Biotechnology 2021;9.
[132] Tsiapalis D, O'Driscoll L. Mesenchymal stem cell derived extracellular vesicles for tissue engineering and regenerative medicine applications'. Cells 2020:9.
[133] Cooper LF, Ravindran S, Huang CC, Kang M. A role for exosomes in craniofacial tissue engineering and regeneration. Frontiers in Physiology 2019;10:1569.
[134] Wang X, Thomsen P. Mesenchymal stem cell-derived small extracellular vesicles and bone regeneration. Basic and Clinical Pharmacology and Toxicology 2021;128:18–36.
[135] Liu Y, Ma Y, Zhang J, Yuan Y, Wang J. Exosomes: a novel therapeutic agent for cartilage and bone tissue regeneration. Dose-Response 2019;17. 1559325819892702.
[136] Kim YG, Choi J, Kim K. Mesenchymal stem cell-derived exosomes for effective cartilage tissue repair and treatment of osteoarthritis. Biotechnology Journal 2020;15:e2000082.
[137] Wang W, Liang X, Zheng K, Ge G, Chen X, Xu Y, et al. Horizon of exosome-mediated bone tissue regeneration: the all-rounder role in biomaterial engineering. Materials Today Bio 2022;16:100355.
[138] Foo JB, Looi QH, How CW, Lee SH, Al-Masawa ME, Chong PP, et al. Mesenchymal stem cell-derived exosomes and MicroRNAs in cartilage regeneration: biogenesis, efficacy, miRNA enrichment and delivery'. Pharmaceuticals 2021:14.
[139] Horita M, Hsu SN, Raper A, Farquharson C, Stephen LA. miR-29b inhibits TGF-β1-induced cell proliferation in articular chondrocytes. Biochemistry and Biophysics Reports 2022;29:101216.
[140] Hong L, Sun H, Amendt BA. MicroRNA function in craniofacial bone formation, regeneration and repair. Bone 2021;144:115789.
[141] Ramos-Zaldívar HM, Polakovicova I, Salas-Huenuleo E, Corvalán AH, Kogan MJ, Yefi CP, et al. Extracellular vesicles through the blood-brain barrier: a review. Fluids and Barriers of the CNS 2022;19:60.
[142] Younas N, Fernandez Flores LC, Hopfner F, Höglinger GU, Zerr I. A new paradigm for diagnosis of neurodegenerative diseases: peripheral exosomes of brain origin. Translational Neurodegeneration 2022;11:28.
[143] Cano A, Muñoz-Morales Á, Sánchez-López E, Ettcheto Arriola M, Souto EB, Camins A, et al. Exosomes-based nanomedicine for neurodegenerative diseases: current insights and future challenges. Pharmaceutics 2023;15:298.
[144] Yang Y, Yang H, Yang Y, Ma Y. Exosomal microRNAs have great potential in the neurorestorative therapy for traumatic brain injury. Experimental Neurology 2022;352:114026.
[145] Feng J, Zhang Y, Zhu Z, Gu C, Waqas A, Chen L. Emerging exosomes and exosomal MiRNAs in spinal cord injury. Frontiers in Cell and Developmental Biology 2021;9:703989.

[146] Li S, Bi G, Han S, Huang R. MicroRNAs play a role in Parkinson's disease by regulating microglia function: from pathogenetic involvement to therapeutic potential. Frontiers in Molecular Neuroscience 2021;14:744942.

[147] Tang X, Sun C. The roles of MicroRNAs in neural regenerative medicine. Experimental Neurology 2020;332:113394.

[148] Wang X, Han Z, Cheng R, Zhou X, Hou X, Chen J, et al. Role of miR-326 in neonatal hypoxic-ischemic brain damage pathogenesis through targeting of the δ-opioid receptor. Molecular Brain 2020;13:51.

[149] Guo X, Shen X, Yong Z. MiR-101 protects against the cerebral I/R injury through regulating JAK2/STAT3 signaling pathway'. Neuropsychiatric Disease and Treatment 2021;17:2791–802.

[150] Sun F, Zhang H, Huang T, Shi J, Wei T, Wang Y. miRNA-221 regulates spinal cord injury-induced inflammatory response through targeting TNF-α expression. BioMed Research International 2021;2021:6687963.

[151] Pinchi E, Frati A, Cantatore S, D'Errico S, Russa R, Maiese A, et al. Acute spinal cord injury: a systematic review investigating miRNA Families involved. International Journal of Molecular Sciences 2019;20.

[152] Liu M, Li P, Jia Y, Cui Q, Zhang K, Jiang J. Role of non-coding RNAs in axon regeneration after peripheral nerve injury. International Journal of Biological Sciences 2022;18:3435–46.

[153] Shi G, Wang Y, Wang Z, Thoreson AR, Jacobson DS, Amadio PC, et al. A novel engineered purified exosome product patch for tendon healing: an explant in an ex vivo model. Journal of Orthopaedic Research 2021;39:1825–37.

[154] Qi J, Liu Q, Reisdorf RL, Boroumand S, Behfar A, Moran SL, et al. Characterization of a purified exosome product and its effects on canine flexor tenocyte biology. Journal of Orthopaedic Research 2020;38:1845–55.

[155] Ren Y, Zhang S, Wang Y, Jacobson DS, Reisdorf RL, Kuroiwa T, et al. Effects of purified exosome product on rotator cuff tendon-bone healing in vitro and in vivo. Biomaterials 2021;276:121019.

[156] Ikumi A, Gingery A, Toyoshima Y, Zhao C, Moran S, Livia C, et al. Administration of purified exosome Product in a rat Sciatic serve Reverse Autograft model. Plastic and Reconstructive Surgery 2021 [Publish Ahead of Print].

[157] Kisby CK, Shadrin IY, Peng LT, Stalboerger PG, Trabuco EC, Behfar A, et al. Impact of repeat dosing and mesh exposure chronicity on exosome-induced vaginal tissue regeneration in a porcine mesh exposure model. Female Pelvic Medicine and Reconstructive Surgery 2021;27:195–201.

[158] Zhao B, Zhang Y, Han S, Zhang W, Zhou Q, Guan H, et al. Exosomes derived from human amniotic epithelial cells accelerate wound healing and inhibit scar formation. Journal of Molecular Histology 2017;48:121–32.

[159] Cossetti C, Lugini L, Astrologo L, Saggio I, Fais S, Spadafora C. Soma-to-germline transmission of RNA in mice xenografted with human tumour cells: possible transport by exosomes. PLoS One 2014;9:e101629.

[160] Zhu Q, Li Q, Niu X, Zhang G, Ling X, Zhang J, et al. Extracellular vesicles secreted by human urine-derived stem cells promote ischemia repair in a mouse model of hind-limb ischemia. Cellular Physiology and Biochemistry 2018;47:1181–92.

[161] Parry HA, Mobley CB, Mumford PW, Romero MA, Haun CT, Zhang Y, et al. Bovine milk extracellular vesicles(EVs) modification elicits skeletal muscle growth in rats. Frontiers in Physiology 2019;10:436.

[162] Zhang W, Hu J, Huang Y, Wu C, Xie H. Urine-derived stem cells: applications in skin, bone and articular cartilage repair. Burns and Trauma 2021;9:tkab039.

CHAPTER FOURTEEN

The importance of exosomal RNA in major ocular and respiratory diseases

Hima Makala[1], Sounak Ghosh Roy[2] and Debasish Sarkar[3]
[1]Radiation Oncology Branch, National Cancer Institute, NIH, Bethesda, MD, United States
[2]Henry M Jackson Foundation for the Advancement of Military Medicine, Inc., Silver Spring, MD, United States
[3]Fosun Pharma USA Inc., Lexington, MA, United States

1. Introduction

Exosomes are groups of biological molecules encased by lipid molecules that serve an important role in cell-to-cell communication. They were discovered in vitro-cultured supernatants of sheep reticulocytes by EG Trams in 1981 [1]. These nano-sized (30—100 nm) vesicles are composed of RNAs, DNAs, proteins, and lipids and can influence physiological and pathological processes. Initially, these small vesicles were considered garbage bins for disposing of damaged cells and undesirable cellular components. However, they are now recognized as a vital part of intercellular communication, functioning as a relay molecule with great specificity. They relay information related to antigen presentation, cellular growth and differentiation, immune response in tumors, and the migration and invasion of tumor cells [2].

Exosomes are ubiquitously secreted in several biofluids, including aqueous, saliva, milk, blood, plasma, serum, tears, and all cell supernatants. They are produced by a variety of cell types, including adipocytes, endothelial cells, and mesenchymal stem cells (MSCs). Their tissue-specific roles have gained considerable attention in identifying disease-specific biomarkers and in regenerative medicine [3,4]. Continuous efforts are being made to find a cure for degenerative diseases. For example, exosomes derived from MSCs have shown promising results in the treatment of Parkinson's disease [5,6]. Exosomes do not provoke an immune response and can be designed as drug vehicles for site-specific drug delivery. A study demonstrated that the use of localized drug-delivery exosomes in non-small cell lung cancer aims to increase therapeutic efficacy and reduce side effects [7]. Understanding the patterns of exosome expression in normal and sick tissues can help in the development of biomarkers, and the improvement of medication delivery [8]. Researchers are providing strong evidence that the alternations in exosome cargo, particularly exosomal RNAs, involve gene regulation and functional aspects of recipient cells.

Exosomal RNA
ISBN 978-0-443-14008-2
https://doi.org/10.1016/B978-0-443-14008-2.00006-1

2. Exosomal mRNA

Exosomal mRNAs were discovered using microarray analysis in MC/9 and HMC-1, mouse and human cell lines, respectively [9]. The 270 transcripts of exosomal RNAs were identified in human HMC-1 upon transferring the exosomal mRNA from mouse mast cells [10]. Furthermore, exosomal mRNAs function by translating into functional proteins in the recipient cells, owing to their superior advantage over noncoding RNAs (ncRNAs). Exosomal RNASs are encapsulated in a lipid membrane, which acts as a protective barrier against RNAse degradation. Exosome-derived urine samples can remain stable for as long as 14 days at 4°C [11]. The presence of MSC exosomal mRNA, specifically MCP-1 mRNA, MCP-1 (involved in the escalation of inflammatory reactions, including monocyte recruitment and photoreceptor apoptosis), revealed protective benefits on tissue wound healing in eyes treated with MSC exosomes [12].

3. Exosomal miRNA

MicroRNAs (miRNAs) are nonncRNAs with 18–21 nucleotides in length that regulate gene expression by binding to the 3′ untranslated region of a target gene [13]. Along with exosomal mRNA, miRNAs are in exosome cargo and can be harnessed as biomarkers for diagnostic and therapeutic purposes [14,15]. The detection of miRNAs is easier in comparison with other exosomal RNAs, which is due to their abundant presence in the exosomes. Lipid bilayer membranes provide stability for miRNAs, like the mRNA in exosomes. They can be stored for up to 2 weeks at 4°C and 5 years at −20°C, with their stability unchanged at lower temperatures [16]. Researchers have identified several exosomal miRNAs, like miR-19 in brain cancer metastasis [17], miR-25-3p, miR-130b-3p, and miR-425-5p in colorectal cancers [18]. Exosomal miRNAs can be used as biomarkers to diagnose diseases other than cancer, such as ocular, pulmonary, and respiratory problems, which are listed in Table 14.1 and will be covered in detail in the next sections.

4. Exosomal lncRNA

Long noncoding RNAs (lncRNAs) are a type of ncRNA that has more than 200 nucleotides in length and can bind to mRNA and miRNA to influence gene expression and splicing [19,20]. Exosomal lncRNA prostate cancer antigen 3 (lncPCA3) has been identified in urinary exosomes and is significantly overexpressed up to 70–100-fold in prostate cancer cell lines [21,22]. The US Food and Drug Administration authorized lncPCA3 as a prostate cancer biomarker [23]. There are no studies that have reported the

Table 14.1 Role of exosomal noncoding RNAs in Ocular diseases.

S. No	Name of the ocular disease	Source of exosomes	Cargo evaluated	Biological function in eye in health and disease
Role of Exosomal RNA in human ocular diseases				
1	AMD	Serum	hsa-miR (19a, 126, 410)	Apoptosis and/or angiogenesis
2	Cataract	AH	hsa-miRNAs (miR-486-5p, miR-204, and miR-184miR181a, miR-191, miR-148a, miR-26a, miR-125a-5p, let-7a, let7b)	Communication between AH inflow and outflow tissues to maintain intraocular pressure homeostasis
3	Myopia	AH	has-miR (582-3p, 17-5p, 885-3p, 19b-3p, and 450-5p)	NA
Ocular cell lines/primary cells in ocular diseased condition				
4	AMD	ARPE-19	hsa-miR-494-3p	Potential biomarkers of mitochondrial dysfunction
5	DR	ARPE-19	hsa-miR-543	Induction of EMT in receptor RPE cells
			hsa-miR-202-5p	Inhibition of growth, migration, and tube formation of DR in HUVECs
6	DR	Human retinal pericytes (ACBRI-183)	Circular RNA cPWWP2A	Regulation of retinal microvascular function of miR-579/occludin/sirtuin 1
7	POAG	NPCE primary cells	hsa-miR-29b	Reduced levels of col3A1 and regulates WNT/β-catenin pathway
		Primary human Trabecular meshwork cells	hsa-miR-182	Associated with POAG via miR-182
Therapeutic applications of human stem cell–derived exosomes for ocular disease				
8	DR	Human umbilical cord –derived exosomes	hsa-miR-126	Reduces hyperglycemia-induced retinal inflammation by inhibiting the HMGB1 pathway
9	RD	Neural stem/ progenitor cell	hsa-miR (26-a-5p, 21-5p), hsa-let-7a-5p	Protection of photoreceptor cells by inactivating microglia cells
10	CNV	Human umbilical cord MSCs	hsa-miR-27b	Inhibition of EMT and subretinal fibrosis by targeting HOXC6
11	DR	Adipose MSCs	hsa-miR-192	Negative regulation of ITGA1

AH, aqueous humor; *AMD*, age-related macular degeneration; *CNV*, choroidal neovascularization; *DR*, diabetic retinopathy; *MSC*, mesenchymal stem cell; *NPCE*, nonpigmented ciliary epithelium; *POAG*, primary open-angle glaucoma; *RD*, retinal degeneration; *ITGA1*, integrin subunit alpha 1.

involvement of lncRNA in nondegenerative diseases. However, more research is needed to determine the functional involvement of exosomal lncRNA in pathological illnesses.

5. Exosomal circRNA

Circular RNAs (circRNA) are a newly discovered family of endogenous ncRNAs distinguished by their covalently closed loop structures and lack of a 5′ cap or a 3′ poly-A tail [24]. They play diverse roles in various molecular and cellular functions, including regulating alternative splicing [25], serving as miRNA sponging [26], participating in RNA-protein interactions [27], and transcriptional regulation [28]. CircRNAs have a unique covalently bonded structure that provides higher tolerance to exonucleases compared to linear RNAs [24]. CircRNA has been identified as a possible molecular marker in several disorders, such as tumors, due to its conservation, abundance, and tissue specificity [29].

The exosomal circRNA recipient cells showed two-fold enrichment in exosomes when compared with donor cells [30]. CircRNA has been identified as a biomarker in several diseases, including cancer, nervous system diseases, and diabetic retinopathy (DR) [30,31]. For example, exosomal circRNA (exo-circRNA)-cPWWP2A has been found to modulate endothelial cells via miR-579 sponging in DR [31]. There is an urgent need to uncover potential exosomal circRNA that can be employed as biomarkers for illness diagnosis in the future (Table 14.1).

There has been growing interest in using exosomal RNAs as biomarkers and in regenerative medicine, in particular for ocular, pulmonary, and respiratory diseases. A study in 2015 identified exosomal RNAs in aqueous tumors as a potential biomarker for disease progression and diagnosis [32]. The role of exosomal RNAs is currently under extensive investigation in both degenerative and nondegenerative diseases, as well as signaling pathways. In this review, we aim to explore the biological significance of exosomal RNA in ocular, pulmonary, and respiratory diseases.

6. Exosomal RNAs in ocular diseases

The association of exosomal RNA was observed in the progression of ocular diseases including corneal disease, DR, glaucoma, age-related macular degeneration (AMD), retinoblastoma (Rb), uveal disease, and optic neuropathy.

7. Corneal disease

A recent study demonstrated that tissue regeneration by corneal stromal stem cells (CSSC) is mediated by the exosomal miRNA [33]. In culture, CSSC exosomes fuse with stromal and corneal epithelial cells, thereby delivering miRNA to the target cells. This treatment resulted in the decreased expression of fibrotic genes such as *Col3a1* and *Acta2*, subsequently inhibiting the invasion of neutrophils into the injured area and restoring normal collagen in the corneal tissue [33]. The suppression of Alix, a component of the endosomal sorting complex essential for transport, by siRNA interference had a significant effect on miR-159a packing into exosomes. This resulted in a considerable decrease in the number of miRNAs detected in the released exosomes. Exosomes lacking miRNAs were used to cure injured corneas, which demonstrated evident scarring and the expression of genes linked with fibrosis and inflammatory cell infiltration [33]. This data indicates that exosomes with low miRNA levels were less effective at suppressing corneal scarring and promoting tissue regeneration [33,34].

8. Diabetic retinopathy

DR is one of the most prevalent diabetic microvascular problems, defined by damage to the blood vessels of the retina, and one of the major causes of blindness in the elderly population [35,36]. Retinal pathologies in DR are associated with high levels of cytokines in the retina. According to the International Diabetes Federation, the global diabetes prevalence was predicted to be 10.5% in 2021 and is expected to increase by 12.2% [37].

Nonproliferative DR (NPDR) and PDR are two types of DR [38]. Retinal hemorrhages, microaneurysms, and intraretinal microvascular anomalies are some of the clinical symptoms of NPDR. Neovascularization classifies PDR [38]. The effective therapeutic approach for PDR involves targeting the Vascular Endothelial Growth Factor (VEGF) and corticosteroid receptors. However, this can also damage vascular and neuronal survival and function. Currently, there are no approved drugs available for the treatment of NPDR.

An in vitro and in vivo model study has described the role of exosomal miRNA's treatment of DR [39,40]. When isolated retinal photoreceptors from C57BL/6J mice CD73(+) were exposed to high levels of glucose (30 mM), this led to elevated levels of VEGF protein and reduced levels of four antiangiogenic miRNAs (miR-20a-3p, miR-20a-5p, miR-20b, and miR-106a-5p) in the exosomes released by the retinal cells [39]. In both studies, miR-20b was able to regulate VEGF expression in a rat model of DR [40]. However, knockdown and overexpression of these miRNAs with mimics would be required to understand their specific functions and mechanisms.

A study in 2018 described that exosomal miR-222 could repair the retinal layers at the wounded area of the retina in diabetic rabbit models [41]. In this study, rabbits were treated with streptozotocin (an antibiotic that produces pancreatic islet β-cell destruction) to create the diabetic condition. MSC-derived exosomes were isolated from rabbit adipose tissue and administered intravenously to diabetic rabbits [41]. After 12 weeks of post exosome treatment, miRNA-222 was upregulated in the retinal tissue of the rabbits. According to the authors, MSC exosomes can transport specific miRNAs like miR-222, and this mechanism can stimulate regenerative changes in damaged areas of the retina [38,41].

Another in-vivo study in 2019 showed that MSC-derived exosomal miR-126 could regulate hyperglycemia-induced retinal inflammation in a rat model [42]. Exosomes of human umbilical cord-derived MSCs were identified, and miRNA-126 was subsequently transferred into the exosomes [42]. Exosomal miR-126 was administered intravitreally to diabetic rats and cocultured in vitro with high-glucose-impacted human retinal endothelial cells. MSC exosomal miR-126 in retinal cells decreases inflammation caused by hyperglycemia by reducing NLRP3 inflammasome activity and down-regulating high-mobility group box signaling [42]. This can further lead to the inhibition of NF-jB/p65 activation, a crucial mediator in regulating the expression of inflammatory factors, and is possibly the mechanism behind the enhancement of the high glucose—induced inflammatory response [42]. Further investigation is needed to fully understand the role of exosomal miR-126 in identifying and managing DR targets.

Unlike MSC exosomal RNA, retinal exosomal RNAs like miR-124-3p and miR-202-5p have been associated with the retinal repair of DR [43,44]. A recent study has revealed that GW4869 is capable of preventing miR-124-3p from being transferred from photoreceptors to the inner retina [43]. The inhibition of exosomes also resulted in damage to the retina, leading to inflammation and impairment of the photoreceptors [43]. This experimental evidence indicates the potential functional activity of exosomal miRNA in protecting against retinal damage. According to Gu et al., exosomes from ARPE19 cells (retinal pigment epithelial cells) can decrease pathological fibrosis in PDR by transmitting miR-202-5p to human umbilical vein endothelial cells [44]. The authors highlighted that exosomal miR-202-5p, which is a known target of TGFβR2, negatively regulates TGFβR2 expression, leading to inhibition of cell growth, migration, and tube formation [44].

A study by Liu et al. demonstrated that the inhibition of miR-579 could indirectly modulate the activity of endothelial cells in the presence of exo-circRNA-cPWWP2A [31]. CPWWP2A functions as a natural miR-579 sponge by capturing and suppressing miR-579's activity. This results in an increased expression of angiopoietin 1, occludin, and sirtuin-1 [31]. Through an in vivo model, the overexpression of cPWWWP2A or the suppression of miR-579 has been shown to improve retinal vascular impairment caused by diabetes [31]. By contrast, overexpression of miR-579 or inhibition of cPWWP2A-mediated signaling by cPWWP2A silencing promotes retinal vascular dysfunction [31]. Research has demonstrated that cPWWP2A overexpression can lead to a decrease in retinal endothelial cell damage caused by diabetes-induced stress.

9. Glaucoma

Glaucoma is the most common cause of permanent blindness, affecting approximately 76 million individuals worldwide [45]. The number of glaucoma cases is expected to increase to 112 million by 2040 [46]. Glaucoma is a group of diseases that are identified by optic atrophy, visual field abnormalities, and loss of vision. It is a degenerative optic nerve disease that causes an abnormal increase in pressure within the eye, resulting in the death of retinal ganglion cells (RGCs) and ultimately blindness [47]. Most cases of abnormally high pressure inside the eye are due to problems with the drainage of aqueous humor (AH), which can be caused by various factors such as a narrow anterior chamber, hardening of the trabecular meshwork, abnormal substances in the extracellular matrix, and damage to the trabecular meshwork. Injuries to the RGCs, which are responsible for transmitting visual signals from the retina to the brain, cannot be healed [48]. Recent studies have identified factors that can harm RGCs, including oxidative stress, neuroinflammation, mitochondrial dysfunction, insufficient neurotrophic factors, and failure in axonal transport [48,49].

Exosomal RNAs isolated from ocular fluids could serve as a potential biomarker for diagnostic and therapeutic purposes. A 2015 in vitro study identified two exosomal miRNAs (miR-191 and miR-26a) isolated from uAH and the primary human trabecular meshwork cell [32]. Further studies have described other miRNAs in AH, including let-7a, let-7b, miR-26a, miR-125a-5p, miR-148a, miR-181a, mIR-184, miR-191, miR-204, and miR-486-5p. Researchers proposed that the progression of glaucoma is associated with various pathways [50].

A study conducted in 2016 reported increased levels of exosomal miR-182 in human trabecular meshwork cells. These findings suggest that this exosomal miRNA may be involved in the pathological increase in intraocular pressure [51].

A study by Tabak et al. reported [37] that the exosomes from non-pigmented ciliary epithelial cells (NPCE-Exo) have been found to enhance Wnt signaling and reduce the expression of phosphorylated GSK3 and β-catenin in trabecular meshwork cells. This has an impact on trabecular meshwork cadherin. Cadherin is a crucial part of the extracellular matrix that enlarges the pores in the trabecular meshwork. This can lead to higher intraocular pressure and greater resistance to the outflow of AH [52].

The NPCE exosomal RNAs like let-7a, miR-16, miR-21, miR-100, and miR-638 are well known to regulate Wnt signaling in the trabecular meshwork. The miRNA profiles listed by AH, miR-17, miR-21, and miR-29, are involved in the modulation of collagen synthesis. These observations suggest that exosomal miRNAs derived from NPCE have the potential to influence the Wnt pathway in the human trabecular meshwork.

In the rat optic nerve crush model, bone marrow stromal cells (BMSCs)-derived exosomal miRNAs (specifically miRNA-17, miRNA-21, and miRNA-92) were found to have a significant impact on the survival and restoration of RGCs' axons [53]. Exosomal miRNAs have potential as biomarkers for ocular disease diagnosis and treatment. Exosomes offer an accessible and promising pathway for immune response and nerve repair.

10. Age-related macular degeneration (AMD)

AMD is a condition in which the retinal tissue in the macular area degenerates in older people. AMD can cause vision loss and eventually blindness, affecting 196 million people worldwide, with instances predicted to grow to 288 million by 2040 [54]. AMD is a degenerative, chronic illness of the central retina that is classified as atrophic (dry) or exudative (wet). Dry AMD is identified by the presence of drusen, a substance similar to lipids that builds up on Bruch's membrane in the macula of the retina [55–59]. On the other hand, late-stage exudative or "wet" AMD is characterized by choroidal neovascularization, which causes leakage of retinal endothelial cells and the formation of fibrous scars [57,59–62].

Studies show that certain serum exosomal miRNAs are harmful to AMD patients, leading to neovascularization in retinal cells [63]. These miRNAs, including miR-19a, miR-126, and miR-410, are found at higher levels in AMD patient serum exosomes compared to control samples [63]. Similarly, a different study conducted in 2019 reported elevated levels of miR-626 and miR-485-5p, as well as reduced expression of miR-885-5p in blood serum exosomes from patients with wet AMD [64]. Their presence significantly increased endothelial tubules and demonstrated the association with VEGF and apoptotic pathways using computational analysis [63].

11. Uveal melanoma

Uveal melanoma is a rare malignancy that arises from melanocytes within the uveal tract of the eye [65], which are responsible for the production of eye pigmentation [66]. Ragusa et al. demonstrated the increased expression of serum exosomal miR-146a in uveal melanoma patients [67], and it can be used as a biomarker.

12. Retinoblastoma

Rb is a prevalent form of eye cancer, mainly affecting children, originating in the retina, the very back part of the eye [68]. The exosomal miRNAs miR-216b-5p and miR-301b-3p were both elevated in both Rb cell lines, WERI-Rb-1 and NCC-Rb-51, in comparison with a control cell line, MIO-M1 [69]. Exosomes from WERI-Rb1 cells increased tumor growth by reducing natural killer cells and increasing tumor-associated macrophages [70].

13. Optic neuropathy

Optic neuropathy, or atrophy, is caused by damage to the optic nerve due to blood flow blockage, swelling, or trauma. In Sprague-Dawley rat models with optic-nerve damage, the use of BMSC exosomes containing specific miRNAs, including the miR-17-92 cluster, resulted in reduced levels of phosphatase and tensin homolog (PTEN). PTEN plays a crucial role in suppressing the growth and survival of RGC axons [71]. BMSCs were found to express high levels of miR-10b-5p, miR-21-5p, miR-22-3p, miR-143-3p, and miR-486-5p [72]. Another study demonstrated that exosome-derived umbilical cord MSC (UMSC) significantly promoted RGC survival and glial cell activation in optic nerve-damaged Wistar rats in comparison to saline controls. These miRNAs, including miRNA-21-5p, miRNA-125b-5p, miRNA-23a-3p, miRNA-100-5p, and let-7f-5p, are identified in UMSC exosomes [73]. While BMSC exosomes were effective in promoting axonal regeneration in the optic nerve, UMSC exosomes did not have the same effect [53].

14. Exosomal miRNA in chronic obstructive pulmonary disease (COPD)

Scientific evidence has reported that the exosomal ncRNAs are involved in the pathogenesis and development of chronic obstructive pulmonary disease (COPD) [74]. The altered expression of exosomal miRNA was identified in lung disease patients in comparison with healthy individuals. Like in ocular diseases, exosomal miRNA and lncRNA offer to be novel therapeutic targets and diagnostic markers in pulmonary diseases [75].

According to the World Health Organization, COPD is a common lung disease due to chronic inflammation of the airways, causing restricted airflow and breathing problems. It is sometimes referred to as emphysema or chronic bronchitis. Smoking and air pollution are the main factors that could be associated with chronic inflammation of the airways. Inflammation has a critical role in the initiation and progression of COPD [76]. Along with several cellular and molecular pathways, exosomal ncRNAs are also involved in inflammation related to COPD pathogenesis. A 2015 study identified that exosomal miR-210 is involved in the suppression of autophagy in COPD patients [77]. Based on the coculture assays, they reported that cigarette smoking could induce upregulation of exosomal miR-210 in primary human bronchial epithelial cells (HBECs), which promotes myofibroblast differentiation in lung fibroblasts (LFs). Exosomal miR-210 could directly regulate autophagy processes by affecting ATG7 and is inversely correlated with ATG7 expression in LFs [77]. In COPD patients, autophagy induction was significantly decreased in LFs, and silencing *ATG7* in LFs led to myofibroblast differentiation [77]. Hence, it is proposed that smoking exposure alters the expression of exosomal miR-210 of the HBECs and can act as a paracrine autophagy regulator of myofibroblast differentiation in LFs [77]. In the future, exosomal miR-210 might be a diagnostic biomarker and/or therapeutic target for COPD patients [74,75,77].

Recent research using high-throughput technologies has brought to light holistic transcriptome footprints that are specific to COPD as compared to healthy participants [78,79]. Hierarchical clustering and next-generation sequencing (NGS) were used to detect differentially expressed exosomal miRNAs in bronchoalveolar lavage fluid (BALF) and lung tissues from 8 healthy smokers (s), 8 healthy nonsmokers (ns), and 16 COPD patients. MiR-320b and miR-22-3p are considerably increased in COPD BALF-derived exosomes compared to BALF-derived exosomes from healthy nonsmokers, but miR-423-5p is dramatically downregulated. MiR-100-5p expression, on the other hand, is lower in COPD patients' BALF-derived exosomes compared to healthy smokers' BALF-derived exosomes. In contrast, the same study found that only one miR-122-5p is downregulated in COPD patients' lung-derived exosomes when compared to both healthy smokers and nonsmokers, implying that this miRNA is a promising target for broad-range therapies [78]. In another study, single-cell RNA sequencing on peripheral blood mononuclear cells in conjunction with plasma exosome sequencing revealed even more details on the distinct genomic makeup of COPD patients. Six COPD patients and four healthy controls have varied expression levels of 135 mRNAs, 132 nonlncRNAs, and 359 circRNAs. A link between this exosomal transcriptome profile and local and systemic immunity pathways was discovered using competing endogenous and protein-protein interaction (PPI) networks. Interestingly, the majority of the differently regulated exosomal RNAs (in COPD patients) are associated with important immunological processes such as cytokine-cytokine interaction and antiviral immunity. This research is fundamental because it identifies direct targets from the exosomal RNA milieu involved in innate immunity and antiviral defense, both of which are important in COPD and other respiratory illnesses [79]! Several exosomal lncRNAs (H19, MALAT1, MEG3, FENDRR, CDKN2B-AS1, TUG1, HOTAIR, and GAS5 lncRNAs) have been discovered as being essential not just in COPD but also in asthma, idiopathic pulmonary fibrosis (IPF), and lung cancer. Furthermore, these lncRNAs control immunological function and apoptotic signaling pathways, adding to our understanding of PPI networks regulated by exosomal non-ncRNAs [80].

15. Exosomal RNA in respiratory diseases

Respiratory infections have a significant impact on world health, which has become even more visible in the aftermath of COVID-19 and the ongoing rise in natural disasters. Exosomal nonncRNAs have been demonstrated to influence many signaling pathways, including cell proliferation, migration, inflammation, and fibrosis, which are all hallmarks of lung illnesses. As a result, researchers can identify possible biomarkers by analyzing exosomal ncRNAs extracted from plasma, blood, lung tissues, sputum, and BALF. Exosomal miRNAs and lncRNAs, for example, have already been identified as promising targets in chronic respiratory illnesses and COVID-19 [2,10,74,81−92].

Given the recent wildfire calamity in North America, researchers are looking for exosomal RNA biomarkers that have been altered by environmental conditions. A study published last year on wildfire smoke condensates, in which female CD-1 mice were exposed to smoke samples, is an example of how exosomal RNA analysis can be useful in tackling the most serious global health challenges of the day [93]. The bulk of the plasma exosomal miRNAs that were elevated after mice were treated with smoke condensates were indicators of cardiovascular injury and inflammation.

Continued efforts to uncover more exosomal targets in various disease situations may allow researchers to build better therapeutics for lung cancer, idiopathic pulmonary illnesses, asthma, and other prevalent respiratory ailments. Furthermore, because exosomal RNAs constantly shuttle between organelles, they are employed in simulating protein−protein interactions that are important in pathophysiology, particularly inflammatory pathways. Exosomal RNA research, for example, has paved the door for a better understanding of the interaction between macrophages and other signaling pathways in lung-associated illnesses. Because most respiratory ailments are linked to inflammation and fibrosis, such investigations are important in identifying novel biomarkers for lung diseases [94]. The following is an overview of what is known about the role of exosomal nonncRNAs in the development of major respiratory disorders such as asthma, pulmonary fibrosis, lung cancer, and pneumonia.

16. Asthma

Asthma is one of the most common chronic respiratory disorders, affecting approximately 300 million individuals worldwide. Asthma is characterized by neutrophil-induced airway wall remodeling and the proliferation of airway smooth muscle cells (ASMCs). Because exosomal transport is so important in the immune system and inflammation, asthma has long been a focus of exosomal analysis research. In 2013, one of the first papers associating BALF-exosomal miRNA with asthma was published.

In this work, a microarray-based comparison of ten healthy control people and 10 patients with moderate intermittent asthma demonstrates that 24 miRNAs are significantly changed in asthma patients. These targets contain members of the let-7 and miR-200 families [95].

Another study found that exosomal miR-1470 from MSC upregulates CD4+CD25+FOXP3+ regulatory T cells (Tregs) in asthmatic patients. Furthermore, miR-1470 performs this function by upregulating P27KIP1, and siRNA silencing reverses the miR1470/P27KIP1-induced differentiation of Tregs. As a result, this study demonstrates the feasibility of using exosomal miRNAs as asthma therapy targets [96].

A persuasive study has found a link between the variable expression of exosomal miRNAs and the extent of asthma treatment by scorpion and centipede (SC). Real-time PCR and small RNA-seq analysis show that 328 exosomal miRNAs are expressed differently in SC-treated mice than in asthmatic mice. There are 118 upregulated miRNAs and 210 downregulated miRNAs among the 328 miRNAs. Apoptosis, cell adhesion, Wnt and MAPK signaling, calcium signaling, and transcription are some of the key pathways linked to these miRNAs. This work demonstrates the value of using exosomal miRNA analysis to evaluate the efficacy of prospective medicines and natural products for the treatment of asthma [97].

lncRNAs, in addition to miRNAs, play a significant role in the evolution of asthmatic airway inflammation and the proliferation of ASMCs. LncRNA AK085865 has been identified as a major regulator of macrophage polarization and inflammation, and its deletion has been shown to ameliorate asthmatic phenotypes in mice. Exosomal transfer of the neutrophil-derived lncRNA CRNDE (colorectal neoplasia differentially expressed) has been found in another study to boost ASMC proliferation and migration in mice [98,99].

17. Lung cancer

Since their discovery in 2007, exosomal RNAs (mRNA, lncRNA, circRNA, and miRNA) have been linked to a variety of cancers, including non-small cell lung cancer, glioblastoma, prostate cancer, colorectal adenoma, pancreatic cancer, gastric cancer, breast cancer, renal cancer, liver cancer, colon cancer, and cervical cancer. Because exosomes can transport diverse types of cargo between organelles and cell types, exosomes and exosomal RNAs play a significant role in protumor mechanisms such as angiogenesis, metastasis, and proliferation. Furthermore, the cargo-like properties of exosomal RNAs have been linked to the development of drug resistance and chemoresistance in cancer cells. Exosomes, it turns out, can also help with the development of antitumor medicines, and the discovery of new biomarkers (in the form of exosomal RNAs) is now a major component in the arsenal of antitumor treatment [100].

Exosomes (including antisense RNAs, mRNAs, and miRNAs) generated from three distinct colorectal cancer cell lines were rapidly transferred into lung cancer A549 cell lines in one of the early reports tying exosomal RNAs to lung cancer cells. This study shows that exosomes and their contents can move between cell types, paving the path for more exciting projects later in the decade [101]. In a similar vein, a 2016 study shows that the mesenchymal HBEC model can transfer the chemoresistance phenotype to epithelial HBEC. ZEB1 mRNA, an epithelial-to-mesenchymal transition (EMT) transcription factor, has been identified as the major factor conferring chemoresistance on epithelial HBECs, highlighting the relevance of exosomal RNAs as prospective targets in NSCLC therapy [102].

More high-throughput investigations have aided in the identification of more exosomal RNAs implicated in the pathogenesis of lung cancer. Furthermore, as with COPD, PPI networks that are directly regulated by exosomal lncRNAs have been identified. H19, MALAT1, MEG3, FENDRR, CDKN2B-AS1, TUG1, HOTAIR, and GAS5 lncRNAs target signaling pathways (AKT/MTOR, WNT/B-catenin, P53, NF-KB, PTEN/PI3K/AKT) in small and nonsmall cell lung cancers. Furthermore, these lncRNAs (particularly HOTAIR) have been linked to numerous therapeutic resistances, including gefitinib, cisplatin, and etoposide, among others, expanding the repertoire of targets that can be examined for lung cancer therapy [80]. Furthermore, RNA-sequencing investigations have been used to identify other lnonncRNAs as possible targets in lung cancer. Among the 90 dysregulated lncRNAs that separate NSCLC patients from disease-free and TB controls, serum exosomal linc01125 has emerged as the most significant [103].

It is important to emphasize that efforts to develop exosomal RNA biomarkers in lung cancer are not confined to lncRNAs alone. Exo-circRNAs have been identified as possible biomarkers in lung cancer etiology in multiple publications, including Circ_0047921, 005628, 0001492, 0001346, 0000690, and 0001439. CircRNAs have a key role in lung cancer growth because they regulate protumor pathways such as EMT, cell proliferation, angiogenesis, metastasis, and apoptosis. CircRNAs can potentially increase medication resistance in some cases. Thus, identifying exosomal circRNAs is critical for lung cancer therapy [104—108].

18. Pneumonia

Pneumonia is one of the leading causes of childhood death and is intimately associated with DNA virus infections, most notably human adenovirus. Reverse transcription-quantitative polymerase chain reaction examination of serum samples collected from 59 patients revealed four exosomal serum miRNAs (miR-450a-5p, miR-103a-3p, miR-103b-5p, and miR-98-5p) that are significantly changed in adenovirus-infected children [109].

19. Pulmonary fibrosis

IPF, the most common kind of pulmonary fibrosis, is a chronic inflammatory lung disease defined by scar tissue accumulation that is seen in men over the age of 65. There is a higher occurrence in the populations of North America and Europe, and it is associated with significant mortality (3–5 years after diagnosis). Although the US Food and Drug Administration has approved a few medications (pirfenidone and nintedanib), an effective treatment is still lacking, necessitating a more aggressive search for novel biomarkers [110–117].

Hierarchical clustering and NGS were used to discover differentially expressed exosomal miRNAs in bronchoalveolar lavage fluid (BALF) and lung tissues from eight healthy smokers (s), eight healthy nonsmokers (ns), and eight IPF patients. When BALF and lung exosomes from healthy participants were compared to those from IPF patients, distinct miRNA signatures were discovered, as previously described in the case of COPD patients. In the case of BALF-derived exosomes, five miRNAs (miR-375-3p, miR-200a-3p, miR-200b-3p, miR-141-3p, and miR-423-5p) were considerably downregulated, while four (miR-22-3p, miR-320a-3p, miR-320b, and miR-24-3p) were dramatically elevated in IPF compared to healthy nonsmokers. In contrast, a study of lung-derived exosomes indicates a substantially larger range of differentially expressed exosomal miRNAs in IPF patients compared to healthy nonsmokers: 26 upregulated and 29 downregulated miRNAs in IPF compared to healthy nonsmokers! When BALF and lung exosomes from COPD and IPF patients were compared, a similar pattern emerged: just one miRNA (miR-375-3p) is differentially expressed in BALF exosomes, whereas 31 are upregulated and 36 are downregulated in COPD lung exosomes compared to IPF lungs [78].

Exosomal miRNAs have emerged at the forefront of the hunt for IPF biomarkers, with several studies linking these potential miRNAs to important biological pathways such as fibroblast differentiation and proliferation, mitochondrial function, senescence, and inflammation. Interestingly, most of these miRNAs have been examined on several "donor cell" samples, including tissue cultures and fibrotic mouse models. It is noteworthy that differentially regulated exosomal miRNAs have been identified in the blood (miR-141, miR-7, miR-21-5p, miR-16, miR-21, miR-26a, miR-210), BALF (miR-125b, miR-128, miR-21, miR-100, miR-140-3p, miR-374b, let 7d), and sputum (miR-33a-5p, miR-142-3p, miR-192-5, let-7d-5p, miR-26a-5p), thus widening the range of available biomarkers. It is also worth adding that sputum is a less invasive mode of treatment, and these findings could be useful in developing alternative therapeutics for IPF [118,119].

Exosomal miRNAs have also been found to have distinct signatures in disorders that are exacerbated and perpetuated by preexisting lung fibrosis problems. Ant-synthetase syndrome (ASSD) is one such pathological disease. Patients with interstitial lung disease (ILD) have a higher death rate in the case of ASSD. Specific exosomal miRNAs (miR-29c-3p, miR-30a-5p) have been found to be increased in ASSD-ILD patients compared to ASSD patients without lung disease using miRNA PCR arrays. Only miR-29c-3p is elevated in IPF patients (without ASSD) when compared to both ASSD and ASSD-ILD. This study demonstrates the potential of exosomal RNA biomarkers not just in pulmonary fibrosis but also in recognizing the danger of secondary diseases in lung patients! [120].

References

[1] Trams EG, Lauter CJ, Salem Jr N, Heine U. Exfoliation of membrane ecto-enzymes in the form of micro-vesicles. Biochimica et Biophysica Acta (BBA) - Biomembranes 1981;645(1):63–70.

[2] Li C, Ni YQ, Xu H, Xiang QY, Zhao Y, Zhan JK, et al. Roles and mechanisms of exosomal non-coding RNAs in human health and diseases. Signal Transduction and Targeted Therapy 2021;6(1):383.

[3] Roberson CD, Atay S, Gercel-Taylor C, Taylor DD. Tumor-derived exosomes as mediators of disease and potential diagnostic biomarkers. Cancer Biomarkers 2010;8(4–5):281–91.

[4] Fais S, O'Driscoll L, Borras FE, Buzas E, Camussi G, Cappello F, et al. Evidence-based clinical use of nanoscale extracellular vesicles in nanomedicine. ACS Nano 2016;10(4):3886–99.

[5] Vilaça-Faria H, Salgado AJ, Teixeira FG. Mesenchymal stem cells-derived exosomes: a new possible therapeutic strategy for Parkinson's disease? Cells 2019;8(2).

[6] Zhao T, Sun F, Liu J, Ding T, She J, Mao F, et al. Emerging role of mesenchymal stem cell-derived exosomes in regenerative medicine. Current Stem Cell Research and Therapy 2019;14(6):482–94.

[7] Frydrychowicz M, Kolecka-Bednarczyk A, Madejczyk M, Yasar S, Dworacki G. Exosomes - structure, biogenesis and biological role in non-small-cell lung cancer. Scandinavian Journal of Immunology 2015;81(1):2–10.

[8] Khalid A, Persano S, Shen H, Zhao Y, Blanco E, Ferrari M, et al. Strategies for improving drug delivery: nanocarriers and microenvironmental priming. Expert Opinion on Drug Delivery 2017;14(7):865–77.

[9] Valadi H, Ekström K, Bossios A, Sjöstrand M, Lee JJ, Lötvall JO. Exosome-mediated transfer of mRNAs and microRNAs is a novel mechanism of genetic exchange between cells. Nature Cell Biology 2007;9(6):654–9.

[10] Wang J, Yue BL, Huang YZ, Lan XY, Liu WJ, Chen H. Exosomal RNAs: novel potential biomarkers for diseases-A review. International Journal of Molecular Sciences 2022;23(5).

[11] El Fekih R, Hurley J, Tadigotla V, Alghamdi A, Srivastava A, Coticchia C, et al. Discovery and validation of a urinary exosome mRNA signature for the diagnosis of human kidney transplant rejection. Journal of the American Society of Nephrology 2021;32(4):994–1004.

[12] Chow LLC, Mead B. Extracellular vesicles as a potential therapeutic for age-related macular degeneration. Neural Regeneration Research 2023;18(9):1876–80.

[13] Bartel DP. MicroRNAs: genomics, biogenesis, mechanism, and function. Cell 2004;116(2):281–97.

[14] Wahid F, Shehzad A, Khan T, Kim YY. MicroRNAs: synthesis, mechanism, function, and recent clinical trials. Biochimica et Biophysica Acta 2010;1803(11):1231–43.

[15] Lu J, Getz G, Miska EA, Alvarez-Saavedra E, Lamb J, Peck D, et al. MicroRNA expression profiles classify human cancers. Nature 2005;435(7043):834–8.

[16] Weber JA, Baxter DH, Zhang S, Huang DY, Huang KH, Lee MJ, et al. The microRNA spectrum in 12 body fluids. Clinical Chemistry 2010;56(11):1733–41.

[17] Zhang L, Zhang S, Yao J, Lowery FJ, Zhang Q, Huang WC, et al. Microenvironment-induced PTEN loss by exosomal microRNA primes brain metastasis outgrowth. Nature 2015;527(7576):100–4.

[18] Wang D, Wang X, Si M, Yang J, Sun S, Wu H, et al. Exosome-encapsulated miRNAs contribute to CXCL12/CXCR4-induced liver metastasis of colorectal cancer by enhancing M2 polarization of macrophages. Cancer Letters 2020;474:36−52.
[19] Guo X, Gao L, Wang Y, Chiu DKY, Wang T, Deng Y. Advances in long noncoding RNAs: identification, structure prediction and function annotation. Briefings in Functional Genomics 2015;15(1):38−46.
[20] Losko M, Kotlinowski J, Jura J. Long noncoding RNAs in metabolic syndrome related disorders. Mediators of Inflammation 2016;2016:5365209.
[21] Bermúdez M, Aguilar-Medina M, Lizárraga-Verdugo E, Avendaño-Félix M, Silva-Benítez E, López-Camarillo C, et al. LncRNAs as regulators of autophagy and drug resistance in colorectal cancer. Frontiers in Oncology 2019;9:1008.
[22] Xue WJ, Ying XL, Jiang JH, Xu YH. Prostate cancer antigen 3 as a biomarker in the urine for prostate cancer diagnosis: a meta-analysis. Journal of Cancer Research and Therapeutics 2014;10(Suppl. l):C218−21.
[23] Hu B, Yang H, Yang H. Diagnostic value of urine prostate cancer antigen 3 test using a cutoff value of 35 μg/L in patients with prostate cancer. Tumour Biol 2014;35(9):8573−80.
[24] Meng S, Zhou H, Feng Z, Xu Z, Tang Y, Li P, et al. CircRNA: functions and properties of a novel potential biomarker for cancer. Molecular Cancer 2017;16(1):94.
[25] Ashwal-Fluss R, Meyer M, Pamudurti NR, Ivanov A, Bartok O, Hanan M, et al. circRNA biogenesis competes with pre-mRNA splicing. Molecular Cell 2014;56(1):55−66.
[26] Li F, Zhang L, Li W, Deng J, Zheng J, An M, et al. Circular RNA ITCH has inhibitory effect on ESCC by suppressing the Wnt/β-catenin pathway. Oncotarget 2015;6(8):6001−13.
[27] Du WW, Yang W, Liu E, Yang Z, Dhaliwal P, Yang BB. Foxo3 circular RNA retards cell cycle progression via forming ternary complexes with p21 and CDK2. Nucleic Acids Research 2016;44(6):2846−58.
[28] Li Z, Huang C, Bao C, Chen L, Lin M, Wang X, et al. Exon-intron circular RNAs regulate transcription in the nucleus. Nature Structural and Molecular Biology 2015;22(3):256−64.
[29] Rybak-Wolf A, Stottmeister C, Glažar P, Jens M, Pino N, Giusti S, et al. Circular RNAs in the mammalian brain are highly abundant, conserved, and dynamically expressed. Molecular Cell 2015;58(5):870−85.
[30] Li Y, Zheng Q, Bao C, Li S, Guo W, Zhao J, et al. Circular RNA is enriched and stable in exosomes: a promising biomarker for cancer diagnosis. Cell Research 2015;25(8):981−4.
[31] Liu C, Ge HM, Liu BH, Dong R, Shan K, Chen X, et al. Targeting pericyte-endothelial cell crosstalk by circular RNA-cPWWP2A inhibition aggravates diabetes-induced microvascular dysfunction. Proceedings of the National Academy of Sciences of the United States of America 2019;116(15):7455−64.
[32] Dismuke WM, Challa P, Navarro I, Stamer WD, Liu Y. Human aqueous humor exosomes. Experimental Eye Research 2015;132:73−7.
[33] Shojaati G, Khandaker I, Funderburgh ML, Mann MM, Basu R, Stolz DB, et al. Mesenchymal stem cells reduce corneal fibrosis and inflammation via extracellular vesicle-mediated delivery of miRNA. Stem Cells Translative Medicine 2019;8(11):1192−201.
[34] Liu A, Hefley B, Escandon P, Nicholas SE, Karamichos D. Salivary exosomes in health and disease: future prospects in the eye. International Journal of Molecular Sciences 2023;24(7).
[35] Lee R, Wong TY, Sabanayagam C. Epidemiology of diabetic retinopathy, diabetic macular edema and related vision loss. Eye and Vision 2015;2(1):17.
[36] Ogurtsova K, da Rocha Fernandes JD, Huang Y, Linnenkamp U, Guariguata L, Cho NH, et al. IDF Diabetes Atlas: global estimates for the prevalence of diabetes for 2015 and 2040. Diabetes Research and Clinical Practice 2017;128:40−50.
[37] Magliano DJB EJ. International diabetes federation: Brussels, Belgium. 10th ed. IDF Diabetes Atlas; 2021.
[38] Sanghani A, Andriesei P, Kafetzis KN, Tagalakis AD, Yu-Wai-Man C. Advances in exosome therapies in ophthalmology−from bench to clinical trial. Acta Ophthalmologica 2022;100(3):243−52.
[39] Maisto R, Trotta MC, Petrillo F, Izzo S, Cuomo G, Alfano R, et al. Corrigendum: resolvin D1 modulates the intracellular VEGF-related miRNAs of retinal photoreceptors challenged with high glucose. Frontiers in Pharmacology 2020;11.

[40] Qin B, Liu J, Liu S, Li B, Ren J. MiR-20b targets AKT3 and modulates vascular endothelial growth factor-mediated changes in diabetic retinopathy. Acta Biochimica et Biophysica Sinica 2016;48(8):732–40.
[41] Safwat A, Sabry D, Ragiae A, Amer E, Mahmoud RH, Shamardan RM. Adipose mesenchymal stem cells-derived exosomes attenuate retina degeneration of streptozotocin-induced diabetes in rabbits. Journal of Circulatory Biomark 2018;7. 1849454418807827.
[42] Zhang W, Wang Y, Kong Y. Exosomes derived from mesenchymal stem cells modulate miR-126 to ameliorate hyperglycemia-induced retinal inflammation via targeting HMGB1. Investigative Ophthalmology and Visual Science 2019;60(1):294–303.
[43] Wooff Y, Cioanca AV, Chu-Tan JA, Aggio-Bruce R, Schumann U, Natoli R. Small-medium extracellular vesicles and their miRNA cargo in retinal health and degeneration: mediators of homeostasis, and vehicles for targeted gene therapy. Frontiers in Cellular Neuroscience 2020;14:160.
[44] Gu S, Liu Y, Zou J, Wang W, Wei T, Wang X, et al. Retinal pigment epithelial cells secrete miR-202-5p-containing exosomes to protect against proliferative diabetic retinopathy. Experimental Eye Research 2020;201:108271.
[45] Flaxman SR, Bourne RRA, Resnikoff S, Ackland P, Braithwaite T, Cicinelli MV, et al. Global causes of blindness and distance vision impairment 1990-2020: a systematic review and meta-analysis. Lancet Global Health 2017;5(12):e1221–34.
[46] Tham YC, Li X, Wong TY, Quigley HA, Aung T, Cheng CY. Global prevalence of glaucoma and projections of glaucoma burden through 2040: a systematic review and meta-analysis. Ophthalmology 2014;121(11):2081–90.
[47] Henson DB, Thampy R. Preventing blindness from glaucoma. BMJ 2005;331(7509):120–1.
[48] Almasieh M, Wilson AM, Morquette B, Cueva Vargas JL, Di Polo A. The molecular basis of retinal ganglion cell death in glaucoma. Progress in Retinal and Eye Research 2012;31(2):152–81.
[49] Syc-Mazurek SB, Libby RT. Axon injury signaling and compartmentalized injury response in glaucoma. Progress in Retinal and Eye Research 2019;73:100769.
[50] Gauthier AC, Liu J. Epigenetics and signaling pathways in glaucoma. BioMed Research International 2017;2017:5712341.
[51] Liu Y, Bailey JC, Helwa I, Dismuke WM, Cai J, Drewry M, et al. A common variant in MIR182 is associated with primary open-angle glaucoma in the NEIGHBORHOOD consortium. Investigative Ophthalmology and Visual Science 2016;57(10):4528–35.
[52] Lerner N, Avissar S, Beit-Yannai E. Extracellular vesicles mediate signaling between the aqueous humor producing and draining cells in the ocular system. PLoS One 2017;12(2):e0171153.
[53] Mead B, Tomarev S. Bone marrow-derived mesenchymal stem cells-derived exosomes promote survival of retinal ganglion cells through miRNA-dependent mechanisms. Stem Cells Translative Medicine 2017;6(4):1273–85.
[54] Wong WL, Su X, Li X, Cheung CM, Klein R, Cheng CY, et al. Global prevalence of age-related macular degeneration and disease burden projection for 2020 and 2040: a systematic review and meta-analysis. Lancet Global Health 2014;2(2):e106–16.
[55] Arya M, Sabrosa AS, Duker JS, Waheed NK. Choriocapillaris changes in dry age-related macular degeneration and geographic atrophy: a review. Eye Vision (London) 2018;5:22.
[56] Wang AL, Lukas TJ, Yuan M, Du N, Tso MO, Neufeld AH. Autophagy and exosomes in the aged retinal pigment epithelium: possible relevance to drusen formation and age-related macular degeneration. PLoS One 2009;4(1):e4160.
[57] Flores R, Carneiro Â, Vieira M, Tenreiro S, Seabra MC. Age-related macular degeneration: pathophysiology, management, and future perspectives. Ophthalmologica 2021;244(6):495–511.
[58] Bowes Rickman C, Farsiu S, Toth CA, Klingeborn M. Dry age-related macular degeneration: mechanisms, therapeutic targets, and imaging. Investigative Ophthalmology and Visual Science 2013;54(14). Orsf68-80.
[59] Bird AC. Therapeutic targets in age-related macular disease. Journal of Clinical Investigation 2010;120(9):3033–41.
[60] Lim LS, Mitchell P, Seddon JM, Holz FG, Wong TY. Age-related macular degeneration. Lancet 2012;379(9827):1728–38.

[61] Penn JS, Madan A, Caldwell RB, Bartoli M, Caldwell RW, Hartnett ME. Vascular endothelial growth factor in eye disease. Progress in Retinal and Eye Research 2008;27(4):331–71.
[62] Shao J, Choudhary MM, Schachat AP. Neovascular age-related macular degeneration. Developments in Ophthalmology 2016;55:125–36.
[63] ElShelmani H, Wride MA, Saad T, Rani S, Kelly DJ, Keegan D. The role of deregulated MicroRNAs in age-related macular degeneration pathology. Translative Visual Science Technology 2021;10(2):12.
[64] Elbay A, Ercan Ç, Akbas F, Bulut H, Ozdemir H. Three new circulating microRNAs may be associated with wet age-related macular degeneration. Scandinavian Journal of Clinical and Laboratory Investigation 2019;79(6):388–94.
[65] Kaliki S, Shields CL. Uveal melanoma: relatively rare but deadly cancer. Eye 2017;31(2):241–57.
[66] Hu D-N, Simon JD, Sarna T. Role of ocular melanin in ophthalmic physiology and pathology. Photochemistry and Photobiology 2008;84(3):639–44.
[67] Qu Z, Liu J, Zhu L, Zhou Q. A comprehensive understanding of choroidal metastasis from lung cancer. OncoTargets and Therapy 2021;14:4451–65.
[68] Ray A, Gombos DS, Vats TS. Retinoblastoma: an overview. Indian Journal of Pediatrics 2012;79(7):916–21.
[69] Ravishankar H, Mangani AS, Phoebe Moses GL, Mani SP, Parameswaran S, Khetan V, et al. Serum exosomal miRNA as biomarkers for Retinoblastoma. Experimental Eye Research 2020;199:108184.
[70] Chen S, Chen X, Qiu J, Chen P, Han X, Wu Y, et al. Exosomes derived from retinoblastoma cells enhance tumour deterioration by infiltrating the microenvironment. Oncology Reports 2021;45(1):278–90.
[71] Zhang Y, Chopp M, Liu XS, Katakowski M, Wang X, Tian X, et al. Exosomes derived from mesenchymal stromal cells promote axonal growth of cortical neurons. Molecular Neurobiology 2017;54(4):2659–73.
[72] Baglio SR, Rooijers K, Koppers-Lalic D, Verweij FJ, Pérez Lanzón M, Zini N, et al. Human bone marrow- and adipose-mesenchymal stem cells secrete exosomes enriched in distinctive miRNA and tRNA species. Stem Cell Research and Therapy 2015;6(1):127.
[73] Fang S, Xu C, Zhang Y, Xue C, Yang C, Bi H, et al. Umbilical cord-derived mesenchymal stem cell-derived exosomal MicroRNAs suppress myofibroblast differentiation by inhibiting the transforming growth factor-β/SMAD2 pathway during wound healing. Stem Cells Translative Medicine 2016;5(10):1425–39.
[74] Li Y, Yin Z, Fan J, Zhang S, Yang W. The roles of exosomal miRNAs and lncRNAs in lung diseases. Signal Transduction and Targeted Therapy 2019;4(1):47.
[75] Salimian J, Mirzaei H, Moridikia A, Harchegani AB, Sahebkar A, Salehi H. Chronic obstructive pulmonary disease: MicroRNAs and exosomes as new diagnostic and therapeutic biomarkers. Journal of Research in Medical Sciences 2018;23:27.
[76] Kaur G, Maremanda KP, Campos M, Chand HS, Li F, Hirani N, et al. Distinct exosomal miRNA profiles from BALF and lung tissue from COPD and IPF patients. medRxiv 2021. 2021.08.24.21262557.
[77] Fujita Y, Araya J, Ito S, Kobayashi K, Kosaka N, Yoshioka Y, et al. Suppression of autophagy by extracellular vesicles promotes myofibroblast differentiation in COPD pathogenesis. Journal of Extracellular Vesicles 2015;4(1):28388.
[78] Kaur G, Maremanda KP, Campos M, Chand HS, Li F, Hirani N, et al. Distinct exosomal miRNA profiles from BALF and lung tissue of COPD and IPF patients. International Journal of Molecular Sciences 2021;22(21).
[79] Pei Y, Wei Y, Peng B, Wang M, Xu W, Chen Z, et al. Combining single-cell RNA sequencing of peripheral blood mononuclear cells and exosomal transcriptome to reveal the cellular and genetic profiles in COPD. Respiratory Research 2022;23(1):260.
[80] Poulet C, Njock MS, Moermans C, Louis E, Louis R, Malaise M, et al. Exosomal long non-coding RNAs in lung diseases. International Journal of Molecular Sciences 2020;21(10).
[81] Alipoor SD, Mortaz E, Garssen J, Movassaghi M, Mirsaeidi M, Adcock IM. Exosomes and exosomal miRNA in respiratory diseases. Mediators of Inflammation 2016;2016:5628404.

[82] Azhdari MH, Goodarzi N, Doroudian M, MacLoughlin R. Molecular insight into the therapeutic effects of stem cell-derived exosomes in respiratory diseases and the potential for pulmonary delivery. International Journal of Molecular Sciences 2022;23(11).
[83] Fujimoto S, Fujita Y, Kadota T, Araya J, Kuwano K. Intercellular communication by vascular endothelial cell-derived extracellular vesicles and their MicroRNAs in respiratory diseases. Frontiers in Molecular Biosciences 2020;7:619697.
[84] Leon-Icaza SA, Zeng M, Rosas-Taraco AG. microRNAs in viral acute respiratory infections: immune regulation, biomarkers, therapy, and vaccines. ExRNA 2019;1(1):1.
[85] Parzibut G, Henket M, Moermans C, Struman I, Louis E, Malaise M, et al. A blood exosomal miRNA signature in acute respiratory distress syndrome. Frontiers in Molecular Biosciences 2021;8:640042.
[86] Popowski KD, Dinh PC, George A, Lutz H, Cheng K. Exosome therapeutics for COVID-19 and respiratory viruses. View (Beijing) 2021;2(3):20200186.
[87] Purghe B, Manfredi M, Ragnoli B, Baldanzi G, Malerba M. Exosomes in chronic respiratory diseases. Biomedicine & Pharmacotherapy 2021;144:112270.
[88] Soccio P, Moriondo G, Lacedonia D, Tondo P, Quarato CMI, Foschino Barbaro MP, et al. EVs-miRNA: the new molecular markers for chronic respiratory diseases. Life (Basel) 2022;12(10).
[89] Vadivalagan C, Shitut A, Kamalakannan S, Chen RM, Serrano-Aroca A, Mishra V, et al. Exosomal mediated signal transduction through artificial microRNA (amiRNA): a potential target for inhibition of SARS-CoV-2. Cellular Signalling 2022;95:110334.
[90] Wang Y, Zhu X, Jiang XM, Guo J, Fu Z, Zhou Z, et al. Decreased inhibition of exosomal miRNAs on SARS-CoV-2 replication underlies poor outcomes in elderly people and diabetic patients. Signal Transduction and Targeted Therapy 2021;6(1):300.
[91] Yang Y, Yuan L, Du X, Zhou K, Qin L, Wang L, et al. Involvement of epithelia-derived exosomes in chronic respiratory diseases. Biomedicine and Pharmacotherapy 2021;143:112189.
[92] Zargar MJ, Kaviani S, Vasei M, Soufi Zomorrod M, Heidari Keshel S, Soleimani M. Therapeutic role of mesenchymal stem cell-derived exosomes in respiratory disease. Stem Cell Research & Therapy 2022;13(1):194.
[93] Carberry CK, Koval LE, Payton A, Hartwell H, Ho Kim Y, Smith GJ, et al. Wildfires and extracellular vesicles: exosomal MicroRNAs as mediators of cross-tissue cardiopulmonary responses to biomass smoke. Environment International 2022;167:107419.
[94] Lai X, Zhong J, Zhang B, Zhu T, Liao R. Exosomal non-coding RNAs: novel regulators of macrophage-linked intercellular communication in lung cancer and inflammatory lung diseases. Biomolecules 2023;13(3).
[95] Levanen B, Bhakta NR, Torregrosa Paredes P, Barbeau R, Hiltbrunner S, Pollack JL, et al. Altered microRNA profiles in bronchoalveolar lavage fluid exosomes in asthmatic patients. The Journal of Allergy and Clinical Immunology 2013;131(3):894–903.
[96] Zhuansun Y, Du Y, Huang F, Lin L, Chen R, Jiang S, et al. MSCs exosomal miR-1470 promotes the differentiation of CD4(+)CD25(+)FOXP3(+) Tregs in asthmatic patients by inducing the expression of P27KIP1. International Immunopharmacology 2019;77:105981.
[97] Tang B, Wu Y, Fang H, Wu Y, Shi K. Small RNA sequencing reveals exosomal miRNAs involved in the treatment of asthma by Scorpio and centipede. BioMed Research International 2020;2020:1061407.
[98] Pei W, Zhang Y, Li X, Luo M, Chen T, Zhang M, et al. LncRNA AK085865 depletion ameliorates asthmatic airway inflammation by modulating macrophage polarization. International Immunopharmacology 2020;83:106450.
[99] Zhang XY, Chen ZC, Li N, Wang ZH, Guo YL, Tian CJ, et al. Exosomal transfer of activated neutrophil-derived lncRNA CRNDE promotes proliferation and migration of airway smooth muscle cells in asthma. Human Molecular Genetics 2022;31(4):638–50.
[100] Zhou R, Chen KK, Zhang J, Xiao B, Huang Z, Ju C, et al. The decade of exosomal long RNA species: an emerging cancer antagonist. Molecular Cancer 2018;17(1):75.
[101] Chiba M, Kimura M, Asari S. Exosomes secreted from human colorectal cancer cell lines contain mRNAs, microRNAs and natural antisense RNAs, that can transfer into the human hepatoma HepG2 and lung cancer A549 cell lines. Oncology Reports 2012;28(5):1551–8.

[102] Lobb RJ, van Amerongen R, Wiegmans A, Ham S, Larsen JE, Moller A. Exosomes derived from mesenchymal non-small cell lung cancer cells promote chemoresistance. International Journal of Cancer 2017;141(3):614–20.
[103] Xian J, Zeng Y, Chen S, Lu L, Liu L, Chen J, et al. Discovery of a novel linc01125 isoform in serum exosomes as a promising biomarker for NSCLC diagnosis and survival assessment. Carcinogenesis 2021;42(6):831–41.
[104] Hussen BM, Abdullah SR, Hama Faraj GS, Rasul MF, Salihi A, Ghafouri-Fard S, et al. Exosomal circular RNA: a signature for lung cancer progression. Cancer Cell International 2022;22(1):378.
[105] Xian J, Su W, Liu L, Rao B, Lin M, Feng Y, et al. Identification of three circular RNA cargoes in serum exosomes as diagnostic biomarkers of non-small-cell lung cancer in the Chinese population. Journal of Molecular Diagnostics 2020;22(8):1096–108.
[106] Huo S, Dou D. Circ_0056285 regulates proliferation, apoptosis and glycolysis of osteosarcoma cells via miR-1244/TRIM44 Axis. Cancer Management and Research 2021;13:1257–70.
[107] Chen F, Huang C, Wu Q, Jiang L, Chen S, Chen L. Circular RNAs expression profiles in plasma exosomes from early-stage lung adenocarcinoma and the potential biomarkers. Journal of Cellular Biochemistry 2020;121(3):2525–33.
[108] Drula R, Braicu C, Harangus A, Nabavi SM, Trif M, Slaby O, et al. Critical function of circular RNAs in lung cancer. Wiley Interdisciplinary Reviews: RNA 2020;11(5):e1592.
[109] Huang F, Bai J, Zhang J, Yang D, Fan H, Huang L, et al. Identification of potential diagnostic biomarkers for pneumonia caused by adenovirus infection in children by screening serum exosomal microRNAs. Molecular Medicine Reports 2019;19(5):4306–14.
[110] Gao J, Kalafatis D, Carlson L, Pesonen IHA, Li CX, Wheelock A, et al. Baseline characteristics and survival of patients of idiopathic pulmonary fibrosis: a longitudinal analysis of the Swedish IPF Registry. Respiratory Research 2021;22(1):40.
[111] Guenther A, Krauss E, Tello S, Wagner J, Paul B, Kuhn S, et al. The European IPF registry (eurIPFreg): baseline characteristics and survival of patients with idiopathic pulmonary fibrosis. Respiratory Research 2018;19(1):141.
[112] Lederer DJ, Martinez FJ. Idiopathic pulmonary fibrosis. New England Journal of Medicine 2018;378(19):1811–23.
[113] Martinez FJ, Lederer DJ. Focus on idiopathic pulmonary fibrosis: advancing approaches to diagnosis, prognosis, and treatment. Chest 2018;154(4):978–9.
[114] Nathan SD, Costabel U, Glaspole I, Glassberg MK, Lancaster LH, Lederer DJ, et al. Efficacy of pirfenidone in the context of multiple disease progression events in patients with idiopathic pulmonary fibrosis. Chest 2019;155(4):712–9.
[115] Pardo A, Selman M. The interplay of the genetic architecture, aging, and environmental factors in the pathogenesis of idiopathic pulmonary fibrosis. American Journal of Respiratory Cell and Molecular Biology 2021;64(2):163–72.
[116] Raghu G, Remy-Jardin M, Myers JL, Richeldi L, Ryerson CJ, Lederer DJ, et al. Diagnosis of idiopathic pulmonary fibrosis. An official ATS/ERS/JRS/ALAT clinical practice guideline. American Journal of Respiratory and Critical Care Medicine 2018;198(5):e44–68.
[117] Rosas IO, Goldberg HJ, Collard HR, El-Chemaly S, Flaherty K, Hunninghake GM, et al. A phase II clinical trial of low-dose inhaled carbon monoxide in idiopathic pulmonary fibrosis. Chest 2018;153(1):94–104.
[118] Negrete-Garcia MC, de Jesus Ramos-Abundis J, Alvarado-Vasquez N, Montes-Martinez E, Montano M, Ramos C, et al. Exosomal micro-RNAs as intercellular communicators in idiopathic pulmonary fibrosis. International Journal of Molecular Sciences 2022;23(19).
[119] Yang T, Wang J, Zhao J, Liu Y. Current and prospective applications of exosomal microRNAs in pulmonary fibrosis (Review). International Journal of Molecular Medicine 2022;49(3).
[120] Bozzini S, Zanframundo G, Bagnera C, Bozza E, Lettieri S, Vertui V, et al. A proof-of-concept analysis of plasma-derived exosomal microRNAs in interstitial pulmonary fibrosis secondary to antisynthetase syndrome. International Journal of Molecular Sciences 2022;23(23).

Index

Note: 'Page numbers followed by *f* indicate figures and *t* indicate tables.'

C

F

G

N

O

P

T

Y

Z

9780443140082